SECOND EDITION

# *Mechanics of Materials*


ARCHIE HIGDON

*Department of Engineering*
*California State Polytechnic College*

EDWARD H. OHLSEN

*Department of Engineering Mechanics*
*Iowa State University*

WILLIAM B. STILES

*Graduate Institute of Technology*
*University of Arkansas*

JOHN A. WEESE

*Department of Mechanical Engineering*
*University of Denver*




JOHN WILEY & SONS, INC. New York London Sydney

SECOND CORRECTED PRINTING, JULY, 1968

Library of Congress Catalog Card Number: 66-25222
Printed in the United States of America

# *Preface*

We sincerely appreciate the numerous helpful suggestions for the improvement of this text, that were submitted by users of the first edition. Every comment was carefully considered in preparing this edition. Some of the recommendations were in direct opposition to others; thus, not all of them could be incorporated in the revision. For example, some of these recommendations were for a more theoretical approach, whereas others called for more emphasis on practical situations. However, we are convinced that the text has been improved as a result of the many suggestions by the users of the first edition, and we solicit constructive criticisms of this revision. Readers are encouraged to submit any errors detected so that corrections can be made in subsequent printings.

The former Chapter 1 has been reorganized and divided into Chapters 1 and 2, with new articles on strain-energy, three-dimensional stress-strain relations, and a more complete presentation of strain analysis.

Chapter 3, on fundamentals of elasticity, has been added to provide an introduction to this important subject area as well as to show that in many cases results from elementary mechanics of materials are in excellent agreement with the more exact mathematical theory of elasticity. This addition has also made it possible to expand the article in Chapter 4 on the torsion of non-circular sections, and to present a more detailed analysis of the stress distribution in beams in Chapter 5. This material is so arranged that it can be omitted without loss of continuity by schools that do not desire, or do not have time, to present it.

Chapter 5, on stresses due to flexural loading, has been reorganized for a more concise presentation. As recommended by a number of readers, the use of singularity functions and the application of energy techniques have been included in Chapter 6 which deals with deflections due to bending. Castigliano's theorem is presented in Chapter 7 in a discussion of statically indeterminate beams. The former Chapters 6 and 9 have been combined into a more concise presentation of dynamic loading and combined static loading; this presentation now appears in Chapter 8. The material on columns, Chapter 9, has been rewritten and updated to conform more closely with current practice. The discussion of stability has been expanded and a treatment of beam columns added. Chapters 10 and 11 are essentially the same as the former Chapters 8 and 10, except for the introduction of new problems and a revision of the presentation of riveted and bolted joints, as recommended by users.

We take this opportunity to express our appreciation to our colleagues and the many users of the first edition for their constructive criticism and suggestions. We wish particularly to acknowledge the assistance of Mrs. Mabel Duykers in the preparation of the manuscript.

*April, 1966*

Archie Higdon
*USAF Academy, Colorado*

Edward H. Ohlsen
*Ames, Iowa*

William B. Stiles
*Little Rock, Arkansas*

John A. Weese
*Denver, Colorado*

# *Preface to the First Edition*

THE primary objectives of a course in mechanics of materials are: (1) to develop a working knowledge of the relations between the loads applied to a non-rigid body made of a given material and the resulting deformations of the body, (2) to develop a thorough understanding of the relations between the loads applied to a non-rigid body and the stresses produced in the body, (3) to develop a clear insight into the relations between stress and strain for a wide variety of conditions and materials, and (4) to develop adequate procedures for finding the required dimensions of a member of a specified material to carry a given load subject to stated specifications of stress and deflection.

The principals and methods used to meet these general objectives are drawn largely from prerequisite courses in mechanics, physics, and mathematics together with the basic concepts of the theory of elasticity and the properties of engineering materials. This book is designed to emphasize the required fundamental principles, with numerous applications to demonstrate and develop logical orderly methods of procedure. Instead of deriving numerous formulas for all types of problems, we have stressed the use of free-body diagrams and the equations of equilibrium, together with the geometry of the deformed body and the observed relations between stress and strain, for the analysis of the force system acting on a body.

The topics to be included in a course in mechanics of materials and the sequence to be followed are somewhat controversial. After careful consideration of the preferences of many teachers of mechanics, as well as consideration of our own convictions, we have used an essentially traditional approach. However, principal stresses are presented in Chapters 1, 2, and 3 and discussed more completely in Chapter 6 to promote a better understanding of the fundamental concepts of stress and to emphasize the significance of principal stresses. This material is contained in Articles 1–7, 1–8, 2–4, and 3–12 and in Chapter 6 so that teachers who prefer to present all of the material at one time can readily do so. The topics on connections are presented in Chapter

10 because many teachers consider that this material should be left for design courses when time is short for the course in mechanics of materials. The presentation of shear and moment diagrams is placed after the development of the flexure formula in Chapter 3 to give the student a better appreciation of the need for these diagrams as he studies them. The material on beam deflections has been organized to permit independent study of the double-integration and area-moment methods. Either method can be stressed or presented exclusively without the necessity of extensive reorganization if the time available precludes the study of both methods.

This book is designed for a first course in mechanics of materials, and the major emphasis is placed on the elastic range of stress. Where feasible, the presentation is extended to include the analysis of the plastic range of stress with appropriate problems.

The illustrative examples and problems have been selected with special attention devoted to problems that require an understanding of the principles of mechanics of materials without demanding excessive time for computational work. A large number of problems are included so that problem assignments may be varied from term to term.

Extensive use has been made of prerequisite course material in statics and calculus. A working knowledge of these subjects is considered an essential prerequisite to a successful study of mechanics of materials as presented in this book.

Ample material is included for a five-semester-hour course, but careful attention has been given to the needs of those with less time available. The book can readily be used for a three-semester-hour course by deleting selected articles and chapters. Suggested assignment sheets for courses of various lengths are available to teachers from the publisher.

The answers to about half the problems are included at the end of this book. We feel that the first assignment on a given topic should include some problems for which the answers are given; and, since the simpler problems are usually reserved for this first assignment, the answers are provided for the first four or five problems of each article and thereafter are given in general for alternate problems. We feel that answers for certain problems should not be given; hence we have not followed the flat rule of answers for all even-numbered problems. Since the convenient designation of problems for which answers are provided is of great value to those who make up assignment sheets, the problems for which answers are provided are indicated by means of an asterisk after the number.

In general, all given data are assumed to be composed of three significant figures regardless of the number of figures shown. Answers are, therefore, given to three significant figures, unless the number lies between 1 and 2 or any decimal multiple thereof, in which case four significant figures are reported. In Chapter 8, the problems involve material properties, which are quite variable, and possibly stress concentration factors, which are difficult to obtain to more than two significant figures from the graphs; therefore, we hesitate to give results to more than two significant figures. However, in order to avoid confusion, the answers are usually given to three significant figures.

We are indebted to present and former colleagues at various universities for ideas and suggestions used in the preparation of this book. We wish to acknowledge the assistance of Miss Mary Stiles and Mrs. Harriet Crum in preparing the manuscript and of our wives for the encouragement and patience they provided. We wish also to express our gratitude for the assistance of the Westinghouse Aviation Gas Turbine Division.

*February, 1960*

Archie Higdon
*USAF Academy, Colorado*

Edward H. Ohlsen
*Ames, Iowa*

William B. Stiles
*University, Alabama*

# *Contents*

## CHAPTER 10
## *Repeated Loading*



## CHAPTER 11
## *Connections*



## *Appendices*

# *Symbols and Abbreviations**

| | |
|---|---|
| $A$ | area |
| Art. | article or articles |
| avg. | average |
| $b$ | breadth, width |
| $C$ | compression |
| $c$ | distance from neutral axis or from center of twist to extreme fiber |
| $E$ | modulus of elasticity in tension or compression |
| $e$ | eccentricity |
| Eq. | equation or equations |
| $F$ | force, or body force |
| $f$ | normal force per unit of length |
| Fig. | figure or figures |
| ft | foot or feet |
| ft-lb | foot-pound |
| $G$ | modulus of rigidity (modulus of elasticity in shear) |
| $g$ | acceleration of gravity |
| $h$ | height, depth of beam |
| hp | horsepower |
| $I$ | moment of inertia or second moment of area |
| in. | inch or inches |
| in-lb | inch-pound |
| $J$ | polar moment of inertia of area |
| $K$ | stress concentration factor |
| $k$ | modulus of spring, factor of safety |
| kip or k | kilopound (1000 lb) |
| ksi | kilopounds (or kips) per square inch |
| $L$ | length |
| lb | pound |
| $M$ | bending moment |
| max | maximum |
| min | minimum |
| $N$ | normal force |
| $n$ | ratio of elastic moduli, number of cycles |
| $P$ | concentrated load, force |
| $p$ | fluid pressure |
| psf | pounds per square foot |

* The symbols and abbreviations used in this book conform essentially with those approved by the American Standards Association.

| | |
|---|---|
| psi | pounds per square inch |
| $Q$ | first moment of area |
| $q$ | shearing force per unit of length |
| $R$ | radius, modulus of rupture, resultant force |
| $r$ | radius of gyration |
| rad | radian |
| rpm | revolutions per minute |
| rps | revolutions per second |
| $S$ | surface force |
| sec | second |
| sq | square |
| $T$ | torque, temperature, tension |
| $t$ | thickness, tangential deviation |
| $U$ | strain energy |
| $u$ | strain energy per unit volume |
| $u, v, w$ | components of displacement |
| $V$ | shearing force |
| $W$ | weight |
| $W_k$ | work |
| $w$ | weight or load per unit of length |
| $y$ | deflection |
| $Z$ | section modulus ($Z = I/c$) |
| $\alpha$ (alpha) | coefficient of thermal expansion |
| $\gamma$ (gamma) | shearing strain (unit), specific weight |
| $\delta$ (delta) | total deformation |
| $\Delta$ (delta) | maximum deflection under energy load |
| $\epsilon$ (epsilon) | normal strain (unit) |
| $\eta$ (eta) | efficiency, portion of energy load which is effective |
| $\theta$ (theta) | total angle of twist in radians, slope of deflected beam |
| $\mu$ (mu) | Poisson's ratio, micro-inches per inch ($10^{-6}$ in. per in.) |
| $\rho$ (rho) | radius |
| $\sigma$ (sigma) | normal stress (unit) |
| $\tau$ (tau) | shearing stress (unit) |
| $\phi$ (phi) | the Airy stress function |
| $\omega$ (omega) | angular velocity |
| 16 WF 88 | a 16-in. wide flange rolled section with a weight of 88 lb per foot. |
| 10 I 35 | a 10-in. American standard rolled section (I-beam) with a weight of 35 lb per foot. |
| 12 ⊔ 30 | a 12-in. rolled channel section with a weight of 30 lb per foot. |
| ° | degree |

CHAPTER 1

# *Stress and Strain: Concepts and Definitions*

**1–1. Introduction.** The primary objective of a course in mechanics of materials is the development of the relations between the loads applied to a non-rigid body and the resulting *internal forces* and *deformations* induced in the body. Ever since the time of Galileo Galilei (1564–1642), men of scientific bent have studied assiduously the problem of the load-carrying capacity of structural and machine elements and the mathematical analysis of the internal forces and deformations induced by applied loads. The experience and observations of these scientists and engineers of the last three centuries are the heritage of the engineer of today, providing him with fundamental knowledge from which he may progress to the development of theories and techniques which permit him to design, with competence and assurance, structures and machines of unprecedented size and complexity.

The subject matter of this book forms the basis for the solution of three general types of problems as follows:

1. Given a certain function to perform (the transporting of traffic over a river by means of a bridge, conveying scientific instruments to Mars in a space vehicle, the conversion of water power into electric power), of what materials should the machine or structure be constructed, and what should be the sizes and proportions of the various elements? This is the designer's task, and obviously there is no single solution to any given problem.

2. Given the completed design, is it adequate? That is, does it perform the function economically and without excessive deformation? This is the checker's problem.

3. Given a completed structure or machine, what is its actual load-carrying capacity? The structure may have been designed for some

purpose other than the one for which it is now to be used. Is it adequate for the proposed use? For example, a building may have been designed for an office building, but is later found to be desirable for use as a warehouse. In such a case, what maximum loading may the floor safely support? This is the rating problem.

Since the complete scope of these problems is obviously too comprehensive for mastery in a single course, this book is restricted to a study of individual members and very simple structures or machines. The design courses which follow will consider the entire structure or machine, and will provide essential background for the complete analysis of the three problems listed.

The principles and methods used to meet the objective stated at the beginning of this chapter depend to a great extent upon prerequisite courses in mathematics and mechanics, supplemented by additional concepts from the theory of elasticity and the properties of engineering materials. The equations of equilibrium from *statics* are used extensively, with one major change in the free-body diagrams; namely, most free bodies are isolated by *cutting through* a member instead of removing a pin or some other connection. The forces transmitted by the cut sections are *internal* forces called *stresses.*

It will frequently be found that the equations of equilibrium (or motion) are not sufficient to determine all the unknown loads or reactions acting on a body. In such cases, it is necessary to consider the geometry (the change in size or shape) of the body after the loads are applied. The *deformation* in any direction or dimension is called *strain.* In some instances, the specified maximum deformation and not the specified maximum stress will govern the maximum load that a member may carry.

Some knowledge of the physical and mechanical properties of materials is required in order to create a design, to properly evaluate a given design, or even to write the correct relation between an applied load and the resulting deformation of a loaded member. Essential information will be introduced as required, and more complete information can be obtained from textbooks and handbooks on properties of materials.

**1–2. Load Classification.** Certain terms are commonly used to describe applied loads; their definitions are given here so that the terminology will be clearly understood.

Loads may be classified with respect to time.

1. *A static load* is a gradually applied load for which equilibrium is reached in a relatively short time.

2. A *sustained load* is a load which is constant over a long period of time, such as the weight of a structure (called *dead load*). This type of load is treated in the same manner as a static load; however, for some materials and conditions of temperature and stress, the resistance to failure may be different under short-time loading and under sustained loading.

3. An *impact load* is a rapidly applied load (an energy load). Vibration normally results from an impact load, and equilibrium is not established until the vibration is eliminated, usually by natural damping forces.

4. A *repeated load* is a load which is applied and removed many thousands of times. The helical springs which close the valves on automobile engines are subjected to repeated loading.

Loads may also be classified with respect to the area over which the load is applied.

1. A *concentrated load* is a load or force which is applied at a point. Any load which is applied to a relatively small area compared with the size of the loaded member is assumed to be a concentrated load; for example, a truck wheel load on the longitudinal members of a bridge.

2. A *distributed load* is a load distributed along a length or over an area. The distribution may be uniform or non-uniform. The weight of a concrete bridge floor of uniform thickness is an example of a uniformly distributed load.

Loads may be classified with respect to the location and method of application.

1. A *centric load* is one in which the resultant force passes through the centroid of the resisting section. If the resultant passes through the centroids of all resisting sections, the loading is termed *axial*. In Fig. 1–1$a$, the line of action of the load $P$ passes through the centroid of section $a$–$a$; therefore, the load is centric for section $a$–$a$. The force $P$ in Fig. 1–1$b$ is applied through the plate to the shank of the rivet and is resisted by the force $V$ on the shaded section $b$–$b$. If the thickness $t$ is not too large (the usual case for riveted joints), the force $P$ can be assumed to be collinear with $V$, thus satisfying the definition of a centric load. The loading illustrated in this figure is called *single shear*, and the force $V$ is a shearing force. If the plate $A$ were sandwiched between two plates $B$ with the rivet extending through all three

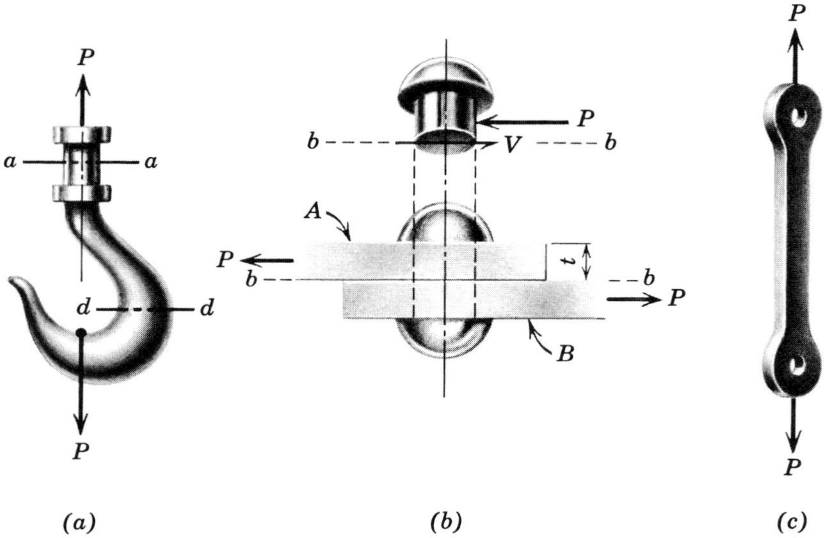


Fig. 1–1

plates, the rivet would be subjected to two shearing forces. Such loading is called *double shear*. Figure 1–1*c* represents an eyebar used for trusses, tie rods, and linkages. This is an example of axial loading.

2. A *torsional load* is one which subjects a shaft or some other member to couples which twist the member. If the couples lie in planes which are transverse to the axis of the member, as in Fig. 1–2, the member is subjected to pure torsion.

3. A *bending* or *flexural load* is one in which the loads are applied

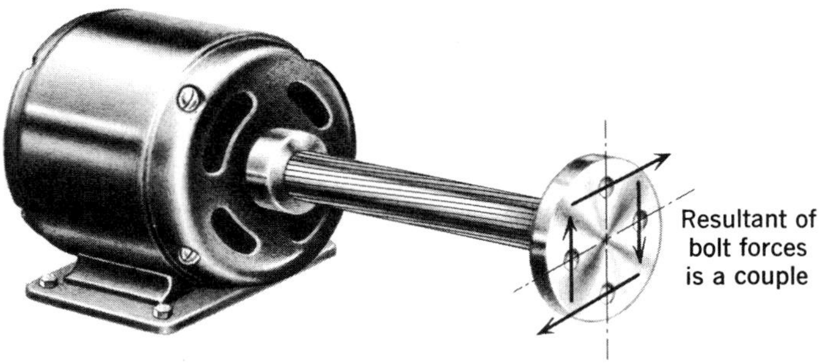


Fig. 1–2

Fig. 1–3

transversely to the longitudinal axis of the member. The applied load may include couples which lie in planes parallel to the axis of the member. A member subjected to bending loads bends or bows along its length. Figure 1–3 illustrates a beam subjected to flexural loading consisting of a concentrated load, a uniformly distributed load, and a couple.

4. A *combined loading* is a combination of two or more of the previously defined types of loading.

**1–3. Concept of Stress.** In every subject area there are certain fundamental concepts of paramount importance to a satisfactory comprehension of the subject matter. For the subject mechanics of materials, a thorough mastery of the physical significance of stress and strain is paramount. The discussion of stress will be undertaken first, and the study of strain will be taken up in later articles.

The eyebar $AB$ from the pile extractor of Fig. 1–4 (used to reclaim steel sheet piles from cofferdams and similar installations) will serve as an example for the introduction of the concept of stress. Although the bar is subjected to dynamic loading in use, an equivalent static load (a static load which will produce the same stresses and strains as the dynamic load) may be used in the design. Since the eyebar is a two-force member, the pin reactions at $A$ and $B$ are collinear with the axis of the bar. A review of two-force members and stress analysis in trusses, as presented in a course in statics, is strongly recommended. When the bar is cut by any plane, such as plane $c$–$c$, and the free-body diagram of Fig. 1–5 is drawn, equilibrium may be established by placing on the cut section a series of parallel forces (stresses), each of which acts on a unit area. The term stress is sometimes used to denote the resultant force on a section; in other instances it is used to indicate the force per unit area. *In this book stress will always refer to the force per unit area, unless otherwise stated.* The equilibrium equation $\Sigma F_y = 0$ gives the expression

$$\text{Applied force} = (\text{Stress})(\text{Area}).$$

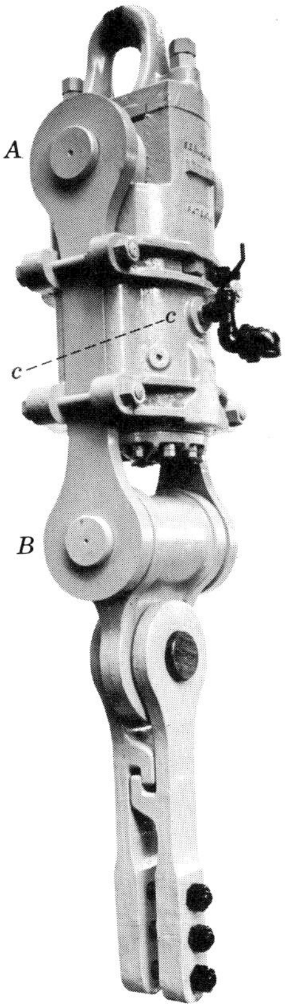


Fig. 1–4. Courtesy: Vulcan Iron Works Inc., Chicago, Ill.

It is immediately apparent that the magnitude of the stress on plane *c–c* is a function of the angle $\theta$, and this fact distinguishes a stress from an ordinary force vector. The aspect of the plane on which the stress acts is an additional characteristic, which means that a stress is a tensor of the second order[1] not a vector, and that the laws of vector addition do not apply to stresses (forces per unit area). This need not be cause for concern if, in the application of the equations of equilibrium (or motion), one always replaces a stress with a total force (stress multiplied by the appropriate area), thus reducing the problem to one involving ordinary force vectors.

Because the stress indicated on plane *c–c* of Fig. 1–5, if evaluated, would convey little information regarding the adequacy of design, it is customary to replace this stress with two components: one perpendicular to the plane, called the *normal stress* (which may be tension or compression) and designated by the Greek letter $\sigma$; and the other parallel to the plane, called the *shearing stress* and designated by the letter $\tau$. These stresses, when evaluated, may be compared with experimentally determined values for the strength of the material in tension, compression, and shear, and thus yield significant information. Such stresses are indicated in Fig. 1–6*a*. Again, the aspect of the plane as an additional characteristic possessed by a stress must be emphasized; therefore, in Fig. 1–6*b* the stresses are replaced by their resultants, the force vectors $N$ and $V$, and the equations of equilibrium may be ap-

[1] A scalar is a tensor of zero order that can be defined by one component. A vector is a tensor of the first order and, in general, has three components. A stress involves a force and the area on which it acts, and is a tensor of the second order requiring six components to specify it completely. See *Tensor Analysis,* I. S. Sokolnikoff, John Wiley and Sons, New York, 1951.

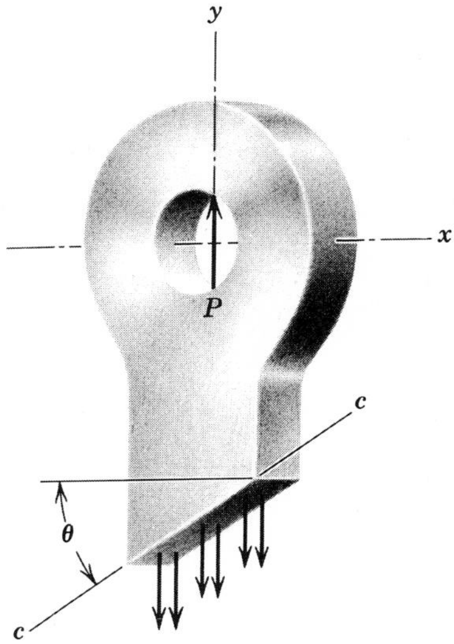


Fig. 1–5

plied to the free-body diagram. In the preceding discussion, the assumption is made that the stress is uniformly distributed over the cut section. A necessary (but not sufficient) condition for the distribution to be uniform is that the loading be *centric*. Non-uniform stress distribution under centric loading will be discussed in Art. 1–8. In this

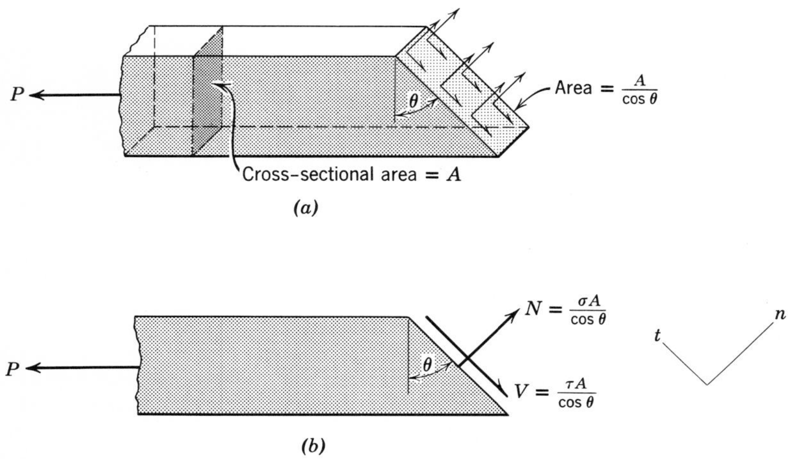


Fig. 1–6

book, unless otherwise indicated, a uniform stress distribution is to be assumed for all centric (or assumed centric) loading.

**1–4. Stresses under Centric Loading.** If the equations of equilibrium are applied to the free-body diagram of Fig. 1–6*b*, the following expressions are obtained:

For $\Sigma F_n = 0$, $$P \cos\theta = N = \frac{\sigma A}{\cos\theta},$$

from which $$\sigma = \frac{P}{A}\cos^2\theta = \frac{P}{2A}(1 + \cos 2\theta);$$

and, for $\Sigma F_t = 0$, $$P \sin\theta = V = \frac{\tau A}{\cos\theta},$$

from which $$\tau = \frac{P}{A}\sin\theta\cos\theta = \frac{P}{2A}(\sin 2\theta).$$

The graph of these expressions in Fig. 1–7 indicates that $\sigma$ is maximum when $\theta$ is 0 or 180°, that $\tau$ is maximum when $\theta$ is 45° or 135°, and also that $\tau_{\max} = \frac{1}{2}\sigma_{\max}$. Therefore the maximum stresses are given by the expressions

$$\sigma_{\max} = P/A \qquad (1\text{–}1)$$

and

$$\tau_{\max} = P/(2A). \qquad (1\text{–}2)$$

It must be emphasized that Eq. (1–2) is valid only for centric tensile or compressive loading in one direction. For convenience, the maximum normal stress under axial loading is called the *axial stress*. Note that the normal stress is either maximum or minimum on planes for which the shearing stress is zero. In Art. 1–6 it is shown that the shear-

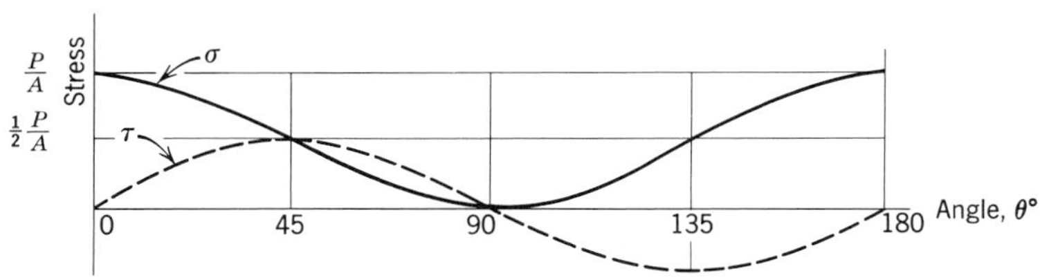


Fig. 1–7

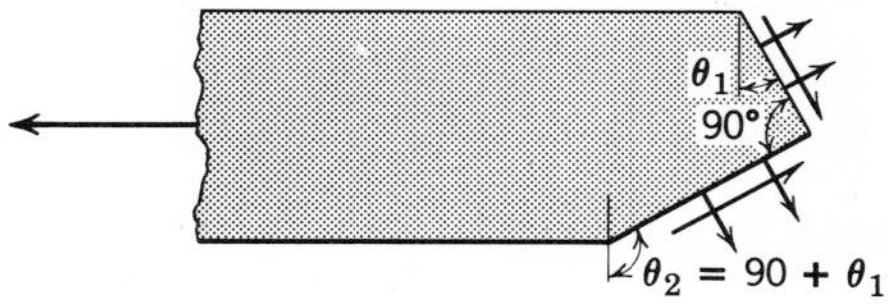


Fig. 1–8

ing stress is always zero on the planes of maximum or minimum normal stress.

Laboratory experiments indicate that both shearing and normal stresses under axial loading are important since a brittle material loaded in tension will fail in tension on the transverse plane, whereas a ductile material loaded in tension will fail in shear on the 45° plane.

One observes that when $\theta$ of Fig. 1–7 is greater than 90°, the sign of the shearing stress changes, but the magnitude of the shearing stress for any angle $\theta$ is the same as that for $90° + \theta$. The stresses are shown in Fig. 1–8. The equality of shearing stresses on orthogonal planes can be demonstrated by applying the equations of equilibrium to the free-body diagram of a small rectangular block of thickness $dz$, shown in Fig. 1–9. If a shearing force[2] $V_x = \tau_{yx}\,dx\,dz$ is applied to the top surface of the block, the equation $\Sigma F_x = 0$ will dictate the application of an oppositely directed force $V_x$ to the bottom of the block, thus leaving the block subjected to a clockwise couple; this

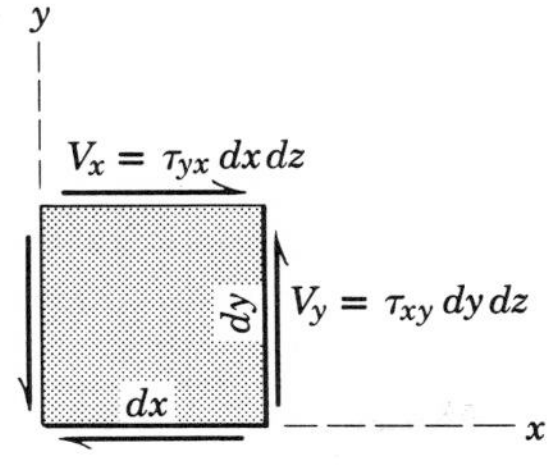


Fig. 1–9

[2] The double subscript on the shearing stress is used to designate both the plane on which the stress acts and the direction of the stress. The first subscript indicates the plane (or rather the normal to the plane) and the second subscript indicates the direction of the stress.

couple must be balanced by a counterclockwise couple composed of the oppositely directed forces $V_y$ applied to the vertical faces of the block. Finally, application of the equation $\Sigma M_z = 0$ yields the following,

$$\tau_{yx}\,(dx\;dz)\;dy = \tau_{xy}\,(dy\;dz)\;dx,$$

from which

$$\tau_{yx} = \tau_{xy}.$$

Therefore, *if a shearing stress exists at a point on any plane, there must also exist at this point a shearing stress of the same magnitude on an orthogonal plane.* This statement is also valid when normal stresses are acting on the planes, since the normal stresses occur in collinear but oppositely directed pairs, and thus have zero moment with respect to any axis.

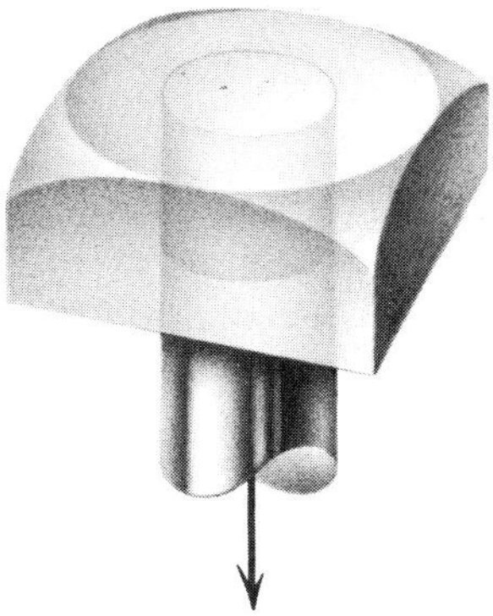

Fig. 1–10

It will be observed that loads on a structure or machine are transmitted to the individual members through some sort of connection, as, for example, the riveted connection of Fig. 1–1*b*, the pins at the ends of the eyebar of Fig. 1–4, and even so simple a detail as the head (or nut) of a bolt (Fig. 1–10) loaded axially in tension. In all these connections, one of the most significant stresses induced is a shearing stress; hence, these members are often said to be subjected to centric shear loading. If the loading is as indicated in Fig. 1–1*b*, the average shearing stress on plane *b–b* is the significant shearing stress (although it is not the maximum stress) and is referred to as the *cross shearing stress.* Another type of shear loading is termed *punching shear,* as, for example, the action of a punch in forming rivet holes in a metal plate, the tendency of building columns to punch through footings, and the tendency of a tensile axial load on a bolt to pull the shank through the head. Under a punching shear load, the significant stress is the average shearing stress on the surface described by the periphery of the punching member and the thickness of the punched member;[3] for example, the shaded cylindrical area of Fig. 1–10.

[3] This is modified in the case of building columns on concrete footings. Reference should be made to reinforced-concrete design specifications.

The following example will serve to illustrate the concepts discussed in Art. 1–3 and 1–4. *Note: In all problems in this textbook the weights of the members will be neglected, unless otherwise specified.*

**Example 1–1:** A hoist is made from a 4 by 4-in. wood post $BC$ and a 1-in. diameter steel eyebar $AB$ as shown in Fig. 1–11$a$. The hoist supports a load $W$ of 7000 lb. Determine the axial stresses in members $AB$ and $BC$ and the cross shearing stress in the 1-in. diameter bolt at $A$ which is in double shear.[4]

*Solution:* The members $AB$ and $BC$ are two-force members subjected to axial loading. When the equations of equilibrium are applied to the free-body diagram of Fig. 1–11$b$ the following results are obtained:

$$\circlearrowleft + \qquad \Sigma MC = 0$$

$$7000(5) - (5/13)(F_A)(7) = 0, \qquad F_A = 13{,}000 \text{ lb } T,$$

$$\sigma_{AB} = \frac{F_A}{A} = \frac{13{,}000}{(\pi/4)(1)^2} = \underline{16{,}550 \text{ psi } T.} \qquad \text{Ans.}$$

[4] Structural timbers specified as rough sawn will be approximately full size. Cross-sectional dimensions of dressed timbers will be between ⅜ and ½ in. smaller than nominal. For convenience, all timbers in this book will be considered as full size.

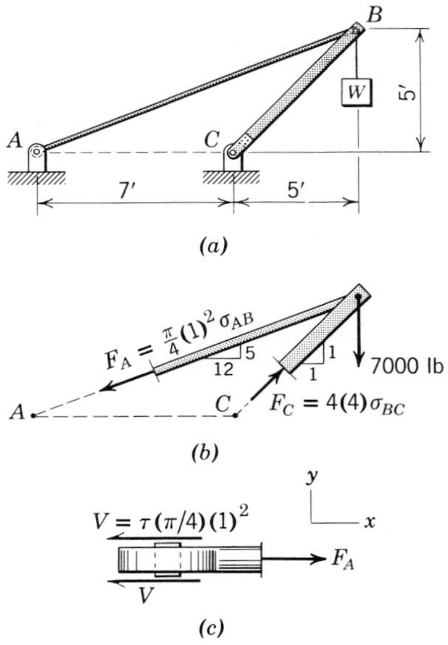


Fig. 1–11

Observe that it is not necessary to solve the equation explicitly for the force $F_A$ in the member if only the stress is required. Hence, for the next equation this step will be omitted; thus,

$$\circlearrowleft + \qquad \Sigma M_A = 0$$

$$7000(12) - \frac{1}{\sqrt{2}}(16)(\sigma_{BC})(7) = 0$$

$$\sigma_{BC} = \underline{1061 \text{ psi } C}. \qquad \text{Ans.}$$

And from the free-body diagram of Fig. 1–11$c$, which is a free-body diagram of the connection at $A$, and using the value of force $F_A$ as determined above,

$$\overset{+}{\rightarrow} \qquad \Sigma F_x = 0$$

$$F_A - 2V = 0$$

$$V = \tau\pi(1)^2/4 = 6500$$

$$\tau = \underline{8280 \text{ psi}}. \qquad \text{Ans.}$$

## PROBLEMS

**1–1.*** Consider the loads of Fig. P1–1 to be axially applied at sections $A$, $B$, $C$, and $D$ to the steel bar, which has a cross-sectional area of 3 sq in. Determine the axial stress in the bar:

($a$) On a section 10 in. to the right of $A$.
($b$) On a section 60 in. to the right of $A$.

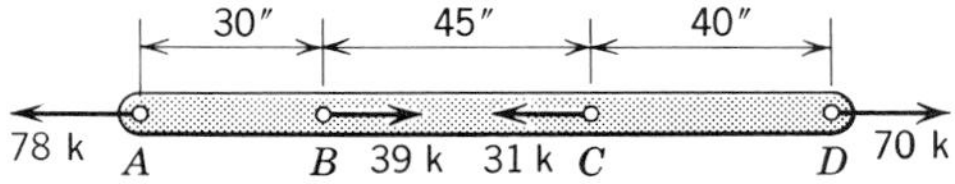


Fig. P1–1

**1–2.*** Consider the loads of Fig. P1–2 to be axially applied at sections $B$, $C$, and $D$ to the concrete structure supported at $A$. The cross-sectional areas are: For $CD$, 12 sq in., for $BC$, 3.6 sq in., and for $AB$, 16 sq. in. Determine the axial stress in the concrete:

(*a*) On a section in the $CD$ interval.
(*b*) On a section in the $BC$ interval.
(*c*) On a section in the $AB$ interval.

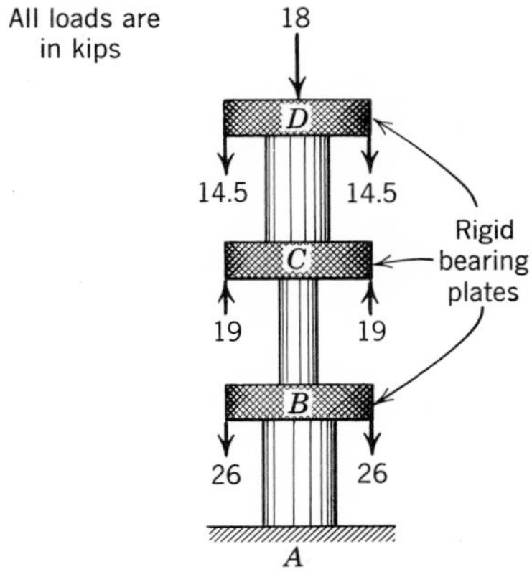


Fig. P1–2

**1–3.*** In Fig. P1–3, the axial stresses are 1150 psi $C$ in the wood post $B$, and 19,000 psi $T$ in the steel bar $A$. Determine the load $P$.

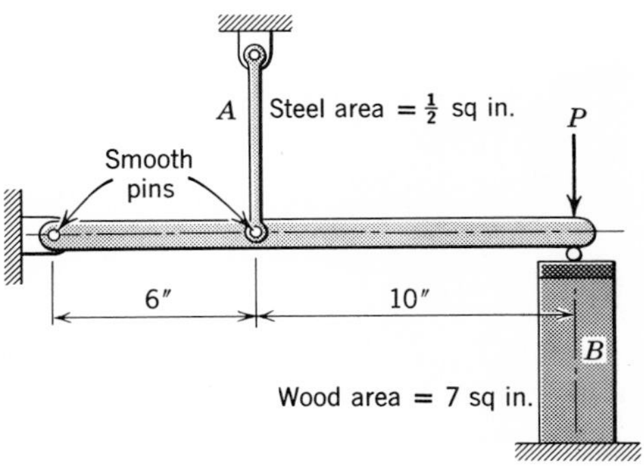


Fig. P1–3

**1–4.** Member $AC$ of the pin-connected frame $ABC$ of Fig. P1–4 is a 2.0-in. diameter rod with an axial stress of 15,000 psi $T$. Determine the force $P$.

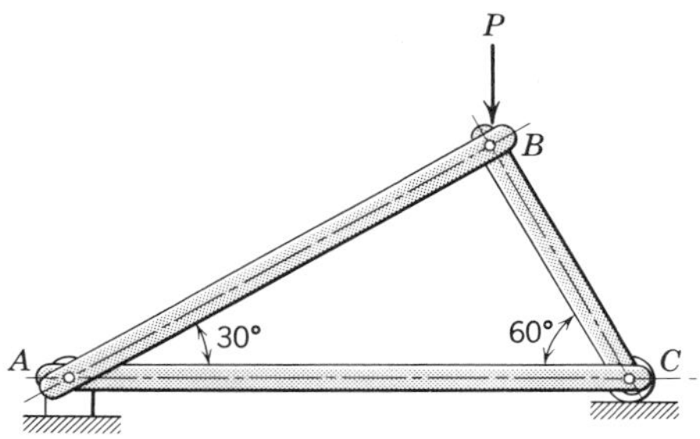


Fig. P1–4

**1–5.*** The tie rod $AB$ of Fig. P1–5 is subjected to an axial stress of 22,000 psi.

(*a*) Determine the cross-sectional area of $AB$.

(*b*) If the cross shearing stress in pin $C$ is 10,000 psi, determine the cross-sectional area of pin $C$.

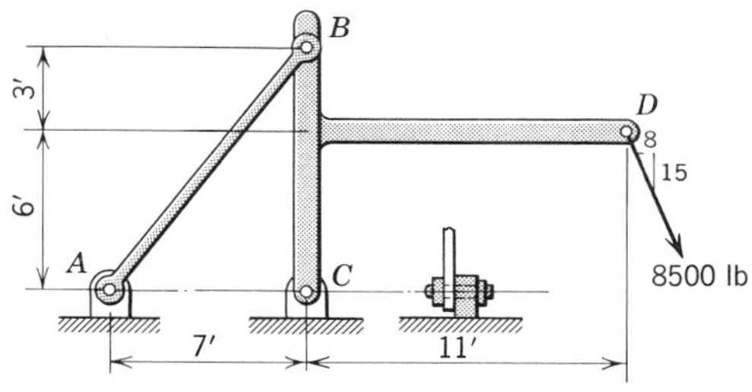


Fig. P1–5

**1–6.*** The concrete cylinder $C$ of Fig. P1–6 has a cross-sectional area of 15 sq in. A pressure-measuring device indicates an average pressure between $C$ and the foundation of 400 psi excluding the weight of the members. Contact with the horizontal member $E$ is made through a ball and steel cap on $C$. The axial stresses in rods $A$ and $B$ are 10,000 psi $T$ and 18,000 psi $T$ respectively. Determine the cross-sectional areas of $A$ and $B$.

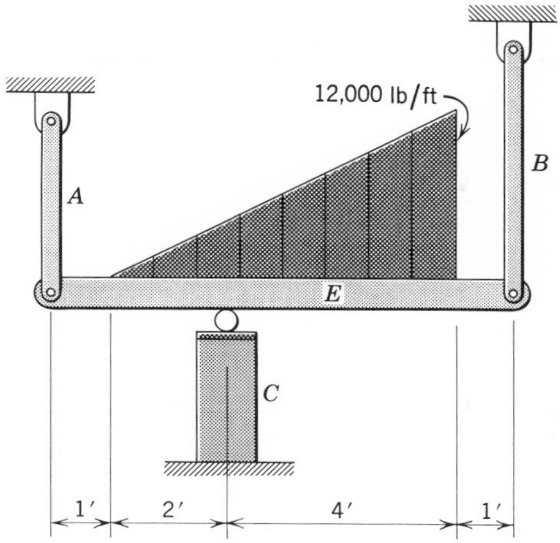


Fig. P1–6

**1–7.** Member $GF$ of the pin-connected truss of Fig. P1–7 has a cross-sectional area of 2.4 sq in. Determine the axial stress in $GF$. The lengths shown are center to center of pins.

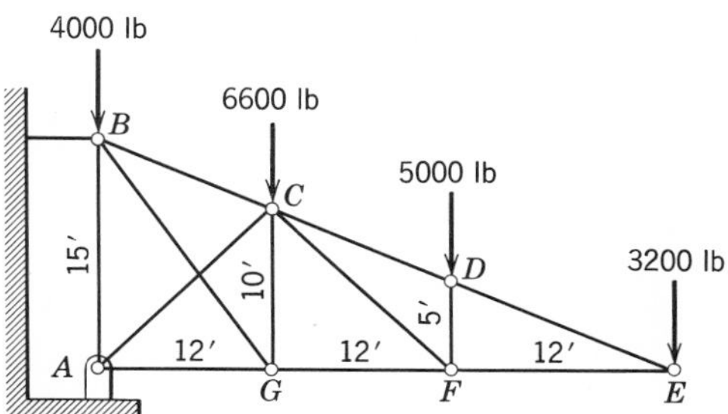


Fig. P1–7

**1–8.*** Member $CF$ of the pin-connected truss of Fig. P1–7 is a ⅞-in. diameter rod. Determine the axial stress in $CF$. The lengths shown are center to center of pins.

**1–9.** Determine the axial stress in member *GF* of the pin-connected truss of Fig. P1–9. The cross-sectional area of *GF* is 4.25 sq in.

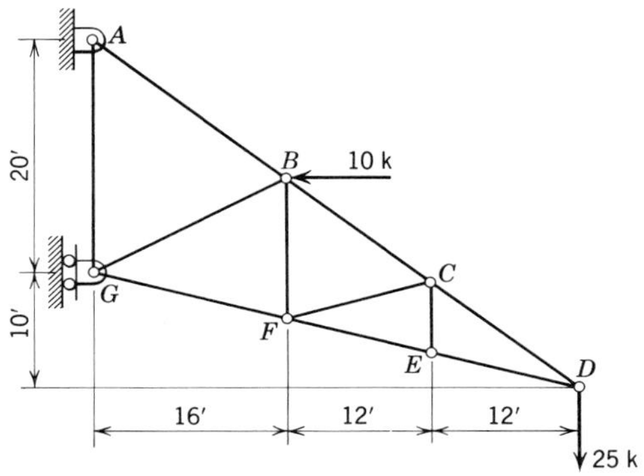


Fig. P1–9

**1–10.*** The axial stress in member *JC* of the pin-connected truss of Fig. P1–10 is 7000 psi *C*. Determine the cross-sectional area of *JC*.

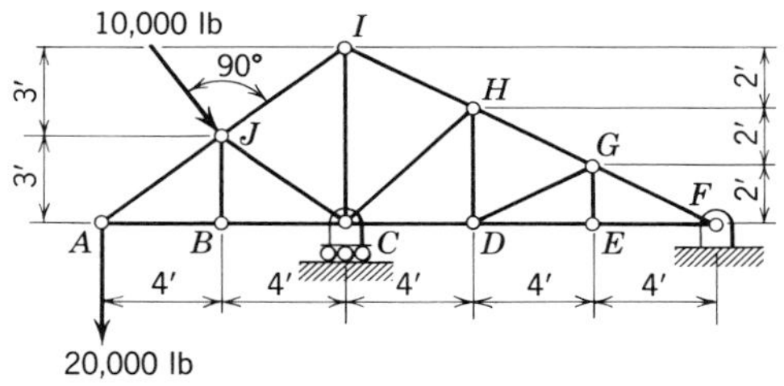


Fig. P1–10

**1–11.** The axial stress in member *IJ* of the pin-connected truss of Fig. P1–10 is 16,000 psi *T*. Determine the cross-sectional area of *IJ*.

**1–12.*** A punch press is used to punch a 1.40-in. diameter hole in a ¼-in. thick steel plate. Determine the force exerted by the press on the plate when the average punching shear resistance of the steel plate is 48,000 psi.

**1–13.** The inclined member *AB* is framed into a 2 by 6-in. western red cedar plank as shown in Fig. P1–13. The plank can withstand a shearing stress parallel to the grain (horizontal) of 108 psi. Determine the axial compressive force in *AB* when the plank fails in horizontal shear.

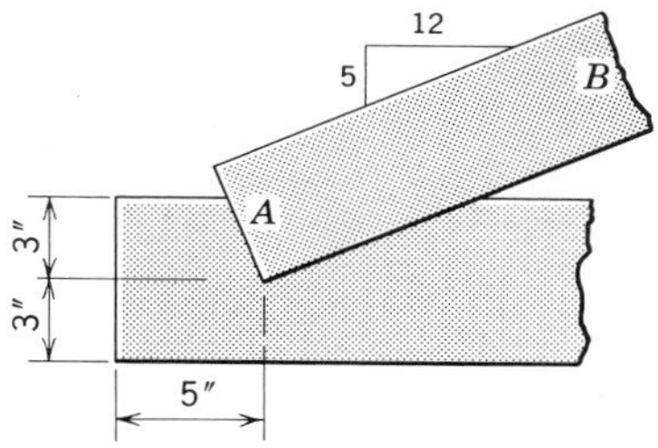


Fig. P1–13

**1–14.*** The Vulcan Iron Works, Inc., furnished the following design data for their pile extractor of Fig. 1–4:

Side bar *AB* body cross section, 7 by 2⅝ in.

Diameter of pins at *A* and *B*, 7½ in.

Design load of 701,640 lb from hammer plus 200,000-lb hoisting line pull.

Material, heat-treated alloy steel.

For the given loading, determine:

(*a*) The axial stress in the side bar.

(*b*) The cross shearing stress in the pin.

**1–15.** For the pin-connected frame of Fig. P1–15, determine:

(*a*) The axial stress in the ¾-in. diameter rod *CD*.

(*b*) The cross shearing stress in the ¾-in. diameter pin at *B*. Pin *B* is in single shear.

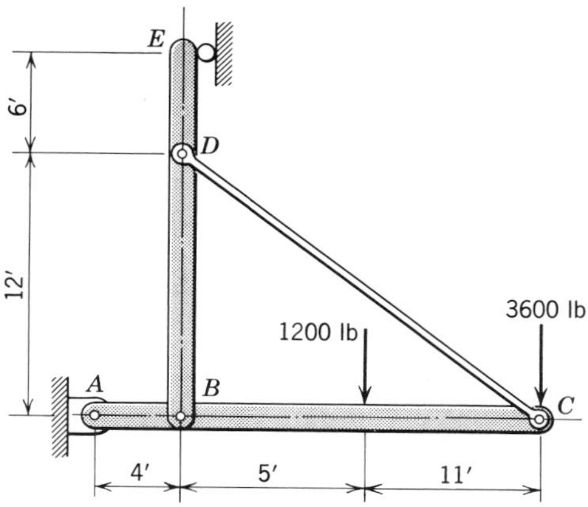


Fig. P1–15

**1–16.*** In the pin-connected structure of Fig. P1–16, a roller at $B$ on member $AB$ is free to slide in a smooth slot in member $CD$. The body $Q$ weighs 1600 lb; the weights of all other members may be neglected; all pins are in single shear, and the cross shearing stress in the pins at $B$ and $C$ is 10,000 psi. Determine the cross-sectional area of:

(*a*) The pin at $B$.

(*b*) The pin at $C$.

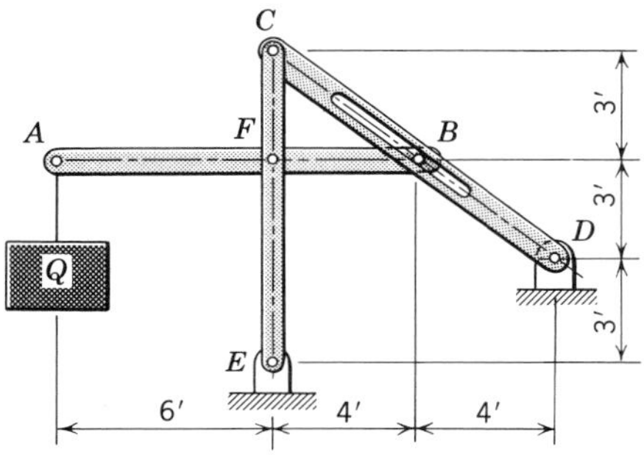


Fig. P1–16

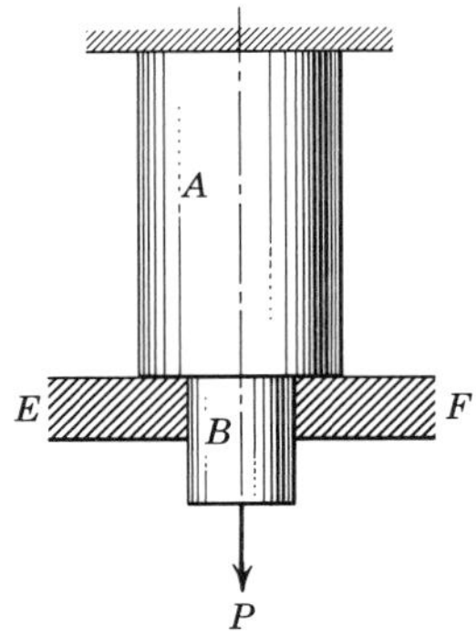


Fig. P1–17

**1–17.** Parts $A$ and $B$ of the vertical circular shaft in Fig. P1–17 have cross-sectional areas of 5.00 sq in. and 2.10 sq in., respectively. The shaft is securely supported at the top and when loaded makes contact with the support at $EF$. When the force $P$ is 125 kips the axial stress in $A$ is 18.0 ksi $T$. Determine the average unit pressure of $A$ on $EF$. Part $B$ fits snugly in a smooth hole in $EF$.

**1–18.** Two identical rods $AB$ are pinned to two identical rods $BC$ with a smooth pin at $B$ as shown in Fig. P1–18. A vertical force of 600 lb and a horizontal force of 1300 lb are applied to the center of the pin $B$ as indicated. The weights of the rods may be neglected. The cross-sectional area of pin $B$ is 0.30 sq in. The total cross-sectional area of the two rods $AB$ is 0.20 sq in. Determine:

(*a*) The highest cross shearing stress in pin $B$, and state at what section it occurs.

(*b*) The axial stress in rods $AB$.

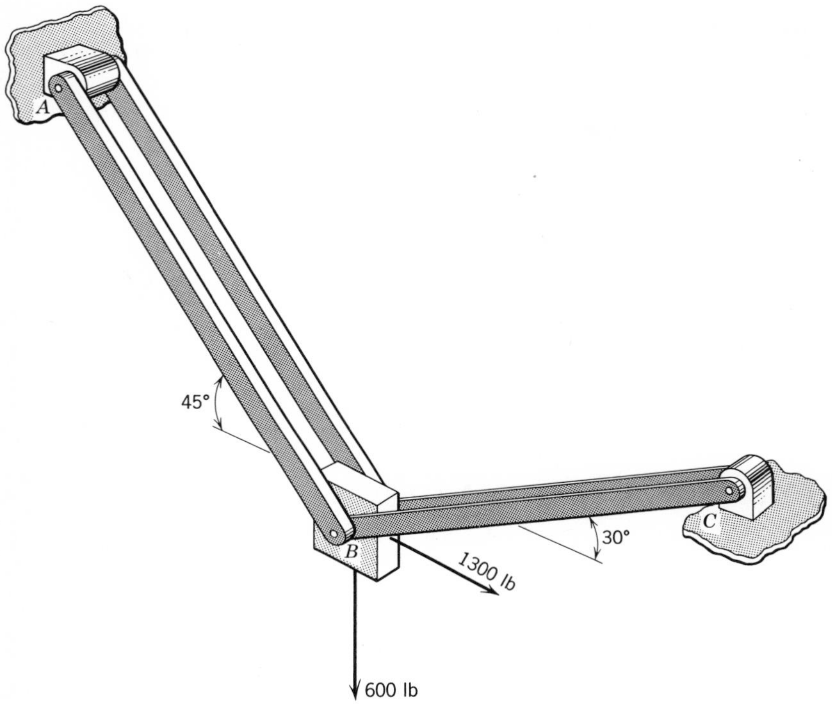


Fig. P1–18

**1–19.*** A riveted joint from a truss is shown in Fig. P1–19. Member $B$ consists of two 2½ by 2½ by ¼-in. angles back to back. (See App. C.) The force $P_1$ is 12,600 lb $T$. Determine the axial stress in the gross section (no holes) of member $B$.

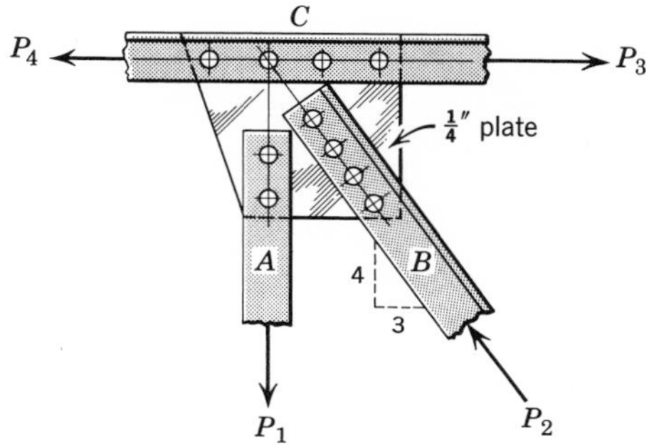


Fig. P1–19

**1–20.** Member $BE$ of the pin-connected frame of Fig. P1–20 is a 1 by 3/8-in. rectangular bar and the pin at $E$ has a diameter of 1.25 in. The force $P$ is applied to the pin outside of members $BE$ and $DE$. Determine:

(*a*) The axial stress in member $BE$.

(*b*) The maximum cross shearing stress in the pin at $E$.

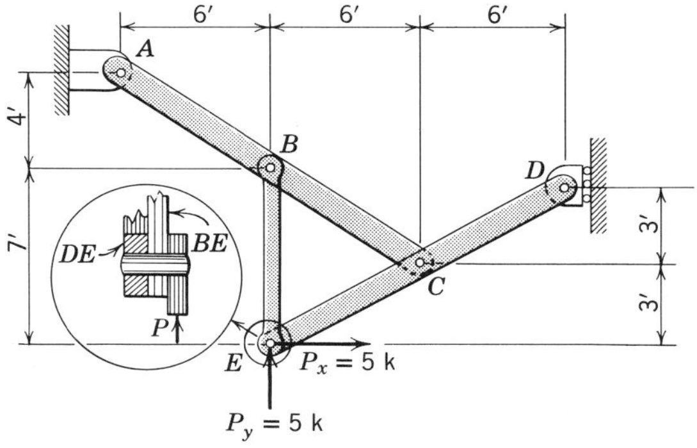


**Fig. P1–20**

**1–21.*** Member $CE$ of Fig. P1–21 is an aluminum alloy tube pinned at $C$ and $E$. The cable is attached to the pins and passes over a smooth pulley at $D$. Determine:

(*a*) The cross-sectional area of the tube if the axial stress is 4 ksi $C$.

(*b*) The diameter of the pin at $C$ if the maximum cross shearing stress is 10 ksi.

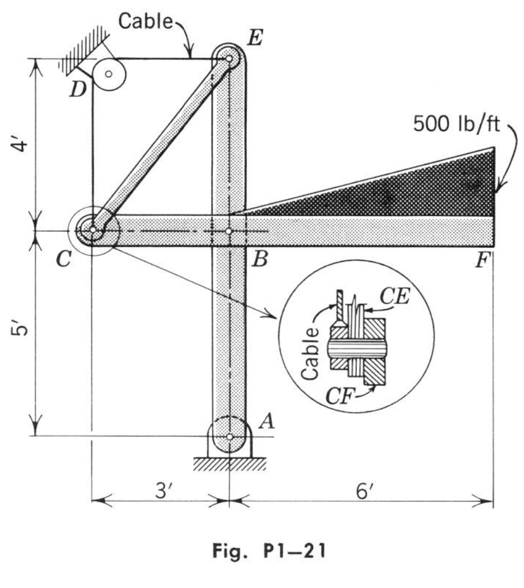


**Fig. P1–21**

**1–5. Stresses Under Biaxial Loading.** Although the expressions developed in Art. 1–4 are of restricted value, being limited to centric loading, the free-body diagram method used to obtain them can be used with equal facility for problems involving loading in two directions (biaxial loading). The solution to the following example illustrates this method of approach.

**Example 1–2:** At a given point in a machine element, the following stresses were evaluated: 8000 psi $T$ and zero shear on a horizontal plane and 4000 psi $C$ on a vertical plane. Determine the stresses at this point on a plane having a slope of 3 vertical to 4 horizontal.

*Solution:* As an aid to visualization of the data, it is suggested that the differential block of Fig. 1–12*a* be drawn (a stress picture, not a free-body diagram). The next step is to draw a free-body diagram subjected to ordinary force vectors. Various satisfactory free-body diagrams can be used such as the wedge-shaped element defined by the three given planes in Fig. 1–12*b*, in which the shaded area indicates the plane on which the stresses to be evaluated are acting. To this area is assigned arbitrarily the magnitude $dA$, and the corresponding areas of the horizontal and vertical faces of the element are 0.8 $dA$ and 0.6 $dA$, respectively. The forces acting on these areas will be as indicated in the diagram. It must be emphasized that *force vectors* act on these areas. For practical purposes, the two-dimensional free-body diagram of Fig. 1–12*c* will be adequate. Summing forces in the $n$ direction

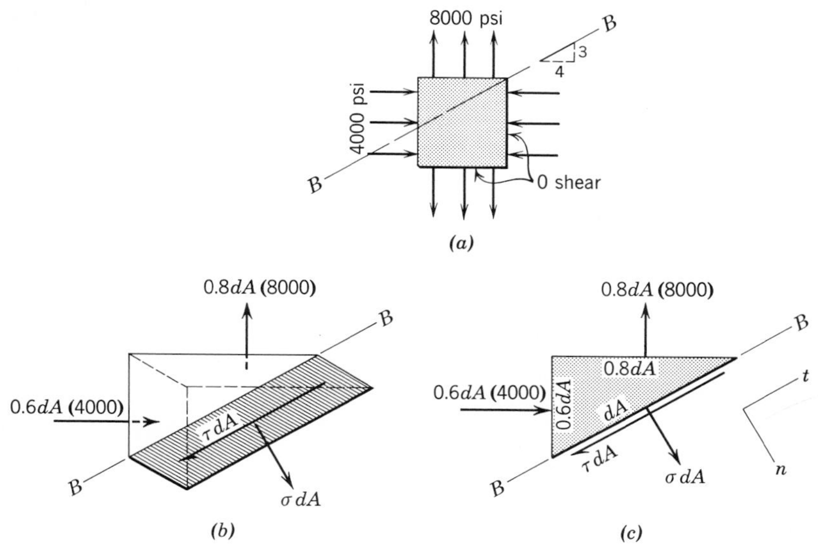


Fig. 1–12

yields the following:

$$(+\searrow)\quad \Sigma F_n = 0, \qquad \sigma\, dA + 0.6\, dA\,(4000)(0.6) - 0.8\, dA\,(8000)(0.8) = 0,$$

from which $$\sigma = \underline{3680 \text{ psi } T}. \qquad \text{Ans.}$$

When forces are summed in the $t$ direction, the shearing stress is found to be

$$\tau = \underline{5760 \text{ psi.}} \qquad \text{Ans.}$$

Note that the normal stress should be designated as *tension* or *compression.* The presence of shearing stresses on the horizontal and vertical planes, had there been any, would merely have required two more forces on the free-body diagram: one parallel to the vertical face and one parallel to the horizontal face. Note, however, that the magnitude of the shearing stresses (not the forces) must be the same on any two orthogonal planes (see Art. 1–4) and that the vectors should be directed as in Fig. 1–9 or both vectors reversed, depending on the given data.

## PROBLEMS

**1–22.*** A structural steel bar 1.20 in. in diameter is subjected to an axial tensile load of $5000\pi$ lb. Determine:

(*a*) The normal and shearing stresses on a plane through the bar that makes an angle of 30° with the direction of the load.

(*b*) The maximum normal and shearing stresses in the bar.

**1–23.*** The normal stress on plane $AB$ of the rectangular block of Fig. P1–23 is 400 psi $C$. Determine:

(*a*) The load $P$.

(*b*) The shearing stress on plane $AB$.

(*c*) The maximum shearing stress on the block.

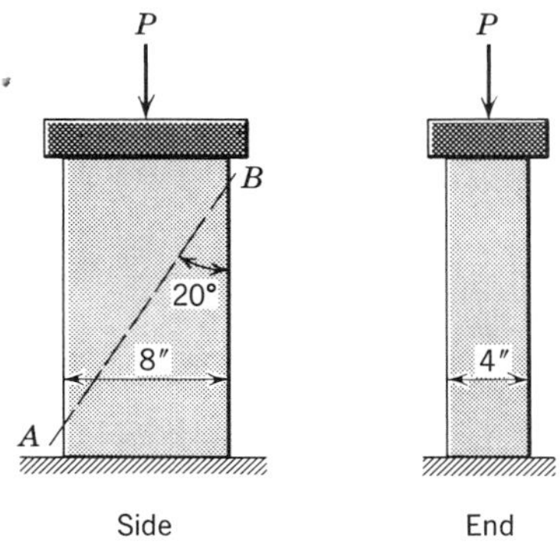


Fig. P1–23

**1–24.*** Specifications for the 3 by 3 by 21-in. block of Fig. P1–24 require that the normal stress and shearing stress on plane *A–A* not exceed 400 psi and 120 psi, respectively. Determine the maximum value of the load *P* that can be applied without exceeding the given requirements.

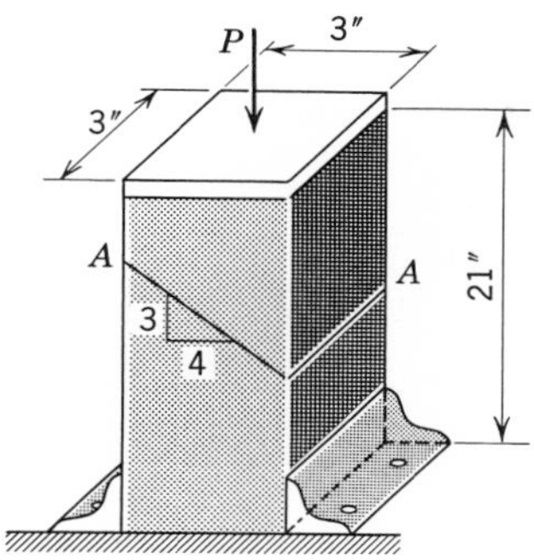


Fig. P1–24

**1–25.*** The coupling of Fig. P1–25 is connected by two ½-in. diameter flange bolts (one on each side). Specifications for the bolts require that the axial tensile stress not exceed 12.0 ksi and the cross shearing stress not exceed 8.0 ksi. Determine the maximum value of the coupling load *P* that can be applied without exceeding either of the given requirements. *Note:* The cross-sectional area of the bolts at the root of the threads is 0.126 sq in.

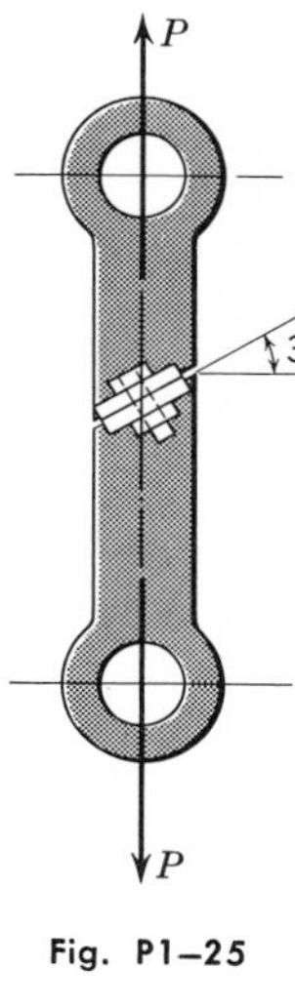


Fig. P1–25

**1–26.** Specifications for the timber block of Fig. P1–26 require that the stresses not exceed the following: shear parallel to the grain 150 psi, shear on any other plane 200 psi, and compression perpendicular to the grain 600 psi. Determine the maximum value of the axial load $P$ that can be applied without exceeding any of the given requirements.

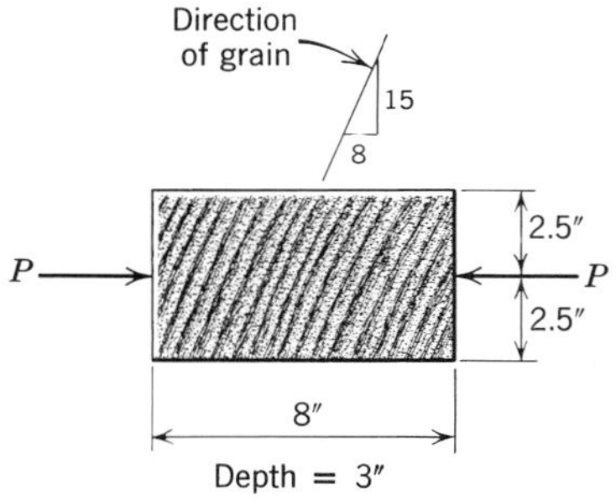


Fig. P1–26

**1–27.*** Because of internal pressure in a boiler, the stresses at a particular point in the boiler plate were found to be as shown in the stress picture of Fig. P1–27. Determine the normal stress at this point on the inclined plane shown.

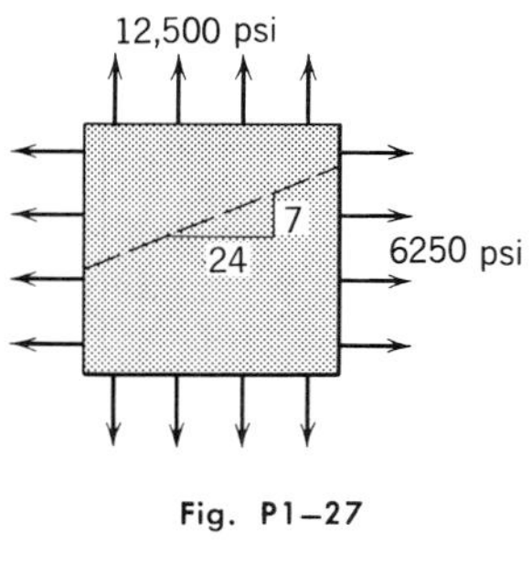


Fig. P1–27

**1–28.** At a point in a stressed body, there are normal stresses of 18 ksi $T$ on a vertical plane and 10 ksi $C$ on a horizontal plane. The shearing stresses on these planes are zero. Determine the normal and shearing stresses on a plane which makes an angle of 30° with the horizontal plane, and an angle of 60° with the vertical plane.

**1–29.** At a point in a structural member, there is a horizontal tensile stress $\sigma_x$ of 3000 psi and an unknown vertical normal stress $\sigma_y$. On a plane having a slope of 12 vertical ($y$) to 5 horizontal ($x$), there is a tensile stress of 4000 psi and an unknown shearing stress. There are no shearing stresses on horizontal and vertical planes through the point. Determine:

(*a*) The vertical normal stress.

(*b*) The shearing stress on the inclined plane.

**1–30.*** At a point in a structural member, there are stresses on horizontal and vertical planes as shown in Fig. P1–30, and the magnitude of $\sigma_c$ is three times the magnitude of $\sigma_t$. Specifications require that the shearing stress on plane $AB$ not exceed 4000 psi, and the normal stress on $AB$ not exceed 7800 psi. Determine the maximum value of $\sigma_c$ within the given specifications.

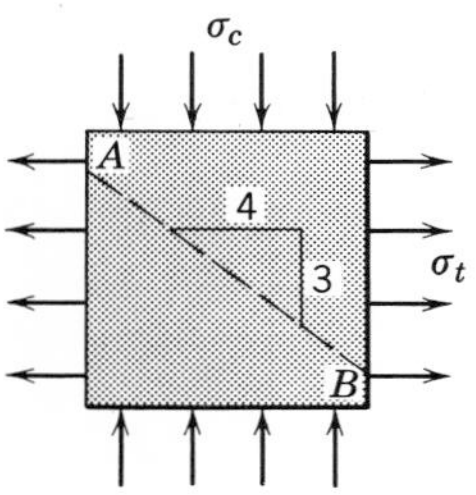


Fig. P1–30

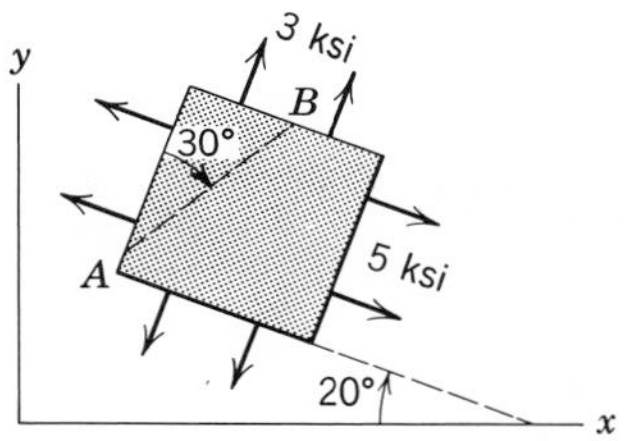


Fig. P1–31

**1–31.** Figure P1–31 represents the stress at a point in a stressed body. Determine the normal and shearing stresses at the point on plane $AB$.

**1–32.*** Figure P1–32 represents the stress situation at a point in a body. The normal stress on the inclined plane $AB$ is 3000 psi $C$. If the shearing stress on the horizontal plane is 1000 psi as shown, determine the shearing stress on plane $AB$.

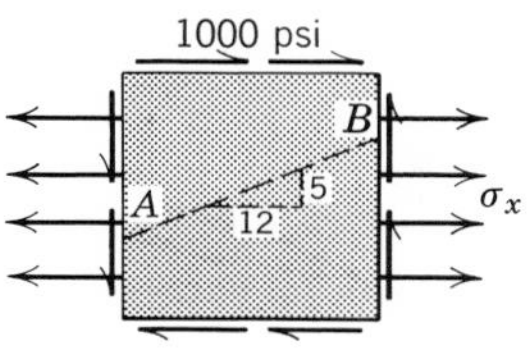


Fig. P1–32

**1–33.*** On a vertical plane through a point in a structural member, there is a shearing stress of 1000 psi as shown in Fig. P1–33. There are no normal stresses on horizontal or vertical planes through the point. Determine the normal stress on plane *B–B*.

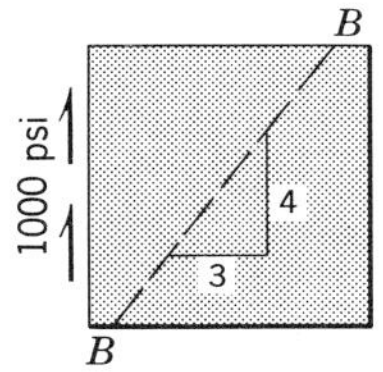


Fig. P1–33

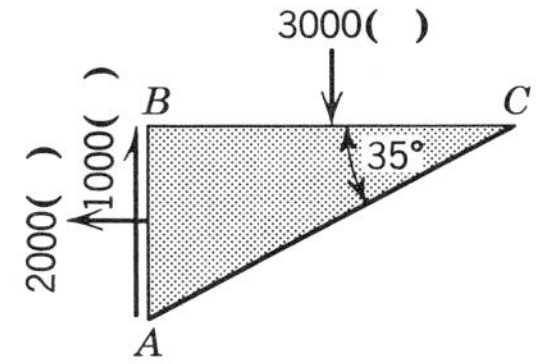


Fig. P1–34

**1–34.** Figure P1–34 is an incomplete free-body diagram for the determination of stresses at a point in a machine element. Draw the complete free-body diagram, and determine the normal stress on plane *AC*. The stresses indicated are in psi.

**1–35.** Solve Prob. 1–34 with the given angle 20° instead of 35°.

**1–36.*** At a point in a body there are stresses as shown in Fig. P1–36. The normal stress on plane *AB* is 4.56 ksi *T*. Determine the stress $\sigma_x$.

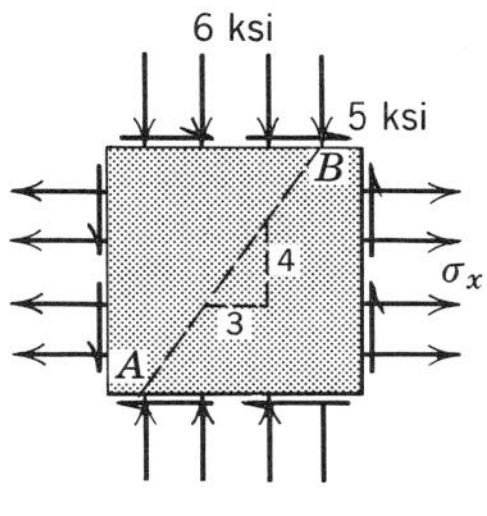


Fig. P1–36

**1–37.** The stresses on horizontal and vertical planes at a point are as indicated in Fig. P1–37. The normal stress at this point on plane *A–A* is 8000 psi *T*. Determine the magnitude of the shearing stresses $\tau_h$ and $\tau_v$.

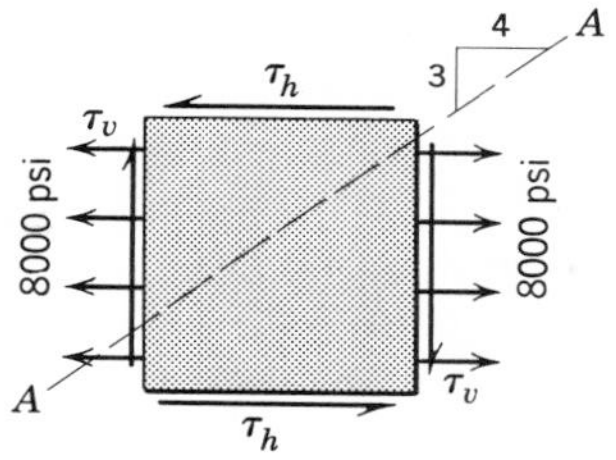


Fig. P1–37

**1–38.*** At point $A$ in the structural element in Fig. P1–38, the stresses on the inclined plane are $\sigma = 3600$ psi $T$ and $\tau = 1400$ psi in the directions indicated, and the normal stress on a vertical plane is zero. Determine the normal stress on a horizontal plane through this point.

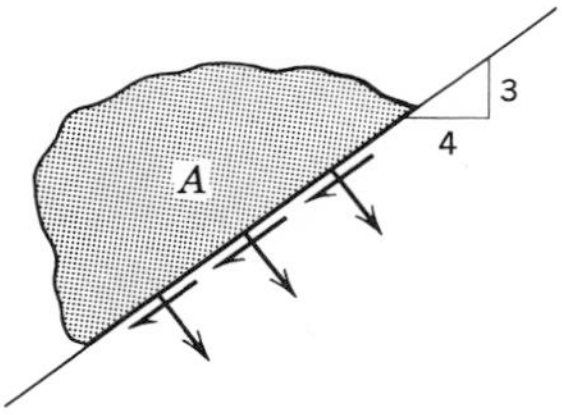


Fig. P1–38

**1–39.** At point $A$ in the structural element of Fig. P1–38, the known stresses are 2400 psi $T$ on the inclined plane and 1800 psi $T$ and zero shear on a horizontal plane. Determine the normal stress on a vertical plane through the point.

**1–40.*** At a point in a stressed region, the known stresses are as shown in Fig. P1–40. Determine the maximum normal stress at the point. From the free-body diagram, write an expression for $\sigma_n$ on plane $AB$ in terms of the angle $\theta$ and by differentiation, determine the angle for which the normal stress is maximum.

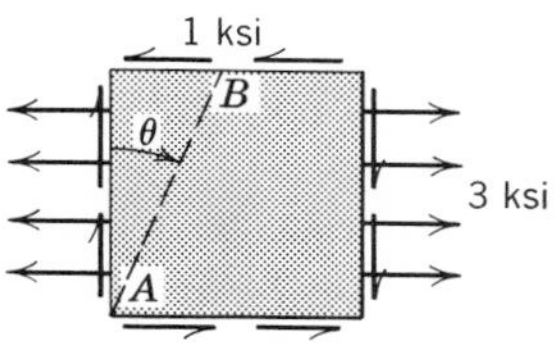


Fig. P1–40

**1–6. Principal Stresses—Plane Stress.** As mentioned in Art. 1–3, a thorough understanding of the physical significance of stress is fundamental in mechanics of materials. The concept of stress (internal force per unit area) at a point in a body on a plane through the point must be kept constantly in mind. The fact that the stresses are, in general, different for different planes through any one point has been emphasized previously.

The sign convention for stresses is given in Fig. 1–13, which illustrates a small element whose faces are parallel to the coordinate planes. The six faces are denoted by the directions of their outward normals so that the positive $x$ face is the one whose outward normal is in the direction of the positive $x$ axis. The coordinate axes $x$, $y$, and $z$ are arranged as a right-hand system. Normal stresses (indicated by the symbol $\sigma$ and a single subscript to indicate the plane on which the stress acts) are positive if they point in the direction of the outward normal. Thus, normal stresses are positive if tensile. Shearing stresses are denoted by the symbol $\tau$ followed by two subscripts; the first subscript designates the plane on which the shearing stress acts and the second the coordinate axis to which it is parallel. Thus $\tau_{xz}$ is the shearing stress on an $x$ plane parallel to the $z$ axis. A positive shearing stress points in the positive direction of the coordinate axis of the second subscript if it acts on a surface with an outward normal in the positive direction. Conversely, if the outward normal of the

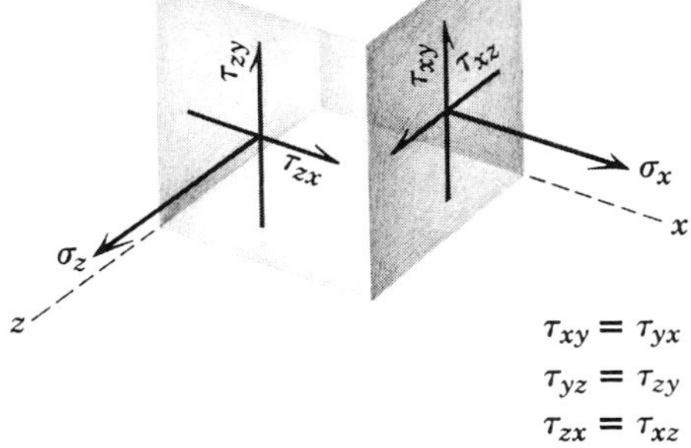


Fig. 1–13

surface is in the negative direction, then the positive shearing stress points in the negative direction of the coordinate axis of the second subscript. The stresses shown on the element in Fig. 1–13 are all positive.

For simplicity, the following discussion will be limited to plane stress, defined as a stress condition such that, of the six possible components of stress indicated in Fig. 1–13, all components in one direction (for example $\tau_{xz}$, $\tau_{yz}$, and $\sigma_z$) are zero.

Three important facts about stresses were developed in Art. 1–4. They are: (1) The maximum normal stress for a centric tensile or compressive load in one direction is in the direction of the load on a plane normal to the load, and its magnitude is $\sigma = P/A$. (2) The maximum shearing stress for a centric tensile or compressive load in one direction is on a plane at an angle of 45° to the load, and its magnitude is $\tau = \sigma/2 = P/(2A)$. (3) For all types of loads, a shearing stress at any point on any plane is accompanied by a shearing stress of equal magnitude on an orthogonal plane through the point.

Once the normal and shearing stresses on orthogonal planes through a point in a body are known, the normal and shearing stresses on any specific oblique plane through the same point can be determined by the free-body diagram method of Art. 1–5. Usually, the critical stresses are the maximum tensile and compressive stresses and the maximum shearing stresses. Therefore, the method of Art. 1–5 is inadequate, unless the slopes of the planes on which the maximum stresses act are known. For a centric load in one direction, the planes on which maximum and minimum normal stresses and maximum shearing stresses act are readily available from the developments of Art. 1–4. As more complicated loads are studied, more general methods for finding the critical stresses become necessary. Two such methods will be presented.

*The maximum and minimum normal stresses at a point in a stressed body are defined as principal stresses, and the planes on which they act are called principal planes. The third principal stress for the case of plane stress in the xy plane is* $\sigma_z$ *which is equal to zero.* The free-body diagram method of Art. 1–5 will be used to derive formulas for the principal stresses and the maximum shearing stress at a point in a stressed member in terms of known stresses on orthogonal planes through the point.

These formulas will be derived for the plane stress situation indicated in Fig. 1–14*a*, where the plane *A–A* represents the trace of any plane through the point (all planes are perpendicular to the plane of zero stress—the plane of the paper). In the following derivation, the angle $\theta$ is positive counterclockwise.

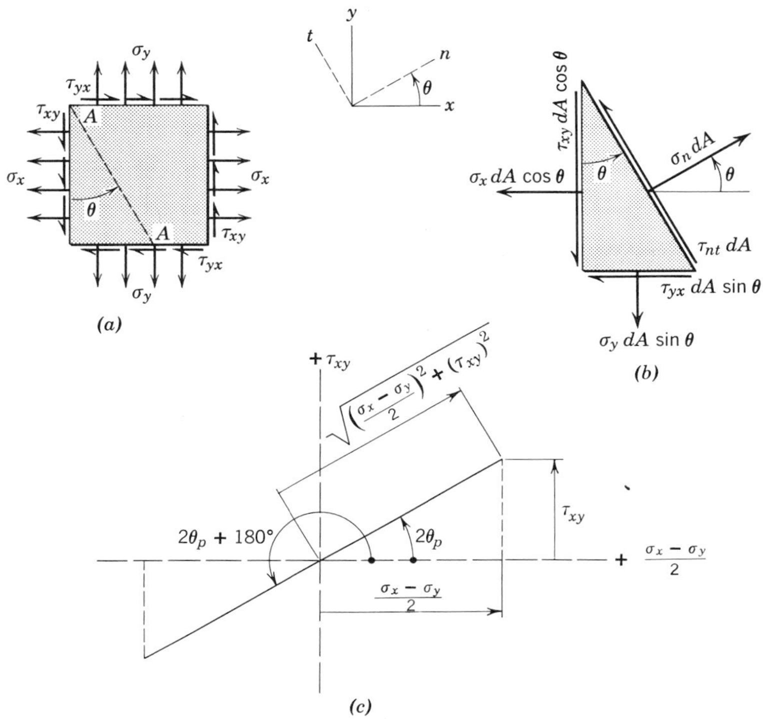


Fig. 1–14

Figure 1–14*b* is a free-body diagram of a wedge-shaped element in which the areas of the faces are $dA$ for the inclined face (plane $A$–$A$), $dA \cos \theta$ for the vertical face, and $dA \sin \theta$ for the horizontal face. The force equation of equilibrium in the $n$ direction gives

$$\sigma_n \, dA - \tau_{yx}(dA \sin \theta) \cos \theta - \tau_{xy}(dA \cos \theta) \sin \theta - \sigma_x(dA \cos \theta) \cos \theta - \sigma_y(dA \sin \theta) \sin \theta = 0,$$

from which, since $\tau_{yx} = \tau_{xy}$,

$$\sigma_n = \frac{\sigma_x(1 + \cos 2\theta)}{2} + \frac{\sigma_y(1 - \cos 2\theta)}{2} + \frac{2\tau_{xy}(\sin 2\theta)}{2}$$

$$= \frac{\sigma_x + \sigma_y}{2} + \frac{\sigma_x - \sigma_y}{2} \cos 2\theta + \tau_{xy} \sin 2\theta. \qquad \text{(a)}$$

The principal stresses (maximum and minimum values of $\sigma_n$) occur at the values of $\theta$ for which $d\sigma_n/d\theta$ is equal to zero.
Differentiation of $\sigma_n$ with respect to $\theta$ gives

$$\frac{d\sigma_n}{d\theta} = -(\sigma_x - \sigma_y)\sin 2\theta + 2\tau_{xy}\cos 2\theta. \tag{b}$$

Setting Eq. (b) equal to zero and solving gives

$$\tan 2\theta_p = \frac{\tau_{xy}}{(\sigma_x - \sigma_y)/2}. \tag{1–3}$$

The values of $\theta_p$ from Eq. (1–3) give the orientation of the principal planes. For a given set of values of $\tau_{xy}$, $\sigma_x$, and $\sigma_y$, there are two values of $2\theta_p$ differing by 180° and, consequently, two values of $\theta_p$ 90° apart. This proves that the principal planes are normal to each other.

When $\tau_{xy}$ and $(\sigma_x - \sigma_y)$ have the same sign, $\tan 2\theta_p$ is positive and one value of $2\theta_p$ is between 0° and 90°, with the other value 180° greater, as shown in Fig. 1–14*c*. Consequently, one value of $\theta_p$ is between 0° and 45°, and the other one is 90° greater. In the first case, both $\sin 2\theta_p$ and $\cos 2\theta_p$ are positive, and in the second case both are negative. When these functions of $2\theta_p$ are substituted into Eq. (a), the principal stresses are found to be

$$\sigma_{p1,p2} = \frac{\sigma_x + \sigma_y}{2} + \frac{\sigma_x - \sigma_y}{2}\left[\pm\frac{(\sigma_x - \sigma_y)/2}{\sqrt{\left(\frac{\sigma_x - \sigma_y}{2}\right)^2 + \tau_{xy}^2}}\right] + \tau_{xy}\left[\pm\frac{\tau_{xy}}{\sqrt{\left(\frac{\sigma_x - \sigma_y}{2}\right)^2 + \tau_{xy}^2}}\right],$$

which reduces to

$$\sigma_{p1,p2} = \frac{\sigma_x + \sigma_y}{2} \pm \sqrt{\left(\frac{\sigma_x - \sigma_y}{2}\right)^2 + \tau_{xy}^2}. \tag{1–4}$$

Equation (1–4) gives the two principal stresses in the $xy$ plane, and the third one is $\sigma_{p3} = \sigma_z = 0$. Equation (1–3) gives the angles $\theta_p$ and $\theta_p + 90°$ between the $x$ (or $y$) axis and the mutually perpendicular planes on which the principal stresses act. When $\tan 2\theta_p$ is positive, $\theta_p$ is positive, and the rotation is counterclockwise from the $x$ and $y$ axes to the planes on which the two principal stresses act. When

$\tan 2\theta_p$ is negative, the rotation is clockwise. Note that one value of $\theta_p$ will always be between positive and negative 45° (inclusive) with the other value 90° greater. The numerically greater principal stress will act on the plane which makes an angle of 45° or less with the plane of the numerically larger of the two given normal stresses. This statement can be confirmed in any case by substitution of the value of $\theta_p$ from Eq. (1–3) into Eq. (a) to obtain the corresponding principal stress.

Note that if one or both of the principal stresses from Eq. (1–4) are negative the algebraic maximum stress can have a smaller absolute value than the "minimum" stress.

For the free-body diagram of Fig. 1–14*b*, the force equation of equilibrium in the $t$ direction gives

$$\tau_{nt}\,dA - \tau_{xy}(dA\cos\theta)\cos\theta + \tau_{yx}(dA\sin\theta)\sin\theta + \sigma_x(dA\cos\theta)\sin\theta - \sigma_y(dA\sin\theta)\cos\theta = 0,$$

from which

$$\tau_{nt} = -\frac{\sigma_x - \sigma_y}{2}\sin 2\theta + \tau_{xy}\cos 2\theta. \tag{c}$$

Note that the expression for $d\sigma_n/d\theta$ from Eq. (b) is numerically twice the value of the expression for $\tau_{nt}$ from Eq. (c). Consequently the *shearing stress is zero on the principal planes.* The maximum shearing stress occurs on planes located by values of $\theta$ where $d\tau_{nt}/d\theta$ is equal to zero. Differentiation of Eq. (c) gives

$$\frac{d\tau_{nt}}{d\theta} = -(\sigma_x - \sigma_y)\cos 2\theta - 2\tau_{xy}\sin 2\theta.$$

When $d\tau_{nt}/d\theta$ is equated to zero, the value of $\theta_\tau$ is given by the expression

$$\tan 2\theta_\tau = -\frac{(\sigma_x - \sigma_y)/2}{\tau_{xy}}. \tag{d}$$

Comparison of Eq. (d) and (1–3) reveals that the two tangents are negative reciprocals. Therefore the two angles $2\theta_p$ and $2\theta_\tau$ differ by 90°, and $\theta_p$ and $\theta_\tau$ are 45° apart. This means that the planes on which the maximum shearing stresses occur are 45° from the principal planes. The maximum shearing stresses are found by substituting values of angle functions obtained from Eq. (d) in Eq. (c). The results are

$$\tau_{\max} = -\frac{\sigma_x - \sigma_y}{2}\left[\mp \frac{(\sigma_x - \sigma_y)/2}{\sqrt{\left(\frac{\sigma_x - \sigma_y}{2}\right)^2 + \tau_{xy}^2}}\right] + \tau_{xy}\left[\pm \frac{\tau_{xy}}{\sqrt{\left(\frac{\sigma_x - \sigma_y}{2}\right)^2 + \tau_{xy}^2}}\right];$$

$$\tau_{\max} = \pm\sqrt{\left(\frac{\sigma_x - \sigma_y}{2}\right)^2 + \tau_{xy}^2}, \tag{1–5}$$

which is observed to have the same magnitude as the second term of Eq. (1–4).

A useful relation between the principal stresses and the maximum shearing stress is obtained from Eq. (1–4) and (1–5) by subtracting the values for the two principal stresses and substituting the value of the radical from Eq. (1–5). The result is

$$\tau_{\max} = (\sigma_{p1} - \sigma_{p2})/2, \tag{1–6}$$

or, in words, the maximum value of $\tau_{nt}$ is equal in magnitude to one-half the difference between the two principal stresses.

In general, when stresses act in three directions it is shown in advanced textbooks[5] that there are three orthogonal planes on which the shearing stress is zero. These planes are known as the principal planes, and the stresses acting on them (the principal stresses) will have three values: one maximum, one minimum, and a third stress between the other two. The maximum shearing stress is one-half the difference between the *maximum* and *minimum* principal stresses, and acts on planes which bisect the angles between the planes of the maximum and minimum normal stresses.

When a state of plane stress exists, one of the principal stresses is zero. If the values of $\sigma_{p1}$ and $\sigma_{p2}$ from Eq. (1–4) have the same sign, then the third principal stress, $\sigma_{p3}$ equals zero, will be either the maximum or the minimum normal stress. Thus, the maximum shearing stress may be $(\sigma_{p1} - \sigma_{p2})/2$, $(\sigma_{p1} - 0)/2$, or $(\sigma_{p2} - 0)/2$, depending on the relative magnitudes and signs of the principal stresses. These three possibilities are illustrated in Fig. 1–15, in which one of the

[5] *Theory of Elasticity,* 2nd ed., S. Timoshenko and J. N. Goodier, McGraw-Hill, New York, 1951, pp. 213–219.

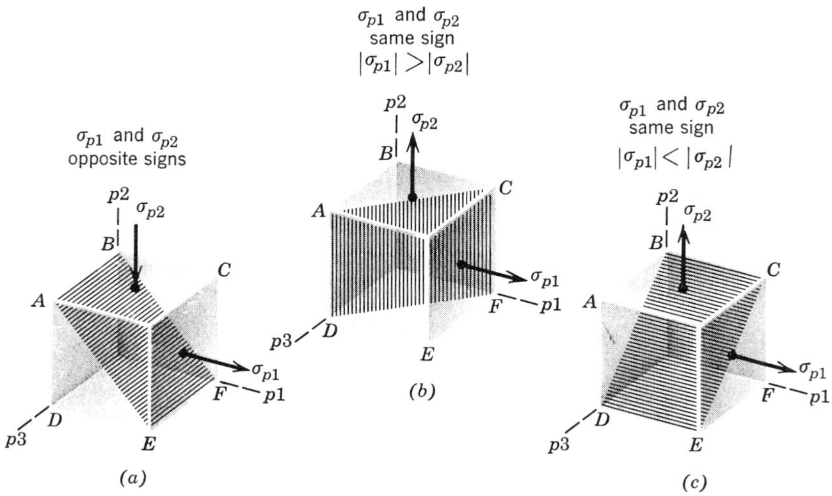


Fig. 1–15

two orthogonal planes on which the maximum shearing stress acts is hatched for each example.

The direction of the maximum shearing stress can be determined by drawing a wedge-shaped block with two sides parallel to the planes having the maximum and minimum principal stresses, and with the third side at an angle of 45° with the other two sides. The direction of the maximum shearing stress must oppose the larger of the two principal stresses.

Another useful relation between the principal stresses and the normal stresses on the orthogonal planes, shown in Fig. 1–14, is obtained by adding the values for the two principal stresses as given by Eq. (1–4). The result is

$$\sigma_{p1} + \sigma_{p2} = \sigma_x + \sigma_y \tag{e}$$

or, in words, *for plane stress, the sum of the normal stresses on any two orthogonal planes through a point in a body is a constant or invariant.*

In the preceding discussion, "maximum" and "minimum" stresses were considered as algebraic quantities, and it has already been pointed out that the minimum algebraic stress may have a larger magnitude than the maximum stress. However, in the application to engineering problems (which includes the problems in this book), the term "maximum" will always refer to the largest absolute value (largest magnitude).

The application of the formulas and procedures developed in this article are illustrated by the following examples.

**Example 1–3:** At a point in a structural member subjected to plane stress there are stresses on horizontal and vertical planes through the point as shown in Fig. 1–16*a*.

(*a*) Determine the principal stresses and the maximum shearing stress at the point.

(*b*) Locate the planes on which these stresses act and show the stresses on a complete sketch.

*Solution:* (*a*) On the basis of the established sign convention, $\sigma_x$ is positive, whereas $\sigma_y$ and $\tau_{xy}$ are negative. For use in Eq. (1–3) and (1–4), the given values are

$$\sigma_x = +10{,}000 \text{ psi}, \quad \sigma_y = -8000 \text{ psi}, \quad \text{and} \quad \tau_{xy} = -4000 \text{ psi}.$$

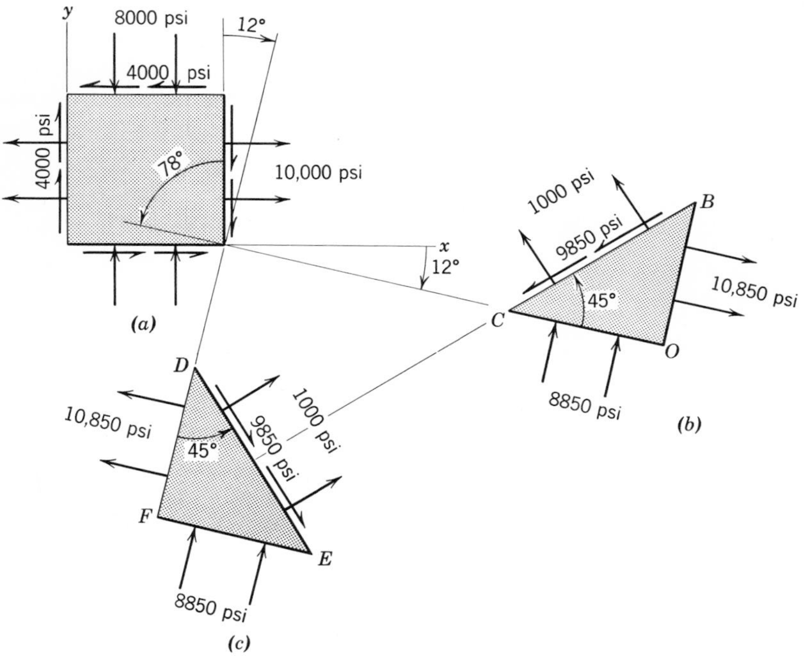


Fig. 1–16

When these values are substituted in Eq. (1–4), the principal stresses are found to be

$$\sigma_p = \frac{10{,}000 - 8000}{2} \pm \sqrt{\left(\frac{10{,}000 + 8000}{2}\right)^2 + (-4000)^2}$$

$$= 1000 \pm 9850$$

$$= \underline{10{,}850 \text{ psi } T \text{ and } 8850 \text{ psi } C} \qquad \text{Ans.}$$

$$\sigma_z = \underline{0}. \qquad \text{Ans.}$$

Since $\sigma_{p1}$ and $\sigma_{p2}$ are of opposite sign, the maximum shearing stress is

$$\tau_{\max} = \frac{10{,}850 - (-8850)}{2} = \underline{9850 \text{ psi.}} \qquad \text{Ans.}$$

(*b*) When the given data are substituted in Eq. (1–3), the results are

$$\tan 2\theta_p = -4000/9000 = -0.4444,$$

from which $2\theta_p = -23.96°$

and $\theta_p = -11.98° \text{ or } 12°\circlearrowright.$

The required sketch is shown in Fig. 1–16*b* or *c*.

From Eq. (e), the sum of the normal stresses on any two orthogonal planes is a constant for plane stress. Planes *CB* and *DE* of Fig. 1–16 are orthogonal and the normal stresses on them are obviously equal. Therefore,

$$2\sigma_n = \sigma_x + \sigma_y = 10{,}000 - 8000 = 2000$$

and $\sigma_n = 1000$ psi on the planes of maximum shear.

**Example 1-4:** Solve Ex. 1–3 when $\sigma_y$ is 8000 psi $T$ with $\sigma_x$ and $\tau_{xy}$ as shown, see Fig. 1–17*a*.

*Solution:* (*a*) The principal stresses, expressed in kips per square inch (ksi), are

$$\sigma_{p1,p2} = \frac{10 + 8}{2} \pm \sqrt{\left(\frac{10 - 8}{2}\right)^2 + 4^2}$$

$$= 9 \pm 4.123$$

$$= \underline{13.12 \text{ ksi } T \text{ and } 4.88 \text{ ksi } T}. \qquad \text{Ans.}$$

Also $\sigma_{p3} = \sigma_z = \underline{0}.$ Ans.

Since $\sigma_{p1}$ and $\sigma_{p2}$ have the same sign, the maximum shearing stress is

$$\tau_{\max} = (13.12 - 0)/2 = \underline{6.56 \text{ ksi.}} \qquad \text{Ans.}$$

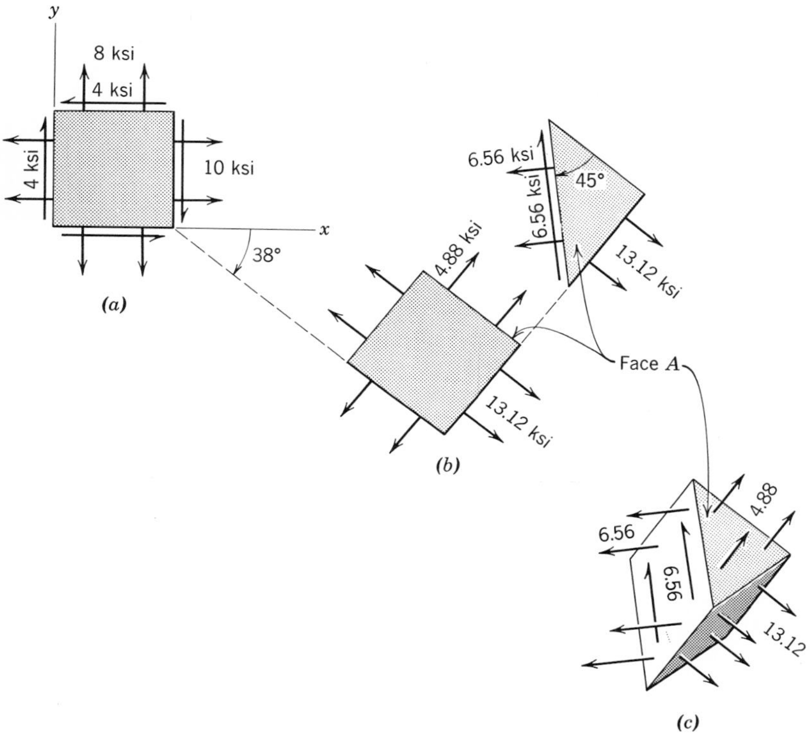


Fig. 1–17

($b$) When the given data are substituted in Eq. (1–3), the results are

$$\tan 2\theta_p = -4/1 = -4,$$

from which

$$2\theta_p = -75.97°$$

and

$$\theta_p = -37.985° \text{ or } 38°\curvearrowright.$$

The maximum shearing stress occurs on the plane making an angle of 45° with the planes of maximum and minimum normal stresses—in this case, 13.12 ksi and zero. The complete sketch is given in Fig. 1–17$b$ where the upper (wedge-shaped) block is the orthographic projection of the lower block. The three-dimensional Fig. 1–17$c$ is presented as an aid to the visualization of Fig. 1–17$b$.

## PROBLEMS

**The following problems are assumed to be plane stress problems unless otherwise noted. The required sketches are to be complete, indicating all pertinent angles and stresses.**

**1–41.*** At a point in a stressed body, the normal stresses are 8000 psi $C$ on a vertical plane and 12,000 psi $T$ on a horizontal plane. A positive shearing stress of 6000 psi acts on a vertical plane at the point. Determine, and show on a sketch, the principal stresses and the maximum shearing stress at the point.

**1–42.*** At a point in a stressed machine part, the known stresses are 20 ksi $T$ and $-8$ ksi shear on a vertical plane, and 10 ksi $C$ on a horizontal plane. Determine, and show on a sketch, the principal and maximum shearing stresses at the point.

**1–43.*** For the stresses given in Prob. 1–34, determine, and show on a sketch, the principal stresses and the maximum shearing stress.

**1–44.** A point within a structural member is subjected to the stresses in Fig. P1–44.

(*a*) Determine, and show on a sketch, the principal and maximum shearing stresses at the point.

(*b*) Determine, by the free-body diagram method, the shearing and normal stresses on plane $AB$.

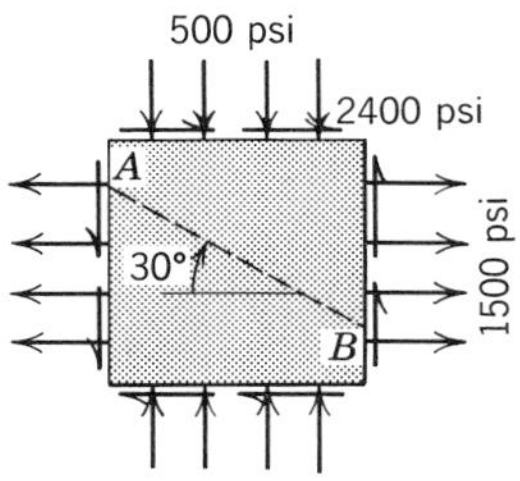


Fig. P1–44

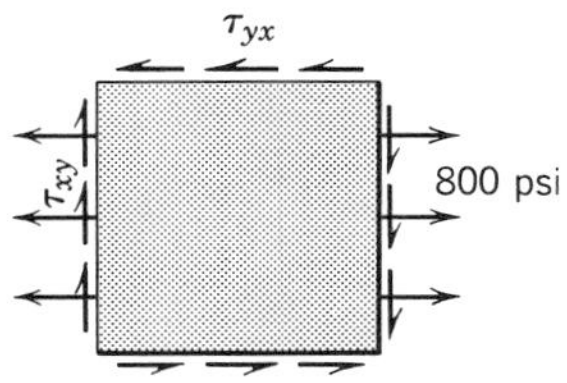


Fig. P1–45

**1–45.*** The stresses shown in Fig. P1–45 act at a point in a structural member. The tensile principal stress is known to be 1200 psi. Determine:

(*a*) The maximum shearing stress.

(*b*) The orientation of the planes on which the stress of part *a* acts.

(*c*) The shearing stress on the horizontal plane.

**1–46.*** At a point in a stressed body, the normal stress and shearing stress acting on a horizontal plane are 5 ksi $C$ and $-3$ ksi respectively. The normal stress acting on a vertical plane through the point is 1 ksi $T$. Determine, and show on a sketch, the principal and maximum shearing stresses.

**1–47.** At a point in a stressed region, there exists on a vertical plane a normal stress of 19 ksi *T*, and an unknown negative shearing stress. The maximum principal stress at the point is 20 ksi *T*, and the maximum shearing stress has a magnitude of 13 ksi. Determine the unknown stresses on the horizontal and vertical planes and show these stresses as well as the principal stresses on a sketch.

**1–48.** At a point in a loaded member, there are stresses of 14 ksi *C* and −5 ksi shear on a vertical plane, and 10 ksi *T* on a horizontal plane. Determine, and show on a sketch, the principal and maximum shearing stresses at the point.

**1–49.** The vertical plane through a point in a member has a normal stress of 6 ksi *T* and a shearing stress of −6 ksi and the horizontal plane through the point has a normal stress of 20 ksi *C*. Determine, and show on a sketch, the principal and maximum shearing stresses at the point.

*Problems 1–50 through 1–56 are problems in plane stress in which certain stresses or angles are given. Determine the unknown quantities for each problem, and draw a sketch indicating the stresses and planes on which they act. Note that in some problems there might be more than one possible value of* $\theta_p$ *depending on the sign of* $\tau_{xy}$.

| | $\sigma_x$ | $\sigma_y$ | $\tau_{xy}$ | $\sigma_{p1}$ | $\sigma_{p2}$ | $\tau_{max}$ | $\theta_p$ |
|---|---|---|---|---|---|---|---|
| **1–50.*** | 5,000 | −11,000 | 6,000 | | | | |
| **1–51.** | −12,000 | 3,000 | −4,000 | | | | |
| **1–52.*** | −16,000 | −8,000 | 3,000 | | | | |
| **1–53.** | 4,000 | −6,000 | | 11,000 | | | |
| **1–54.*** | 6,000 | 4,000 | | | −12,000 | | |
| **1–55.** | 6,000 | | | 9,000 | | 5,600 | |
| **1–56.*** | | | 3,000 | 8,000 | | | 30° |

**1–7. Mohr's Circle for Principal Stresses.** The equations for finding the principal stresses and the maximum shearing stresses at a point in a stressed member were developed in Art. 1–6. The German engineer Otto Mohr (1835–1918) developed a useful pictorial or graphical interpretation of these equations.[6] This method, commonly called Mohr's circle,[7] involves the construction of a circle in such a manner that the coordinates of each point on the circle represent the normal and shear-

[6] The particular case for plane stress was presented by the German engineer K. Culmann (1829–1881) about sixteen years before Mohr's paper.

[7] Mohr's circle is also used as a pictorial interpretation of the formulas for obtaining maximum and minimum second moments of areas and maximum products of inertia. See *Engineering Mechanics*, by Higdon and Stiles, Prentice-Hall, Englewood Cliffs, N.J., 1955, 1962. Mohr's circle is also used for plane strain, three-dimensional stress, and maximum and minimum moments of inertia of mass.

ing stresses on one plane through the stressed point, and the angular position of the radius to the point gives the orientation of the plane. Normal stresses are plotted as horizontal coordinates, with tensile stresses considered positive, and compressive stresses considered negative. Shearing stresses are plotted as vertical coordinates, with shearing forces that give a clockwise couple considered positive, and shearing forces that produce a counterclockwise couple considered negative. Note that this convention for plotting shearing stresses differs from that stated in Art. 1–6. This convention is used *only* with Mohr's circle.

The Mohr's circle for any point subjected to plane stress can be drawn when the stresses on two mutually perpendicular planes through the point are known. Consider, for example, the stress situation of Fig. 1–14*a* with $\sigma_x$ greater than $\sigma_y$, and plot in Fig. 1–18 the points representing the given stresses. The coordinates of point $V$ of Fig. 1–18 are the stresses on the vertical plane through the stressed

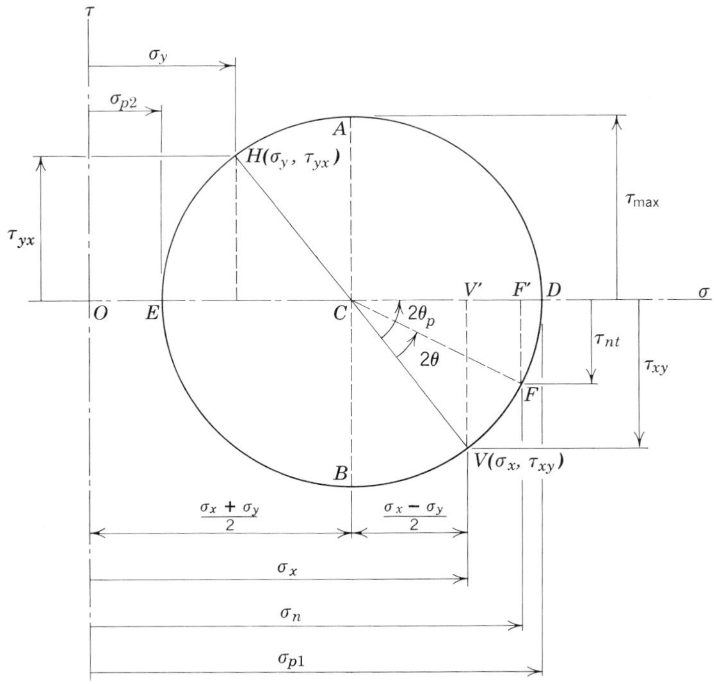


Fig. 1–18

point of Fig. 1–14*a*, and point $H$ is determined by the stresses on the horizontal plane through the point. Line $VH$ is the diameter of Mohr's circle for the stresses at the point. Because $\tau_{yx}$ is equal to $\tau_{xy}$, point $C$, the center of the circle, is on the $\sigma$ axis. Line $CV$ on Mohr's circle represents the plane (the vertical plane of Fig. 1–14*a*) through the stressed point from which the angle $\theta$ is measured. The coordinates of each point on the circle represent $\sigma_n$ and $\tau_{nt}$ for one particular plane through the stressed point, the abscissa representing $\sigma_n$ and the ordinate representing $\tau_{nt}$. To demonstrate this statement, draw any radius $CF$ in Fig. 1–18 at an angle $2\theta$ counterclockwise from radius $CV$. From the figure, it is apparent that

$$OF' = OC + CF \cos(2\theta_p - 2\theta),$$

and since $CF$ equals $CV$, the above equation reduces to

$$OF' = OC + CV \cos 2\theta_p \cos 2\theta + CV \sin 2\theta_p \sin 2\theta.$$

By referring to the figure, the expression for $OF'$ reduces to

$$\begin{aligned} OF' &= OC + CV' \cos 2\theta + VV' \sin 2\theta \\ &= \frac{\sigma_x + \sigma_y}{2} + \frac{\sigma_x - \sigma_y}{2} \cos 2\theta + \tau_{xy} \sin 2\theta. \end{aligned}$$

This expression is identical to Eq. (a) of Art. 1–6. Therefore $OF'$ is equal to $\sigma_n$. In a similar manner,

$$\begin{aligned} F'F &= CF \sin(2\theta_p - 2\theta) \\ &= CV \sin 2\theta_p \cos 2\theta - CV \cos 2\theta_p \sin 2\theta \\ &= V'V \cos 2\theta - CV' \sin 2\theta \\ &= \tau_{xy} \cos 2\theta - \frac{\sigma_x - \sigma_y}{2} \sin 2\theta = \tau_{nt}, \end{aligned}$$

which is Eq. (c) of Art. 1–6.

Since the horizontal coordinate of each point on the circle represents a particular value of $\sigma_n$, the maximum normal stress is represented by $OD$, and its value is

$$\begin{aligned} \sigma_{p1} &= OD = OC + CD = OC + CV \\ &= \frac{\sigma_x + \sigma_y}{2} + \sqrt{\left(\frac{\sigma_x - \sigma_y}{2}\right)^2 + \tau_{xy}^2}, \end{aligned}$$

which agrees with Eq. (1–4).

Likewise, the vertical coordinate of each point on the circle represents a particular value of $\tau_{nt}$, which means that the maximum shearing stresses are represented by $CA$ and $CB$, and their value is

$$\tau_{\max} = CA = CV = \sqrt{\left(\frac{\sigma_x - \sigma_y}{2}\right)^2 + \tau_{xy}^2},$$

which agrees with Eq. (1–5).

The angle $2\theta_p$ from $CV$ to $CD$ is counterclockwise or positive, and its tangent is

$$\tan 2\theta_p = \frac{\tau_{xy}}{(\sigma_x - \sigma_y)/2},$$

which is Eq. (1–3). From the derivation of Eq. (1–4), the angle between the vertical plane and one of the principal planes was $\theta_p$. In obtaining the same equation from Mohr's circle, the angle between the radii which represent these same two planes is $2\theta_p$. In other words, all angles in Mohr's circle are twice the corresponding angles for the actual stressed body. The angle from the vertical plane to the horizontal plane in Fig. 1–14*a* is 90°, but in Fig. 1–18 the angle between line $CV$ (which represents the vertical plane) and line $CH$ (which represents the horizontal plane) in Mohr's circle is 180°.

The results obtained from Mohr's circle have been shown to be identical with the equations derived from the free-body diagram of Art. 1–6. Consequently, Mohr's circle provides an extremely useful aid in the solution for, and visualization of the stresses on various planes through a point in a stressed body in terms of the stresses on two mutually perpendicular planes through the point. Although Mohr's circle can be drawn to scale and used to obtain values of stresses and angles by direct measurements on the figure, it is probably more useful as a pictorial aid to the analytic determination of the desired quantities. The relationships between the stresses and the planes on which they act are readily apparent from the circle. Problems of the type presented in Art. 1–5 as well as those of Art. 1–6 can readily be solved by this semi-graphical method as will be illustrated in the following example.

**Example 1–5:** At a point in a structural member that is subjected to plane stress there are stresses on horizontal and vertical planes through the point, as shown in Fig. 1–19*a*. Determine, and show on a sketch: (*a*) The principal stresses and maximum shearing stress at the point. (*b*) The normal and shearing stresses on plane $A$–$A$ through the point.

*Solution:* Mohr's circle is constructed from the given data as follows: On a set of coordinate axes, Fig. 1–19*b*, plot point $V$ (representing the stresses on the vertical plane) at (8, −4) because the stresses on the vertical plane are 8 ksi $T$ and −4 ksi (counterclockwise) shear. Likewise, point $H$ (representing the stresses on the horizontal plane) has the coordinates (−6, 4). Draw

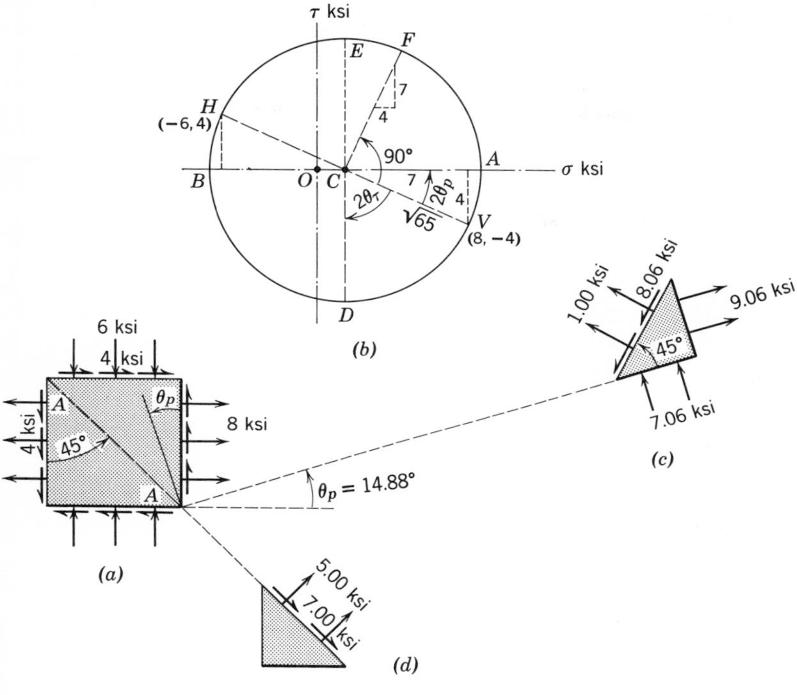


Fig. 1–19

line $HV$, which is a diameter of Mohr's circle, and note that the center of the circle is at (1, 0). The radius of the circle is

$$CV = \sqrt{7^2 + 4^2} = 8.06 \text{ ksi}.$$

(*a*) The principal stresses and the maximum shearing stress at the point are

$$\sigma_{p1} = OA = 1 + 8.06 = \underline{9.06 \text{ ksi } T}, \qquad \text{Ans.}$$

$$\sigma_{p2} = OB = 1 - 8.06 = -7.06 = \underline{7.06 \text{ ksi } C}, \qquad \text{Ans.}$$

$$\sigma_{p3} = \sigma_z = \underline{0}. \qquad \text{Ans.}$$

Since $\sigma_{p1}$ and $\sigma_{p2}$ have opposite signs, the maximum shearing stress is

$$\tau_{max} = CE = CD = \underline{8.06 \text{ ksi.}} \qquad \text{Ans.}$$

The principal planes are represented by lines $CA$ and $CB$ where

$$\tan 2\theta_p = 4/7 = 0.5714,$$

which gives

$$2\theta_p = 29.75° \circlearrowleft.$$

Since the angle $2\theta_p$ is counterclockwise, the principal planes are counterclockwise from the vertical and horizontal planes of the stress block, as shown in Fig. 1–19*c*. To determine which principal stress acts on which plane, note that as the radius of the circle rotates counterclockwise, the end of the radius $CV$ moves from $V$ to $A$, indicating that as the initially vertical plane rotates through 14.88°, the stresses change to 9.06 ksi $T$ and zero. Note also that the end $H$ of radius $CH$ moves to $B$, indicating that as the initially horizontal plane rotates through 14.88°, the stresses change to 7.06 ksi $C$ and zero. The required sketch is Fig. 1–19*c*.

(*b*) Plane $A$–$A$ is 45° counterclockwise from the vertical plane; therefore, the corresponding radius of Mohr's circle is 90°, $2\theta$, counterclockwise from the line $CV$, and is shown as $CF$ on Fig. 1–19*b*. The coordinates of the point $F$ are seen to be (5, 7) which means that the stresses on plane $A$–$A$ are 5.00 ksi $T$ and 7.00 ksi negative shear as shown on Fig. 1–19*d*. Note that Fig. 1–19*d* does not show all stresses acting on the element.

## PROBLEMS

**Note: The bodies in the following problems are assumed to be in a state of plane stress.**

**1–57.*** Solve Prob. 1–41 by means of Mohr's circle.

**1–58.*** Solve Prob. 1–42 by means of Mohr's circle.

**1–59.*** Solve Prob. 1–43 by means of Mohr's circle.

**1–60.** Solve Prob. 1–48 by means of Mohr's circle.

**1–61.*** Solve Prob. 1–52 by means of Mohr's circle.

**1–62.** Solve Prob. 1–28 by means of Mohr's circle.

**1–63.*** Solve Prob. 1–27 by means of Mohr's circle.

**1–64.** Solve Prob. 1–31 by means of Mohr's circle.

**1–65.*** Solve Prob. 1–33 by means of Mohr's circle.

**1–66.** Solve Prob. 1–34 by means of Mohr's circle.

**1–67.** Solve Prob. 1–44 by means of Mohr's circle.

**1–68.** For the data of Prob. 1–46, determine, and show on a sketch: (*a*) normal and shearing stresses on a plane located 10 deg. clockwise from the horizontal plane. (*b*) The principal and maximum shearing stresses.

**1–69.** At a point in a stressed body, the normal stresses are 9.60 ksi $C$ on a vertical plane and 10.40 ksi $T$ on a horizontal plane. A negative shearing stress of 8.00 ksi acts on a vertical plane at the point. Determine, and show on a sketch, the stresses on a plane 20° clockwise from the vertical plane.

**1–70*** At a point in a machine element, the normal stresses are 10,000 psi $T$ on a vertical plane and 14,000 psi C on a horizontal plane. A positive shearing stress of 6000 psi acts on the vertical plane at the point. Determine and show on a sketch:

(*a*) The stresses on a plane 30° counterclockwise from the vertical plane.
(*b*) The principal and maximum shearing stresses.

**1–71.** At a point in a stressed body, the stresses on a vertical plane are 8000 psi *C* and + 1000 psi shear. The normal stress on the horizontal plane is 2000 psi *T*. Determine, and show on a sketch:

(*a*) The stresses on a plane 60° counterclockwise from the vertical plane through the point.

(*b*) The principal and maximum shearing stresses.

**1–72.*** At a point in a stressed body, there are stresses of 1 ksi *T* on a vertical plane and 7 ksi *C* on a horizontal plane. A positive unknown shearing stress acts on the vertical plane and the maximum shearing stress at the point has a magnitude of 5 ksi. Determine the shearing stress on the vertical plane and also the principal stresses. Show the results on an appropriate sketch.

**1–73.** Solve Prob. 1–47 by means of Mohr's circle.

**1–74.*** At a point in a stressed member, there exists on the horizontal plane a normal stresss of 10 ksi *C* and an unknown positive shearing stress. One principal stress at the point is 2 ksi *C*, and the maximum shearing stress has a magnitude of 13 ksi. Determine the unknown stresses on the horizontal and vertical planes and show these stresses as well as the principal stresses on a sketch.

**1–75.** At a point in a stressed body, the principal and maximum shearing stresses are oriented as indicated in Fig. P1–75. Determine, and show on a sketch, the stresses acting on the horizontal and vertical planes at this same point.

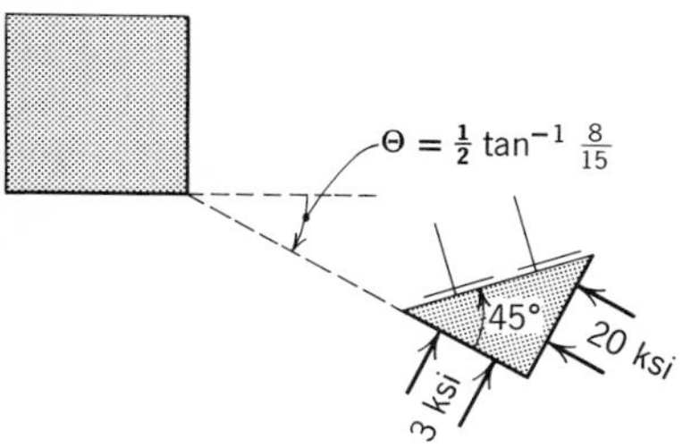


Fig. P1–75

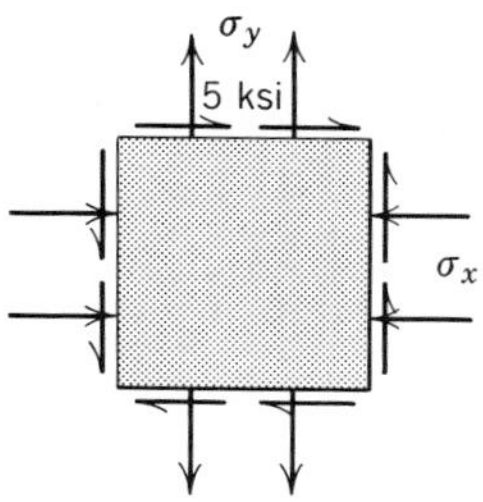


Fig. P1–76

**1–76.*** At a point in a stressed body, there are stresses on horizontal and vertical planes as shown in Fig. P1–76. The principal stresses at the point are 21 ksi *C* and 5 ksi *T*. Determine $\sigma_x$ and $\sigma_y$ and show on a complete sketch the principal and maximum shearing stresses at the point.

**1–77.** The principal tensile stress on a vertical plane through a point in a machine element is equal to four times the principal tensile stress on a horizontal plane. On a plane 30° clockwise from the vertical plane, the normal stress shall not exceed 4000 psi *T*, and the shearing stress shall not exceed 1800 psi. Using Mohr's circle, determine the maximum permissible stress on the horizontal plane.

**1–78.*** The known stresses at a point in a stressed body are 1000 psi compression and 1800 psi shear on a horizontal plane. The maximum compressive stress at the point is 1200 psi. There is an unknown tensile stress acting on the vertical plane at the point. Compute this unknown tensile stress, and show on a complete sketch the principal and maximum shearing stresses at this point.

**1–8. Stress Concentration.** In the foregoing articles, it was assumed that the average stress, as determined by the expression $\sigma$ or $\tau$ equals $P/A$, is the significant or critical stress. For many problems, this is true; for other problems, however, the maximum normal stress on a given section may be considerably greater than the average; and, for certain combinations of loading and material, the maximum rather than the average is the important stress. If there exists in the structural or machine element a discontinuity which interrupts the stress path (called stress trajectory),[8] the stress at the discontinuity may be considerably greater than the nominal (average, in the case of centric loading) stress on the section; thus, there is a *stress concentration* at the discontinuity. This is illustrated in Fig. 1–20 in which a type of discontinuity is shown in the upper figure and the approximate distribution of normal stress on a transverse plane is shown in the accompanying lower figure. The ratio of the maximum stress to the nominal stress on the section is known as the *stress concentration factor.* Thus the expression for the maximum normal stress in a centrically loaded member becomes

$$\sigma = K(P/A)$$

where $A$ is either the gross or the net area (area at the reduced section) depending on the value used for $K$, the stress concentration factor. Curves for stress concentration factors will be found in Chap. 10. It is important that the user of such curves (or tables of factors) ascertain whether the factors are based on the gross or net section.

[8] A stress trajectory is a line which everywhere is parallel to the maximum normal stress.

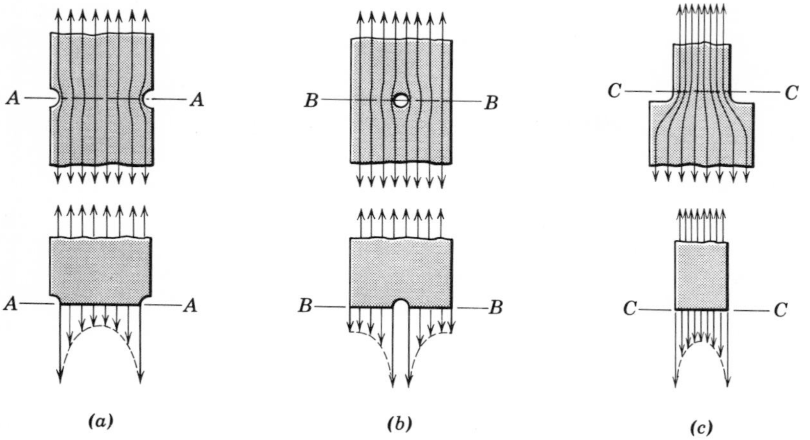


Fig. 1–20

Stress concentration is not significant in the case of static loading of a ductile material (defined in Art. 2–1) because the material will yield inelastically in the region of high stress and, with the accompanying redistribution of stress, equilibrium may be established and no harm done. However, if the load is an impact or repeated load, instead of the action described above, the material may fracture. Also, if the material is brittle, even a static load may cause fracture. Therefore, in the case of impact or repeated loading on any material or static loading on a brittle material, the presence of stress concentration should not be ignored. Before leaving the subject of stress concentration, it should be noted that in regions of support and load application, the stress distribution varies from the nominal (defined as the stress obtained from elementary theories of stress distribution—uniform for centric loading). This fact was discussed in 1864 by Barre de Saint-Venant (1797–1886), a French mathematician. Saint-Venant observed that although localized distortions in such regions produced stress distributions different from the theoretical, these localized effects disappeared at some distance (the implication being that the distance is not of great magnitude[9]) from such locations. This statement is known as

[9] For example, it can be shown mathematically that the localized effect of a concentrated load on a beam may disappear at a section slightly greater than the depth of the beam away from the load. See *Theory of Elasticity,* 2nd Ed., S. Timoshenko and J. N. Goodier, McGraw-Hill, New York, 1951.

*Saint-Venant's principle,* and is constantly used in engineering design. A more complete discussion of stress concentration is included in Chapter 10.

**1–9. Normal and Shearing Strain.** When the size or shape of a body is altered, the change in any dimension is termed a *deformation,* $\delta$, and the *normal strain,* $\epsilon$, is defined as the deformation per unit length. The *shearing strain,* $\gamma$, is defined as the change in angle between two planes which in the unstressed material are orthogonal. The deformation or strain may be the result of a change of temperature or stress or of other physical phenomena such as grain growth or shrinkage. In this book, only strains resulting from changes in temperature and stress are considered. The term strain is sometimes used to denote total deformation, $\delta$; however, *in this book, strain will always refer to deformation per unit length for normal strains or change in angle for shearing strains.*

The normal strain is given by the expressions

$$\epsilon = \delta_n/L \quad \text{or} \quad \epsilon = d\delta_n/dL,$$

where the first expression denotes average strain over the length, $L$, and the second denotes strain at a point. The shearing strain will be defined here for small deformations only (i.e. $\tan \gamma = \gamma$); then

$$\gamma = \delta_s/L \quad \text{or} \quad \gamma = d\delta_s/dL,$$

where again, the first expression denotes average strain over the length, $L$, and the second denotes strain at a point. Figure 1–21*a* represents a normal strain (change in length), and Fig. 1–21*b* shows a

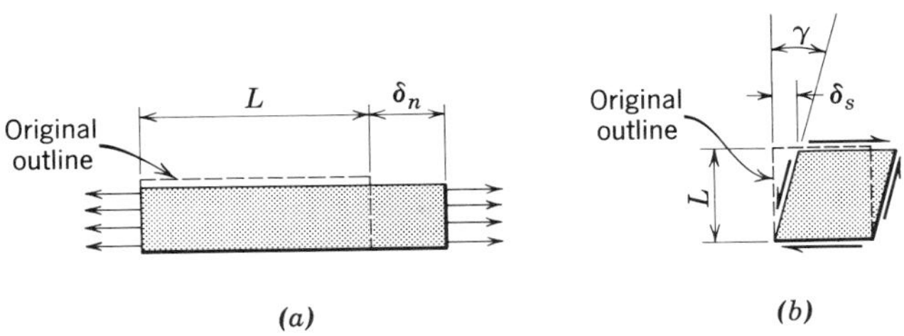


Fig. 1–21

shearing strain (change in shape). Note that the shearing strain is the tangent of the angle of distortion, which for small strains is equal to the angle in radians. Both normal and shearing strains are dimensionless.

When a member is subjected to an axial stress, the normal strain in the direction of the load is defined as an axial strain. The axial strain will also be accompanied by a smaller normal strain perpendicular to the direction of the load. In general, if the axial stress is tensile, the axial strain will be an elongation, and the perpendicular strain will be a decrease in dimension; the reverse will be true for compressive stress.

## PROBLEMS

**1–79.*** A ½-in. round structural steel rod is loaded in tension to fracture. The 8.00-in. gage length of the rod was marked off in 1.00-in. lengths before loading. After the rod broke, the strain in the 2.00-in. length containing the fracture was found to be 0.450, and the total elongation of the other six inches was found to be 1.50 in. Determine the average strain for the 8.00-in. gage length.

**1–80.*** Mutually perpendicular axes in an unstressed member were found to be oriented at 89.54° when the member was stressed. Determine the shearing strain in the stressed member in the direction of the axes.

**1–81.*** A rigid (does not bend) 14,300-lb plate is supported as shown in Fig. P1–81. Posts *B* and *D* are hard rubber 4 by 5 by 8 in. Post *C* is a 2 by 2 by 7.84-in. timber. Given that all three posts are compressed when the load (plate) is applied, determine the relation between the axial strains in the two materials.

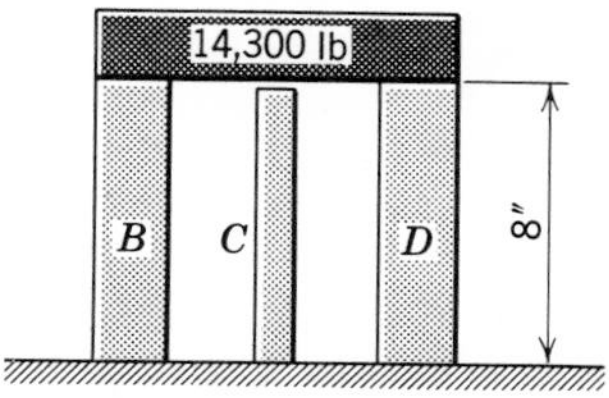


Fig. P1–81

**1–82.** In Fig. P1–82, neglect the bending of bar $C$. There is no strain in the vertical bars before load $P$ is applied. After load $P$ is applied, the axial strain in bar $B$ is found to be 0.0006. Determine the axial strain produced in bar $A$.

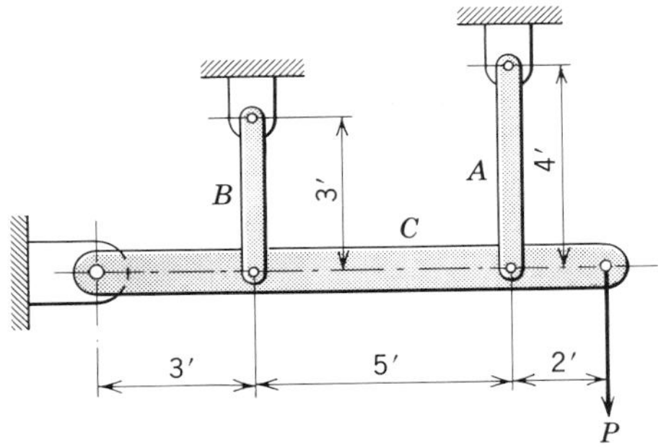


Fig. P1–82

**1–83.*** Solve Prob. 1–82 if there is a 0.001-in. slack in the connection between $A$ and $C$ before the load is applied.

**1–84.** The load $P$ produces an axial strain in the brass post $B$ of Fig. P1–84 of 0.0014. Determine the axial strain in the aluminum alloy rod $A$.

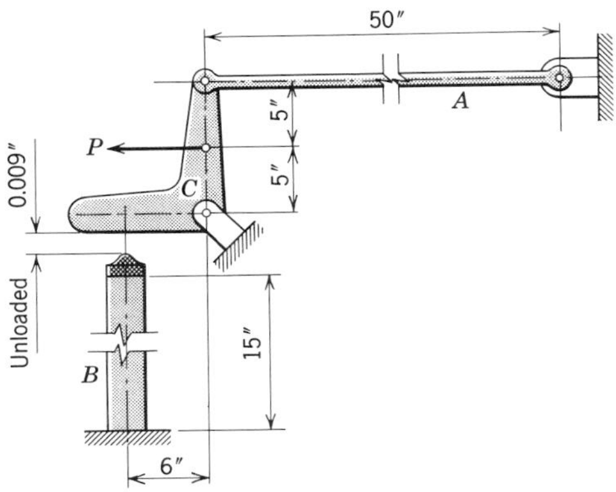


Fig. P1–84

**1–85.*** Solve Prob. 1–84 if in addition to the 0.009-in. clearance between $B$ and $C$, there is 0.005 in. of slack in the connection between $A$ and $C$.

**1–10. Principal Strains—Plane Stress.** The material of Art. 1–9 serves to convey the concept of strain as a unit deformation, but it is inadequate for other than one-directional loading. The extension of the concept to biaxial loading is essential because of the important role played by the strain in experimental methods of stress evaluation. In many practical problems involving the design of structural or machine elements, the configuration and loading are too complicated to permit stress determination solely by mathematical analysis; hence, this technique is supplemented by laboratory measurements.

Strains can be measured by several methods but, except for the simplest cases, stresses cannot be obtained directly. Consequently, the usual procedure in experimental stress analysis is to measure strains and calculate the state of stress from the stress-strain equations which will be presented in Art. 2–3.

Strains are always three-dimensional, except for the case of a member restrained from deforming in one or two directions. However, in most experimental work involving strain measurement, the strains are measured on a free surface (restraint from supports is assumed to be negligible), and the strain normal to the surface has no effect on the stresses. For such problems, the expressions for strain associated with plane stress (biaxial loading) are sufficient; hence, the following discussion will emphasize the strains accompanying plane stress.

Consider Fig. 1–22*a* in which the shaded rectangle represents a small unstrained element of material having the configuration of a rectangular parallelepiped. When subjected to the stresses of Fig.

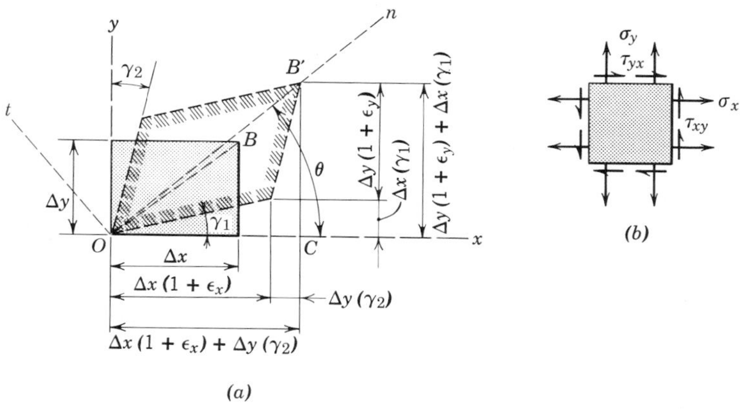


Fig. 1–22

1–22*b*, the element assumes the shape indicated by the dashed lines.

The strains due to the normal stresses are denoted by $\epsilon_x$ and $\epsilon_y$. The strain $\gamma_{xy}$ due to the shearing stress is the change in the right angle and is the sum of the angles $\gamma_1$ and $\gamma_2$, that is

$$\gamma_{xy} = \gamma_1 + \gamma_2. \tag{a}$$

The dimensions of the deformed element in terms of the strains are indicated in Fig. 1–22*a*. Since the deformations are assumed to be very small, $OB'$ has essentially the same direction as $OB$, and the angles resulting from the deformations may be replaced by their tangents.

The normal strain in the $n$ direction, denoted by $\epsilon_n$, is related to the length of the diagonal $OB$ by

$$OB' = OB(1 + \epsilon_n) \tag{b}$$

and $OB'$ can be expressed in terms of the strains $\epsilon_x$, $\epsilon_y$, and $\gamma_{xy}$ by applying the Pythagorean theorem to the triangle $OCB'$ as follows:

$$(OB')^2 = (OC)^2 + (CB')^2$$

$$(OB)^2(1 + \epsilon_n)^2 = [(1 + \epsilon_x)\,\Delta x + \gamma_2\,\Delta y]^2 + [(1 + \epsilon_y)\,\Delta y + \gamma_1\,\Delta x]^2. \tag{c}$$

Since the strains are small, it follows that $\epsilon_x^2 \ll \epsilon_x$, etc., and all second-degree terms such as $\epsilon_x^2$, $\epsilon_x\gamma_2$, $\gamma_1^2$, etc., will be neglected in comparison with the first degree terms as Eq. (c) is expanded. The equation becomes

$$(OB)^2(1 + 2\epsilon_n) = (\Delta x)^2(1 + 2\epsilon_x) + 2\gamma_2\,\Delta x\,\Delta y + (\Delta y)^2(1 + 2\epsilon_y) + 2\gamma_1\,\Delta x\,\Delta y. \tag{d}$$

Since $\Delta x = OB \cos\theta$, $\Delta y = OB \sin\theta$, and $\gamma_1 + \gamma_2 = \gamma_{xy}$, from Eq. (a), Eq. (d) becomes

$$1 + 2\epsilon_n = (1 + 2\epsilon_x)\cos^2\theta + 2\gamma_{xy}\sin\theta\cos\theta + (1 + 2\epsilon_y)\sin^2\theta. \tag{e}$$

When the double-angle identities for $\cos^2\theta$, $\sin^2\theta$, and $\sin\theta\cos\theta$ are applied, Eq. (e) reduces to

$$\epsilon_n = \frac{\epsilon_x + \epsilon_y}{2} + \frac{\epsilon_x - \epsilon_y}{2}\cos 2\theta + \frac{\gamma_{xy}}{2}\sin 2\theta. \tag{f}$$

The similarity between Eq. (f) and Eq. (a) for stresses in Art. 1–6, p. 30, is apparent; therefore, one can conclude that the principal normal and shearing strains are given by the expressions

$$\epsilon_{p1,p2} = \left(\frac{\epsilon_x + \epsilon_y}{2}\right) \pm \sqrt{\left(\frac{\epsilon_x - \epsilon_y}{2}\right)^2 + \left(\frac{\gamma_{xy}}{2}\right)^2} \tag{1-7}$$

and

$$\gamma_{\max} = 2\sqrt{\left(\frac{\epsilon_x - \epsilon_y}{2}\right)^2 + \left(\frac{\gamma_{xy}}{2}\right)^2} \tag{1-8}$$

where the signs are plus for tensile normal strains and plus for an angle $\gamma$ (shearing strain) which decreases the angle between the faces of the element at the origin of coordinates as indicated in Fig. 1–22*a*.

The Mohr's circle for the strains of Fig. 1–22*a* (with $\epsilon_x > \epsilon_y$) is given in Fig. 1–23. It is apparent that the sign convention for shearing strain needs to be extended to cover the construction of Mohr's circle. Observe that for positive shearing strain (indicated in Fig. 1–22*a*), the edge of the element parallel to the $x$ axis rotates counterclockwise while the edge parallel to the $y$ axis rotates clockwise. For the Mohr's circle construction, the clockwise rotation will be designated positive, and the counterclockwise designated negative. This is consistent with the sign convention for shearing stress given in Art. 1–6. One observes

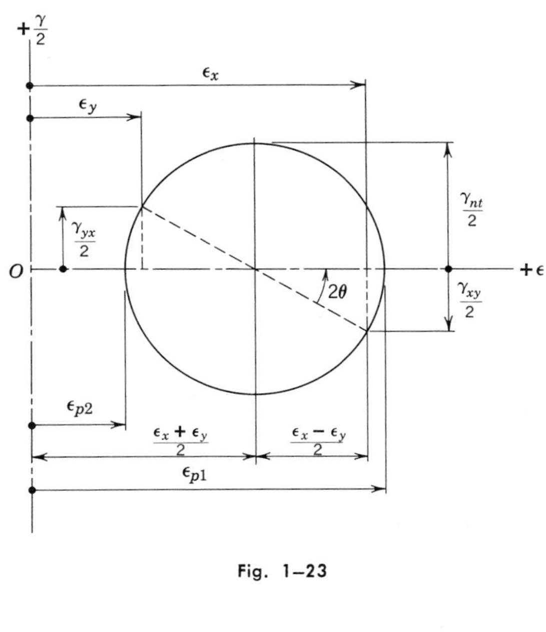


Fig. 1–23

that so far as this diagram is concerned, the maximum shearing strain would be given by the expression

$$\frac{\gamma_{nt}}{2} = \frac{\epsilon_{p1} - \epsilon_{p2}}{2}.$$

However, for this case of plane stress, $\sigma_{p1}$ and $\sigma_{p2}$ are tensile stresses and both will decrease the element dimension in the direction of the third principal axis (see Art. 1–9); hence, $\epsilon_{p3}$ is negative and the maximum shearing strain (where $\epsilon_{p1}$ and $\epsilon_{p2}$ have the same sign and $\epsilon_{p1} > \epsilon_{p2}$) is $2(\epsilon_{p1} - \epsilon_{p3})/2$. Although the stated scope of this article is strains in bodies subjected to plane stress, the maximum shearing strain when plane strain exists will be indicated. Plane strain is defined as the condition in which the only non-zero strains are $\epsilon_x$, $\epsilon_y$, and $\gamma_{xy}$; in general, this requires a non-zero value for $\sigma_z$. For this state of strain, the maximum shearing strain is $2(\epsilon_{p1} - 0)/2 = \epsilon_{p1}$.

In general, an investigator is interested in principal stresses; hence, he wishes to measure principal strains. However, the principal directions are frequently unknown.[10] In such cases, the strains can be measured in three directions and from Eq. (f) the three unknowns $\epsilon_x$, $\epsilon_y$, and $\gamma_{xy}$ can be determined. From these strains, the principal strains can be evaluated by means of Eq. (1–7) or Mohr's circle and the principal stresses computed using the equations of Art. 2–3 (provided Hooke's law applies). Mohr's circle for strains can be constructed graphically from three measured normal strains using a simple construction.[11] Since this is graphical rather than analytical, it will not be described further. Of course, one must be aware of the fact that strains (hence, stresses) cannot be evaluated experimentally at a point, since strain-measuring devices (gages) must have a finite length. Therefore, strains and stresses are determined experimentally in a region rather than at a point.

**Example 1–6:** With the aid of a strain rosette composed of three wire resistance gages making angles of 0°, 60°, and 120° with the $x$ axis, the following strains were measured:

$$\epsilon_0 = \epsilon_x = +100(10^{-5}), \qquad \epsilon_{60} = -65(10^{-5}), \qquad \epsilon_{120} = +75(10^{-5}).$$

Determine the principal strains and the maximum shearing strain.

[10] There are techniques available, such as the brittle coat method, for ascertaining principal directions.

[11] Glenn Murphy, *J. Appl. Mech.* (Trans. A.S.M.E.), Vol. 12, p. 209, 1945; N. J. Hoff, *ibid.*

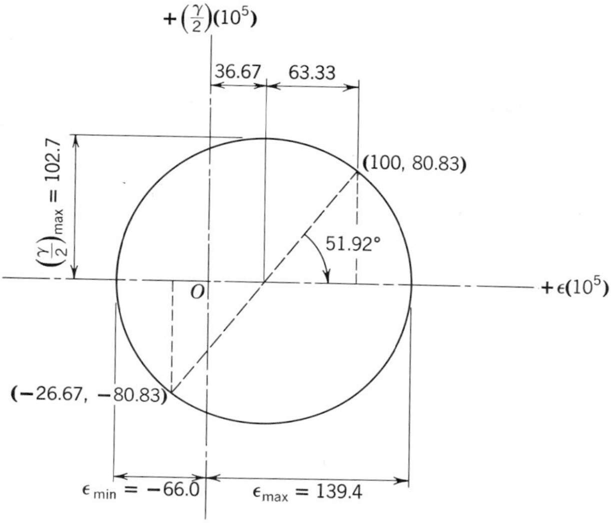


Fig. 1–24

*Solution:* The given data are substituted in Eq. (f) to yield the following (in units of $10^{-5}$ in. per in.):

$$\epsilon_x = +100 \text{ (measured)},$$

$$-65 = \frac{100 + \epsilon_y}{2} + \frac{100 - \epsilon_y}{2}(\cos 120°) + \frac{\gamma_{xy}}{2}(\sin 120°)$$

$$= \frac{100 + \epsilon_y}{2} + \frac{100 - \epsilon_y}{2}\left(\frac{-1}{2}\right) + \frac{\gamma_{xy}}{2}\left(\frac{\sqrt{3}}{2}\right),$$

$$75 = \frac{100 + \epsilon_y}{2} + \frac{100 - \epsilon_y}{2}(\cos 240°) + \frac{\gamma_{xy}}{2}(\sin 240°)$$

$$= \frac{100 + \epsilon_y}{2} + \frac{100 - \epsilon_y}{2}\left(\frac{-1}{2}\right) + \frac{\gamma_{xy}}{2}\left(\frac{-\sqrt{3}}{2}\right);$$

from which $\epsilon_y = -26.67$ and $\gamma_{xy} = -161.66$.
These values are then used to construct the Mohr's circle of Fig. 1–24. Since $\gamma_{xy}$ is negative, the angle between the element faces at the origin increases; hence, the $xz$ face rotates clockwise (+) and the $yz$ face rotates counterclockwise (−). The results taken from the circle are

$$\epsilon_{max} = \underline{+139.4(10^{-5})},$$

$$\epsilon_{min} = \underline{-66.0(10^{-5})}, \qquad \text{Ans.}$$

$$\gamma_{max} = \underline{\pm 205.4(10^{-5})}.$$

## PROBLEMS

**Note: In the following plane stress problems, the symbol $\mu$ is used to represent micro-inches per inch ($10^{-6}$ in./in.).**

**1–86.*** An investigator measured the following average normal strains at a point on the surface of a stressed body. The strains, measured at 30°, 60°, and 135° counterclockwise from the horizontal $x$ axis, are $+400\mu$, $+150\mu$, and $-200\mu$, respectively. Determine the principal and maximum shearing strains.

**1–87.** Strains measured with a strain rosette composed of three foil gages at angles of 0°, 60°, and 120° with the $x$ axis are $-400\mu$, $-800\mu$, and $-1200\mu$, respectively. Determine the principal and maximum shearing strains.

**1–88.*** Solve Prob. 1–87 when measured strains are $+100\mu$, $+200\mu$, and $+120\mu$, respectively.

**1–89.** Strains measured with a strain rosette composed of three resistance gages at angles of 0°, 45°, and 90° with the $x$ axis are $-200\mu$, $-300\mu$, and $+100\mu$, respectively. Determine the principal and maximum shearing strains.

**1–90.*** Solve Prob. 1–89 when the measured strains are $400\mu$, $-240\mu$, and $-80\mu$, respectively.

**1–91.** If the strains in the region of Prob. 1–90 had been measured with a (0–60–120)-degree rosette, what should the strain readings have been?

CHAPTER 2

# *Material Properties and Associated Topics*

**2–1. Properties of Materials—Static.** The satisfactory performance of a structure frequently is determined by the amount of deformation or distortion which can be permitted. A deflection of a few thousandths of an inch might make a boring machine useless, whereas the boom on a dragline might deflect several inches without impairing its usefulness. It is often necessary to relate the loads on a structure, or on a member in a structure, to the deflection which the loads will produce. Such information can be obtained by plotting diagrams showing loads and deflections for each member and type of loading in a structure, but such diagrams will vary with the dimensions of the members, and it would be necessary to draw new diagrams each time the dimensions were varied. A more useful diagram is one showing the relation between the stress and strain. Such diagrams are called stress-strain diagrams.

Data for stress-strain diagrams are usually obtained by applying an axial load to a test specimen and measuring the load and deformation simultaneously. A testing machine, see Fig. 2–1, is used to strain the member and to measure the load required to produce the strain. The stress is obtained by dividing the load by the initial cross-sectional area of the member. The area will change somewhat during the loading, and the stress obtained using the initial area is obviously not the exact stress occurring at higher loads. It is the stress most commonly used, however, in designing structures. The stress obtained by dividing the load by the actual area is frequently called the *true stress* and is useful in explaining the fundamental behavior of materials. Strains are usually relatively small in materials used in engineering structures, often less than 0.001, and their accurate determination requires special measuring equipment. Normal strain can be obtained by measuring the

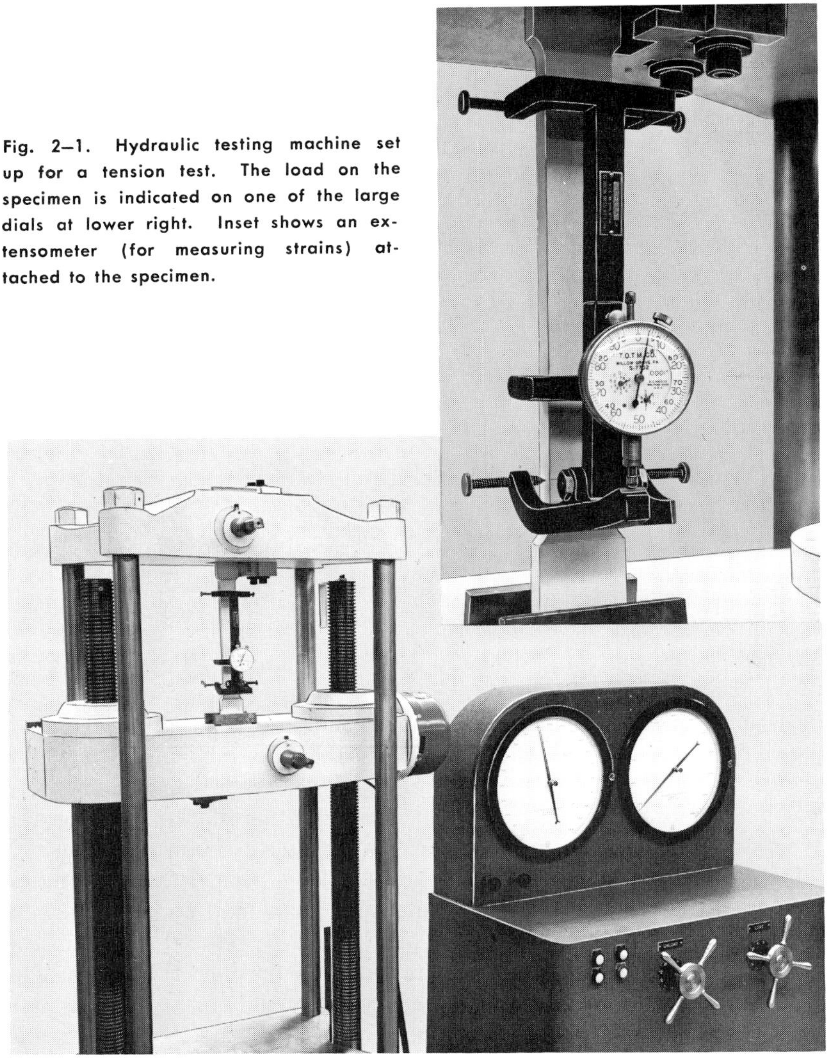

Fig. 2–1. Hydraulic testing machine set up for a tension test. The load on the specimen is indicated on one of the large dials at lower right. Inset shows an extensometer (for measuring strains) attached to the specimen.

deformation, $\delta$, in a length, $L$, and dividing $\delta$ by $L$. Instruments for measuring the deformation $\delta$ are called strain gages, or sometimes extensometers or compressometers, and obtain the desired accuracy by multiplying levers, dial indicators, beams of light, or other means. Resistance strain gages are among the most convenient strain gages, and are widely used. The resistance gage may be a wire gage consisting of a length of 0.001-in. diameter wire arranged as shown in Fig. 2–2*a*, and cemented between two pieces of paper; or it may be an etched foil conductor mounted on epoxy or Bakelite backing, see Fig. 2–2*b*. The wire or foil gage is cemented to the material for which the strain is to be determined. As the material is strained, the wires are lengthened or shortened; this changes the electrical resistance of the gage. The change in resistance is measured by means of a Wheatstone bridge, which may be calibrated to read strain directly. Figure 5–1 shows SR-4 wire strain gages mounted on an aluminum beam. Shearing strains are more difficult to measure directly than normal strains and are often obtained by measuring normal strains in two or three different directions and computing the shearing strains by methods which were described in Art. 1–10.

*True strain,* like true stress, is computed on the basis of the actual length of the test specimen during the test and is used primarily to

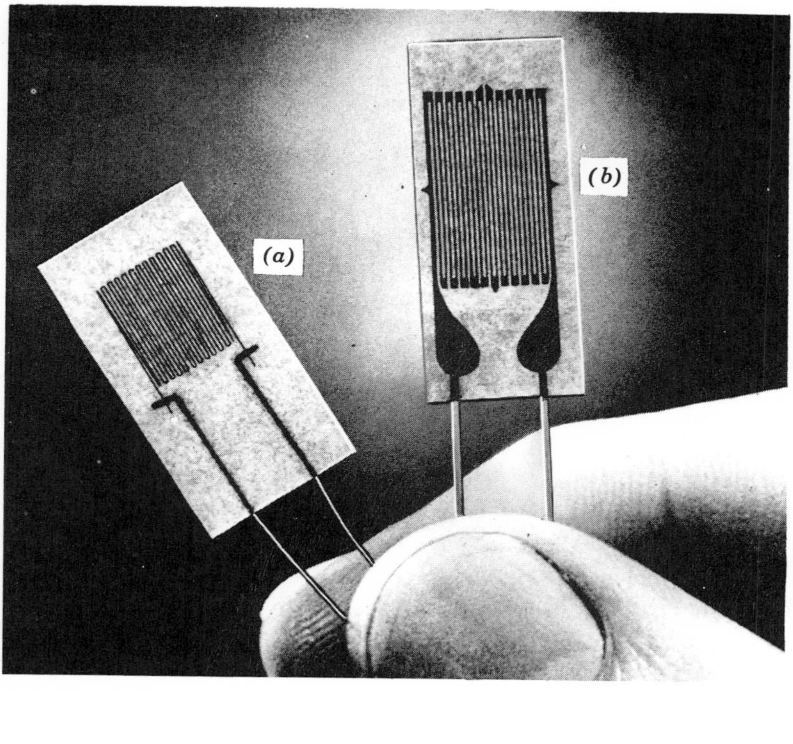


Fig. 2–2

study the fundamental properties of materials. The difference between nominal stress and strain, computed from initial dimensions of the specimen, and true stress and strain is negligible for stresses usually encountered in engineering structures, but sometimes the difference becomes important with larger stresses and strains.

A more complete discussion of the experimental determination of stress and strain will be found in various books on experimental stress analysis.[1]

Figures 2–3*a* and *b* show tensile stress-strain diagrams for structural steel (a low-carbon steel) and for a magnesium alloy. These diagrams will be used to explain a number of properties useful in the study of mechanics of materials.

The initial portion of the stress-strain diagram for most materials used in engineering structures is a straight line. The stress-strain diagrams for some materials, such as gray cast iron and concrete, show a slight curve even at very small stresses, but it is common practice to draw a straight line to average the data for the first part of the diagram and neglect the curvature. The proportionality of load to deflection was first recorded by Robert Hooke who observed in 1678, "*Ut tensio sic vis*" (as the stretch so the force); this is frequently referred to as Hooke's law. Thomas Young, in 1807, suggested what amounts to using the ratio of stress to strain to measure the stiffness of a material. This ratio is called *Young's modulus* or the *modulus of elasticity* and is the slope of the straight line portion of the stress-strain diagram. Young's modulus is written as

$$E = \sigma/\epsilon \quad \text{or} \quad G = \tau/\gamma$$

where $E$ is used for normal stress and strain and $G$ (sometimes called the modulus of rigidity) is used for shearing stress and strain. The maximum stress for which stress and strain are proportional is called the *proportional limit* and is indicated by the ordinates at points $A$ on Fig. 2–3*a* or *b*.

The action is said to be *elastic* if the strain which results from loading disappears when the load is removed. The *elastic limit* is the maximum stress for which the material acts elastically. For most materials it is found that the stress-strain diagram for unloading is approximately parallel to the loading portion; see line $BC$ in Fig. 2–3*b*. If the specimen is again loaded, the stress-strain diagram will usually fol-

[1] *Handbook of Experimental Stress Analysis,* M. Hetenyi, John Wiley and Sons, New York, 1950.

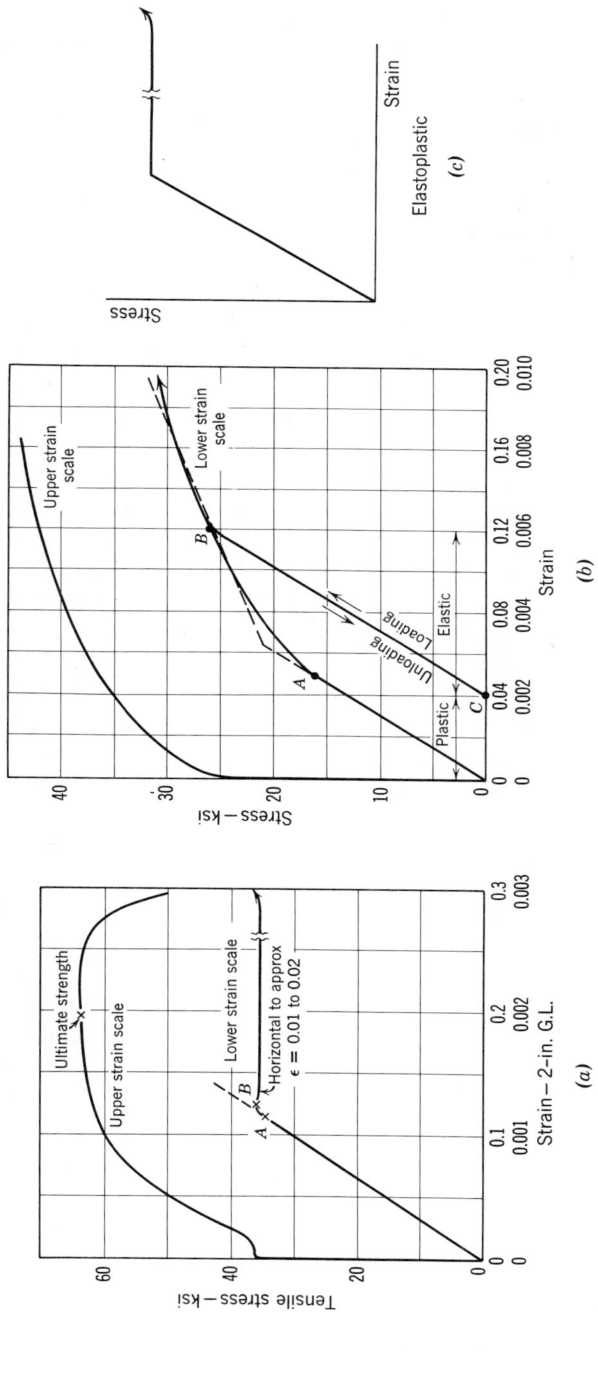

Tensile stress – ksi
Ultimate strength
Upper strain scale
Lower strain scale
Horizontal to approx
ε = 0.01 to 0.02
A
B
Strain – 2-in. G.L.
(a)
Stress – ksi
Upper strain scale
Lower strain scale
Loading
Unloading
Elastic
Plastic
A
B
C
Strain
(b)
Stress
Strain
Elastoplastic
(c)


Fig. 2–3

low the unloading curve until it reaches stress a little less than the maximum stress attained during the initial loading, at which time it will start to curve in the direction of the initial loading curve. As indicated in Fig. 2–3*b* the proportional limit for the second loading is greater than that for the initial loading. This phenomenon is called *strain hardening* or *work hardening.*

When the stress exceeds the elastic limit (or proportional limit for practical purposes), it is found that a portion of the deformation remains after the load is removed. The deformation which remains after an applied load is removed is called *plastic deformation.* Plastic deformation which is independent of the time duration of the applied load is known as *slip. Creep* is plastic deformation which continues to increase under a constant stress. In many instances, creep continues until fracture occurs; however, in other instances, the rate of creep decreases and approaches zero as a limit. Some materials are much more susceptible to creep than are others, but most materials used in engineering exhibit creep at elevated temperatures. The total strain is thus made up of elastic strain, possibly combined with plastic strain which results from slip, creep, or both. When the load is removed, the elastic portion of the strain is recovered, but the plastic part (slip and creep) remains as permanent set.[2]

A precise value for the proportional limit is difficult to obtain, particularly when the transition of the stress-strain diagram from a straight line to a curve is gradual. For this reason, other measures of the maximum stress to which a material may be subjected with little or no plastic strain are required. The yield point and yield strength for a specified offset are frequently used for this purpose.

The *yield point* is the stress at which there is an appreciable increase in strain with no increase in stress, with the limitation that, if straining is continued, the stress will again increase. This latter specification indicates that there is a kink or "knee" in the stress-strain diagram as indicated in Fig. 2–3*a*. The yield point is easily determined without the aid of strain-measuring equipment as the beam of the testing machine drops (or if a dial indicator is used, the needle halts) at the yield point. Unfortunately, few materials possess this property, the most common examples being the low-carbon steels.

The *yield strength* is defined as the stress which will induce a specified permanent set, usually 0.05 to 0.3 per cent which is equiva-

---

[2] In some instances a portion of the strain which remains immediately after the stress is removed may disappear after a period of time. This reduction of strain is sometimes called *recovery.*

lent to a strain of 0.0005 to 0.003. The yield strength is particularly useful for materials which have no yield point. The yield strength can be conveniently determined from a stress-strain diagram by laying off the specified offset (permanent set) on the strain axis as $OC$ in Fig. 2–3*b* and drawing a line $CB$ parallel to $OA$. The stress indicated by the intersection of $CB$ and the stress-strain diagram is the yield strength.

The maximum stress, based on the original area, which is developed in a material before rupture is called the *ultimate strength* of the material, and the term may be modified as the ultimate tensile, compressive, or shearing strength of the material. Ductile materials are those which undergo considerable plastic tensile or shearing deformation before rupture. When the ultimate strength of a ductile material is reached, the cross-sectional area of the test specimen starts to decrease or neck down, and the resultant load which can be carried by the specimen decreases. Thus, the stress based on the original area decreases beyond the ultimate strength of the material although the true stress continues to increase until rupture.

Many engineering structures are designed so that the stresses are less than the proportional limit and Young's modulus provides a simple and convenient relationship between the stress and strain. When the stress exceeds the proportional limit, the deformation becomes partially plastic and no simple relation exists between the stress and strain. Various empirical equations have been proposed relating the stress and strain beyond the proportional limit. The following equation, known as the Ramberg-Osgood[3] formula, is one which has been used in plastic analysis:

$$\epsilon = \epsilon_0 \left[ \frac{\sigma}{\sigma_0} + \frac{3}{7} \left( \frac{\sigma}{\sigma_0} \right)^n \right]$$

in which $\epsilon$ and $\sigma$ are the strain and stress to be related and $\epsilon_0$, $\sigma_0$, and $n$ are experimentally determined constants. Corresponding values of stress and strain can also be read from stress-strain diagrams. A stress-strain diagram similar to the one in Fig. 2–3*c* is frequently assumed for mild steel or other material with similar properties in order to simplify the calculations. In this book, material with such a stress-strain diagram will be designated *elastoplastic*, and the proportional limit and yield points will be assumed to be the same. For mild steel the plastic

[3] "Description of Stress-Strain Curves by Three Parameters," W. Ramberg and W. R. Osgood, *NACA* TN 902, 1943.

strain that occurs at the yield point with no increase in stress is fifteen to twenty times the elastic strain at the proportional limit.

Strength and stiffness are not the only properties of interest to a design engineer. Another important property is *ductility,* defined as the capacity for plastic deformation in tension or shear. This property controls the amount of cold forming to which a material may be subjected. The forming of automobile bodies, the bending of concrete-reinforcing bars, and the manufacture of fencing and other wire products all require ductile materials. Ductility is also an important property of materials used for fabricated structures because under static loading the presence of stress concentration in the region of rivet holes or welds may be ignored since the ductility permits considerable plastic action to take place in the region of high stresses with a resulting redistribution of stress and the establishment of equilibrium. Two commonly used quantitative indices of ductility are the *ultimate elongation* (expressed as a percentage elongation of the gage length at rupture) and the *reduction of cross-sectional area* at the section where rupture occurs (expressed as a percentage of the original area).

The property which indicates the resistance of a material to failure by creep is known as the *creep limit* and is defined as the maximum stress for which the plastic strain will not exceed a specified amount during a specified time interval at a specified temperature. This property is important when designing metal parts subjected to high temperatures and sustained loads, for example, the turbine blades in a turbojet engine.

A material loaded in one direction will undergo strains perpendicular to the direction of the load as well as parallel to it. The ratio of the lateral or perpendicular strain to the longitudinal or axial strain is called *Poisson's ratio,* after Simeon D. Poisson who identified this constant in 1811, and it is a constant for stresses below the proportional limit. Poisson's ratio varies from $\frac{1}{4}$ to $\frac{1}{3}$ for most metals. The symbol $\mu$ is used for Poisson's ratio which is given by the equation

$$\mu = \epsilon_{\text{lateral}}/\epsilon_{\text{longitudinal}}.$$

It will be shown in Art. 2–3 that Poisson's ratio is related to $E$ and $G$ by the formula

$$E = 2(1 + \mu)G.$$

The properties discussed in this article are primarily concerned with static or continuous loading. Properties which are important when considering energy or impact loads and repeated loads are discussed in Art. 2–2 and Chapter 10. A more complete discussion of properties of materials is contained in several excellent properties books.[4, 5]

**Note: For simplicity of calculations in the solution of problems in this book, use the approximate values of $30(10^6)$ psi and $12(10^6)$ psi for the moduli of elasticity and rigidity respectively for steel without identification as to kind. If the steel is identified (eg: structural steel, etc.) the values given in App. A will be used. Unless the proportional limit is given or referred to in the problem, it is assumed that the action is elastic. If the proportional limit is given without a stress-strain diagram, it is assumed that the diagram is elastoplastic (see Fig. 2–3c). When the stress-strain diagram is given or referred to, it is to be used to check the possibility of plastic action.**

## PROBLEMS

**2–1.*** The rigid yokes *B* and *C* of Fig. P2–1 are securely fastened to the 2-in. square steel bar *AD*. Determine the total axial deformation in the 5-ft length of bar between *B* and *C*.

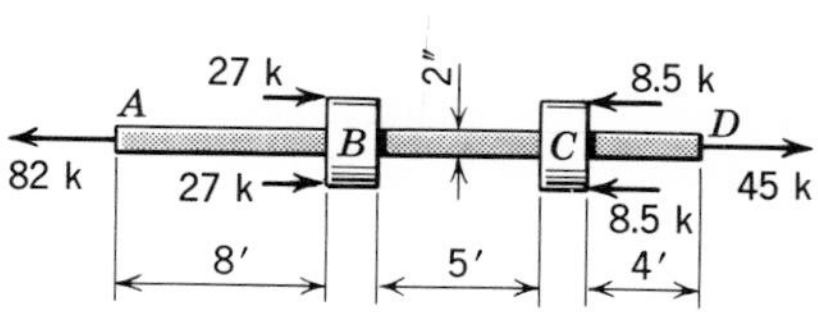


**Fig. P2–1**

**2–2.*** The steel bar in Prob. 2–1 is replaced by wrought aluminum 2014-*T*4. Determine the total axial deformation in the 8-ft length *AB*.

**2–3.*** At the proportional limit, an 8-in. gage length of a ½-in. diameter alloy bar has elongated 0.012 in., and the diameter has been reduced 0.00025 in. The total axial load carried was 4800 lb. Determine the following properties of this material:

(*a*) Modulus of elasticity.
(*b*) Poisson's ratio.
(*c*) Proportional limit.

---

[4] *The Structure and Properties of Materials, Vol. III,* "Mechanical Behavior," H. W. Hayden, W. G. Moffatt, and John Wulff; John Wiley and Sons, New York, 1965.

[5] *Elements of Material Science,* 2nd Ed., L. H. VanVlack, Addison-Wesley Publishing Company, Reading, Mass., 1964.

**2–4.*** A 100-kip axial load is slowly applied to a 1 by 4 by 90-in. rectangular bar. When loaded (the action is elastic), the 4-in. side measures 3.9986 in., and the length has increased 0.09 in. Determine Poisson's ratio and Young's modulus for the material.

**2–5.** A $\frac{1}{4}$ by 2-in. flat alloy bar elongates 0.06 in. in a length of 5 ft under a total axial load of 13,000 lb. The proportional limit of the material is 44,000 psi.

(*a*) Determine the axial stress in the bar.

(*b*) What is the modulus of elasticity of this material?

(*c*) If Poisson's ratio for the material is 0.32, what will be the total change in each lateral dimension?

**2–6.*** A 1.5-in. diameter rod 20 ft long elongates 0.48 in. under a load of 53 kips. The diameter decreased 0.001 in. during the loading. Determine the following properties of the material:

(*a*) Poisson's ratio.

(*b*) Young's modulus.

(*c*) The modulus of rigidity.

**2–7.** A steel and an aluminum bar are coupled together end to end and loaded axially at the extreme ends. Both bars are 2 in. in diameter; the steel bar is 6 ft long, and the aluminum bar is 4 ft long. When the load is applied, it is found that the steel bar elongates 0.0052 in. in a gage length of 8 in. Poisson's ratio for this steel is $\frac{1}{4}$, and the modulus of elasticity of the aluminum is $10.6(10^6)$ psi. Determine:

(*a*) The load.

(*b*) The total change in length from one end to the other.

(*c*) The change in diameter of the steel bar.

**2–8.*** A bar of uniform cross section and homogeneous material hangs vertically while suspended from one end. Determine, in terms of $W$, $L$, $A$, and $E$, the extension of the bar due to its weight.

**2–9.** Determine the extension, due to its weight, of the homogeneous conical bar of Fig. P2–9. Give the result in terms of $L$, $E$, and the specific weight, $\gamma$. Assume that the taper of the bar is slight enough for the assumption of uniform axial stress distribution over a cross section to be valid.

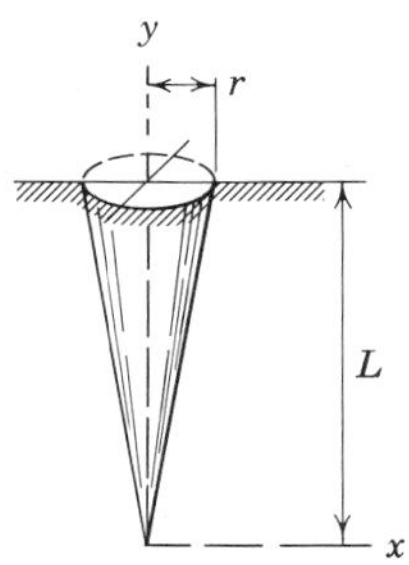


Fig. P2–9

**2–10.*** A homogeneous bar of length $2L$ is in the shape obtained by rotating the parabola, $y = (L/r^2)x^2$ about the $y$ axis. This bar is supported at its upper end ($y$ axis vertical and origin at the lower end of the bar). Determine the elongation of that portion of the bar between $y = L/4$ and $y = 5L/4$ due to the weight of the bar. Give the result in terms of $L$, $E$, and the specific weight, $\gamma$. Assume that $L$ is much larger than $r$.

**2–11.** Determine the extension, due to its own weight, of the homogeneous bar of Fig. P2–11. Give the result in terms of $L$, $E$, and the specific weight, $\gamma$. Assume that the taper of the bar is slight enough that the assumption of uniform axial stress distribution over a cross section is valid.

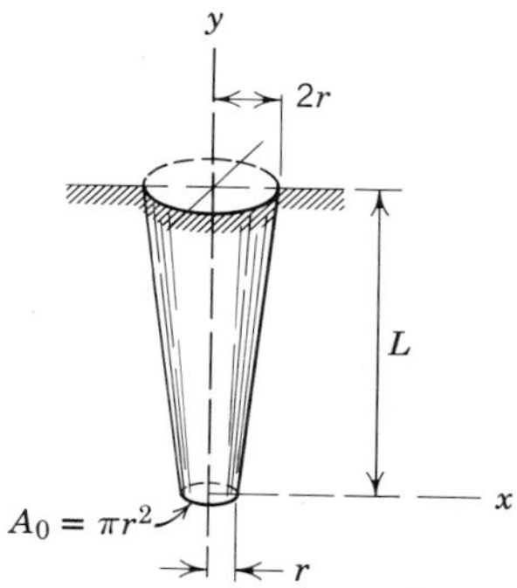


Fig. P2–11

**2–12.*** Solve Prob. 2–11 with an additional axial tensile load of $\gamma A_0 L$ applied at the lower end.

**2–2. Strain Energy.** A significant topic involving stress and strain is the strain energy concept, which is not only important to an understanding of the behavior of materials under dynamic loading but has useful applications in the solution of certain types of problems involving static loading.

The concept of strain energy is illustrated by Fig. 2–4, in which ($a$) represents a bar of uniform cross section subjected to a *slowly applied* axial load, $P$, and held at the upper end by a support which is assumed to be rigid. From the load-deformation diagram, Fig. 2–4$b$, the work done in elongating the bar an amount $\delta_2$ is

$$W_k = \int_0^{\delta_2} P \, d\delta, \qquad \text{(a)}$$

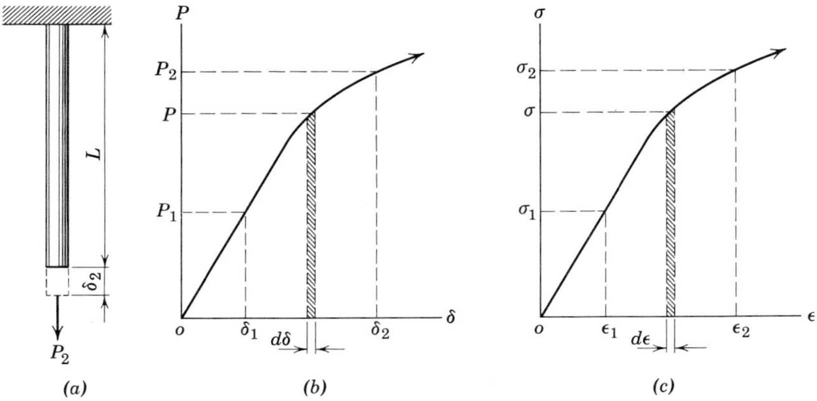


Fig. 2–4

where $P$ is some function of $\delta$. The work done on the bar must equal the change in energy of the material,[6] and this energy change, because it involves the strained configuration of the material, is termed *strain energy,* $U$. If $\delta$ is expressed in terms of axial strain ($\delta = L\epsilon$) and $P$ in terms of axial stress ($P = A\sigma$), Eq. (a) becomes

$$W_k = U = \int_0^{\epsilon_2} (\sigma)(A)(L)\,d\epsilon = AL \int_0^{\epsilon_2} \sigma\,d\epsilon, \tag{b}$$

where $\sigma$ is a function of $\epsilon$ (see Fig. 2–4*c*). If Hooke's law applies,

$$\epsilon = \sigma/E, \qquad d\epsilon = d\sigma/E;$$

and Eq. (b) becomes

$$U = \left(\frac{AL}{E}\right) \int_0^{\sigma_1} \sigma\,d\sigma$$

or

$$U = AL\left(\frac{\sigma_1{}^2}{2E}\right). \tag{c}$$

Equation (c) gives the *elastic strain energy* (which is, in general, recoverable[7]) for axial loading of a material obeying Hooke's law. The quantity in parentheses, $\sigma_1{}^2/(2E)$, is the elastic strain energy, $u$,

[6] Known as *Clapeyron's Theorem,* after the French engineer B. P. E. Clapeyron (1799–1864).

[7] Elastic hysteresis is neglected here as an unnecessary complication.

in tension or compression per unit volume for a particular value of $\sigma$. For shear loading, the expression would be identical except that $\sigma$ would be replaced by $\tau$ and $E$ by $G$.

The integral, $\int \sigma \, d\epsilon$, of Eq. (b) represents the area under the stress-strain curve, and, if evaluated from zero to the elastic limit (for practical purposes, the proportional limit), yields a property known as the *modulus of resilience*. The modulus is defined as the maximum strain energy per unit volume that a material will absorb without inelastic deformation. Customary units are inch-pounds per cubic inch. The area under the entire stress-strain curve from zero to rupture gives the property known as the *modulus of toughness* and denotes the energy per unit volume necessary to rupture the material. The values of these moduli for the steel of Fig. 2–3*a* are

$$u_R = (35{,}000)(0.001{,}17)/2 = 22 \text{ in-lb per cu in.}$$

and

$$u_T = 57{,}000(0.3) = 17{,}000 \text{ in-lb per cu in.},$$

where the 57,000 is an estimated average value of stress for a rectangle having approximately the same area as that under the upper curve of Fig. 2–3*a*.

Equation (c) or its equivalent will be found useful later in the development of certain expressions for deformation or deflection.

**2–3. Generalized Hooke's Law for Isotropic Materials.** Hooke's law [see Art. (2–1)] may be extended to include biaxial and triaxial states of stress which are frequently encountered in engineering applications. Consider, for example, the differential cube of Fig. 2–5 subjected to *biaxial* principal stresses in the elastic range (obeying Hooke's law). The shaded square indicates the unstrained configuration of the cube. Under the action of the stress, $\sigma_x$, the cube extends in the $x$ direction and contracts in the $y$ direction to the configuration indicated by the dotted lines (deformations greatly exaggerated). Then under the action of the stress, $\sigma_y$, superimposed on $\sigma_x$, the cube assumes the outline shown by the hatched dashed lines. If the material is isotropic, Young's modulus has the same value for all directions and the final deformation in the $x$ direction is

$$\delta_x = \epsilon_x \, dx = \frac{\sigma_x}{E} \, dx - \mu \frac{\sigma_y}{E} \, dx,$$

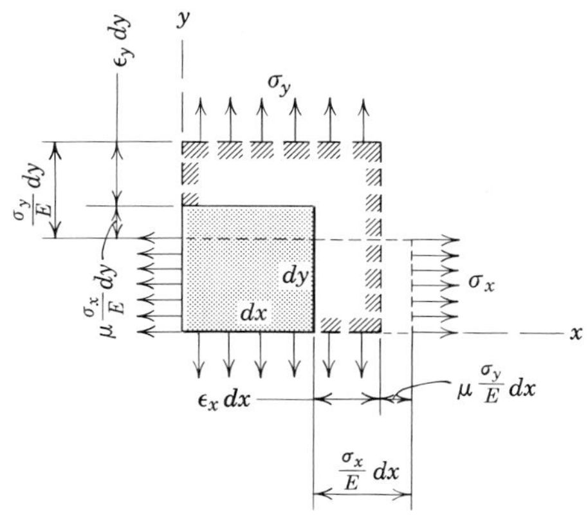


Fig. 2–5

and the three principal strains are

$$\begin{aligned} \epsilon_x &= (\sigma_x - \mu\sigma_y)/E, \\ \epsilon_y &= (\sigma_y - \mu\sigma_x)/E, \\ \epsilon_z &= -\mu(\sigma_x + \sigma_y)/E. \end{aligned} \tag{2–1}$$

The analysis above is readily extended to triaxial principal stresses and the expressions for strain become

$$\begin{aligned} \epsilon_x &= [\sigma_x - \mu(\sigma_y + \sigma_z)]/E, \\ \epsilon_y &= [\sigma_y - \mu(\sigma_x + \sigma_z)]/E, \\ \epsilon_z &= [\sigma_z - \mu(\sigma_x + \sigma_y)]/E. \end{aligned} \tag{2–2}$$

In these expressions, tensile stresses and strains are considered positive, and compressive stresses and strains are considered negative.

When Eq. (2–1) are solved for the stresses in terms of the strains, they give

$$\begin{aligned} \sigma_x &= \frac{E}{1-\mu^2}(\epsilon_x + \mu\epsilon_y), \\ \sigma_y &= \frac{E}{1-\mu^2}(\epsilon_y + \mu\epsilon_x). \end{aligned} \tag{2–3}$$

Equations (2–3) can be used to calculate normal stresses from measured or computed normal strains (see Art. 1–10). When Eq. (2–2) are solved for stresses in terms of strains, they give

$$\sigma_x = \frac{E}{(1+\mu)(1-2\mu)}[(1-\mu)\epsilon_x + \mu(\epsilon_y + \epsilon_z)],$$

$$\sigma_y = \frac{E}{(1+\mu)(1-2\mu)}[(1-\mu)\epsilon_y + \mu(\epsilon_z + \epsilon_x)], \qquad (2\text{–}4)$$

$$\sigma_z = \frac{E}{(1+\mu)(1-2\mu)}[(1-\mu)\epsilon_z + \mu(\epsilon_x + \epsilon_y)].$$

In order to simplify the physical concept (see Fig. 2–5), the derivation of the equations above was based on a cube of material subjected to principal stresses. However, the resulting expressions are not changed by the presence of shearing stresses on the cube faces. For example, if one superimposes shearing stresses $\tau_{xy}$ and $\tau_{yx}$ on the final configuration of Fig. 2–5 and rotates the cube so that the lower edge remains parallel to the $x$ axis, it will be observed that the face of the cube assumes the shape of a parallelogram with the initially vertical edges making the angle $\gamma$ with the $y$ axis. In effect, then, all horizontal filaments merely translate parallel to the $x$ axis without changing length.

Shearing strains are independent of normal strains and Hooke's law extended to shearing stresses and strains is simply

$$\tau = G\gamma. \qquad \text{(a)}$$

However, it will be found convenient in later developments to replace $G$ with an equivalent expression in terms of $E$ and $\mu$. A relationship between the elastic constants $E$, $G$, and $\mu$ can be developed as follows: When only shearing stresses (no normal stresses) act on a pair of orthogonal ($x$, $y$) faces of an element, it is said to be loaded in *pure shear*, and Mohr's circle for the stresses at the point has its center at the origin of coordinates. In this case, the stresses are related by the expression

$$\sigma_{p1} = -\sigma_{p2} = \tau_{\max} \qquad \text{(b)}$$

and act on elements as indicated in Fig. 2–6$a$ and 2–6$b$. The element of Fig. 2–6$a$ will be distorted (changed in shape), whereas the element in Fig. 2–6$b$ will undergo normal strains (a change in volume) with no distortion. The relation between these strains can be obtained by noting that the strain energy per unit volume (see Art. 2–2) must

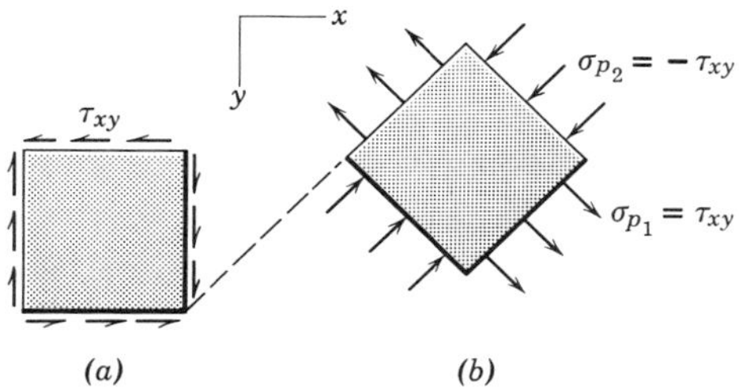


Fig. 2–6

have the same value regardless of whether it is expressed in terms of the shearing stresses or the normal stresses; thus

$$\frac{1}{2}\tau_{xy}\gamma_{xy} = \frac{1}{2}\sigma_{p1}\epsilon_{p1} + \frac{1}{2}\sigma_{p2}\epsilon_{p2},$$

and from Eq. (2–1),

$$\epsilon_{p1} = \frac{\sigma_{p1} - \mu\sigma_{p2}}{E}; \qquad \epsilon_{p2} = \frac{\sigma_{p2} - \mu\sigma_{p1}}{E}.$$

Since $\sigma_{p1} = \tau_{xy} = -\sigma_{p2}$, and $\gamma_{xy} = \tau_{xy}/G$,

$$\frac{1}{2}\frac{\tau_{xy}^2}{G} = \frac{1}{2}\frac{\tau_{xy}^2(1+\mu)}{E} + \frac{1}{2}\frac{\tau_{xy}^2(1+\mu)}{E},$$

from which

$$E = 2(1+\mu)G. \tag{2–5}$$

Equation (2–5) substituted in Eq. (a) yields the generalized Hooke's law for shearing stress and strain (isotropic material),

$$\begin{aligned} \tau_{xy} &= \frac{E}{2(1+\mu)}\gamma_{xy}, \\ \tau_{xz} &= \frac{E}{2(1+\mu)}\gamma_{xz}, \\ \tau_{yz} &= \frac{E}{2(1+\mu)}\gamma_{yz}. \end{aligned} \tag{2–6}$$

Since Eq. (2–3) are so widely used in experimental stress determination, the following example is offered to illustrate two methods of application.

**Example 2–1:** Assume the strains of Ex. 1–6 were measured on a model constructed of steel for which the modulus of elasticity and Poisson's ratio may be taken as $30(10^6)$ psi and 0.300, respectively. Determine the principal and maximum shearing stresses, and show them on a complete sketch.

*Solution:* The principal strains from Ex. 1–6 are

$$\epsilon_{max} = +139.4(10^{-5}), \qquad \epsilon_{min} = -66.0(10^{-5}), \qquad \gamma_{max} = 205.4(10^{-5}).$$

From Eq. (2–3)

$$\sigma_{max} = \frac{30(10^6)}{1 - 0.09}[139.4 + 0.3(-66.0)](10^{-5}) = \frac{300}{0.91}(119.6)$$

$$= \underline{39{,}400 \text{ psi } T.} \qquad \text{Ans.}$$

$$\sigma_{min} = \frac{300}{0.91}[-66.0 + 0.3(139.4)] = \underline{7980 \text{ psi } C.} \qquad \text{Ans.}$$

From Eq. (2–6)

$$\tau_{max} = \frac{30(10^6)}{2(1 + 0.3)}(205.4)(10^{-5}) = \underline{23{,}700 \text{ psi.}} \qquad \text{Ans.}$$

Note that the value for $\tau_{max}$ checks that given by Eq. (1–6) p. 33, thus;

$$\tau_{max} = \frac{\sigma_{p1} - \sigma_{p2}}{2} = \frac{39{,}400 + 7{,}980}{2} = 23{,}690 \text{ psi.}$$

The directions of the principal stresses will be the same as those for the principal strains; hence, from Fig. 1–24, p. 55, the directions are obtained for the sketch of Fig. 2–7.

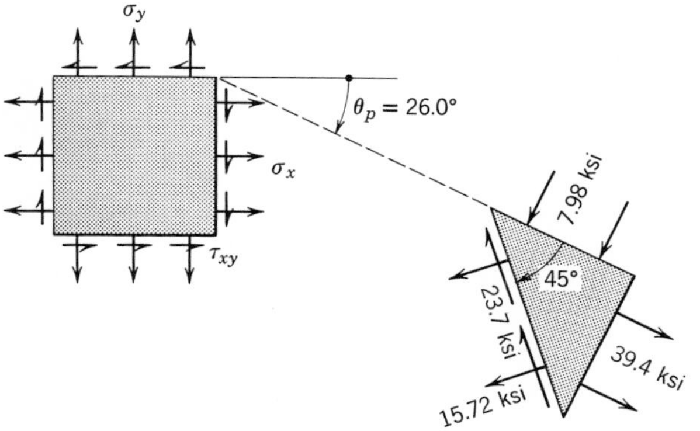


Fig. 2–7

*Alternate solution:* Additional computed strains from Ex. 1–6, p. 55, are

$$\epsilon_y = -26.67(10^{-5}); \qquad \gamma_{xy} = -161.66(10^{-5}).$$

From Eq. (2–3)

$$\sigma_x = \frac{30(10^6)}{1-0.09}[100 + 0.3(-26.67)](10^{-5}) = \frac{300}{0.91}(92) = 30{,}330 \text{ psi},$$

$$\sigma_y = \frac{300}{0.91}[-26.67 + 0.3(100)] = 1{,}099 \text{ psi},$$

and from Eq. (2–6)

$$\tau_{xy} = \frac{30(10^6)}{2(1+0.3)}(-161.66)(10^{-5}) = -18{,}650 \text{ psi}.$$

The Mohr's circle of Fig. 2–8 is constructed from these data and the following results are obtained:

$$\underline{\sigma_{\max} = 39.4 \text{ ksi } T.} \qquad \text{Ans.}$$

$$\underline{\sigma_{\min} = 7.98 \text{ ksi } C.} \qquad \text{Ans.}$$

$$\underline{\tau_{\max} = 23.7 \text{ ksi}.} \qquad \text{Ans.}$$

$$\underline{\theta_p = 25.96°} \circlearrowright. \qquad \text{Ans.}$$

The complete answer is given in Fig. 2–7.

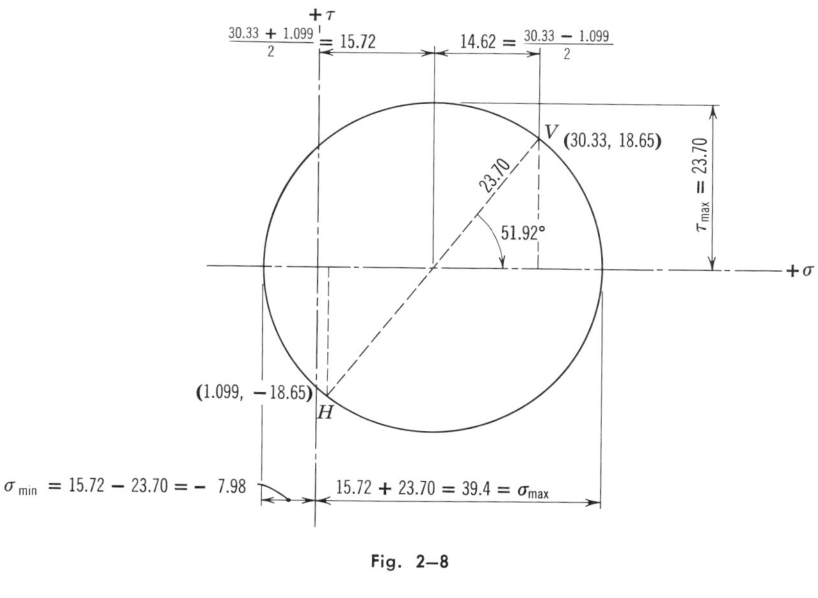


Fig. 2–8

**2–4. Design Loads, Working Stresses, and Factor of Safety.** As mentioned in Art. 1–1, the designer is required to select a material and properly proportion a member therefrom to perform a specified function without failure. Failure is defined as the state or condition in which a member or structure no longer functions as intended. Several types of failure warrant discussion at this point. *Elastic failure* is characterized by excessive elastic deformation—a bridge may deflect elastically under traffic to an extent which may result in discomfort to the vehicle passengers—a close-fitting machine part may deform sufficiently under load to prevent proper operation of the machine. When a structure is designed to avoid elastic failure, the stiffness of the material, indicated by Young's modulus, is the significant property. *Slip failure* is characterized by excessive plastic deformation due to slip. Yield strength, yield point, and proportional limit are used as indices of strength with respect to failure by slip for members subjected to static loads. *Creep failure* is characterized by excessive plastic deformation over a long period of time under constant stress. For machines or structures that are to be subjected to relatively high stress, high temperature, or both over a long period of time, creep is a design consideration, and the creep limit is the strength index to be used. The creep limit normally decreases as the temperature increases. *Failure by fracture* is a complete separation of the material. The ultimate strength of a material is the index of resistance to failure by fracture under static loads where creep is not involved.

Most design problems involve many unknown variables. The load that the structure or machine must carry is usually estimated; the actual load may vary considerably from the estimate, especially when loads at some future time must be considered. Since testing usually damages the material, the properties of the material used in the structure cannot be evaluated directly but are normally determined by testing specimens of a similar material. Furthermore, the actual stresses which will exist in a structure are unknown because the calculations are based on assumptions as to the distribution of stresses in the material. Because of these and other unknown variables, it is customary to design for the load required to produce failure, which is larger than the estimated actual load, or to use a working (or design) stress which is below the stress required to produce failure.

Figure 2–9 is a photograph showing the tail section of a Boeing 707 transport plane being prepared for a vibration test. Such tests can be used to measure the ultimate load-carrying capacity of the structure.

A *working stress* or *allowable stress* is defined as the maximum stress permitted in the design computation. The *factor of safety* may

Fig. 2–9. Vibration test of tail section of airplane. Courtesy Boeing Airplane Co., Seattle, Washington.

be defined as the ratio of a failure-producing load to the estimated actual load. Factor of safety may also be defined as the ratio of the strength of a material to the maximum computed stress in the material. The latter factor of safety may be used to determine an allowable (or working) stress by dividing the strength of a material by the factor of safety; however, because actual sections of structural and machine elements in general are not the same as the minimum computed sections, the actual (computed) factor of safety is the strength divided by the maximum computed stress based on actual sections of members in the design.

Factor of safety is also defined as the ratio of the strain energy per unit volume required to cause failure (slip or fracture) to the maximum computed strain energy per unit volume in a material under dynamic loading.

It will be shown in Chap. 5 that the factor of safety based on load may be greater than that based on stress. In other words, the term is meaningless without a statement of the criterion upon which it is based. In this book, unless otherwise indicated, the term *factor of safety* will denote the *ratio of the strength of the material to the maximum computed stress (unit strain energy for energy loading)*. For a given design, the factor of safety based on the ultimate strength is, in general, quite different from that based on the elastic strength (proportional limit, yield point, or yield strength); hence, the term *strength* is ambiguous without a qualifying statement regarding the pertinent mode of failure. The following example illustrates the use of a factor of safety.

**Example 2–2:** The improvised system of cables in Fig. 2–10*a* is used to remove the pile $Q$. A tractor with a maximum drawbar pull of 16 kips is available to apply the force $P$. Cable $A$ is a 1-in. diameter iron hoisting cable which has a breaking load of 29.0 kips. Cables $B$, $C$, and $D$ are parts of a continuous $3/4$-in. diameter carbon steel aircraft cable with a breaking strength of 49.6 kips. Cable $E$ is a $7/8$-in. diameter steel hoisting cable with a nominal area of 0.60 sq in. and an ultimate tensile strength, based on the nominal area, of 71.6 ksi. The minimum factor of safety for each cable, based on fracture, is to be 2.0. Determine:

(*a*) The maximum allowable force that can be exerted on the pile.

(*b*) The factor of safety for each of the five cables when the load in part *a* is applied.

*Solution:* (*a*) The two free-body diagrams needed to solve the problem are shown in Fig. 2–10*b* and *c*. The maximum allowable tension in each cable can be obtained from the strength of the cable and the factor of safety be-

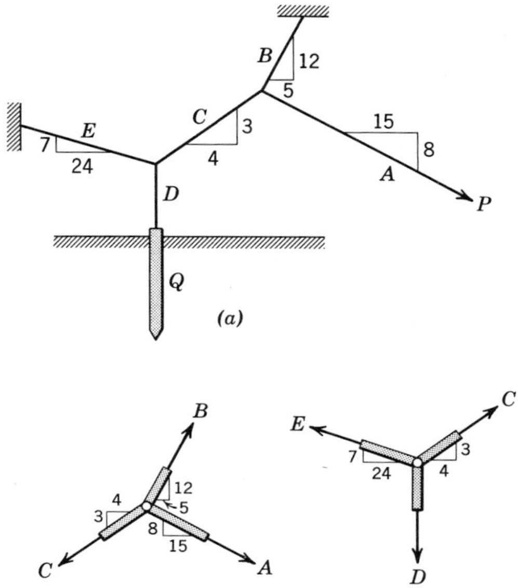


Fig. 2–10

cause for axial loading the factor of safety based on load is the same as that based on stress. The allowable tensions are:

$$A = 29.0/2 = 14.5 \text{ kips}$$

$$B = C = D = 49.6/2 = 24.8 \text{ kips}$$

$$E = 0.60(71.6)/2 = 21.5 \text{ kips.}$$

The maximum tractor pull (16.0 kips) is greater than the allowable force in cable $A$, therefore the maximum tractor pull cannot be utilized without exceeding the specifications. Assuming that the force in $A$ is 14.5 kips, the equations of equilibrium applied to Fig. 2–10*b* give

$$\frac{5}{13}B + \frac{15}{17}14.50 - \frac{4}{5}C = 0$$

and

$$\frac{12}{13}B - \frac{8}{17}14.50 - \frac{3}{5}C = 0$$

from which $B = 25.9$ kips and $C = 28.4$ kips. Since the maximum allowable load on cables $B$ and $C$ is only 24.8 kips, it is apparent that the force in $A$ must be less than the allowable load. The force in $B$ is less than that in $C$; therefore cable $C$ is the controlling member for Fig. 2–10*b*. The equations of equilibrium for this free-body diagram with $C = 24.8$ kips give the results

$$A = 12.65 \text{ kips} \quad \text{and} \quad B = 22.6 \text{ kips.}$$

The equations of equilibrium for Fig. 2–10*c* are

$$\frac{4}{5}C - \frac{24}{25}E = 0$$

and

$$\frac{3}{5}C + \frac{7}{25}E - D = 0.$$

When $C = 24.8$ kips is substituted in these equations, the results are

$$D = E = 20.7 \text{ kips.}$$

Both these forces are less than the maximum allowable loads in the corresponding cables and are therefore safe loads. The strength of cable $C$ is thus the limiting factor. The maximum pull on the pile is the force in cable $D$ and is

$$\underline{20.7 \text{ kips.}} \qquad \text{Ans.}$$

(*b*) The factor of safety in each of the cables is

$$k_A = 29.0/12.65 = \underline{2.29,} \qquad \text{Ans.}$$

$$k_B = 49.6/22.6 = \underline{2.20,} \qquad \text{Ans.}$$

$$k_C = 49.6/24.8 = \underline{2.00,} \qquad \text{Ans.}$$

$$k_D = 49.6/20.7 = \underline{2.40,} \qquad \text{Ans.}$$

$$k_E = 0.60(71.6)/20.7 = \underline{2.08.} \qquad \text{Ans.}$$

## PROBLEMS

**Note: Unless otherwise specified, all members are assumed to have negligible weight, and all pins used for connections are assumed to be smooth.**

**2–13.*** Specifications for a steel machine part require that the axial stress not exceed 20,000 psi and that the total elongation not exceed 0.015 in. If the part is a bar 2 in. in diameter and 3 ft long, determine the greatest permissible axial tensile load to which it may be subjected.

**2–14.*** The load $W$ in Fig. P2–14 is supported by two tie rods as shown. Rod $A$ has an area of 0.18 sq in. and a working stress of 12 ksi, and rod $B$ has an area of 0.10 sq in. and a working stress of 17 ksi. Determine the maximum safe load $W$ that can be supported.

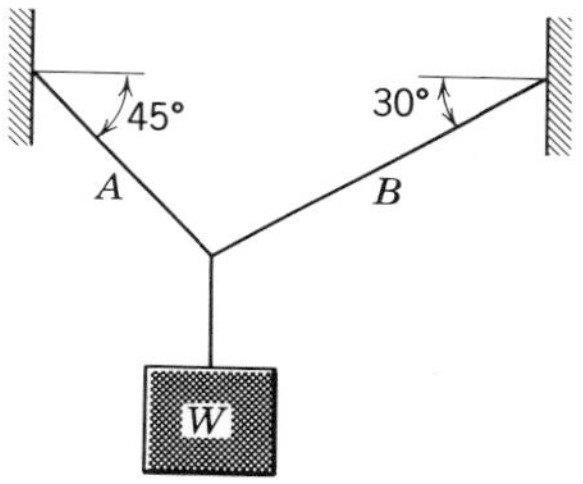


Fig. P2–14

**2–15.*** A structural steel rod (see App. A) 10 ft long is subjected to an axial stress of 25,000 psi. Determine:

(*a*) The maximum shearing stress in the rod.
(*b*) The axial strain in the rod.
(*c*) The factor of safety with respect to failure by fracture.

**2–16.*** A structural member made of wrought aluminum alloy 2024-T4 (see App. A) has a cross-sectional area of 5 sq in. and is subjected to an axial tensile load $P$. The member is to have a strain of not more than 0.0018 and a minimum factor of safety of 4 with respect to failure by fracture. Determine the greatest value $P$ may have without violating any one of these stipulations.

**2–17.** A material has the following properties: ultimate strength 65.0 ksi, yield strength 50.0 ksi, creep limit 25.0 ksi. What working stress should be used when designing with a factor of safety of 2 with respect to failure by slip under static load?

**2–18.*** A 5-ft structural steel bar ½ by 4 in. in cross section is to support an axial tensile load with allowable normal and shearing stresses of 18,000 psi and 8000 psi respectively and with an allowable elongation of 0.035 in. Determine the maximum permissible load.

**2–19.** The rod *DC* and rivet *B* in Prob. 1–15 are made of aluminum alloy 6061-T6 (see App. A). Determine the factor of safety with respect to failure by fracture for:

(*a*) Rod *CD*.

(*b*) Rivet *B*.

**2–20.*** The load *P* in Fig. P1–26 is 5200 lb. The timber block shown has the following ultimate strengths: shear parallel to grain 300 psi, compression perpendicular to the grain 700 psi, and compression parallel to the grain 3000 psi. Determine the factor of safety with respect to failure by fracture.

**2–21.** A 0.2 per cent carbon hardened steel (see App. A) rod will be required to carry an axial tensile load of 8000 lb with a factor of safety of 4 with respect to failure by slip. Determine the necessary cross-sectional area of the rod.

**2–22.*** In Fig. P1–82, bar *B* is made of 0.4 per cent carbon hot-rolled steel and bar *A* is made of aluminum alloy 6061-T6. Other data are as given in Prob. 1–82. Determine the factor of safety with respect to failure by slip for both bars. (See App. A.)

**2–23.** The rigid bar *DC* of Fig. P2–23 is horizontal under no load, and bars *A* and *B* are unstressed. When the load *P* is applied, the strain in bar *B* is found to be 0.0015. Determine the factor of safety with respect to failure by slip for each bar. Bar *A* is 0.8 per cent carbon hot-rolled steel (see App. A) with a cross-sectional area of 2.0 sq in. Bar *B* is cold-rolled red brass with a cross-sectional area of 1.0 sq in.

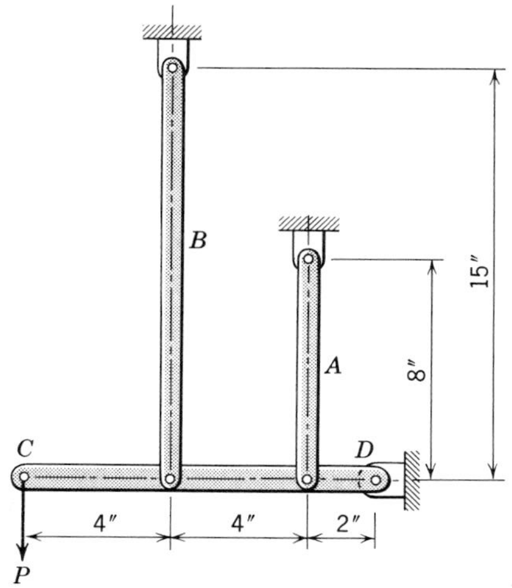


Fig. P2–23

**2–5. Temperature Effects.** Most engineering materials when unrestrained expand when heated and contract when cooled. The strain due to a one-degree temperature change is designated by $\alpha$ and is known as the coefficient of thermal expansion. The strain due to a temperature change of $\Delta T$ degrees is

$$\epsilon_T = \alpha(\Delta T).$$

The coefficient of thermal expansion is approximately constant for a considerable range of temperatures (in general, the coefficient increases with an increase of temperature). For a homogeneous, isotropic material, the coefficient applies to all dimensions (all directions). Values of the coefficient of expansion for several materials are included in App. A.

When a member is restrained (free movement prevented) while a temperature change takes place, stresses (referred to as *thermal stresses*) are induced in the member. If the action is elastic, thermal stresses are easily computed by, first, assuming that the restraining influence has been removed and the member permitted to expand or contract freely, and, second, applying forces that cause the member to assume the configuration dictated by the restraining influence. For example, the bar $AB$ of Fig. 2–11$a$ is securely fastened to rigid supports at the ends. If the end $B$ is assumed to be cut loose from the wall and the temperature drops, the end will move to $B'$ a distance $\delta_T$ as indicated in Fig. 2–11$b$. Next a force $P$, shown in Fig. 2–11$c$, is applied to $B$ of sufficient magnitude to move end $B$ through a distance $\delta_P$ so that the length of the bar is again $L$, the distance between the walls. Since

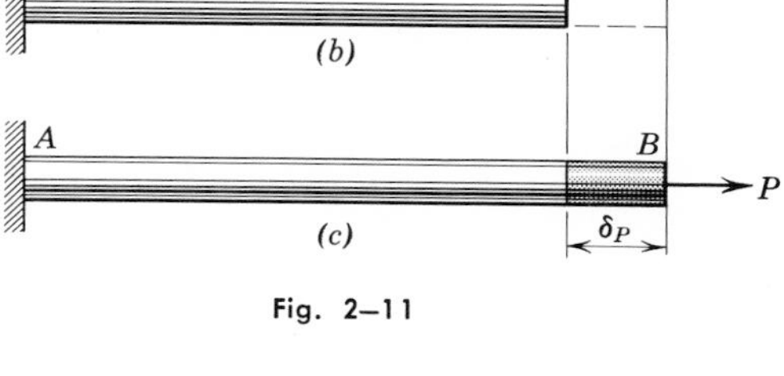


Fig. 2–11

the walls do not move, $\delta_T = \delta_P$, where $\delta_T = \alpha L \, \Delta T$, and $\delta_P = L\epsilon$; and, if Hooke's law applies, $\epsilon = \sigma/E$, where $\sigma$ is the induced axial stress.

In case a stress exists before the temperature change takes place (the rod of Fig. 2-11*a* might have been tightened by means of a turnbuckle or nuts at the ends), the stress due to the temperature change may be added algebraically to the original stress by using the *principle of superposition.* The principle states that stresses (at a point on a given plane) due to different loads may be computed separately and added algebraically, provided that the sum of the stresses does not exceed the proportional limit of the material and that the structure remains stable. In explanation of this last limitation, consider a member with a length considerably greater than the lateral dimension subjected to axial compression loads. This member may successfully carry either of two applied loads, and the sum of the axial stresses for both loads may be within the proportional limit; yet, upon application of both loads, the member may collapse (this action is discussed further in Chap. 9).

## PROBLEMS

**Note: Unless otherwise specified, all members are assumed to have negligible weight, and all pins used for connections are assumed to be smooth.**

**2–24.*** Determine the horizontal movement of point $A$ in Fig. P2–24 due to a temperature rise of 50 F. Assume member $AE$ to have an insignificant coefficient of thermal expansion. The steel bar is structural steel (see App. A) and the aluminum alloy is 6061-T6.

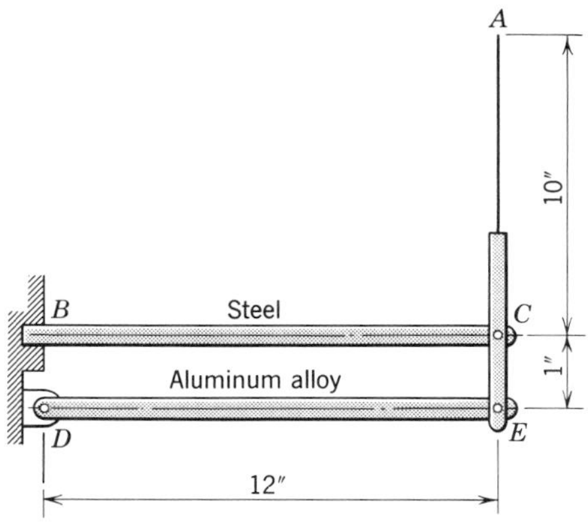


Fig. P2–24

**2–25.*** If the temperature of the mechanism shown in Fig. P2–25 increases 100 F, determine the movement of the pointer in inches with respect to the scale zero. The steel member is made of 0.2 per cent carbon steel. See App. A.

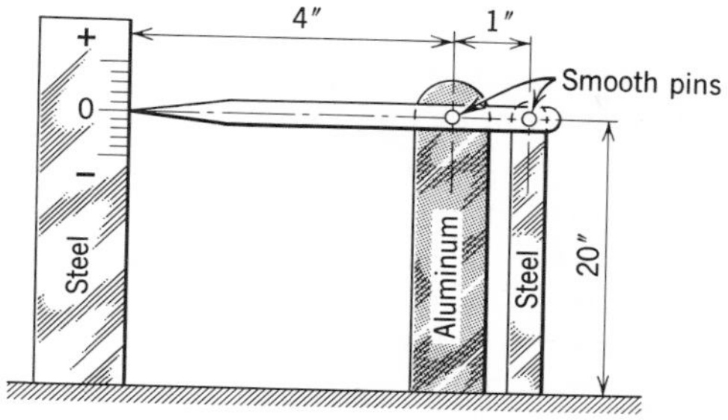


Fig. P2–25

**2–26.** A cold-rolled bronze rod and a hot-rolled wrought Monel rod (see App. A) are coupled together end to end and loaded axially at the extreme ends with a tensile load of $20\pi$ kips. Both rods are 2 in. in diameter; the bronze rod is 5 ft long, and the Monel rod is 4 ft long. Determine the total change in length due to the load, plus a temperature drop of 60 F.

**2–27.*** A 20-ft horizontal railroad rail with a cross-sectional area of 11.6 sq in. is supported against lateral movement. Under an increase in temperature, one end can move 1/16 in. relative to the other end before both ends are tight against adjacent rails which, for this problem, can be assumed to be rigid. The coefficient of thermal expansion of this steel is $6.6(10^{-6})$ per degree F. For an increase in temperature of 80 F, determine:

(*a*) The internal force on a cross section of the rail.

(*b*) The shearing stress developed in the rail on a plane having an angle of 30° with the horizontal.

**2–28.** The ends of a 12-ft long steel rod are securely fastened to rigid supports. At 75 F there is no stress in the rod, and the diameter is 2 in. The steel has the following properties: yield point 38 ksi, ultimate strength 65 ksi, ultimate elongation 28 per cent in 8 in., Poisson's ratio 0.25, and coefficient of thermal expansion $6.5(10^{-6})$ per degree F. For a temperature of 0 F, determine:

(*a*) The factor of safety with respect to failure by slip.

(*b*) The change in the diameter of the bar.

**2–29.*** A 2-in. diameter 20-ft long duralumin rod is secured at the ends to supports that permit a change in length of the bar of 0.06 in. The following properties apply: modulus of elasticity $10(10^6)$ psi, coefficient of thermal expansion $13(10^{-6})$ per degree F, and Poisson's ratio 0.333. When the temperature is 80 F, there is no stress in the rod. For a temperature of $-20$ F, determine:

(*a*) The maximum normal stress in the rod.
(*b*) The maximum shearing stress in the rod.
(*c*) The change in diameter of the rod.

**2–30.** An annealed stainless steel (18-8), (see App. A) bolt through a 3-ft long gray cast iron tube has the nut tightened until the axial tensile stress in the bolt is 6000 psi at 80 F. Determine the change in axial stress in the bolt when the temperature is $-10$ F. Neglect the change in length of the tube due to the change in stress in the bolt.

**2–31.*** A steel tire is to be heated and placed on a 72.00-in. diameter locomotive drive wheel. At 80 F the tire has an inside diameter of 71.94 in. The coefficient of thermal expansion for the steel is $6.5(10^{-6})$ per degree F, and its yield strength for 0.2 per cent offset is 60 ksi.
Determine:

(*a*) The temperature increase required for the inside diameter of the tire to be 72.04 in. to facilitate placing it on the wheel.

(*b*) The maximum tensile stress in the tire when it has cooled to $-20$ F, assuming that the diameter of the wheel does not change.

(*c*) The factor of safety with respect to failure by slip for the stress obtained in part (*b*).

**2–6. Statically Indeterminate Axially Loaded Members.** In the work so far, it has always been possible to find the internal forces in any member of a structure by means of the equations of equilibrium; that is, the structures were statically determinate. If, in any structure, the number of unknown forces and distances exceeds the number of independent equations of equilibrium which are applicable, the structure is said to be *statically indeterminate*, and it is necessary to write additional equations involving the geometry of the deformations in the members of the structure. The following outline of procedure will be helpful in the analysis of problems involving statically indeterminate structures.

1. Draw a free-body diagram.
2. Note the number of unknowns involved (magnitudes and positions).

3. Recognize the type of force system on the free-body diagram, and note the number of independent equations of equilibrium available for this system.

4. If the number of unknowns exceeds the number of equilibrium equations, a deformation equation must be written for each extra unknown.

5. When the number of independent equilibrium equations and deformation equations equals the number of unknowns, the equations can be solved simultaneously. Deformations and forces must be related in order to solve the equations simultaneously. Hooke's law and the definitions of stress and strain can be used when all stresses are less than the corresponding proportional limits. If some of the stresses exceed the proportional limits of the materials, the stress-strain diagrams can be used to relate the loads and deformations or simplifying assumptions (such as perfect plasticity, see Fig. 2–3*c*) can be made to obtain satisfactory results.

It is recommended that a displacement diagram be drawn showing deformations to assist in obtaining the correct deformation equation. The displacement diagram should be as simple as possible (a line diagram) with the deformations indicated with exaggerated magnitudes and clearly dimensioned. Loading members (members assumed to be rigid), especially, should be indicated as single lines. In all problems and examples which follow, the loading members, pins, and supports are assumed to be rigid, and the mechanism is so constructed that the force system is coplanar. The procedure outlined above is illustrated in the following examples.

**Example 2–3:** The rigid plate $C$ in Fig. 2–12*a* is fastened to the 0.50-in. diameter steel rod $A$ and to the aluminum pipe $B$. The other ends of $A$ and $B$ are fastened to rigid supports. When the force $P$ is zero, there are no stresses in $A$ and $B$. Rod $A$ is made of low-carbon steel which is assumed to be elastoplastic (see Fig. 2–3*c*) and has a proportional limit and yield point of 40 ksi ($E$ = 30,000 ksi). Pipe $B$ has a cross-sectional area of 2.0 sq in. and is made of an aluminum alloy which has the stress-strain diagram shown in Fig. 2–12*b*. A load $P$ of 30 kips is applied to $C$ as shown. Determine the axial stresses in $A$ and $B$ and the displacement of plate $C$.

*Solution:* Figure 2–12*c* is a free-body diagram of plate $C$ and portions of the members $A$ and $B$. The free-body diagram contains two unknown forces, and only one equation of equilibrium is available; therefore the problem is statically indeterminate. As the load is applied to the plate $C$, it moves downward an amount $\delta$ which represents the total deformation in members $A$ and $B$; see Fig. 2–12*d*. This observation provides a displacement equation which can be used with the equilibrium equation to solve the problem. The two

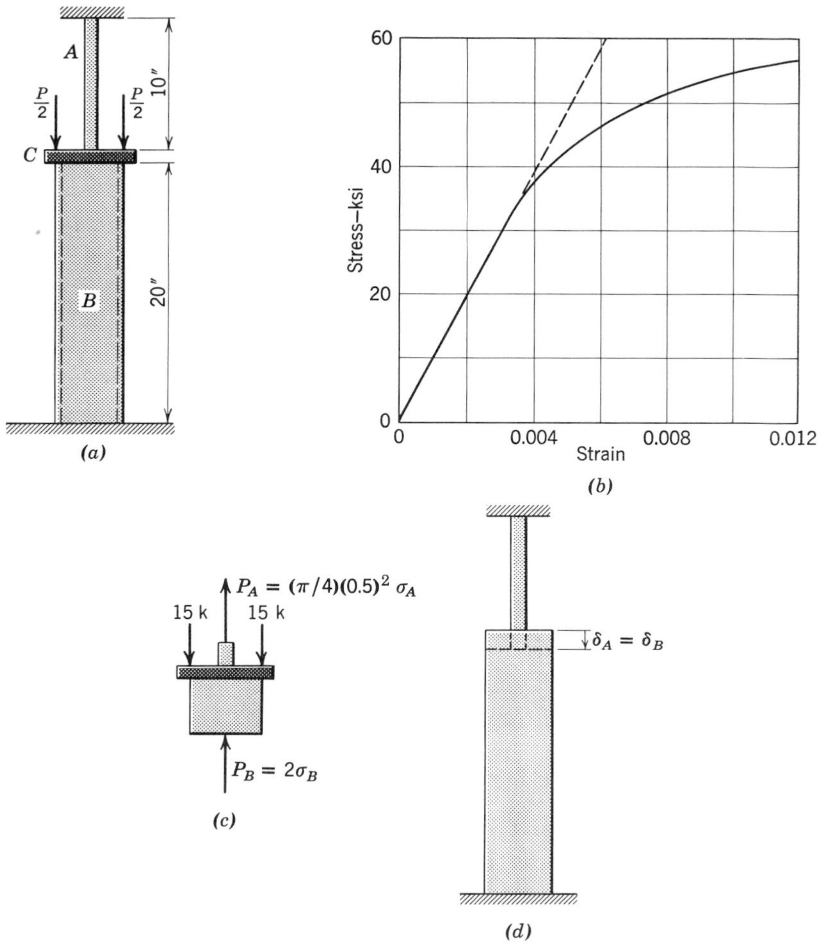


Fig. 2–12

equations are

$$\left(\uparrow +\right) \qquad \Sigma F_y = 0 = P_A + P_B - 30$$

or

$$(\pi/4)(0.50)^2\sigma_A + 2\sigma_B = 30 \tag{a}$$

and

$$\delta_A = \delta_B$$

or

$$10\epsilon_A = 20\epsilon_B.$$

In order to solve these equations, the stresses and strains must be related. If the stresses are less than the corresponding proportional limits, Hooke's law can be used. If Hooke's law is assumed to be valid, the strain equation becomes

$$\frac{10\sigma_A}{30(10^3)} = \frac{20\sigma_B}{10(10^3)},$$

from which

$$\sigma_A = 6\sigma_B. \tag{b}$$

The value of $E_B = 10(10^3)$ ksi is obtained as the slope of the stress-strain diagram in Fig. 2–12*b*. The simultaneous solution of Eq. (a) and (b) gives $\sigma_A = 56.6$ ksi $T$ and $\sigma_B = 9.45$ ksi $C$.

These results indicate that bar $A$ is stressed beyond the proportional limit of the steel and that Hooke's law does not apply. Since the material is assumed to be perfectly plastic beyond the proportional limit, the stress in bar $A$ must be 40 ksi. When this value is substituted in the equilibrium equation (a), the equation becomes

$$(\pi/16)(40) + 2\sigma_B = 30,$$

from which $\sigma_B = \underline{11.07 \text{ ksi } C}$ and $\sigma_A = \underline{40 \text{ ksi } T.}$ Ans.

The stress in $B$ is less than the proportional limit for aluminum; therefore Hooke's law can be used to determine the movement of the plate which is

$$\delta = \delta_B = 20(11.07)/(10^4) = \underline{0.0221 \text{ in. downward.}}$$ Ans.

If the stress in $B$ had been more than the proportional limit of the aluminum, it would have been necessary to determine the strain from the curve in Fig. 2–12*b*. If the material in bar $A$ were not perfectly plastic, a trial-and-error solution would be required, since the stresses in both $A$ and $B$ depend on the movement of $C$.

**Example 2–4:** The steel post $B$ of Fig. 2–13*a* is 2 by 2 in. square, and the aluminum alloy bar $C$ is 1.5 in. wide by 1 in. thick. Bar $A$ and the bearing block on $B$ are to be considered rigid; all connections are treated as smooth pins; the temperature is assumed to remain constant; and the modulus of elasticity of the aluminum alloy is taken as $(10^7)$ psi. Under no load, the clearance between $A$ and $B$ is 0.002 in. Determine the maximum permissible load, $P$, which may be applied as shown if the axial stresses are not to exceed 15,000 psi for the steel and 18,000 psi for the aluminum alloy.

*Solution:* The free-body diagram of body $A$ in Fig. 2–13*b* shows five unknown forces; however, one of the axial stresses, either $\sigma_B$ or $\sigma_C$, will equal the allowable value (which is given), and the other one will be less than the allowable stress (or equal to it in an exceptional case). The force system is coplanar for which three independent equations of equilibrium are available. With four unknowns and three equations, the system is statically indeterminate, and an additional equation must be obtained from the deformations of $B$ and $C$. The displacement diagram of Fig. 2–13*c* shows the positions of the members, after the load $P$ is applied, as dashed lines. The deformations of members $B$ and $C$ are related by the equation (from similar triangles)

$$\frac{\delta_B + 0.002}{4} = \frac{\delta_C}{12}$$

or

$$3\delta_B + 0.006 = \delta_C.$$

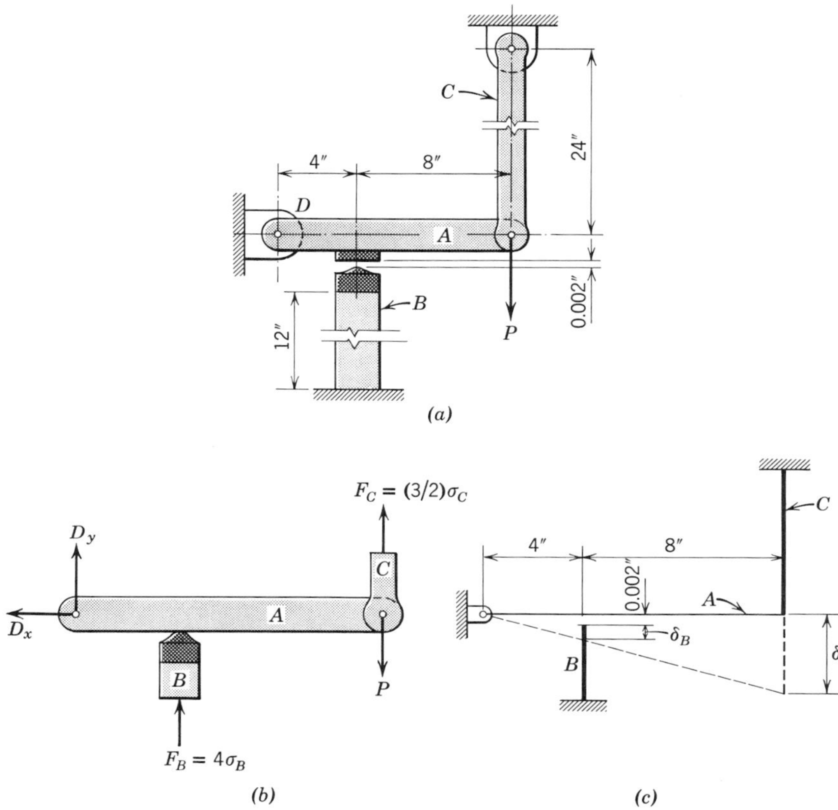


Fig. 2–13

In terms of strain ($\delta = L\epsilon$), the relation is

$$3(12)\epsilon_B + 0.006 = 24\epsilon_C.$$

Since the stresses are limited to values less than the proportional limit (see App. A), Hooke's law ($\sigma = E\epsilon$) applies, and the equation becomes

$$\frac{36\sigma_B}{30(10^6)} + 0.006 = \frac{24\sigma_C}{10(10^6)}$$

or

$$\sigma_B + 5000 = 2\sigma_C.$$

When $\sigma_B$ equals its allowable value of 15,000 psi $\sigma_C$ becomes 10,000 psi as determined from this strain relation. It is apparent that higher values for $\sigma_C$ would yield a stress in the steel above the specified limit. Therefore, the axial stresses are 15,000 psi in the steel and 10,000 psi in the aluminum. From

Fig. 2–13*b* and the equilibrium equation $\Sigma M_D = 0$,

$$4(15{,}000)4 + 12(10{,}000)(3/2) - 12P = 0,$$

from which $$P = \underline{35{,}000 \text{ lb as shown.}}$$ Ans.

**Example 2–5:** If the mechanism of Ex. 2–4 is loaded with a force $P$ of 10,000 lb as shown and the temperature increases 20 F, what are the axial stresses in members $B$ and $C$? For the coefficients of thermal expansion per degree F, use $13(10^{-6})$ for the aluminum and $6.5(10^{-6})$ for the steel.

*Solution:* The free-body diagram in Fig. 2–13*b* is valid for this problem. Following the suggestion of Art. 2–5, the thermal deformation is first permitted to take place by removing all restraints (bar $A$ is imagined to be removed). The deformations of $B$ and $C$ due to the change of temperature are indicated as $\delta_T$ values in Fig. 2–14*a*. Loads are then applied to $B$ and $C$ which change their lengths so that the rigid bar $A$ can be connected to them as indicated in Fig. 2–14*b*. The deformations due to the forces are indicated as $\delta_P$ values. The deformation equation is

$$\frac{\delta_{BP} - \delta_{BT} + 0.002}{4} = \frac{\delta_{CP} + \delta_{CT}}{12},$$

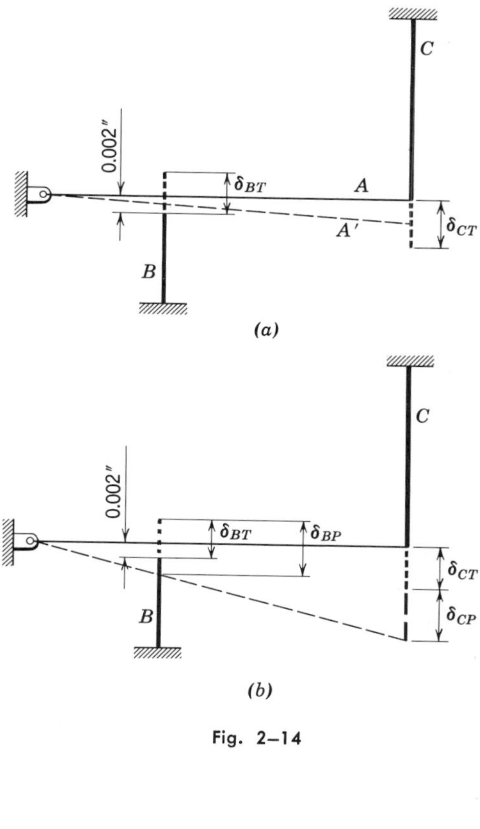


Fig. 2–14

or
$$3\delta_{BP} - 3\delta_{BT} + 0.006 = \delta_{CP} + \delta_{CT}.$$

When deformations are replaced by their values in terms of strains, the equation becomes

$$3(12)\epsilon_{BP} - 3(12)\epsilon_{BT} + 0.006 = 24\epsilon_{CP} + 24\epsilon_{CT}.$$

When the strains due to stress are replaced by their values in terms of stress (assuming Hooke's law is valid) and the strains due to the temperature change by their numerical values, the displacement equation becomes

$$\frac{36\sigma_B}{30(10^6)} - \frac{36(6.5)20}{(10^6)} + \frac{6}{(10^3)} = \frac{24\sigma_C}{10(10^6)} + \frac{24(13.0)20}{(10^6)}$$

or
$$\sigma_B = 2\sigma_C + 4100. \tag{c}$$

From Fig. 2–13*b* and the equilibrium equation $\Sigma M_D = 0$,

$$4\sigma_B(4) + (3/2)\sigma_C(12) - 10{,}000(12) = 0. \tag{d}$$

The simultaneous solution of Eq. (c) and (d) gives

$$\sigma_C = \underline{1088 \text{ psi } T} \qquad \text{Ans.}$$

and
$$\sigma_B = \underline{6280 \text{ psi } C.} \qquad \text{Ans.}$$

These stresses are well within the proportional limits of the materials; therefore Hooke's law is valid.

The use of colored pencils may be found helpful in the construction of the deformation diagram by using one color for temperature deformations and another for load deformations.

From a study of Fig. 2–14*a* it will be observed that with no applied load ($P = 0$), the bar $A$ might have occupied some position such as that shown dashed (marked $A'$) and would, therefore, have induced compression in both bars $B$ and $C$. In the solution given, the assumption was made that the force $P$ was of sufficient magnitude to induce tension in bar $C$. The positive sign on the value for $\sigma_C$ indicates that the assumption was correct *if the deformation diagram and the free-body diagram are compatible;* that is, where a tensile deformation is assumed in the deformation diagram, a tensile stress must be indicated on the free-body diagram. This very important consideration must not be overlooked.

## PROBLEMS

**Note: Unless otherwise specified, all members are assumed to have negligible weight, and all pins used for connections are assumed to be smooth. The temperature remains constant unless otherwise specified.**

**2–32.*** A steel plate 1 by 6 by 20 in. is bolted to an oak block 6 by 6 by 20 in. The structure is shortened 0.02 in. by the unknown load $P$ applied as shown in Fig. P2–32. The modulus of elasticity of the timber is $1.5(10^6)$ psi.

(*a*) Determine the axial stress in each material.

(*b*) Determine the magnitude of the load $P$.

(*c*) Locate the position of the line of action of $P$ with respect to the outer edge of the steel plate.

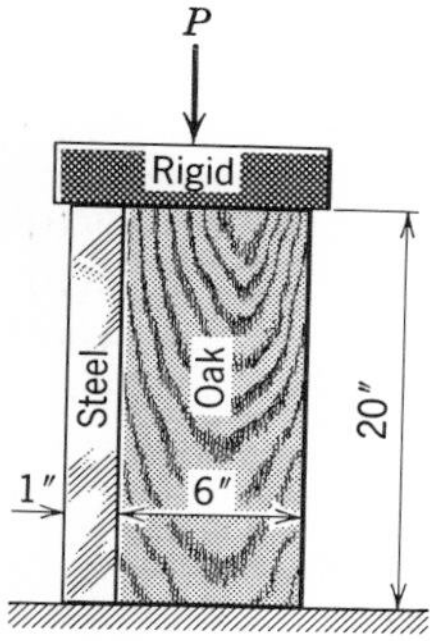


Fig. P2–32

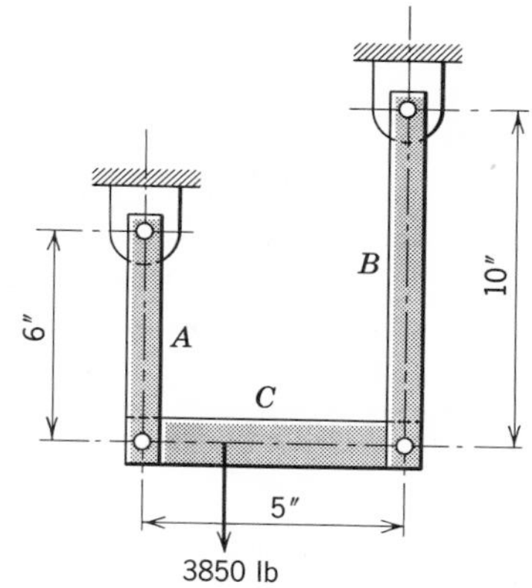


Fig. P2–33

**2–33.*** In Fig. P2–33, *A* is a steel bar 0.5 by 0.8 in. and *B* is a brass bar 1 by 0.5 in. The bar *C* may be regarded as rigid. A load of 3850 lb is hung from *C* at such a place that the bar *C* remains horizontal. The modulus of elasticity of brass is $15(10^6)$ psi. Determine:

(*a*) The axial stress in each material.

(*b*) The vertical displacement of bar *C*.

(*c*) The position of the line of action of the 3850-lb force.

**2–34.*** In Fig. P2–34, *A* is a wrought iron bar and *C* is a 6061-T6 aluminum alloy bar. Bar *B* may be regarded as rigid. A load of 32,000 lb is hung from *B* as shown, and *B* remains horizontal. Determine minimum cross-sectional areas of *A* and *C* such that the axial stresses in them do not exceed working stresses with a factor of safety of 2.5 with respect to failure by slip. (See App. A.)

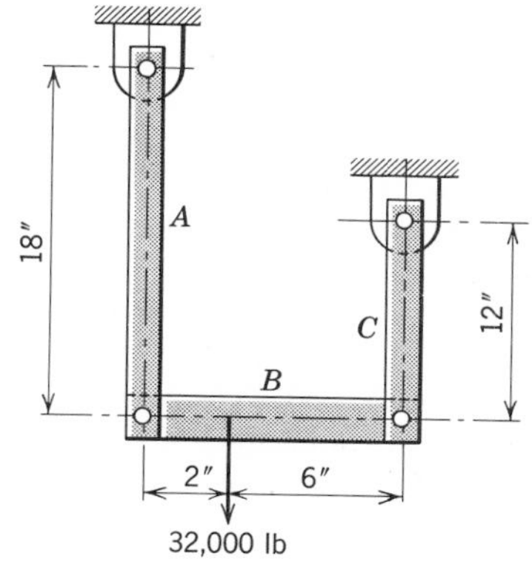


Fig. P2–34

**2–35.*** The system in Fig. P2–35 is initially unstressed. Bar $C$ is rigid and is 0.002 in. above the cast iron member $D$. Determine the value of $P$ such that the stresses in members $A$, $B$, and $D$ have the same magnitude. The following data apply:

| Member | Material | Area, sq in. | $E$, psi |
|---|---|---|---|
| $A$ | Aluminum alloy | 1 | $10(10^6)$ |
| $B$ | Aluminum alloy | 1 | $10(10^6)$ |
| $D$ | Cast iron | 3 | $14(10^6)$ |

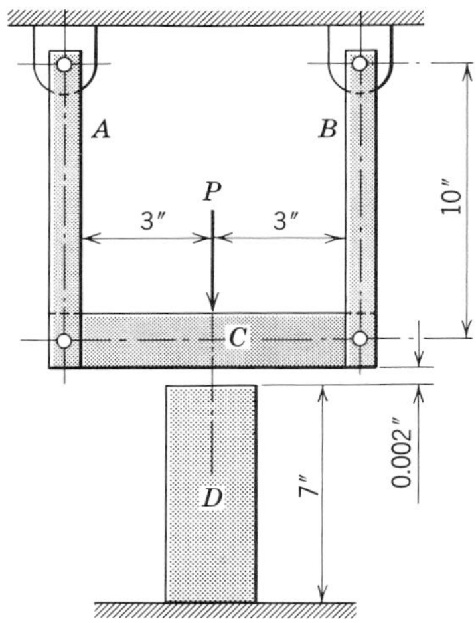


Fig. P2–35

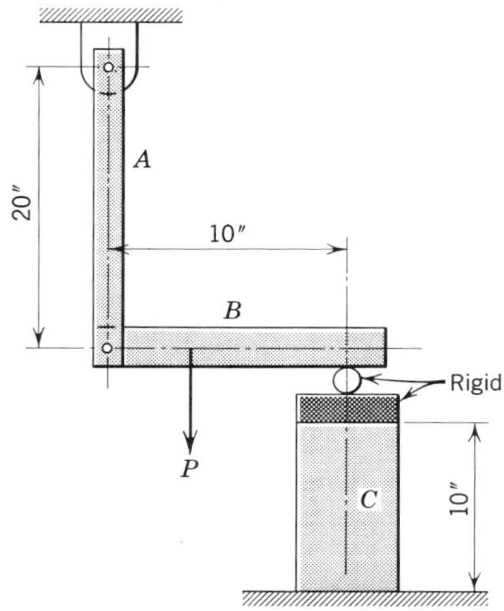


Fig. P2–36

**2–36.** As shown in Fig. P2-36, a downward load $P$ is applied to the rigid member $B$ so that the member remains horizontal. Member $B$ is supported at the left end by the Monel metal rod $A$ and at the right end by the concrete post $C$. The modulus of elasticity of Monel metal is $26(10^6)$ psi and that of the concrete is $4(10^6)$ psi. The cross-sectional areas of $A$ and $C$ are 2 sq in. and 15 sq in., respectively.

($a$) Determine the greatest permissible load $P$ if the allowable tensile stress in the Monel metal is 8400 psi and the allowable compressive stress in the concrete is 1200 psi.

($b$) What is the distance from the center of bar $A$ to the load?

($c$) If the restriction that $B$ remain horizontal were removed, what would be the value of the load $P$, and what would be the distance from the center of the bar $A$ to the load?

**2–37.*** Solve Ex. 2–3 with $P = 50$ kips and the cross-sectional area of $B = 1.00$ sq in.

**2–38.** In Fig. P2–38 *CD* is a rigid bar which sustains a 5000-lb load at the left end and is pin connected to the support bracket *D* and the two aluminum alloy bars *A* and *B*, as shown. The bar *CD* is horizontal under no load. Determine the vertical displacement of point *C* and the axial stress in bar *A*. The following data apply to each of the bars *A* and *B*: Cross-sectional area—0.5 sq in., Young's modulus—$10(10^6)$ psi.

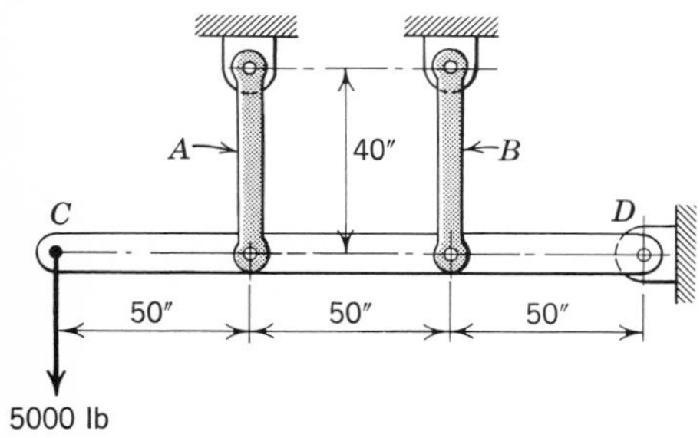


Fig. P2–38

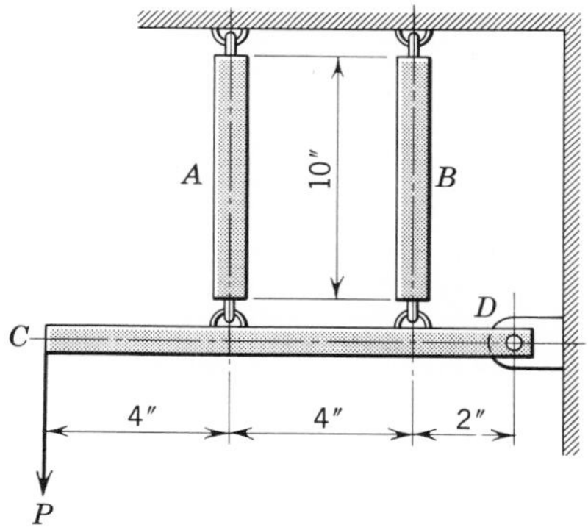


Fig. P2–39

**2–39.*** In Fig. P2–39 bars *A* and *B* are connected by rigid links to a fixed position at the top and to a rigid bar *CD* at the bottom. Assume that the bar *CD* remains straight during loading and that the smooth pin at *D* maintains a fixed position. The following data apply to the bars:

| Bar | Material | Area, sq in. | *E*, psi |
|---|---|---|---|
| *A* | Aluminum alloy | 2.5 | $10(10^6)$ |
| *B* | Monel | 2 | $25(10^6)$ |

Determine the load *P* necessary to produce an axial stress of 18,000 psi in bar *A*.

**2–40.** If the load, *P*, in Fig. P2–39 is 20,000 lb, as shown, what are the axial stresses in bars *A* and *B*? Refer to Prob. 2–39 for data.

**2–41.*** Solve Prob. 2–39 after making the following modifications: Change the distance between bars from 4 in. to 2 in., change $B$ from Monel to steel, and change the stress in $A$ from 18 ksi to 26 ksi. The proportional limit for both materials is 35 ksi. All other dimensions remain as shown.

**2–42.** Solve Prob. 2–40 after making the following modifications: Change the horizontal dimensions from 4, 4, and 2 to 2, 2, and 4; change $B$ from Monel to steel; and change $P$ from 20,000 lb to 68,000 lb. The proportional limit for both materials is 35,000 psi.

**2–43.*** The steel bar $C$ of Fig. P2–43 has a cross-sectional area of 0.25 sq in., and the brass ($E$ = 15,000 ksi) bar $D$ has a cross-sectional area of 3 sq in. The supports and the bar $AB$ are to be considered rigid. Determine the value of the load, $P$, which will produce an axial stress in bar $D$ of 4.6 ksi $C$.

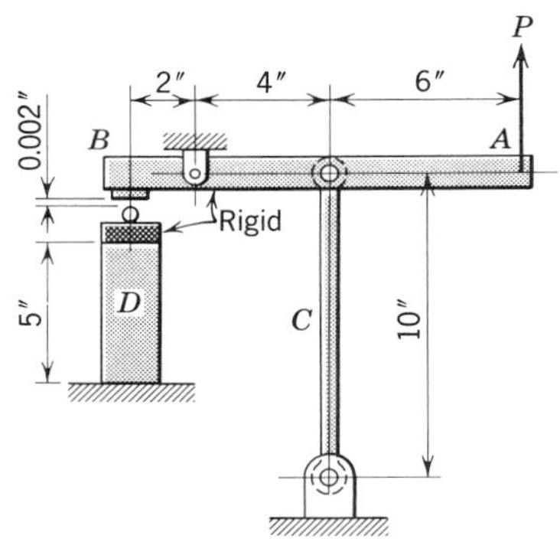


Fig. P2–43

**2–44.** The bar $C$ in Fig. P2–43 is made of structural steel, and $D$ is cold-rolled red brass (see App. A). The areas are 1.2 sq in. for $C$ and 3.0 sq in. for $D$. The clearance above $D$ is changed from 0.002 in. to 0.004 in., and the load $P$ is 30.0 kips. (The axial stress in $D$ is unknown.) Determine the movement of point $A$ caused by the load.

**2–45.*** Bar $B$ of the pin-connected mechanism of Fig. P2–45 is made of aluminum alloy 6061-T6, and bar $A$ is made of 0.2 per cent carbon steel. The cross-sectional areas are 1.5 sq in. for $A$ and 0.25 sq in. for $B$. Bar $CDE$ is to be considered rigid. When the mechanism is unloaded, there is a 0.004-in. clearance in the connection at $D$ which must be taken up before bar $A$ will be stressed. If the allowable axial stresses are 21,000 psi for $A$ and 16,000 psi for $B$, determine the maximum permissible load that may be applied downward at a point 6 in. to the right of $D$.

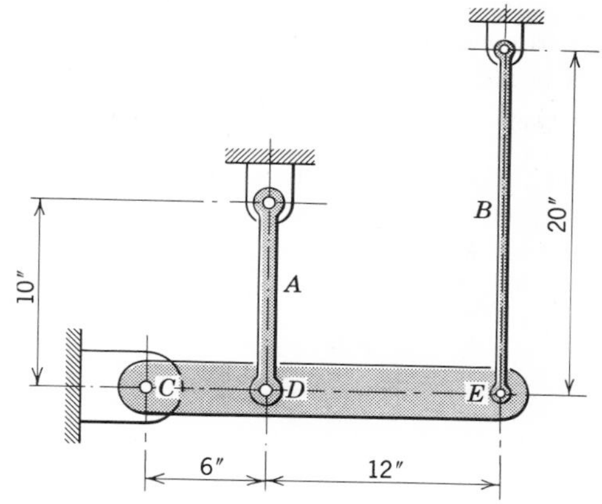


Fig. P2–45

**2–46.*** The pin-connected structure in Fig. P2–46 occupies the position shown under no load. When the loads $D$ (= 2000 lb) and $E$ are applied, the rigid bar $C$ becomes horizontal. Determine the weight of $E$. The following data apply:

| Bar | Material | Area, sq in. | $E$, psi |
|---|---|---|---|
| $A$ | Aluminum alloy | 2 | $10(10^6)$ |
| $B$ | Monel | 4 | $24(10^6)$ |

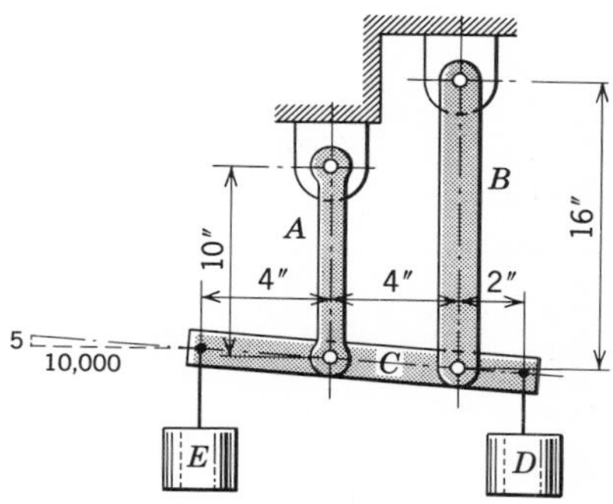


Fig. P2–46

**2–47.** Bar $C$ of Fig. P2–47 is to be considered rigid. Determine the force $P$ which will cause $D$ to move downward 0.011 in. more than point $E$ will move. The following data apply:

| Member | Material | Modulus of Elasticity, psi | Cross-Sectional Area, sq in. |
|---|---|---|---|
| $A$ | Copper | $16(10^6)$ | 0.40 |
| $B$ | Steel | $30(10^6)$ | 0.50 |

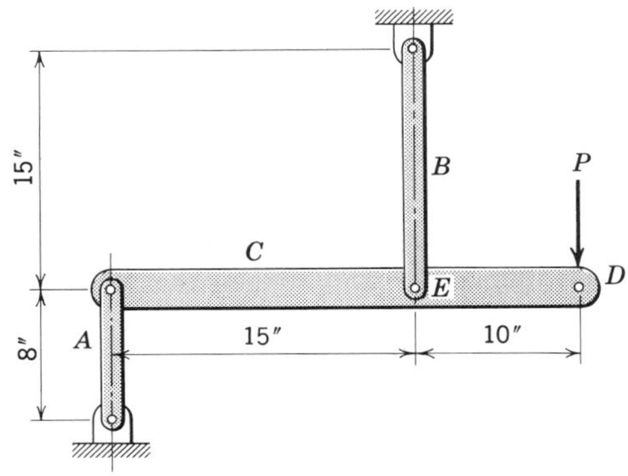


Fig. P2–47

**2–48.*** For the mechanism of Fig. P2–48, assume bar $C$ and all supports to be rigid. Determine the axial stress induced in the gross section of rod $B$ by advancing the nut at the top of rod $B$ 0.1 in. (one turn). The following data apply:

| Rod | Material | Gross Cross-Sectional Area, sq in. | Modulus of Elasticity, psi |
|---|---|---|---|
| $A$ | Brass (C.R.) | 0.500 | $15(10^6)$ |
| $B$ | Steel | 0.400 | $30(10^6)$ |

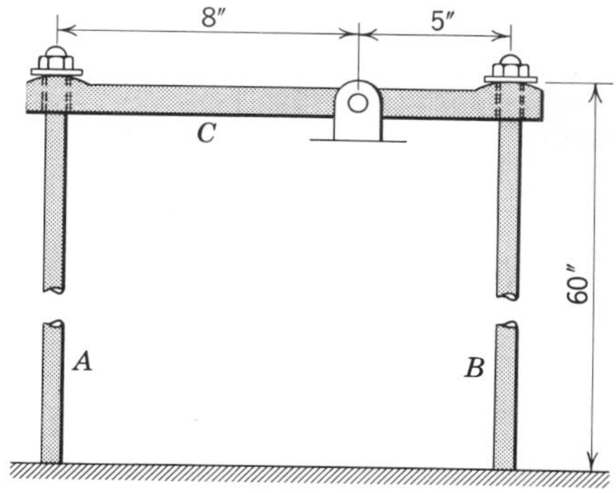


Fig. P2–48

**2–49.** Determine the vertical force $P$ that must be applied at $B$ to make the initially horizontal rigid bar $C$ of Fig. P2–49 horizontal when the temperature rises 100 F. The following data apply:

| Material | $E$, psi | $\alpha$, per deg F | Cross-Sectional Area, sq in. |
|---|---|---|---|
| $A$ (Aluminum alloy) | $10(10^6)$ | $13.0(10^{-6})$ | 4 |
| $D$ (Steel) | $30(10^6)$ | $6.5(10^{-6})$ | 1 |

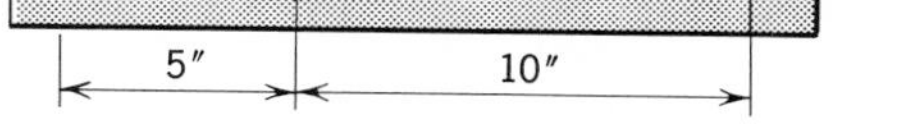


Fig. P2–49

**2–50.*** The rigid body $W$ of Fig. P2–50 is supported by a steel wire 20 ft long. The body hangs in the position shown when the temperature is 90 F. Determine the weight of $W$ if the maximum normal stress in the steel wire is 18.0 ksi $T$ when the temperature is 10 F. The coefficient of thermal expansion is $6.5(10^{-6})$ per degree F.

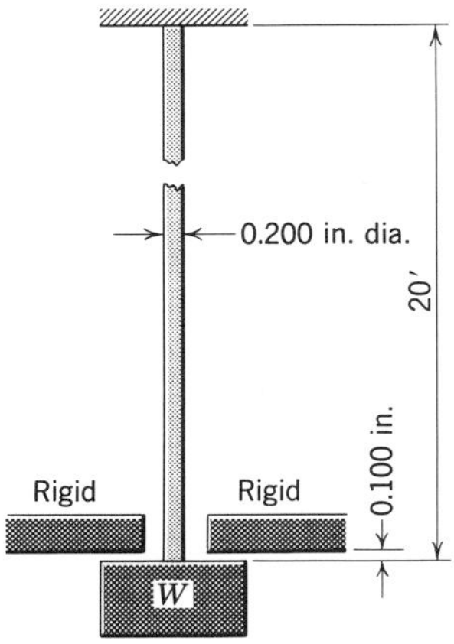


Fig. P2–50

**2–51.** The cross-sectional area of each of the outer brass rods in Fig. P2–51 is ¾ sq in. The cross-sectional area of the center steel (0.2% C hardened) rod is 1 sq in. Bar $A$ remains horizontal. (See App. A.)

(*a*) Determine the load carried by each rod.

(*b*) What change in temperature will cause the steel rod to carry the entire load?

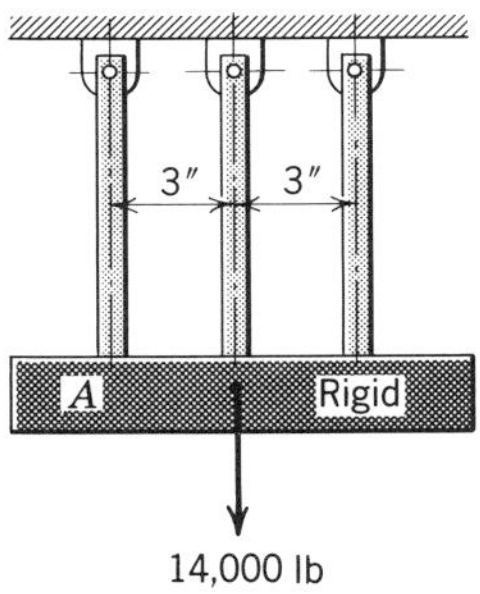


Fig. P2–51

**2–52.** The assembly in Fig. P2–52 consists of a steel bar $A$, a rigid block $C$, and a bronze bar $B$ securely fastened together and fastened to rigid supports at the ends. Initially there is no stress in the members. The temperature drops 40 F, and the load $P$ of 30 kips is applied. Determine the maximum normal stresses in $A$ and $B$ for this condition. The following data apply:

| Bar | Cross-Sectional Area, sq in. | Length, in. | Young's Modulus, ksi | Coef. of Expansion, per deg F |
|---|---|---|---|---|
| $A$ | 1.0 | 90 | $30(10^3)$ | $6.5(10^{-6})$ |
| $B$ | 3.0 | 15 | $15(10^3)$ | $10.0(10^{-6})$ |

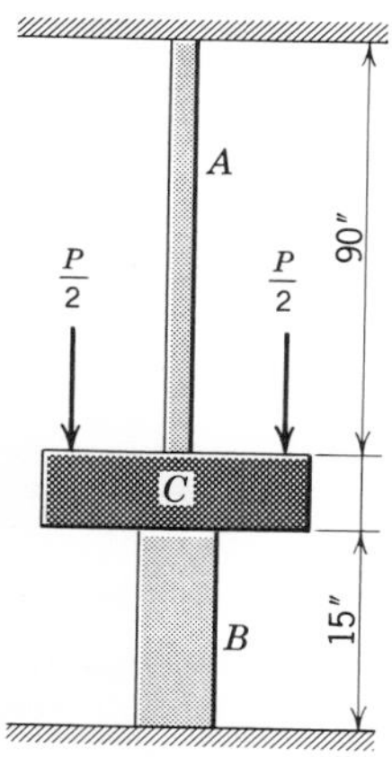


Fig. P2–52

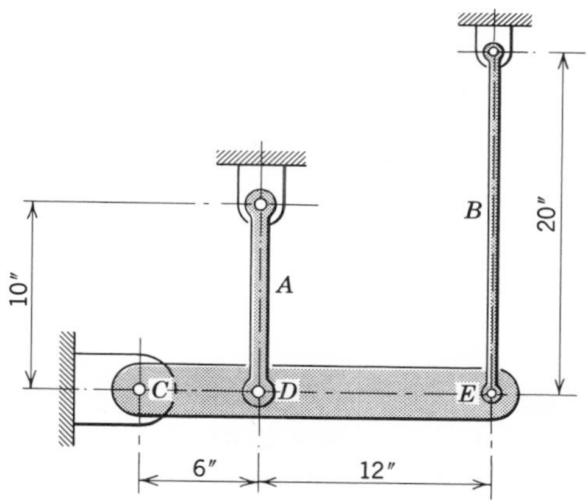


Fig. P2–53

**2–53.*** Bar $B$ of the pin-connected mechanism of Fig. P2–53 is made of aluminum alloy 6061-T6, and bar $A$ is made of 0.2 per cent carbon steel. The cross-sectional areas are 1.5 sq in. for $A$ and 0.25 sq in. for $B$. Bar $CDE$ is to be considered rigid. Initially, the bars are unstressed. Determine the load that must be applied to bar $CDE$ 4 in. to the right of $D$ to make the magnitude of the stress in bar $A$ equal to that in bar $B$ when the temperature drops 100 F. Refer to App. A.

**2–54.*** A hot-rolled wrought Monel bolt through a malleable cast iron (see App. A) tube has the nut tightened until the axial stress in the bolt is 12,000 psi $T$ at 80 F. The length of the tube is 2 ft, and the cross-sectional areas of the bolt and tube are equal. Determine the axial stress in the bolt when the temperature is $-20$ F.

**2–55.** In Fig. P2–55 post $A$ is brass ($E = 15{,}000$ ksi) with a cross-sectional area of 4 sq in. and $B$ is steel with a cross-sectional area of 1.5 sq in. Bars $CD$, $DE$, and $EF$ are to be considered rigid. Determine the load $P$ which will produce a maximum normal stress of 10,000 psi $C$ in post $A$.

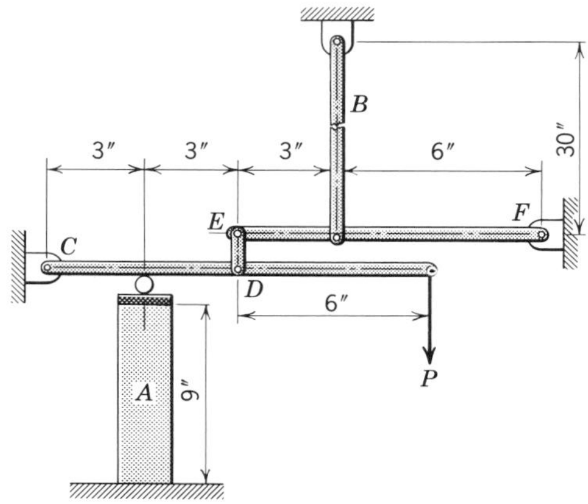


Fig. P2–55

**2–56.*** Members $A$, $B$, and $C$ of Fig. P2–56 are elastic and the bell cranks $D$ and $E$ are to be considered rigid. Determine the maximum permissible value of the load $P$ if the axial stresses in $A$ and $B$ are not to exceed 16 and 3 ksi, respectively. The following data apply:

| Bar | Material | Cross-Sectional Area, sq in. | Modulus of Elasticity, ksi |
|---|---|---|---|
| $A$ | Steel | 0.5 | 30,000 |
| $B$ | Concrete | 4.0 | 4,000 |
| $C$ | Brass | 1.5 | 15,000 |

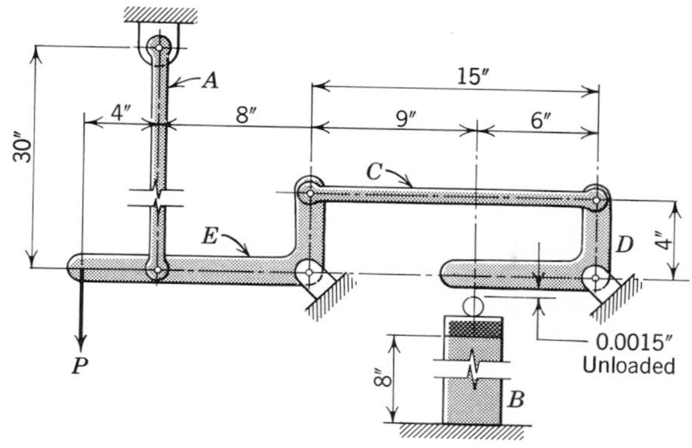


Fig. P2–56

**2–57.** When the structure of Fig. P2–57 is unloaded, there is a 0.001-in. gap between the rigid bearing plate *D* and the rigid bearing cap on bar *B*. Determine the magnitude of the force *P* which will cause the axial stresses in bars *A*, *B*, and *C* to have the same value. The following data apply:

| Bar | Material | Cross-Sectional Area, sq in. | Modulus of Elasticity, psi |
|---|---|---|---|
| *A* | Aluminum Alloy | 4.0 | $10(10^6)$ |
| *B* | Steel | 2.0 | $30(10^6)$ |
| *C* | Brass | 1.0 | $15(10^6)$ |

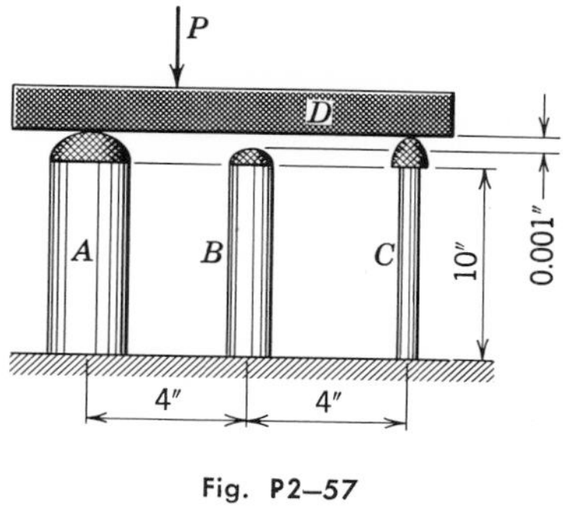


Fig. P2–57

**2–58.*** Bar $A$ of Fig. P2–58 is made of aluminum alloy 7075-T6 with a proportional limit of 55.0 ksi and a modulus of elasticity of $10.5(10^3)$ ksi. Bars $B$ are of structural steel (proportional limit = yield point = 36.0 ksi). All bars have a rectangular cross section 2.2 by 1.2 in. For the indicated load, determine the axial stresses in the bars and the displacement of the pin $C$.

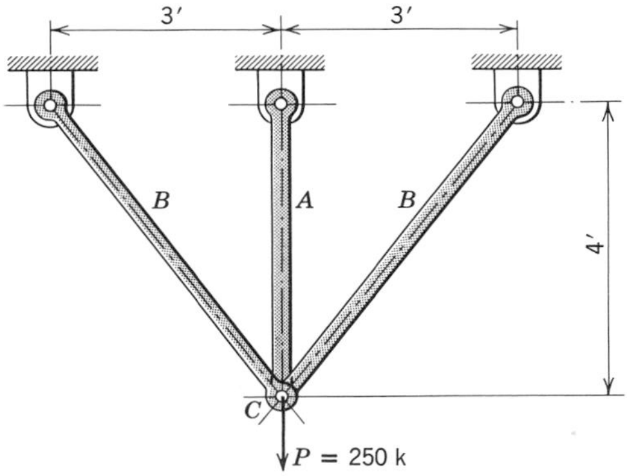


Fig. P2–58

**2–59.** In Fig. P2–58 all three bars are of the structural steel described in Prob. 2–58. If plastic action is permissible in all bars but strain hardening is not to occur, determine:

(*a*) The maximum load $P$ which the structure will support (sometimes called the "limit load").

(*b*) The displacement of pin $C$ when the proportional limit is reached in the bar which is the last to change from elastic to plastic action.

(*c*) The maximum deflection of pin $C$ if the maximum strain before strain hardening occurs is 0.018.

**2–60.*** Solve Prob. 2–58 when there is a clearance of 0.05 in. between the pin at $C$ and bar $A$.

**2–61.** Referring to the data of Prob. 2–58, determine the axial stresses in the bars when the load $P$ is zero and the temperature increases 100 F. The coefficient of thermal expansion for the aluminum alloy is $12.5(10^{-6})$ per degree F.

**2–62.*** In Fig. P2–62, bar $B$ is steel and bar $A$ is aluminum alloy ($E$ = 10,000 ksi). Each bar has a cross-sectional area of 1.5 sq in. Determine the maximum permissible value for the load, $P$, applied as shown if the allowable axial stresses are 20 ksi for the steel and 8 ksi for the aluminum alloy.

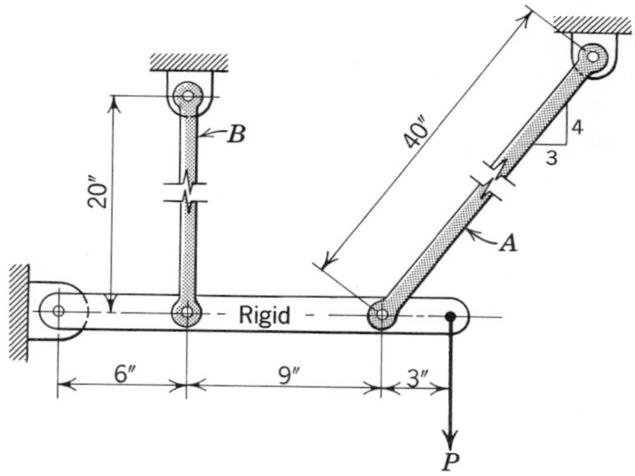


Fig. P2–62

**2–63.** All members of the structure of Fig. P2–63 are elastic. The deflection of the center of the steel beam $C$ with respect to the pins at the ends is given by the expression, $\delta = F/160$, where $\delta$ is in inches and $F$ is the resultant force at the center of the beam in kips. Before loading, beam $C$ just touches the rigid bearing block on top of member $A$. Determine the maximum permissible magnitude for the load $P$. The following data apply:

| Member | Material | Allowable Axial Stress, ksi | Modulus of Elasticity, ksi | Cross-Sectional Area, sq in. |
|---|---|---|---|---|
| $A$ | Aluminum Alloy | 15 | 10,000 | 4.0 |
| $B$ | Brass | 7 | 16,000 | 0.5 each |

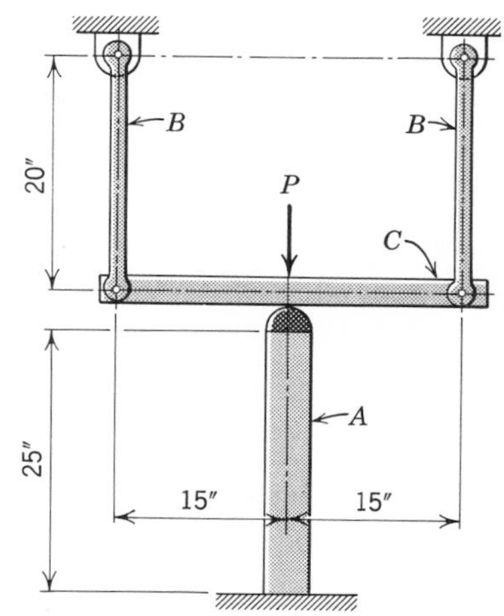


Fig. P2–63

**2–64.*** In Fig. P2–64, the steel bars $A$ and $C$ are pinned to the rigid member $D$, and the rigid yoke $E$ is securely attached to bar $C$ and the cold-rolled brass bar $B$ (see App. A). The cross-sectional areas are 4 sq in. for $A$, 2 sq in. for $C$, and 4 sq in. for $B$. The axial stresses are not to exceed 20 ksi for the steel and 30 ksi for the brass, and the deflection at $E$ is not to exceed 0.05 in. downward. Determine the maximum allowable value of $P$.

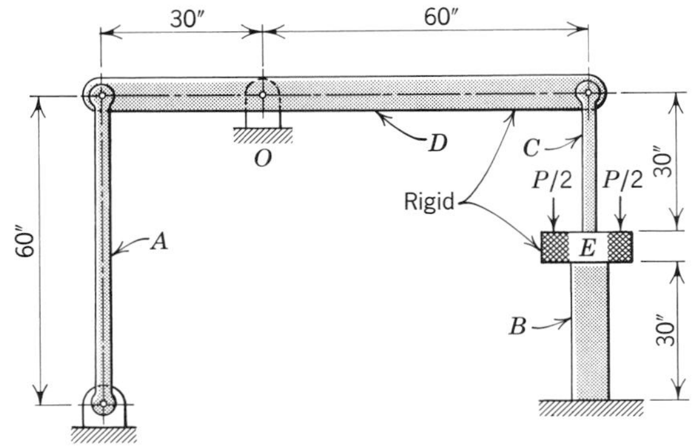


Fig. P2–64

**2–65.** Solve Ex. 2–3 with P = 50 kips, the lengths of $A$ and $B$ changed to 60 in. and 5.0 in. respectively, and the cross-sectional areas of $A$ and $B$ changed to 0.20 sq in. and 1.0 sq in., respectively.

**2–7. Thin-walled Pressure Vessels.** A pressure vessel is described as *thin walled* when the ratio of the wall thickness to the radius of the vessel (an approximately cylindrical or spherical shape is implied) is so small that the distribution of normal stress on a plane perpendicular to the surface of the shell is essentially uniform throughout the thickness of the shell. Actually this stress varies from a maximum value at the inside surface to a minimum value at the outside surface of the shell, but it can be shown[8] that if the ratio of the wall thickness to the inner radius of the vessel is less than 0.1, the maximum normal stress is not more than 5 per cent greater than the average. Boilers,

[8] See, for example, *Advanced Mechanics of Materials,* G. Murphy, McGraw-Hill, New York, 1946.

gas storage tanks, pipe lines, metal tires, and hoops are normally analyzed as thin-walled elements. Gun barrels, certain high-pressure vessels in the oil refining industry, and cylinders and piping for heavy hydraulic presses need to be treated as thick-walled vessels.

Problems involving thin-walled vessels subjected to internal fluid pressure are readily solved with the aid of free-body diagrams of sections of the vessels and the fluid contained therein. Figure 2–15 shows a section of a cylindrical pressure vessel including the fluid. The weights of the fluid and vessel are assumed to be negligible. The force, $P$, is the resultant fluid force acting on the fluid remaining within the portion of the container shown, and the forces, $Q$, are the resultant internal forces on the sections of the cylinder indicated. For simplicity, the pressures and stresses in the $z$ direction have been omitted. The free-body diagram of Fig. 2–15 can be used to evaluate the normal stress on a longitudinal plane (sometimes referred to as the circumferential or hoop stress). The stresses on a transverse plane can be evaluated from a free-body diagram showing stresses and pressures in the $z$ direction. The free-body diagram will be similar to Fig. 2–16*b*. Analysis of the stresses in a cylindrical vessel, subjected to uniform internal pressure only, indicates that the stress on the longitudinal plane is twice the stress on a transverse plane. Consequently, a longitudinal joint needs to be twice as strong as a transverse (or girth) joint.

Note that there are no shearing stresses on the planes shown because there are no loads to induce such shearing stresses. Therefore, the maximum (and minimum) normal stresses occur on the planes selected.

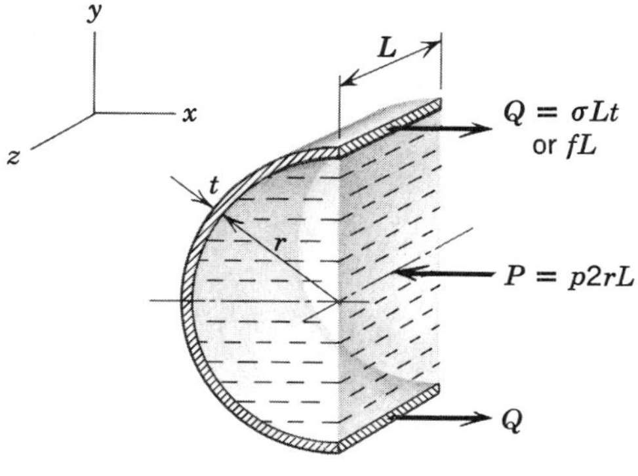


Fig. 2–15

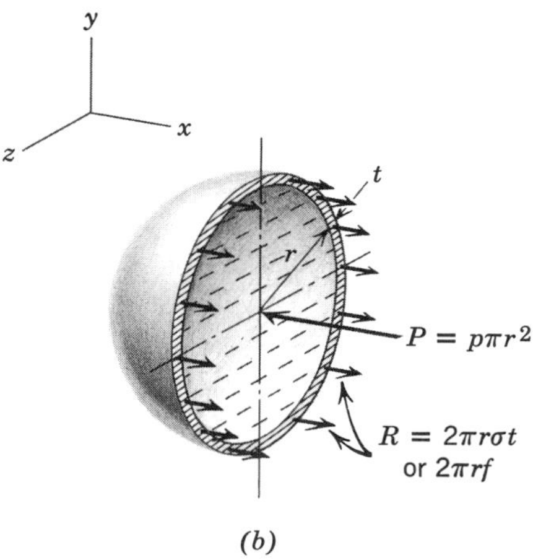


**Fig. 2–16.** Hortonsphere for gas storage. Superior, Wisconsin. Courtesy: Chicago Bridge and Iron Co., Chicago, Illinois.

**Example 2–6:** The 60-ft diameter Hortonsphere of Fig. 2–16*a* is constructed of 1.13-in. steel plates with welded joints and contains gas at a pressure of 75 psi gage (meaning above atmospheric pressure). Determine:

(*a*) The force transmitted across a 10-in. length of a vertical meridian joint.

(*b*) The maximum (average) normal stress in the shell.

*Solution:* (*a*) Using the free-body diagram[9] of Fig. 2–16*b* and summing forces in the $x$ direction gives

$$p\pi r^2 = 2\pi r f,$$

$$75\pi(30^2)(12^2) = 2\pi(30)(12)(f),$$

from which

$$f = 13{,}500 \text{ lb per in.},$$

and the force on a 10-in. length is

$$F = 13{,}500(10) = \underline{135{,}000 \text{ lb.}} \qquad \text{Ans.}$$

(*b*) The normal stress can be obtained from the force per inch as

$$\sigma = \frac{f}{t} = \frac{13{,}500}{1.13} = \underline{11{,}950 \text{ psi } T.} \qquad \text{Ans.}$$

If only the stress were wanted, it could have been obtained directly from the first equation written by substituting $\sigma t$ for $f$.

## PROBLEMS

**2–66.*** The cylindrical propane tanks of Fig. P2–66 are 124¾ in. in diameter and are constructed of 1-in. steel plates with all joints welded. The working pressure for these tanks is 250 psi. Determine:

(*a*) The maximum (average) normal stress in the shell.

(*b*) The factor of safety with respect to failure by fracture. The ultimate tensile strength of the steel is 70 ksi.

Fig. P2–66. All-welded horizontal pressure tanks for propane storage. Northlake, Illinois. Courtesy: Chicago Bridge and Iron Co., Chicago, Illinois.

[9] As a permissible approximation the inside circumference rather than the mean is used in the expression for $R$.

**2–67.*** At Hoover Dam water may be discharged under a maximum head of almost 600 ft (pressure approximately 260 psi). What thickness of plate would be required for a 5-ft diameter pressure pipe at a working tensile stress of 13,000 psi if the joints are welded?

**2–68.** A cylindrical boiler is 6 ft in diameter and is constructed of $\frac{3}{4}$-in. plates. What intensity of internal pressure will produce a maximum tensile stress of 10,800 psi in the plate? The joints are welded.

**2–69.*** A 50-in. diameter spherical vessel is subjected to an internal pressure of 210 psi. Determine the force transmitted across 6 in. of the joint uniting the two halves.

**2–70.*** Two 36-in. diameter hemispheres are to be joined to form a spherical pressure vessel. If the internal pressure is to be 220 psi, determine the force that must be transmitted across an 8-in. length of the joint.

**2–71.** A spherical gas holder 30 ft in diameter contains gas at a pressure of 40 psi. Determine the force transmitted across a 6-in. length of the joint connecting the two halves of the sphere.

**2–72.** (*a*) Prove that the normal stress on a transverse plane of a thin-walled cylindrical pressure vessel is $\sigma = pD/(4t)$.

(*b*) Prove that the normal stress on a longitudinal plane of symmetry of a thin-walled cylindrical pressure vessel is $\sigma = pD/(2t)$.

**2–73.*** A cylindrical thin-walled pressure vessel 25 in. in diameter has a welded spiral seam as indicated in Fig. P2–73. If the vessel is subjected to an internal pressure of 200 psi, determine the normal and shearing forces which must be carried per linear inch of the weld.

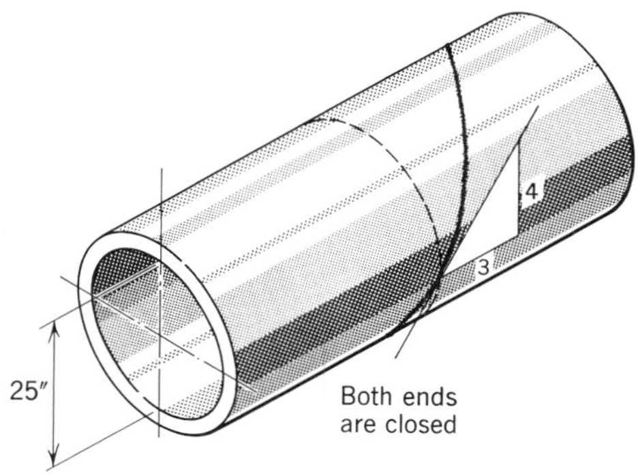


Fig. P2–73

**2–74.** The cylindrical pressure tank of Fig. P2–74 is made of $\frac{1}{4}$-in. steel plate. The normal stress at $A$ on plane $B$–$B$ (which is perpendicular to the surface of the plate at $A$) is 8000 psi. Determine the air pressure in the tank.

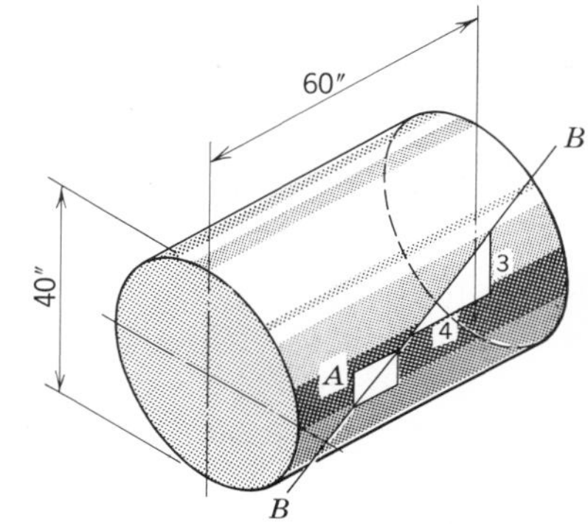


Fig. P2–74

**2–75.*** A cylindrical pressure tank is fabricated by butt welding $\frac{1}{2}$-in. plate with a spiral seam as shown in Fig. P2–75. The pressure in the tank is 200 psi, and an axial load of $4800\pi$ lb is applied to the end of the tank through a rigid bearing plate. Determine the normal and shearing stresses in the weld on the plane of the weld.

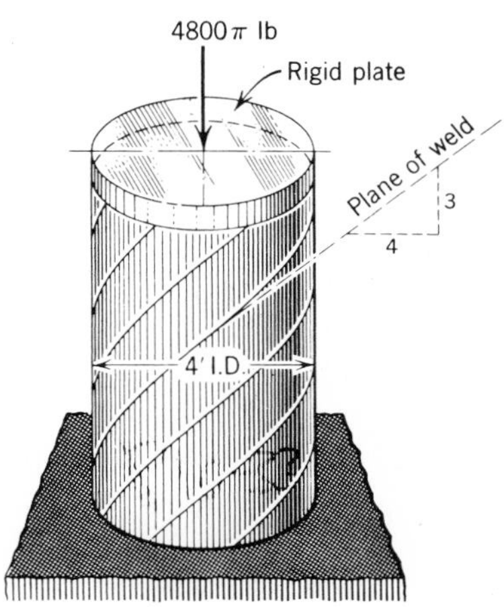


Fig. P2–75

CHAPTER 3

# *Fundamentals of Elasticity*

**3–1. Introduction.** In the study of mechanics of materials it is sometimes necessary to make simplifications to render the analysis more manageable. Experiments indicate that these simplifications give results which are quite satisfactory for most engineering applications. However, severe design limitations such as minimized weights and high reliability may dictate a more detailed analysis.

The purpose of this chapter is to introduce some of the fundamentals of a more refined analytical method of stress analysis, known as the *theory of elasticity*. The results of elasticity studies can be used as standards by which theories used in mechanics of materials may be judged. In some of the succeeding chapters, comparison with the theory of elasticity will be made to provide a better understanding of the merits and limitations of mechanics of materials results.

**3–2. Notation and Sign Convention.** Bodies may be deformed by the following two types of external loads:

Surface forces—pressures distributed over the boundary arising from such sources as the action of one body on another, aerodynamic pressure, or hydrostatic pressure.

Body forces—forces acting on elements of volume arising from gravitational attraction, inertial loading, magnetic fields, and electrical charges.

For convenience, both surface forces and body forces will be resolved into components parallel to the coordinate axes. Surface forces, expressed as force per unit area, will be denoted by $S_x$, $S_y$, $S_z$; body forces, expressed as force per unit volume, will be denoted by $F_x$, $F_y$, $F_z$.

The sign convention for stresses is that defined in Art. 1–6, see Fig. 1–13.

Due to the equality of the shearing stresses $\tau_{xy}$ and $\tau_{yx}$, $\tau_{xz}$ and $\tau_{zx}$, and $\tau_{yz}$ and $\tau_{zy}$, there are only six different stress components to consider, $\sigma_x$, $\sigma_y$, $\sigma_z$, $\tau_{xy}$, $\tau_{yz}$, and $\tau_{xz}$.

A state of plane stress exists when all of the forces acting on a thin flat plate are in the plane of the plate. (See Art. 1–6). The surface forces applied at the boundary and the body forces applied to the volume: (1) have no components perpendicular to the plane of the plate and (2) are distributed uniformly across the thickness of the plate.[1] The faces of the plate are stress free. If the mid-plane of the plate is taken to be the $xy$ plane as shown in Fig. 3–1, the stresses $\sigma_z$, $\tau_{xz}$, and $\tau_{yz}$ are all zero, so that the state of stress is described by $\sigma_x$, $\sigma_y$, and $\tau_{xy}$. This state of stress is known as plane stress since all of the stresses are parallel to one plane; in this case, the $xy$ plane.

The relationships between the normal stresses and strains for plane stress are given by Eq. (2–1) and (2–3) and the relationship between the shearing stresses and strains is the first of Eq. (2–6). Note that although $\sigma_z$ is zero the normal strain in the $z$ direction is

$$\epsilon_z = -\mu(\sigma_x + \sigma_y)/E.$$

A topic closely related to plane stress is plane strain. When a prismatic bar is loaded and supported in such a manner that it is not permitted to expand or contract in the axial direction, there can be no resultant axial strain. If the external load is uniformly distributed

[1] A more detailed examination of plane stress dictates that the loading be distributed parabolically rather than uniformly. For thin plates, however, the error is insignificant. See *Theory of Elasticity,* 2nd Ed., S. Timoshenko and J. N. Goodier, McGraw-Hill, New York, 1951, pp. 241–244.

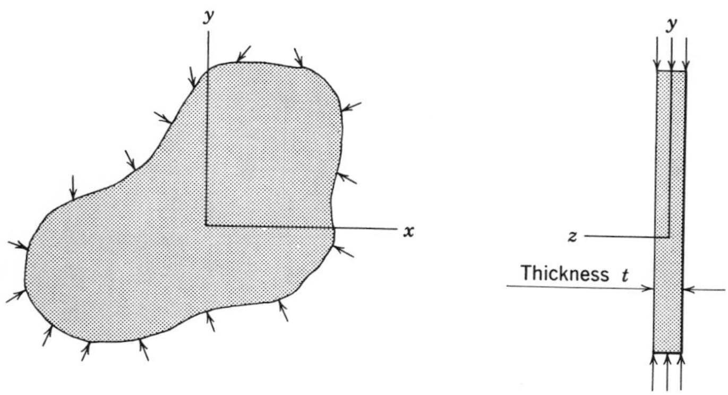


Fig. 3–1

along the axis of the member, all cross sections will be in the same condition of stress and strain and the axial strains will be zero at each section along the body. If the axis of the body is selected as the $z$ axis, it is seen that $\epsilon_z$, $\gamma_{xz}$, and $\gamma_{yz}$ are all zero. This condition is called *plane strain.* It should be noted, however, that the axial stress is not necessarily zero. From Eq. (2–2) it is apparent that when $\epsilon_z = 0$

$$\sigma_z = \mu(\sigma_x + \sigma_y).$$

**3–3. Differential Equations of Equilibrium for Plane Stress.** When a plate is subjected to plane stress the stresses $\sigma_x$, $\sigma_y$, and $\tau_{xy}$ will, in general, vary from point to point throughout the plate and therefore are functions of both $x$ and $y$. It is essential that the stress distribution be such that all elements of the body are in equilibrium. The conditions for equilibrium are obtained by examining the forces acting on a differential element of the plate such as the one shown in Fig. 3–2. The dimensions of the element are $\Delta x$, $\Delta y$, and $t$ where $t$ is the thickness. Consequently, the areas of the $x$ and $y$ faces are $t\,\Delta y$ and $t\,\Delta x$, respectively, and the volume of the element is $t\,\Delta x\,\Delta y$. Since the stresses vary throughout the plate it should be noted that the stress $\sigma_x$ on the faces $x$ and $x + \Delta x$ differ by writing the stresses as $\sigma_x(x, y)$ and $\sigma_x(x + \Delta x, y)$ with similar expressions for the other stress components.

The conditions for equilibrium can be obtained from the free-body diagram shown in Fig. 3–2. For equilibrium in the $x$ direction,

$$[\sigma_x(x + \Delta x, y) - \sigma_x(x, y)]t\,\Delta y + [\tau_{xy}(x, y + \Delta y) - \tau_{xy}(x, y)]t\,\Delta x + F_x t\,\Delta x\,\Delta y = 0.$$

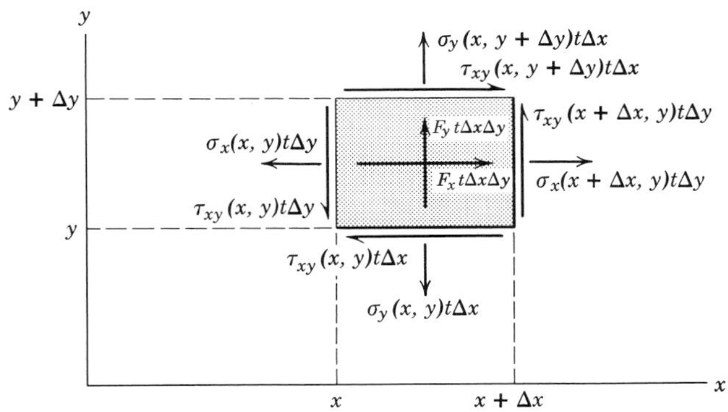


Fig. 3–2

When this equation is divided by $t\,\Delta x\,\Delta y$, it becomes

$$\left[\frac{\sigma_x(x+\Delta x, y) - \sigma_x(x, y)}{\Delta x}\right] + \left[\frac{\tau_{xy}(x, y+\Delta y) - \tau_{xy}(x, y)}{\Delta y}\right] + F_x = 0.$$

If $\Delta x$ and $\Delta y$ approach zero independently the quantities in the two brackets become $\partial\sigma_x/\partial x$ and $\partial\tau_{xy}/\partial y$ respectively. The same method is used to treat equilibrium in the $y$ direction and the differential equations of equilibrium are found to be

$$\begin{aligned} \frac{\partial\sigma_x}{\partial x} + \frac{\partial\tau_{xy}}{\partial y} + F_x &= 0, \\ \frac{\partial\tau_{xy}}{\partial x} + \frac{\partial\sigma_y}{\partial y} + F_y &= 0. \end{aligned} \qquad (3\text{–}1)$$

Since these equations involve no material properties, they can be applied to any matter in a state of plane stress and are not limited in application to elastic bodies.

**3–4. Boundary Conditions.** The differential equations of equilibrium, Eq. (3–1), govern the variation of stress from point to point in the interior of the body. At the boundaries the stresses must be in equilibrium with the surface forces. Consider the shaded differential element which contains a small section of the boundary $\Delta S$ as shown in Fig. 3–3*a*. This element is subjected to surface pressure with components $S_x$ and $S_y$. Let $n$ be the normal to the boundary and $\theta_n$ the angle between this normal and the $x$ axis. Figure 3–3*b* is a free-body

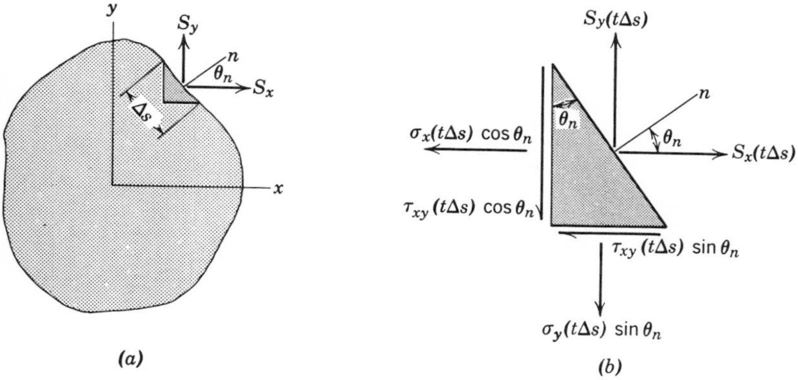


Fig. 3–3

diagram of the element which is in equilibrium. The force equations in the $x$ and $y$ directions result in

$$S_x = \sigma_x \cos\theta_n + \tau_{xy} \sin\theta_n,$$
$$S_y = \sigma_y \sin\theta_n + \tau_{xy} \cos\theta_n. \tag{3–2}$$

Equations (3–2) are the *boundary equations* for the body. If the boundary is a plane surface with its normal in the $x$ direction, Eq. (3–2) become

$$S_x = \sigma_x,$$
$$S_y = \tau_{xy}. \tag{3–2a}$$

Similar equations arise if the boundary is a plane surface which has a normal in the $y$ direction.

**3–5. Strain-Displacement Equations.** For most engineering materials, the deformations under load are comparatively small. This results in a considerable simplification in the development of the strain-displacement equations. It is assumed that the body is so restrained that rigid body movements are prevented and the displacements of the particles of the body produce deformation.

The differential element $OABC$ in Fig. 3–4 illustrates the deformation of an elastic body under plane stress. The displacement of point $O$ to $O'$ is resolved into two components—$u$ and $v$ in the $x$ and $y$ direc-

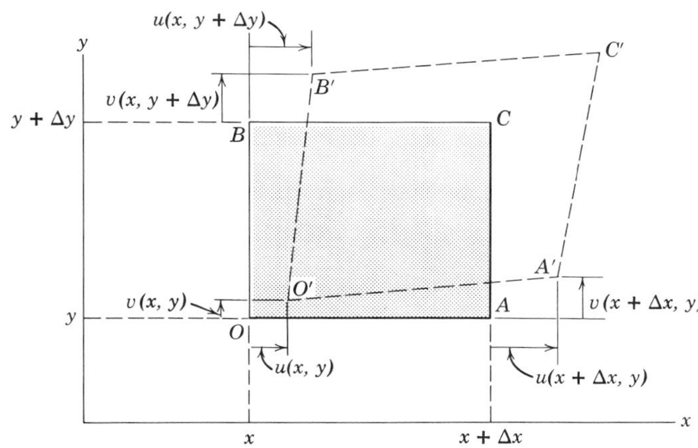


Fig. 3–4

tions, respectively. As $O$ moves to $O'$, $A$ moves to $A'$ and $B$ moves to $B'$. The displacement of $A$ is resolved into two components—$u(x + \Delta x, y)$ in the $x$ direction and $v(x + \Delta x, y)$ in the $y$ direction, as shown on Fig. 3–4. Similar expressions describe the displacement of $B$ to $B'$.

The average normal strain in the $x$ direction is

$$\epsilon_x = (O'A' - OA)/OA$$

as defined in Art. 1–9. The angle between $OA$ and $O'A'$ is small; therefore,

$$O'A' \simeq \Delta x + u(x + \Delta x, y) - u(x, y)$$

and the average normal strain is

$$\epsilon_x = \frac{\Delta x + u(x + \Delta x, y) - u(x, y) - \Delta x}{\Delta x}.$$

As $\Delta x$ approaches zero the average strain approaches the strain at point $O$ and becomes

$$\epsilon_x = \lim_{\Delta x \to 0} \frac{u(x + \Delta x, y) - u(x, y)}{\Delta x} = \frac{\partial u}{\partial x}.$$

A similar analysis for element $OB$ gives the strain in the $y$ direction as $\epsilon_y = \partial v/\partial y$.

The shearing strain $\gamma_{xy}$, which by definition is the change in the right angle $AOB$, is the sum of the angles between $O'A'$ and $OA$ and between $O'B'$ and $OB$. For small deformations, these angles may be replaced by their tangents so that at the point $O$ the shearing strain is

$$\gamma_{xy} = \lim_{\substack{\Delta x \to 0 \\ \Delta y \to 0}} \left[\frac{v(x + \Delta x, y) - v(x, y)}{\Delta x} + \frac{u(x, y + \Delta y) - u(x, y)}{\Delta y}\right]$$

$$= \frac{\partial v}{\partial x} + \frac{\partial u}{\partial y}.$$

Thus, for plane stress the strain-displacement equations are

$$\epsilon_x = \frac{\partial u}{\partial x},$$

$$\epsilon_y = \frac{\partial v}{\partial y}, \qquad (3\text{–}3)$$

$$\gamma_{xy} = \frac{\partial v}{\partial x} + \frac{\partial u}{\partial y}.$$

## PROBLEMS

**3–1.*** The components of the displacement at any point in the rectangular plate in Fig. P3–1 are given by the equations

$$u = -\frac{Mxy}{EI}, \qquad v = \frac{\mu M y^2}{2EI} + \frac{Mx^2}{2EI}$$

where $M$, $E$, $I$, and $\mu$ are constants. Determine the strains and stresses in the plate and on its boundaries.

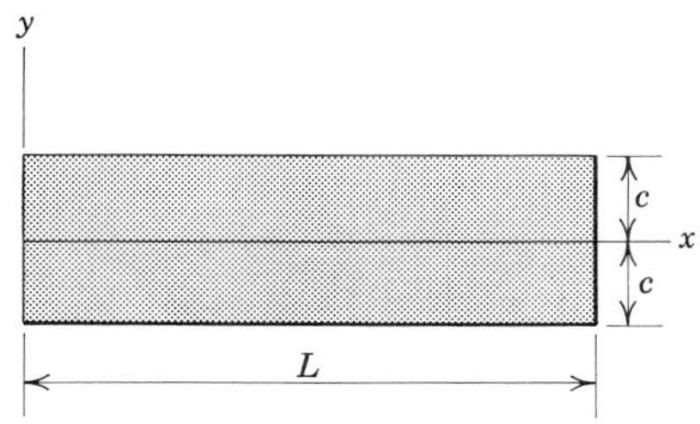


Fig. P3–1

**3–2.*** The components of the displacement at any point in the rectangular plate in Fig. P3–2 are given by the equations

$$u = \frac{\gamma}{E}\left(xL - \frac{x^2}{2} - \frac{\mu y^2}{2}\right), \qquad v = -\frac{\mu\gamma}{E}\, y(L - x)$$

where $\gamma$ is the weight per unit volume of the plate, a constant. Determine the strains and stresses in the plate and state the type of loading. The plate hangs in a vertical plane.

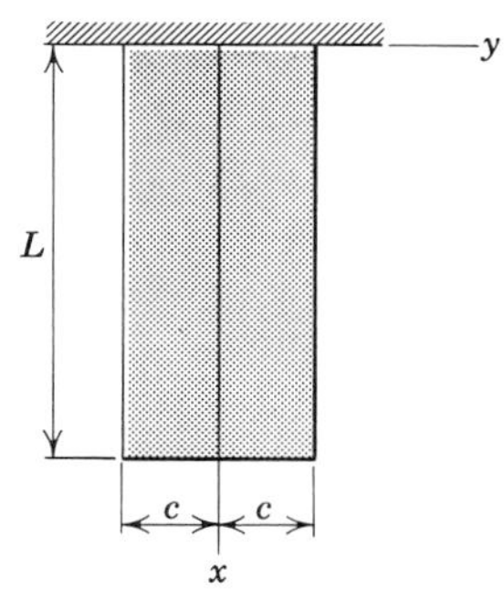


Fig. P3–2

**3–3.** By substituting the strain-displacement equations into the stress-strain equations and substituting the resulting stress-displacement equations into the equilibrium equations, show that, in the absence of body forces, the displacements must satisfy the differential equations

$$\frac{\partial^2 u}{\partial x^2} + \frac{\partial^2 u}{\partial y^2} + \frac{1+\mu}{1-\mu}\frac{\partial}{\partial x}\left(\frac{\partial u}{\partial x} + \frac{\partial v}{\partial y}\right) = 0$$

and

$$\frac{\partial^2 v}{\partial x^2} + \frac{\partial^2 v}{\partial y^2} + \frac{1+\mu}{1-\mu}\frac{\partial}{\partial y}\left(\frac{\partial u}{\partial x} + \frac{\partial v}{\partial y}\right) = 0.$$

These are known as the displacement equations for plane stress.

**3–4.** The displacements at any point in a flat plate are given by the functions

$$u = \frac{pa^2x}{E(b^2-a^2)}\left[(1-\mu^2) + \frac{(1+\mu)b^2}{x^2+y^2}\right],$$

$$v = \frac{pa^2y}{E(b^2-a^2)}\left[(1-\mu^2) + \frac{(1+\mu)b^2}{x^2+y^2}\right].$$

Show that these displacements satisfy the displacement equations in Prob. 3–3.

**3–6. Compatibility.** The differential equations of equilibium of Art. 3–3 and the boundary equations of Art. 3–4 are not sufficient by themselves for the solution of the general stress problem, since the problem is statically indeterminate. To obtain a solution, an additional equation involving deformations is needed. The strain equations of Art. 3–5 are also insufficient by themselves, since not only must the deformations be continuous functions of position, but the first and second derivatives must also be continuous functions to insure that the displacements will be single valued at a particular point, that there will be no abrupt changes in the slope of a filament of material, and that there will be only one solution regardless of the direction in which the point in question is approached. In other words, there will be no jogs or kinks in the continuum.

The strain-deformation equations of Art. 3–5 express three strain components ($\epsilon_x$, $\epsilon_y$, and $\gamma_{xy}$) in terms of two displacement components ($u$, $v$); therefore these three equations are not independent.

By differentiating Eq. (3–3) as follows,

$$\frac{\partial^2 \epsilon_x}{\partial y^2} = \frac{\partial^3 u}{\partial x\, \partial y^2},$$

$$\frac{\partial^2 \epsilon_y}{\partial x^2} = \frac{\partial^3 v}{\partial x^2\, \partial y},$$

$$\frac{\partial^2 \gamma_{xy}}{\partial x\, \partial y} = \frac{\partial^3 v}{\partial x^2\, \partial y} + \frac{\partial^3 u}{\partial x\, \partial y^2},$$

the displacements $u$ and $v$ may be eliminated since the sum of the first two equations equals the third. Thus,

$$\frac{\partial^2 \epsilon_x}{\partial y^2} + \frac{\partial^2 \epsilon_y}{\partial x^2} = \frac{\partial^2 \gamma_{xy}}{\partial x\, \partial y}. \tag{3–4}$$

Equation (3–4) is called the equation of compatibility, and this equation, together with the boundary and equilibrium equations, must be satisfied by the solution to any given problem.

The compatibility equation frequently is more useful when it is written in terms of stresses by means of Hooke's law and the equilibrium equations.

When Eq. (2–1) and the first of Eq. (2–6) are substituted in Eq. (3–4) it becomes

$$\frac{1}{E}\frac{\partial^2(\sigma_x - \mu\sigma_y)}{\partial y^2} + \frac{1}{E}\frac{\partial^2(\sigma_y - \mu\sigma_x)}{\partial x^2} = \frac{2(1+\mu)}{E}\frac{\partial^2 \tau_{xy}}{\partial x\, \partial y}. \tag{a}$$

Additional simplification can be obtained by using Eq. (3–1). If the first equation is differentiated with respect to $x$ and the second with respect to $y$, then the sum of these expressions gives

$$2\frac{\partial^2 \tau_{xy}}{\partial x\, \partial y} = -\left[\frac{\partial^2 \sigma_x}{\partial x^2} + \frac{\partial^2 \sigma_y}{\partial y^2} + \frac{\partial F_x}{\partial x} + \frac{\partial F_y}{\partial y}\right]. \tag{b}$$

When Eq. (b) is substituted in Eq. (a) it becomes, after simplification,

$$\left(\frac{\partial^2}{\partial x^2} + \frac{\partial^2}{\partial y^2}\right)(\sigma_x + \sigma_y) = -(1+\mu)\left(\frac{\partial F_x}{\partial x} + \frac{\partial F_y}{\partial y}\right). \tag{3–5}$$

In a significant number of cases, the body forces $F_x$ and $F_y$ either are zero or vary in such a manner that the right-hand side of Eq. (3–5) is zero. In this case

$$\left(\frac{\partial^2}{\partial x^2} + \frac{\partial^2}{\partial y^2}\right)(\sigma_x + \sigma_y) = 0. \tag{3–5a}$$

The differential operator in Eq. (3–5) and (3–5a) is known as the Laplacian operator and is frequently written as

$$\frac{\partial^2}{\partial x^2} + \frac{\partial^2}{\partial y^2} = \nabla^2.$$

In three-dimensional cartesian coordinates the operator is

$$\nabla^2 = \frac{\partial^2}{\partial x^2} + \frac{\partial^2}{\partial y^2} + \frac{\partial^2}{\partial z^2}.$$

Equation (3–5a) is called Laplace's equation, and solutions to this equation are known as harmonic functions. It is seen from Eq. (3–5a) that in the absence of body forces the sum of the normal stresses at each point must be a harmonic function.

When plane stress exists in an elastic plate, the equations of equilibrium, Eq. (3–1), together with the boundary and compatibility conditions, Eq. (3–2) and Eq. (3–5), are sufficient to solve the problem. The computational difficulties may be considerable, however. These are significantly reduced if it is possible to use a stress function approach such as the one described in the next article.

## PROBLEMS

Problems 3–5 through 3–8 are examples of plane stress in which the stresses $\sigma_x$, $\sigma_y$, and $\tau_{xy}$ are given. In each case, check to see if the compatibility equations and both equilibrium equations are satisfied. All quantities except $x$ and $y$ are constants.

**3–5.** $\sigma_x = \dfrac{-P}{I} y(L - x);\ \sigma_y = 0;\ \tau_{xy} = \dfrac{P}{2I}(c^2 - y^2).$

**3–6.** $\sigma_x = \dfrac{W}{2I} y(L - x)^2;\ \sigma_y = 0;\ \tau_{xy} = \dfrac{-W}{2I}(L - x)(c^2 - y^2).$

**3–7.** $\sigma_x = \dfrac{W}{2I}\left(x^2 y - \dfrac{2}{3} y^3\right);\ \sigma_y = \dfrac{W}{2I}\left(\dfrac{1}{3} y^3 - c^2 y + \dfrac{2}{3} c^3\right);\ \tau_{xy} = \dfrac{W}{2I} x(c^2 - y^2).$

**3–8.** $\sigma_x = \dfrac{pa^2}{(b^2 - a^2)}\left[1 - \dfrac{b^2(x^2 - y^2)}{(x^2 + y^2)^2}\right];$

$$\sigma_y = \frac{pa^2}{(b^2 - a^2)}\left[1 + \frac{b^2(x^2 - y^2)}{(x^2 + y^2)^2}\right];$$

$$\tau_{ry} = \frac{-2pa^2b^2xy}{(b^2 - a^2)(x^2 + y^2)^2}.$$

**3–7. Airy Stress Function.** In the absence of body forces, the differential equations which govern the distribution of stress in a body under plane stress are

$$\frac{\partial \sigma_x}{\partial x} + \frac{\partial \tau_{xy}}{\partial y} = 0,$$
$$\frac{\partial \tau_{xy}}{\partial x} + \frac{\partial \sigma_y}{\partial y} = 0, \tag{3–6}$$
$$\nabla^2(\sigma_x + \sigma_y) = 0.$$

The first two equations are satisfied identically if the stresses are written as derivatives of a function $\varphi$ as

$$\sigma_x = \frac{\partial^2 \varphi}{\partial y^2}, \qquad \sigma_y = \frac{\partial^2 \varphi}{\partial x^2}, \qquad \tau_{xy} = \frac{-\partial^2 \varphi}{\partial x\, \partial y}. \tag{3–7}$$

The sum of the normal stresses is

$$\sigma_x + \sigma_y = \frac{\partial^2 \varphi}{\partial y^2} + \frac{\partial^2 \varphi}{\partial x^2} = \nabla^2 \varphi$$

and the last of Eq. (3–6) becomes

$$\nabla^2(\nabla^2 \varphi) = \frac{\partial^4 \varphi}{\partial x^4} + 2\frac{\partial^4 \varphi}{\partial x^2\, \partial y^2} + \frac{\partial^4 \varphi}{\partial y^4} = 0. \tag{3–8}$$

This is known as the biharmonic equation.

The function $\varphi$ defined by Eq. (3–7) and (3–8) was first introduced by C. B. Airy in 1862, and is generally known as the Airy stress function. It reduces a plane-stress problem to a boundary-value problem; that of finding a biharmonic function which, through Eq. (3–7) and the boundary conditions Eq. (3–2), yields the prescribed loading on the boundary of the plate.

**Example 3–1:** The function $\varphi = a_1x^3 + a_2x^2y + a_3xy^2 + a_4y^3$ is proposed as an Airy stress function.

(*a*) Show that it satisfies the equation of compatibility.

(*b*) A rectangular plate is bounded by $-b \leq x \leq +b$ and $-c \leq y \leq +c$. Draw sketches showing the contributions of the coefficients $a_3$ and $a_4$ to the boundary forces on the plate.

(*c*) Determine the resultant (force or couple) arising from the boundary stresses normal to the face $x = b$ when the only non-zero coefficient is (1) $a_3$ and (2) $a_4$. The plate thickness is $t$.

*Solution:* (*a*) The compatibility equation is

$$\nabla^2(\nabla^2 \varphi) = \frac{\partial^4 \varphi}{\partial x^4} + 2\frac{\partial^4 \varphi}{\partial x^2\, \partial y^2} + \frac{\partial^4 \varphi}{\partial y^4} = 0.$$

Since each of these terms is zero for the given function, the compatibility equation is satisfied.

(*b*) The stresses are, see Eq. (3–7),

$$\sigma_x = \frac{\partial^2 \varphi}{\partial y^2} = 2a_3x + 6a_4y,$$

$$\sigma_y = \frac{\partial^2 \varphi}{\partial x^2} = 6a_1x + 2a_2y,$$

$$\tau_{xy} = -\frac{\partial^2 \varphi}{\partial x\, \partial y} = -2a_2x - 2a_3y.$$

These results indicate that $\sigma_y$ is not a function of $a_3$ and $a_4$. The variation of $\sigma_x$ and $\tau_{xy}$ on the boundary of the plate when $a_3$ is the only non-zero coefficient is shown on Fig. 3–5*a*. Note that $\sigma_x$ is constant along the boundaries $x = \pm b$. The value of $\tau_{yx}$ is constant along the boundaries $y = \pm c$ and varies linearly from $2a_3c$ to $-2a_3c$ along the edges $x = \pm b$. When $a_4$ is the only non-zero coefficient, the only stress is $\sigma_x$ and its variation along the edges $x = \pm b$ is shown in Fig. 3–5*b*.

(*c*) The differential force $dR$ in Fig. 3–5*a* and *b* is the normal force acting on the area $t\, dy$ and is equal to $\sigma_x t\, dy$. The resultant force acting normal to the surface $x = b$ is

$$R_x = \int_{-c}^{+c} \sigma_x \Big]_{x=b} t\, dy.$$

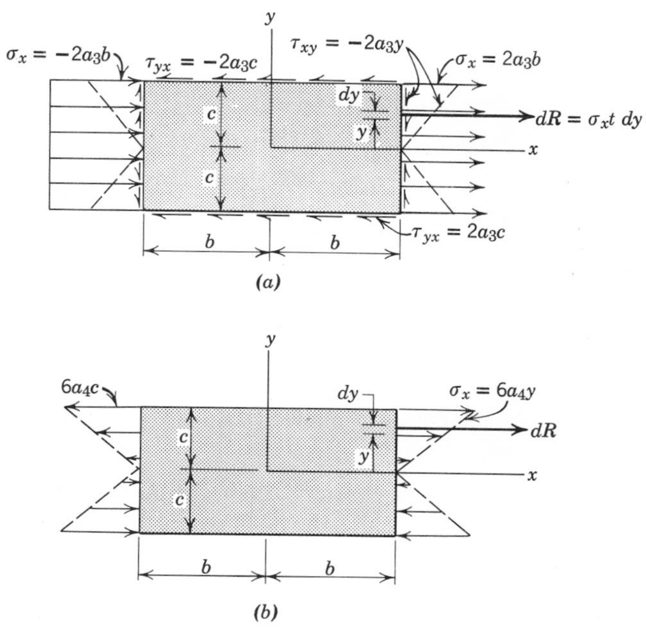


Fig. 3–5

The moment of the differential force with respect to the origin is $y\,dR = y\sigma_x t\,dy$ and the moment of the resultant is

$$M_0 = \int_{-c}^{+c} y\sigma_x \Big]_{x=b} t\,dy.$$

The resultant force corresponding to $a_3$ (with $a_1$, $a_2$, and $a_4$ zero) is

$$R_x = \int_{-c}^{+c} (2a_3b)t\,dy = 2a_3bty \Big]_{-c}^{+c} = \underline{4a_3btc} \qquad \text{Ans.}$$

and the moment is

$$M_0 = \int_{-c}^{+c} y(2a_3b)t\,dy = a_3bty^2 \Big]_{-c}^{+c} = \underline{0}. \qquad \text{Ans.}$$

In the same manner, the resultant boundary force when $a_1$, $a_2$, and $a_3$ are zero is

$$R_x = \int_{-c}^{+c} (6a_4y)t\,dy = 3a_4ty^2 \Big]_{-c}^{+c} = \underline{0}, \qquad \text{Ans.}$$

and the moment (couple) is

$$M = \int_{-c}^{+c} y(6a_4y)t\,dy = 2a_4ty^3 \Big]_{-c}^{+c} = \underline{4a_4tc^3}. \qquad \text{Ans.}$$

## PROBLEMS

**3–9.** Determine the relation which must exist between the coefficients of the fourth-order polynomial,

$$\varphi_4 = a_1x^4 + a_2x^3y + a_3x^2y^2 + a_4xy^3 + a_5y^4$$

in order for it to be biharmonic.

**3–10.*** Determine the stresses $\sigma_x$, $\sigma_y$, and $\tau_{xy}$ for the stress function in Prob. 3–9. Assume that $a_1$, $a_3$, $a_4$, and $a_5$ are all zero. Draw a sketch showing the boundary stresses on the plate bounded by $0 \le x \le b$ and $0 \le y \le c$. Determine the resultant normal and shearing boundary *forces* on each of the faces.

**3–11.*** Solve Prob. 3–10 if only $a_1$ and $a_3$ are non-zero. Note that these two coefficients cannot be independent if the compatibility equation is to be satisfied.

**3–12.** Determine the relations which must exist between the coefficients of the fifth-order polynomial,

$$\varphi_5 = a_1x^5 + a_2x^4y + a_3x^3y^2 + a_4x^2y^3 + a_5xy^4 + a_6y^5$$

in order for it to be biharmonic.

**3–13.*** Determine the stresses $\sigma_x$, $\sigma_y$, and $\tau_{xy}$ for the stress function in Prob. 3–12 assuming that only $a_2$ and $a_4$ are not zero. Determine the resultant boundary forces (both normal and tangential) on each of the faces of the plate bounded by $0 \le x \le b$ and $0 \le y \le c$ where $b$ is equal to $2c$.

**3–14.** The cantilever beam shown in Fig. P3–14 is loaded by a uniform shearing stress, $\tau$, along the surface $y = +c$ while the surface $y = -c$, and $x = L$ are free of stress. In what way does the stress function

$$\varphi = \frac{\tau}{4}\left(xy - \frac{xy^2}{c} - \frac{xy^3}{c^2} + \frac{Ly^2}{c} + \frac{Ly^3}{c^2}\right)$$

match the conditions of the problem, and in what respects does it fail?

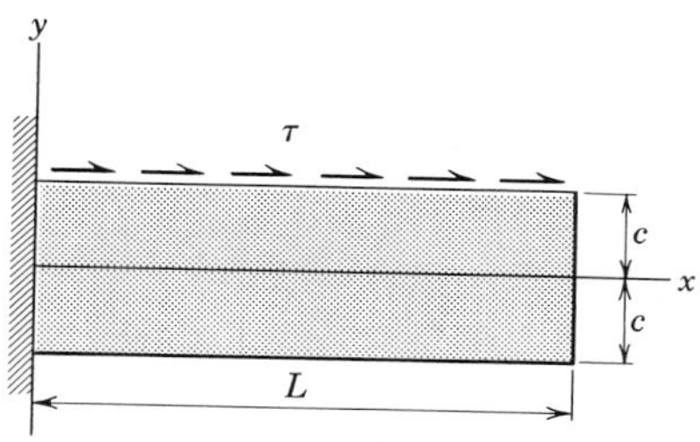


Fig. P3–14

**3–15.** The annular ring in Fig. P3–15 with inside and outside radii of $a$ and $b$, respectively, is subjected to a pressure $p$ on the inside surface. Verify that the biharmonic function

$$\varphi = \frac{pa^2}{2(b^2 - a^2)}\left[(x^2 + y^2) - b^2 \ln (x^2 + y^2)\right]$$

is the stress function for this problem.

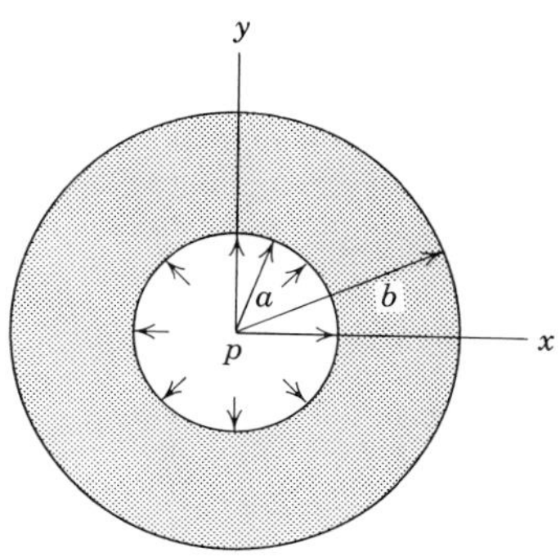


Fig. P3–15

**3–16.*** The biharmonic function

$$\varphi = \frac{Px}{\pi} \arctan \frac{y}{x}$$

is constructed for the problem of a concentrated load on an infinite half plane as shown in Fig. P3–16.

(*a*) Determine the stresses $\sigma_x$, $\sigma_y$, and $\tau_{xy}$ and show that, except at the origin, the stresses on the boundary vanish.

(*b*) Show that the stresses in cylindrical coordinates are

$$\sigma_r = -\frac{2P}{\pi}\frac{\sin\theta}{r}, \qquad \sigma_\theta = 0, \qquad \tau_{r\theta} = 0.$$

(*c*) On a surface $r$ equal to a constant, show that the resultant vertical force is given by

$$F_y = \int_{\theta=0}^{\theta=\pi} \sigma_r(r \sin\theta)\, d\theta$$

and that $F_y$ balances the force $P$. The plate is assumed to be one unit thick.

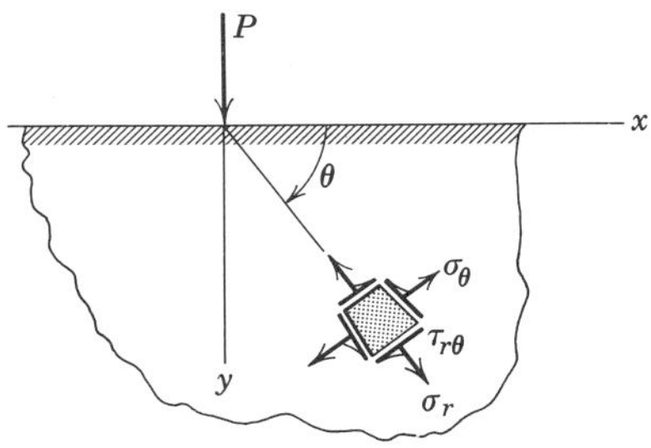


Fig. P3–16

**3–8. Elasticity Equations in Three Dimensions.** The preceding articles were concerned with the analysis of problems in plane stress. A similar analysis for more general problems can be developed by using the same general procedure. In this case, there are three differential equations of equilibrium, three boundary equations, and six compatibility equations. The normal and shearing strains are related to the three displacement components $u$, $v$, and $w$ by the following strain-deformation equations.

$$\begin{aligned} \epsilon_x &= \frac{\partial u}{\partial x}, & \gamma_{xy} &= \frac{\partial v}{\partial x} + \frac{\partial u}{\partial y}, \\ \epsilon_y &= \frac{\partial v}{\partial y}, & \gamma_{yz} &= \frac{\partial w}{\partial y} + \frac{\partial v}{\partial z}, \\ \epsilon_z &= \frac{\partial w}{\partial z}, & \gamma_{zx} &= \frac{\partial u}{\partial z} + \frac{\partial w}{\partial x}. \end{aligned} \tag{3–9}$$

The equations of equilibrium are derived from a free-body diagram of a cube showing three normal and three shearing stresses together with the stress variations similar to Fig. 3–2. The resulting equations are

$$\begin{aligned}
\frac{\partial \sigma_x}{\partial x} + \frac{\partial \tau_{xy}}{\partial y} + \frac{\partial \tau_{xz}}{\partial z} + F_x &= 0,\\
\frac{\partial \tau_{yx}}{\partial x} + \frac{\partial \sigma_y}{\partial y} + \frac{\partial \tau_{yz}}{\partial z} + F_y &= 0,\\
\frac{\partial \tau_{zx}}{\partial x} + \frac{\partial \tau_{zy}}{\partial y} + \frac{\partial \sigma_z}{\partial z} + F_z &= 0,
\end{aligned} \qquad (3\text{–}10)$$

in which $F_x$, $F_y$, and $F_z$ are body forces per unit of volume.

The boundary equations are

$$\begin{aligned}
l\sigma_x + m\tau_{xy} + n\tau_{xz} &= S_x,\\
l\tau_{yx} + m\sigma_y + n\tau_{yz} &= S_y,\\
l\tau_{zx} + m\tau_{zy} + n\sigma_z &= S_z,
\end{aligned} \qquad (3\text{–}11)$$

where $l$, $m$, and $n$ are the direction cosines of the external normal to the surface and $S_x$, $S_y$, and $S_z$ are the components of the surface forces per unit area. Note that $\tau_{yx} = \tau_{xy}$, etc. The equations are written using a cyclic notation for ease of checking subscripts.

The six components of strain $\epsilon_x$, $\epsilon_y$, $\epsilon_z$, $\gamma_{xy}$, $\gamma_{yz}$, and $\gamma_{zx}$ are expressed in terms of the three components of displacement $u$, $v$, and $w$ in Eq. (3–9). The displacement components can be eliminated from the expressions for strains by a proper combination of differentiation and addition. These operations result in the following six equations of compatibility in terms of strains.

$$\begin{aligned}
\frac{\partial^2 \epsilon_x}{\partial y^2} + \frac{\partial^2 \epsilon_y}{\partial x^2} &= \frac{\partial^2 \gamma_{xy}}{\partial x\, \partial y},\\
\frac{\partial^2 \epsilon_y}{\partial z^2} + \frac{\partial^2 \epsilon_z}{\partial y^2} &= \frac{\partial^2 \gamma_{yz}}{\partial y\, \partial z},\\
\frac{\partial^2 \epsilon_z}{\partial x^2} + \frac{\partial^2 \epsilon_x}{\partial z^2} &= \frac{\partial^2 \gamma_{zx}}{\partial z\, \partial x},\\
2\frac{\partial^2 \epsilon_x}{\partial y\, \partial z} &= \frac{\partial}{\partial x}\left(-\frac{\partial \gamma_{yz}}{\partial x} + \frac{\partial \gamma_{zx}}{\partial y} + \frac{\partial \gamma_{xy}}{\partial z}\right),\\
2\frac{\partial^2 \epsilon_y}{\partial z\, \partial x} &= \frac{\partial}{\partial y}\left(\frac{\partial \gamma_{yz}}{\partial x} - \frac{\partial \gamma_{zx}}{\partial y} + \frac{\partial \gamma_{xy}}{\partial z}\right),\\
2\frac{\partial^2 \epsilon_z}{\partial x\, \partial y} &= \frac{\partial}{\partial z}\left(\frac{\partial \gamma_{yz}}{\partial x} + \frac{\partial \gamma_{zx}}{\partial y} - \frac{\partial \gamma_{xy}}{\partial z}\right).
\end{aligned} \qquad (3\text{–}12)$$

These equations must be satisfied by the strain components if the displacements are to exist as continuous functions.

The equations of compatibility can be expressed in terms of the stresses by means of Hooke's law, Eq. (2–2) and (2–6), and the equations of equilibrium, Eq. (3–10). If there are no body forces, the equations become

$$
\begin{aligned}
(1 + \mu)\nabla^2\sigma_x + \frac{\partial^2}{\partial x^2}(\sigma_x + \sigma_y + \sigma_z) &= 0,\\
(1 + \mu)\nabla^2\sigma_y + \frac{\partial^2}{\partial y^2}(\sigma_x + \sigma_y + \sigma_z) &= 0,\\
(1 + \mu)\nabla^2\sigma_z + \frac{\partial^2}{\partial z^2}(\sigma_x + \sigma_y + \sigma_z) &= 0,\\
(1 + \mu)\nabla^2\tau_{yz} + \frac{\partial^2}{\partial y\,\partial z}(\sigma_x + \sigma_y + \sigma_z) &= 0,\\
(1 + \mu)\nabla^2\tau_{zx} + \frac{\partial^2}{\partial z\,\partial x}(\sigma_x + \sigma_y + \sigma_z) &= 0,\\
(1 + \mu)\nabla^2\tau_{xy} + \frac{\partial^2}{\partial x\,\partial y}(\sigma_x + \sigma_y + \sigma_z) &= 0.
\end{aligned}
\tag{3–13}
$$

The derivation of these equations is given in various textbooks on elasticity.[2]

The preceding equations can be used to obtain solutions or to check assumed solutions to problems. They will be used in later chapters with known solutions to compare results from elementary solutions with the more exact solutions based on these expressions.

It can be proved that[3] if a solution is obtained which satisfies the boundary conditions, the equilibrium equations, and the compatibility equations the solution is unique. This is known as the uniqueness theorem, and guarantees that there is only one possible distribution of stress.

In many instances, it may be extremely difficult to solve the general equations of elasticity together with particular boundary conditions. The *inverse* and the *semi-inverse* methods are therefore often used to solve problems. The inverse method consists of examining functions which satisfy the differential equations in order to see what boundary conditions they give. The semi-inverse method consists of

---

[2] *Theory of Elasticity,* 2nd Ed., S. Timoshenko and J. N. Goodier, McGraw-Hill, New York, 1951, pp. 228–232.

[3] *Mathematical Theory of Elasticity,* 2nd Ed., I. S. Sokolnikoff, McGraw-Hill, New York, 1956, pp. 86–88.

making simplifying assumptions regarding the stresses, strains, or displacements, based on the mechanics of the class of problems under study, so that the differential equations are reduced to a more manageable form. Once a solution is found by either method, the uniqueness theorem provides assurance that the stress distribution so obtained is the only one possible for that example.

The developments presented in this chapter deal primarily with the problems of prescribed stresses on the boundary, but it is also common to encounter situations in which displacements on the boundary are prescribed. These problems can also be treated with the same equations, but the application of the boundary conditions requires considerations beyond the scope of this chapter.

## PROBLEM

**3–17.** A prismatic bar has a length $L$ and a rectangular cross section $a$ by $b$. The bar has a specific weight of $\gamma$ lb per unit volume, and is suspended in a vertical position from its upper end (see Fig. P3–17). Show that the displacements given by the equations

$$u = -\mu\gamma xz/E,$$

$$v = -\mu\gamma yz/E,$$

$$w = \gamma[(z^2 - L^2) + \mu(x^2 + y^2)]/(2E)$$

give stresses and strains which satisfy the equations of equilibrium and compatibility. The symbol $\mu$ is Poisson's ratio, and $E$ is Young's modulus. What are the boundary stresses on the surfaces of the bar? What is the shape of the surface at the lower end?

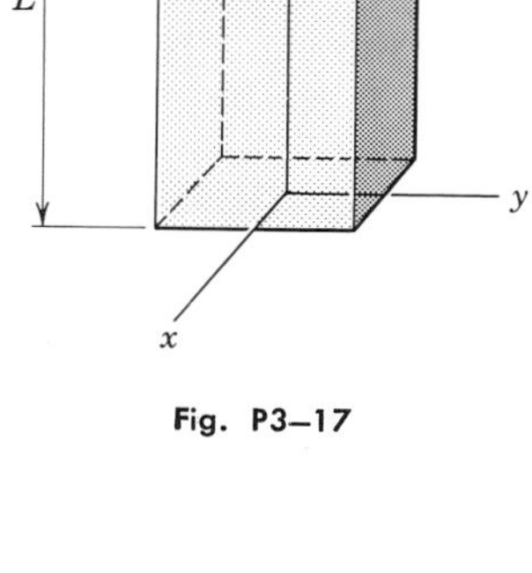


Fig. P3–17

CHAPTER 4

# *Torsional Loading*

**4–1. Torsional Shearing Stress.** In the design of machinery (and some structures) the problem of transmitting a torque (a couple) from one plane to a parallel plane is frequently encountered. The simplest device for accomplishing this function is a circular shaft such as that connecting an electric motor with a pump, compressor, or other mechanism. For example, the shaft $AB$ of Fig. 4–1 transmits a torque from the driving motor $A$ to the coupling $B$. A modified free-body diagram (the weight and bearing reactions are not shown because they contribute no useful information to the torsion problem) of a typical installation is shown in Fig. 4–2. The resultant of the electromagnetic forces applied to the armature $A$ of the motor is a couple which is resisted by the resultant of the bolt forces (another couple) acting on the flange coupling $B$, and the circular shaft transmits the torque from armature to coupling. The problem is the determination of significant stresses in and the deformation of the shaft.

As a point of departure, a segment of the shaft between transverse planes $a$–$a$ and $b$–$b$ of Fig. 4–2 will be studied. In so doing, the complicated stress patterns at the locations of torque applying devices may be avoided and Saint-Venant's principle applies. A free-body diagram of the segment of the shaft is shown in Fig. 4–3 with the torque applied by the armature indicated on the left end as $T$. The resisting torque $T_r$, indicated at the right end, is the resultant of the differential forces $dF$ acting on the transverse plane $b$–$b$. The force $dF$ is equal to $\tau\, dA$ where $\tau$ is the shearing stress *on the transverse plane* at a distance $\rho$ from the shaft center, and $dA$ is a differential area. *In keeping with the notation outlined in Art. 1–6, the shearing stress should be designated* $\tau_{z\theta}$ *to indicate that it acts on the z face in the direction of increasing* $\theta$. For the elementary theory of torsion of circular sections, however, the shearing stress on any transverse plane is always perpendicular to the radius to the point, and the stress will

Fig. 4–1. A 1000-Hp. 180-rpm motor coupled to ball and tube mills. Photo courtesy of Electric Machinery Mfg. Co., Minneapolis, Minn.

be referered to as $\tau$ or $\tau_c$ for convenience. In Art. 4–8, for *non-circular members in torsion, the more precise designation for shearing stress and strain is essential.* If the shaft is in equilibrium, the summation of moments about the axis of the shaft indicates that

$$T = T_r = \int^{\text{Area}} \rho\tau \, dA. \qquad \text{(a)}$$

The law of variation of the shearing stress on the transverse plane must be known before the integral of Eq. (a) can be evaluated. The approach to the problem of stress distribution will be prefaced by a brief reference to the past.

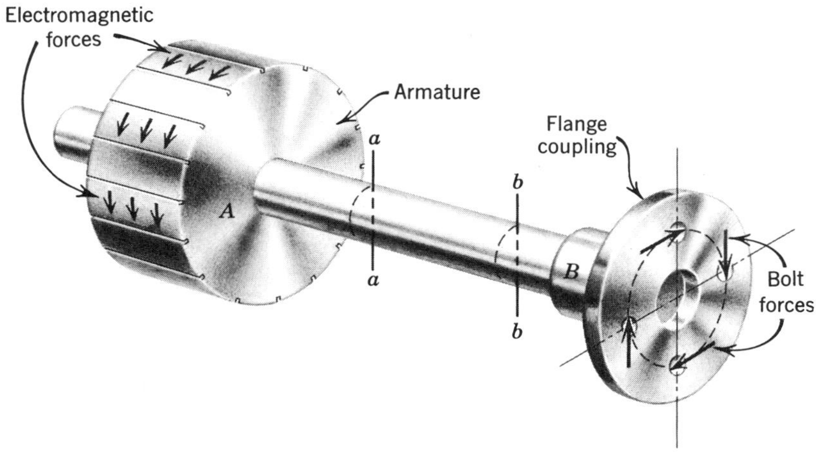


Fig. 4—2

C. A. Coulomb, a French engineer, in 1784 developed experimentally the relationship between the applied torque and the angle of twist for circular bars,[1] and A. Duleau, another French engineer, in a paper published in 1820[1] derived analytically the same relationship by mak-

[1] From *History of Strength of Materials,* S. P. Timoshenko, McGraw-Hill, New York, 1953.

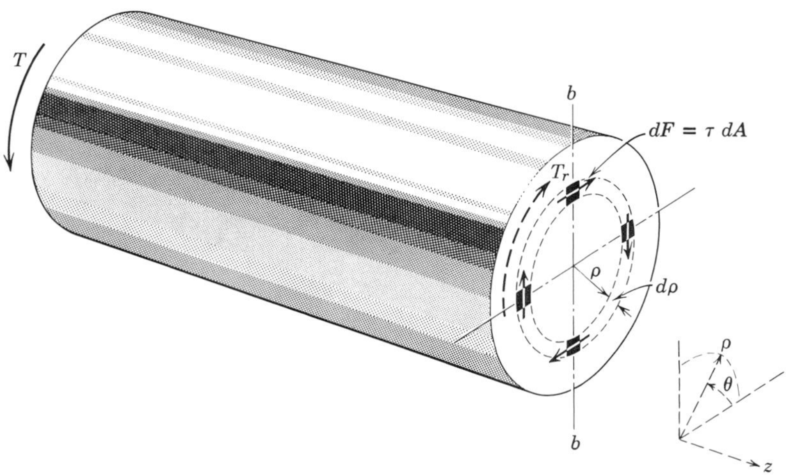


Fig. 4—3

Fig. 4–4a

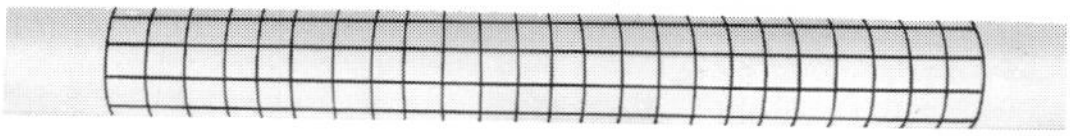

Fig. 4–4b

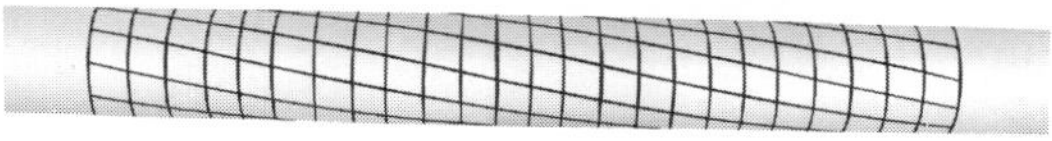

Fig. 4–4c

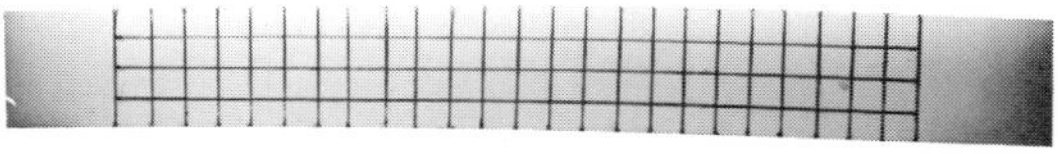

Fig. 4–4d

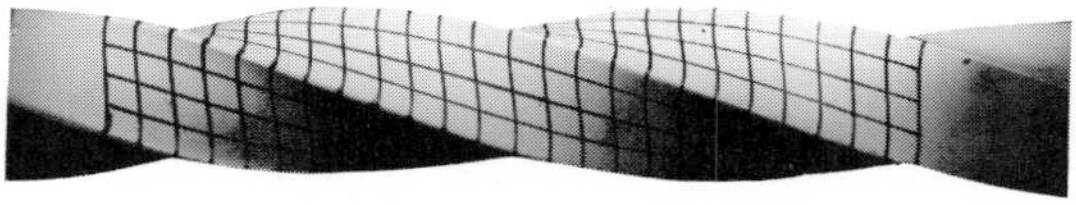

ing the assumption that *a plane section before twisting remains plane after twisting and a diameter remains a straight line.* Visual examination of twisted models indicates that these assumptions are apparently correct for circular sections either solid or hollow (provided the hollow section is circular and symmetrical with respect to the axis of the shaft), but incorrect for any other shape. Compare, for example, the distortions of the rubber models with circular and rectangular cross sections shown in Fig. 4–4.

If the assumptions noted above are made, the distortion of the shaft will be as indicated in Fig. 4–5*a*, wherein points $B$ and $D$ on a common radius in a plane section move to $B'$ and $D'$ in the same plane and still on the same radius. The angle $\theta$ is called the *angle of twist.* The surface $ABB'$ of Fig. 4–5*a* is shown developed in Fig. 4–5*b* and a differential cube of the material at $B$ assumes at $B'$ the familiar distortion due to shearing stress. At this point, the assumption will be made that *all longitudinal elements have the same length* (which limits the results to straight shafts of constant diameter). From Fig. 4–5, the following strain relations may be written:

$$\tan \gamma_c = \frac{BB'}{L} = \frac{c\theta}{L},$$

and

$$\tan \gamma = \frac{DD'}{L} = \frac{\rho\theta}{L},$$

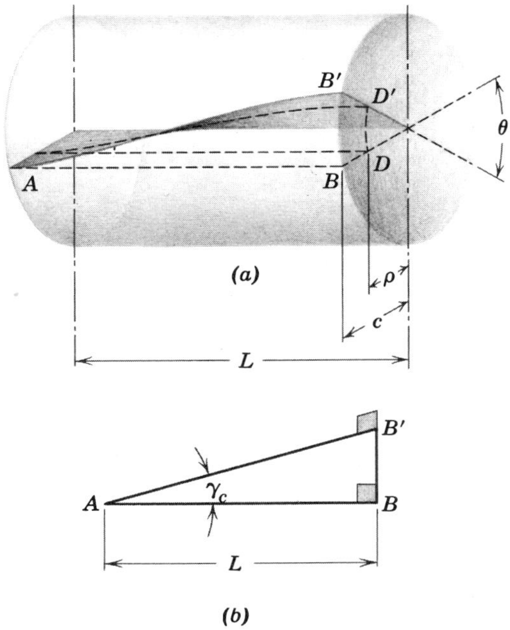


Fig. 4–5

or, *if the strain is small* (the case when the action is elastic and the material is stiff),

$$\gamma_c = \frac{c\theta}{L} \quad \text{and} \quad \gamma = \frac{\rho\theta}{L}. \tag{b}$$

From Eq. (b)

$$\frac{\gamma_c L}{c} = \frac{\gamma L}{\rho},$$

and as the longitudinal elements have the same length, the expression reduces to

$$\frac{\gamma_c}{c} = \frac{\gamma}{\rho},$$

which means that the shearing strain is proportional to the distance from the axis of the shaft. The assumption is now made that *Hooke's law applies* (the accompanying limitation is that the stresses must be within the proportional limit), and the expression becomes

$$\frac{\tau_c}{G_c c} = \frac{\tau}{G_\rho \rho}.$$

The final assumption is that $G$ *has the same value at all points in the shaft cross section,* which limits the resulting expressions to homogeneous and isotropic materials. The last expression then becomes

$$\frac{\tau_c}{c} = \frac{\tau}{\rho} = \text{a constant}, \tag{c}$$

or the shearing stress is proportional to the distance from the axis of the shaft. Equation (c) provides the law of variation of shearing stress required to evaluate the integral of Eq. (a). When the expression

$$\tau = \frac{\rho \tau_c}{c} = \frac{\rho \tau}{\rho}$$

from Eq. (c) is substituted in Eq. (a), the result is

$$T_r = \frac{\tau_c}{c} \int^{\text{Area}} \rho^2 \, dA = \frac{\tau}{\rho} \int^{\text{Area}} \rho^2 \, dA. \tag{d}$$

Equation (d) may also be written as

$$T_r = (\tau/\rho) \int_r^c \int_0^{2\pi} \rho^2 (\rho \, d\rho \, d\theta),$$

which, when integrated with respect to $\theta$ becomes

$$T_r = (\tau/\rho) \int_r^c 2\pi \rho^3 \, d\rho,$$

where $r$ is the inner radius of the resisting area (for example, the inner radius of a hollow shaft) and $c$ is the outer radius of the area.

The integral in Eq. (d) is the polar second moment of area, $J$ (called polar moment of inertia), with respect to the longitudinal axis of the shaft.[2] Equation (d) can then be written as

$$T_r = \tau J / \rho,$$

or more briefly,

$$\tau = T\rho / J. \tag{4–1}$$

Equation (4–1) is known as the *torsion formula,* in which $\tau$ is the *shearing stress on a transverse plane at a distance* $\rho$ *from the axis of the shaft* and $T$ is the resisting torque, which, in general, is obtained from a free-body diagram and an equilibrium equation. In Art. 4–3, it will

[2] See App. B for formulas; and Chap. 6 in *Engineering Mechanics,* Vector ed., A. Higdon and W. B. Stiles, Prentice-Hall, Englewood Cliffs, N.J., 1962, for fundamentals.

be shown that the transverse plane is a plane of maximum shearing stress for pure torsion.

**4–2. Torsional Displacements.** Frequently the amount of twist in a shaft (or structural element) is of paramount importance; making the computation of the angle of twist a necessary accomplishment of the machine or structural designer. The fundamental approach to the problem is provided by the following equations:

$$\gamma = \rho\theta/L, \tag{4–2}$$

$$\tau = \rho T/J, \tag{4–1}$$

and

$$G = \tau/\gamma.$$

Equation (4–2) is Eq. (b) of Art. 4–1; Eq. (4–1) was also derived in Art. 4–1. The third expression is Hooke's law for shearing stresses. The last two expressions are limited to stresses below the proportional limit of the material. These three equations can be combined to give several different relationships; for example,

$$\theta = \frac{\gamma L}{\rho} = \frac{\tau L}{\rho G} = \frac{TL}{JG}.$$

The angle of twist determined from the above expressions is for a length of shaft of constant diameter sufficiently removed from sections where pulleys, couplings, or other mechanical devices are attached to make Saint-Venant's principle applicable. However, for practical purposes, it is customary to neglect local distortion at all connections and compute angles as though there were no discontinuities, and as though the torques applied were distributed over the cross sections in exactly the same manner in which the stresses given by the torsion formula are distributed. The following example illustrates the application of the principles developed.

**Example 4–1:** The steel shaft of Fig. 4–6*a* is in equilibrium under the torques shown. Determine:

(*a*) The maximum shearing stress in the shaft.
(*b*) The angle of twist of end *B* of the 6-in. segment with respect to end *A*.
(*c*) The angle of twist of the end *C* with respect to end *A*.

*Solution:* (*a*) In general, free-body diagrams should be drawn in order to evaluate correctly the resisting torque. Such diagrams are shown in Fig. 4–6*b* and *c*, where in part *b*, the shaft is cut by any transverse plane through the 6-in. segment, and $T_6$ is the resisting torque on this section. Similarly, in Fig. 4–6*c*, the plane is passed through the 4-in. section, and $T_4$ is the resisting torque on this section. The location of the maximum shearing stress is not

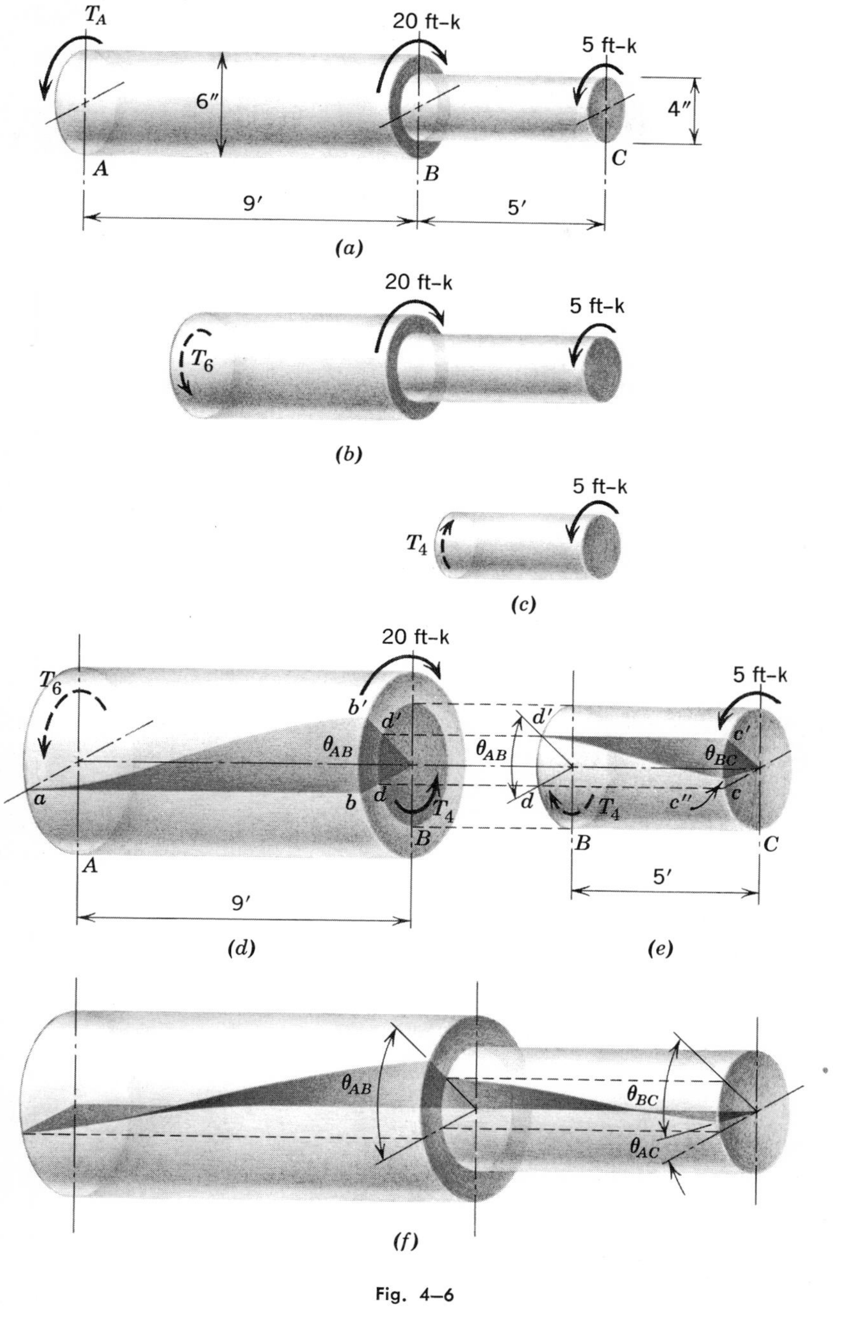

$T_A$
20 ft-k
5 ft-k
6″
4″
A
B
C
9′
5′
(a)
20 ft-k
5 ft-k
$T_6$
(b)
5 ft-k
$T_4$
(c)
20 ft-k
5 ft-k
$T_6$
b′
d′
$\theta_{AB}$
a
b
d
$T_4$
B
A
9′
(d)
d′
$\theta_{AB}$
c′
$\theta_{BC}$
d
$T_4$
c″
c
B
C
5′
(e)
$\theta_{AB}$
$\theta_{BC}$
$\theta_{AC}$
(f)


Fig. 4–6

apparent; hence, the stress must be checked at both sections. Thus, from Fig. 4–6*b*,

$$\Sigma M = 0, \quad T_6 = 20 - 5 = 15 \text{ ft-kips} = 15{,}000(12) \text{ in-lb},$$

and
$$\tau_{\max} = \frac{Tc}{J} = \frac{15{,}000(12)(3)}{(\pi/2)(3^4)} = \frac{40{,}000}{3\pi}.$$

From Fig. 4–6*c*, $\Sigma M = 0, \quad T_4 = 5 \text{ ft-kips} = 5000(12) \text{ in-lb},$

and
$$\tau_{\max} = \frac{5000(12)(2)}{(\pi/2)(2^4)} = \frac{15{,}000}{\pi}; \quad > \frac{40{,}000}{3\pi};$$

therefore, the maximum shearing stress is

$$15{,}000/\pi = \underline{4770 \text{ psi.}} \qquad \text{Ans.}$$

This stress is less than the shearing proportional limit of any steel; hence, the torsion formula applies. Note that had the larger torque been carried by the smaller section, the maximum stress would obviously occur in the small section and only one stress determination would have been required.

(*b, c*) As an aid to visualizing the distortion of the shaft, the segments *AB* and *BC* and the torques acting on them are drawn separately in Fig. 4–6*d* and *e* with the distortions greatly exaggerated. As the resultant torque of 15 ft-kips twists segment *AB* through the angle $\theta_{AB}$, points *b* and *d* move to *b′* and *d′* respectively, and segment *BC* may be considered as a rigid body rotating through the same angle, point *c* moving to *c′*, after which the torque of 5 ft-kips acting on *BC* twists this part of the shaft back through the angle $\theta_{BC}$, point *c′* moving back to *c″*. The resultant distortion for the entire shaft is shown in Fig. 4–6*f*.

When computing distortions, it is recommended that the two fundamental relationships of Eq. (4–1) and (4–2) plus Hooke's law be used; thus,

$$\gamma = \frac{c\theta}{L} = \frac{\tau}{G} = \frac{Tc}{JG}.$$

When this expression is applied to Fig. 4–6*d*, the twist in segment *AB* is obtained as,

$$\theta_{AB} = \frac{(20-5)(12)(9)(12)}{(\pi/2)(3^4)(12)(10^3)} = \frac{40}{\pi(10^3)} = \underline{0.012{,}73 \text{ rad}}. \qquad \text{Ans. } (b)$$

From Fig. 4–6*e* and *f*,

$$\theta_{AC} = \theta_{AB} - \theta_{BC} = \frac{40}{\pi(10^3)} - \frac{5(12)(5)(12)}{(\pi/2)(2^4)(12)(10^3)}$$

$$= \frac{40 - 37.5}{\pi(10^3)} = \underline{0.000{,}796 \text{ rad.}} \qquad \text{Ans. } (c)$$

## PROBLEMS

**4–1.*** In a 10-in. diameter solid circular steel shaft subjected to pure torsion, the shearing strain, $\gamma$, 2 in. from the surface of the shaft, is 0.00054. Determine the maximum shearing strain in the shaft.

**4–2.*** A pure torque acting on a 3-in. diameter circular shaft produces a maximum shearing stress of 5600 psi in the shaft. Determine:

(*a*) The magnitude of the shearing force on a transverse area of 0.15 sq in. considered concentrated at a point 1.2 in. from the center of the shaft.

(*b*) The magnitude of the torsional resistance developed by the area of part (*a*).

**4–3.*** The maximum shearing stress in the solid circular steel shaft of Fig. P4–3 is 9000 psi. Determine the magnitude of the resisting torque offered by the shaded area. Assume a uniform stress distribution over the shaded area.

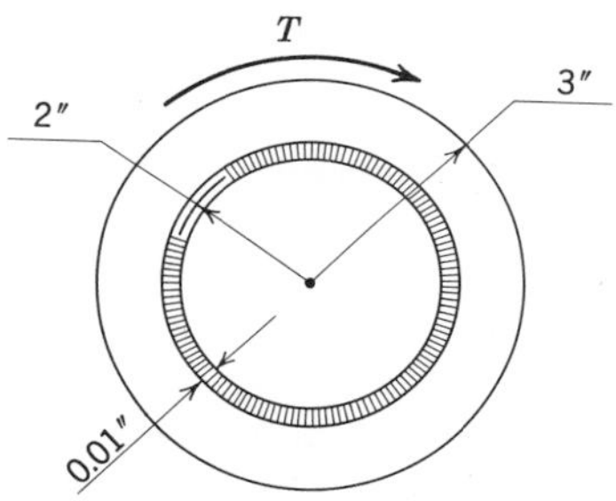


Fig. P4–3

**4–4.*** A pure torque acting on a 16-in. diameter solid circular shaft produces a maximum shearing stress of 20,000 psi. Determine by integration the magnitude of the resisting torque offered by the stress on the shaded band in Fig. P4–4.

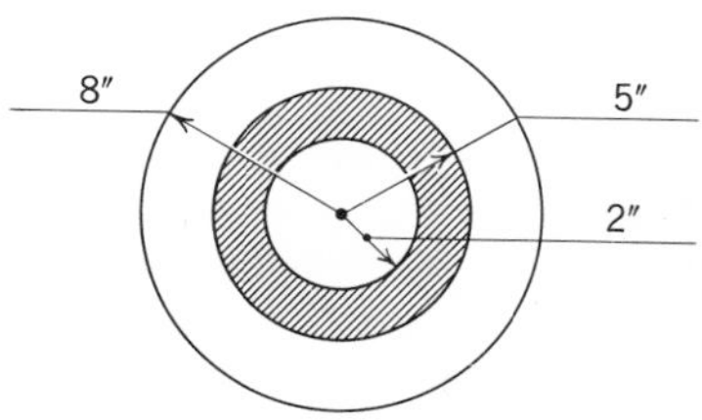


Fig. P4–4

**4–5.** Given that the torsional shearing stress in a circular shaft is proportional to the distance from the center of the cross section of the shaft, *derive* an expression for the resisting torque developed by the band of material which lies between $c/3$ and $2c/3$ of a solid circular shaft having a radius $c$. Express the result in terms of $c$ and $\tau_{max}$.

**4–6.*** (*a*) A hollow circular steel shaft 10 ft long has inside and outside diameters of 3 in. and 4 in., respectively. Determine the shearing stress on a transverse plane at the inner surface of the shaft when it is 10,000 psi at the outer surface.

(*b*) If the shaft were solid instead of hollow, what would be the percentage increase in the area, and what would be the percentage increase in torsional resistance developed by the shaft at the same maximum stress?

**4–7.** Determine the minimum diameter for a solid circular steel shaft required to resist a torque of 15,000 ft-lb without exceeding a shearing stress of 6000 psi.

**4–8.*** A hollow circular steel shaft, 5 in. outside diameter and 3 in. inside diameter, is subjected to a torque of 12,000 ft-lb.

(*a*) Through what angle (in radians) will the shaft be twisted in a length of 20 ft?

(*b*) Determine the shearing stress on a transverse plane at the inner surface of the shaft.

(*c*) Determine the shearing strain at the outside surface.

**4–9.*** A hollow circular steel shaft is 6 ft long and has an outside diameter of 4 in.

(*a*) If the maximum shearing stress is 12,000 psi when the shaft is subjected to a torque of $3200\pi$ ft-lb, what is the inside diameter of the shaft?

(*b*) Determine the magnitude of the angle of twist in the shaft when the maximum shearing stress is 12,000 psi and the torque $2500\pi$ ft-lb.

**4–10.** The member shown in Fig. P4–10 is a solid circular steel shaft 4 in. in diameter. Rod *AB* is a rigid pointer fastened to the end of the shaft. Determine:

(*a*) The maximum shearing stress in the shaft.

(*b*) The distance point *A* moves, measured from its no load position.

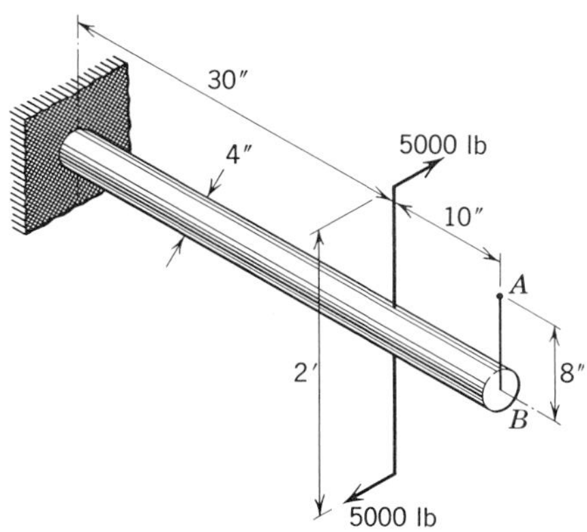


Fig. P4–10

**4–11.*** Determine the maximum allowable torque to which a solid circular wrought iron bar 10 ft long and 4 in. in diameter can be subjected when it is specified that the shearing stress must not exceed 10,500 psi and the angle of twist must not exceed 0.06 rad. The shearing modulus of elasticity for the wrought iron is assumed to be $10^7$ psi.

**4–12.** A solid circular steel shaft 10 ft long is to resist a torque of 2000 ft-lb. The allowable shearing stress is 12,000 psi and the maximum allowable angle of twist is 7°. Determine the minimum permissible diameter.

**4–13.** Specifications for a solid circular steel rod 36 in. long require that it shall be adequate to resist a torque of 15,700 in-lb without twisting more than 2° or exceeding a shearing stress of 10,000 psi. What minimum diameter is required?

**4–14.*** In the factory drive shaft of Fig. P4–14*a* and *b*, a torque of 12,000 in-lb is supplied to the shaft by a belt which drives pulley *A*. A torque of 7000 in-lb is taken off by pulley *B* and the remainder by pulley *C*.

(*a*) Determine the maximum torque in each portion of each drive shaft.

(*b*) Which arrangement (*a* or *b*) produces the smaller maximum stress in the drive shaft?

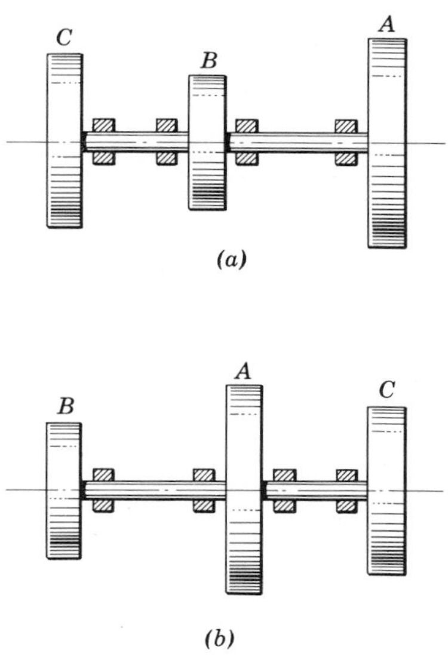


Fig. P4–14

**4–15.** For the hollow circular shaft shown in Fig. P4–15 determine:

(*a*) The magnitude of the total resisting torque at section *A–A* of the shaft.

(*b*) The maximum shearing stress on section *A–A*.

(*c*) The minimum shearing stress on section *A–A*.

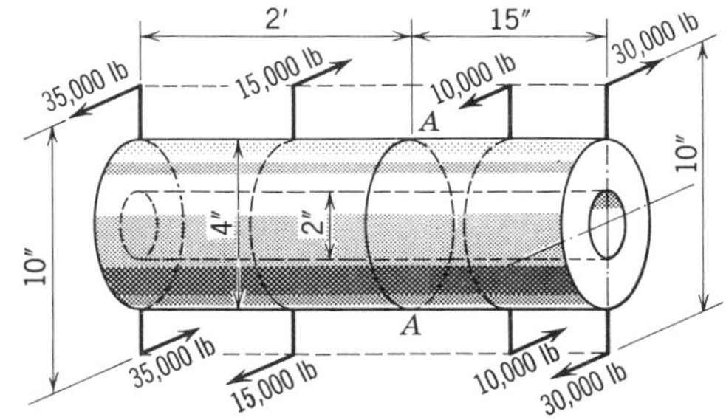


Fig. P4–15

**4–16.** A torque of 80,000 in-lb is supplied to the driving pulley *B* of Fig. P4–16 by a motor. Pulley *A* takes off 30,000 in-lb of torque and the remainder is taken by pulley *C*. Determine:

(*a*) The maximum shearing stress in each shaft.

(*b*) The magnitude of the angle of twist between pulleys *A* and *C*. Both shafts are made of steel and are rigidly connected to pulley *B*.

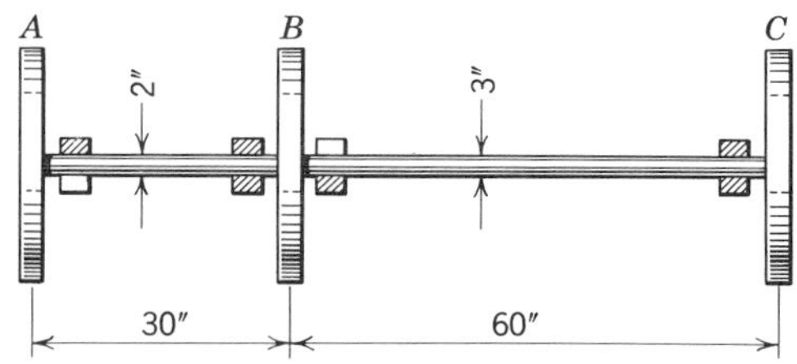


Fig. P4–16

**4–17.*** Under the torques shown in Fig. P4–17, the maximum stress in segment *BC* of the aluminum alloy ($G$ = 4000 ksi) shaft is 12 ksi.

(*a*) Determine by integration the magnitude of the resisting torque carried by the portion of the cross section between $\rho = 1$ and $\rho = 2$ in. in segment *BC*.

(*b*) Determine the rotation of the section at *B* when the torques $T_1$ and $T_2$ are applied as shown.

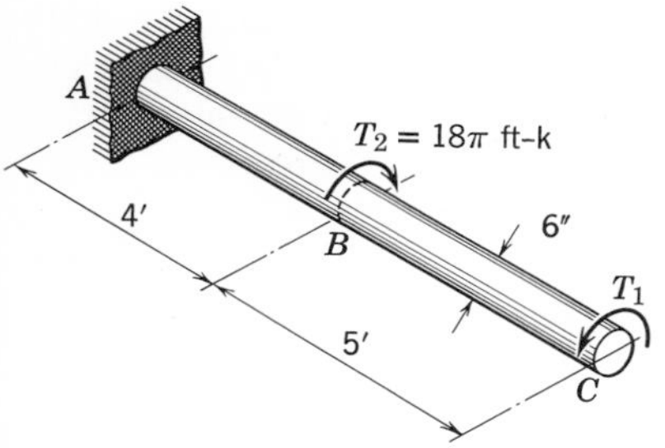


Fig. P4–17

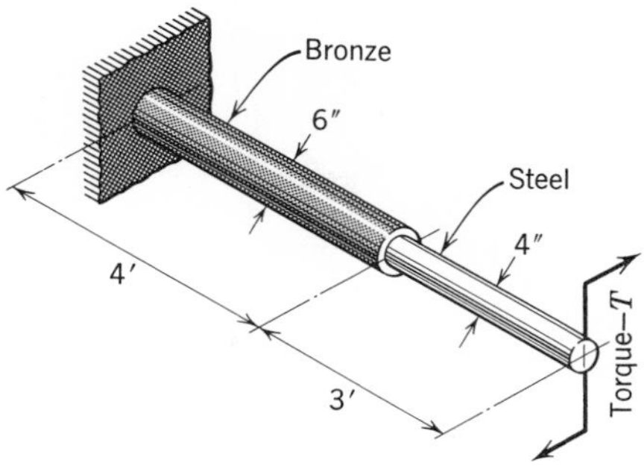


Fig. P4–18

**4–18.*** A solid circular steel shaft is fastened securely to a solid circular bronze shaft as indicated in Fig. P4–18. Determine the value of the couple $T$ that will produce an angle of twist of 0.016 rad in the 7-ft length. The shearing modulus of elasticity of the bronze is assumed to be 6000 ksi.

**4–19.** Solve Prob. 4–18 with an additional torque of $3T$ applied at the right end of the bronze portion; both torques having the same sense.

**4–20.*** A solid circular steel shaft with diameters as shown in Fig. P4–20 is subjected to a torque $T$. The allowable shearing stress is 14,000 psi and the maximum allowable angle of twist in the 7-ft length is 0.06 rad. Determine the maximum allowable value of $T$.

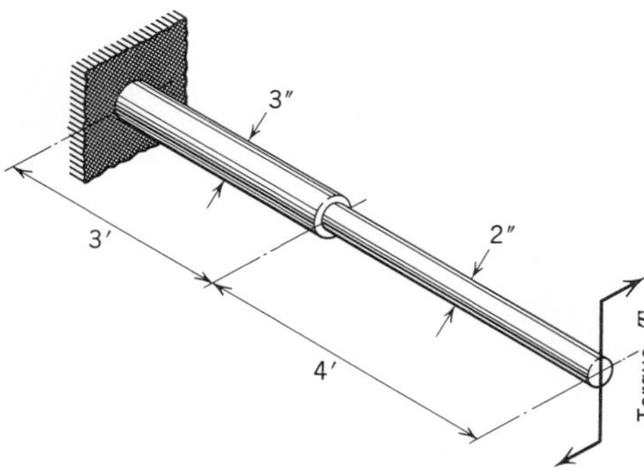


Fig. P4–20

**4–21.** Solve Prob. 4–20 with an additional torque of 4000 ft-lb applied at the right end of the 3-in. portion with the same sense as that of $T$.

**4–22.*** A shaft is composed of two 6-in. diameter circular sections coupled together as indicated in Fig. P4–22. Part $B$ is bronze with a modulus of rigidity of 6000 ksi and part $A$ is steel. A torque is applied at the free end which causes a maximum shearing stress in the shaft $B$ of 10 ksi. Determine:

(*a*) The angle of twist of the free end of the shaft in degrees.

(*b*) The maximum shearing stress in the steel portion.

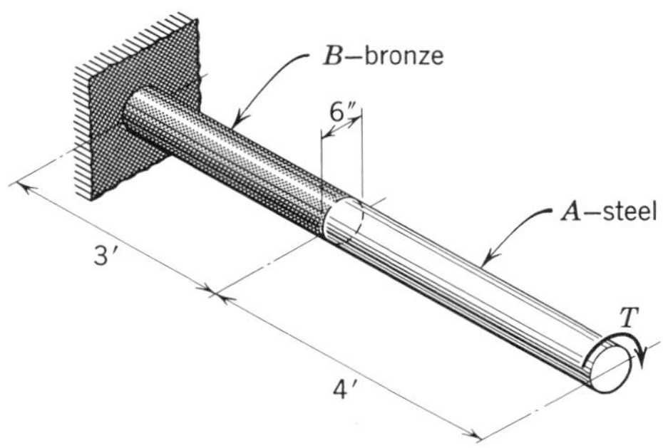


Fig. P4–22

**4–23.** The 4-in. diameter shaft shown in Fig. P4–23 is composed of brass ($G = 5000$ ksi) and steel sections rigidly connected. Determine the maximum allowable torque, applied as shown, if the shearing stresses in the brass and the steel are not to exceed 7500 psi and 10,000 psi respectively, and the distance $AC$, through which the end of the 10-in. pointer $AB$ moves, is not to exceed 0.72 in.

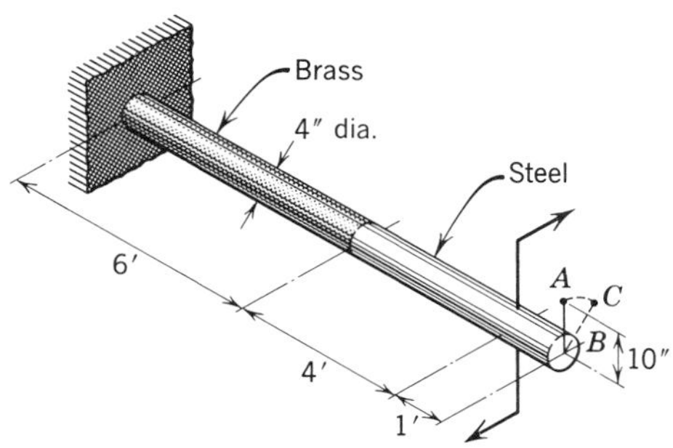


Fig. P4–23

**4–24.*** For the steel shaft loaded and supported as shown in Fig. P4–24, determine:

(*a*) The maximum shearing stress.

(*b*) The angle of twist of the right end with respect to its unloaded position.

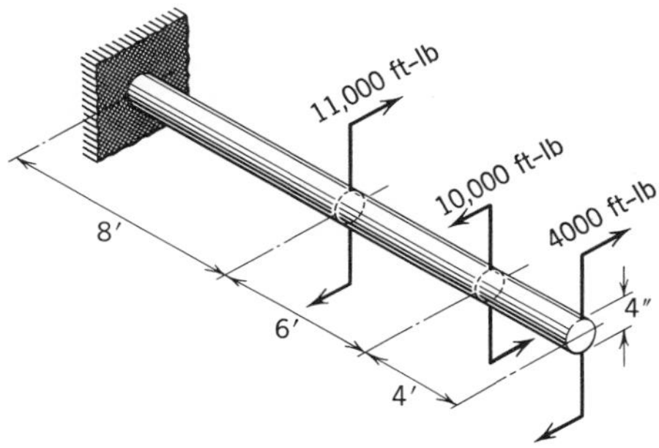


Fig. P4–24

**4–25.** (*a*) Determine the maximum shearing stress in the hollow steel shaft of Fig. P4–25 which is in equilibrium under the torques indicated.

(*b*) Determine the relative rotation of end *D* of the shaft of part *a* with respect to the section at *B*.

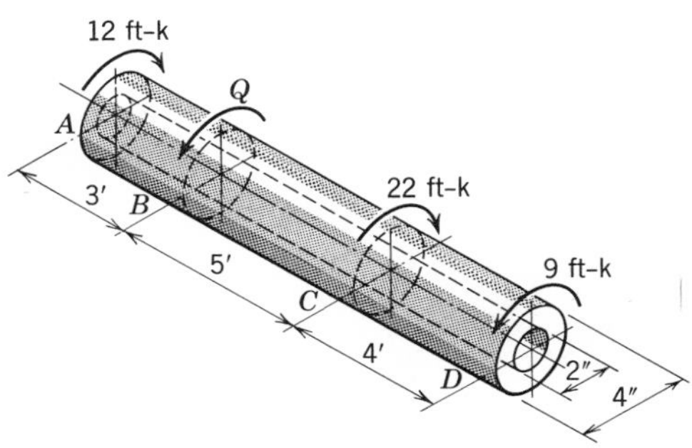


Fig. P4–25

**4–26.*** A steel shaft has the dimensions shown in Fig. P4–26. Determine:

(*a*) The maximum shearing stress on a section 8 ft from the left end.

(*b*) The angle of twist at the section 5 ft from the left end with respect to its "no load" position.

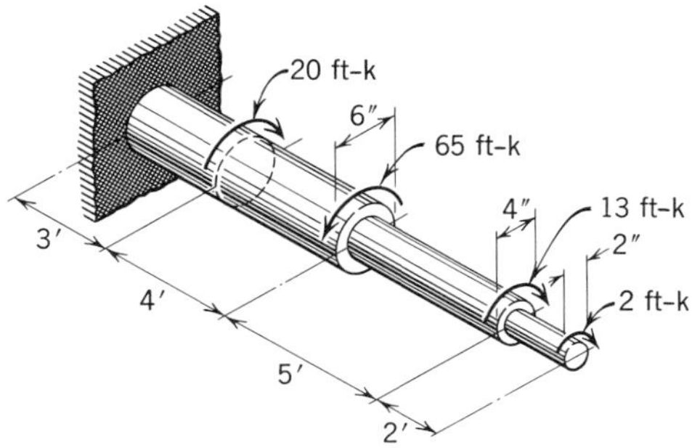


Fig. P4–26

**4–27.** For the steel shaft loaded and supported as shown in Fig. P4–27, determine:

(*a*) The maximum permissible value for the torque $T_3$ which may be applied as shown without exceeding a shearing stress of 10,000 psi in the segment $AB$.

(*b*) The rotation of a diameter at $B$ due to the application of torques $T_1$ and $T_2$ only.

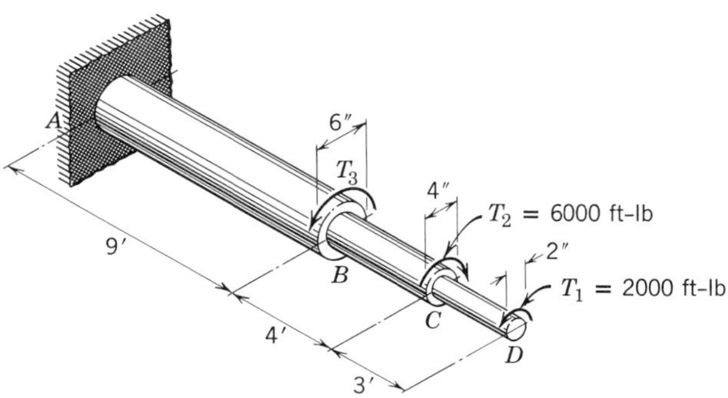


Fig. P4–27

**4–28.*** The hollow circular steel shaft $ABCD$ of Fig. P4–28 is in equilibrium under the torques indicated. The inside diameter is one-half the outside diameter.

($a$) Set up, in terms of $d$, the expression for the rotation of section $C$ with respect to section $B$. Do not solve, but indicate units and sense for the answer.

($b$) Determine the minimum permissible value for the outside diameter for a maximum shearing stress of 12,000 psi.

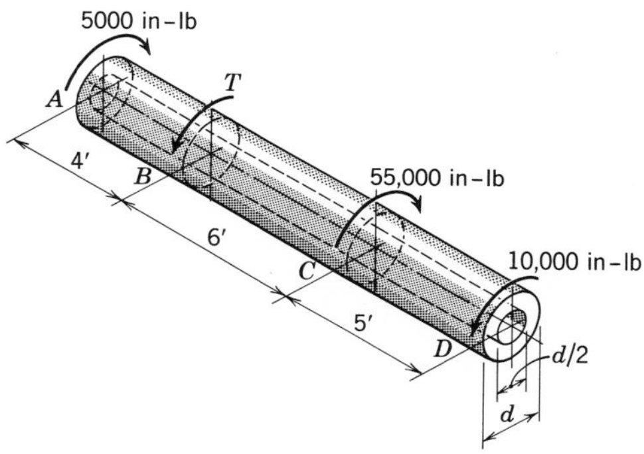


Fig. P4–28

**4–29.** The torsional loads applied as shown to the steel shaft of Fig. P4–29 produce a maximum shearing stress of 11,400 psi and twist the free end through 0.0135 rad. Determine the torques $T_1$ and $T_2$.

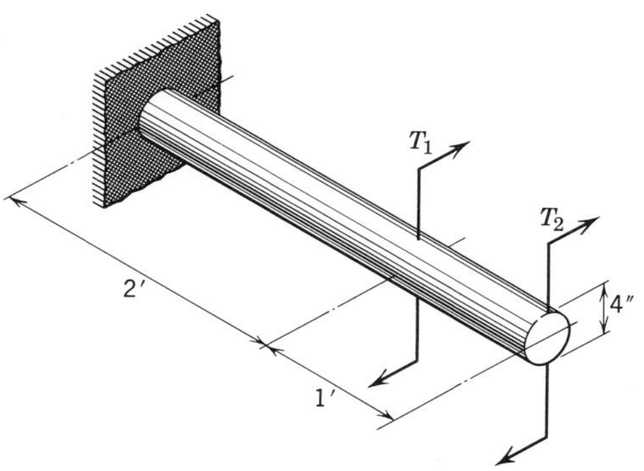


Fig. P4–29

**4–30.*** The steel shaft of Fig. P4–30 is in equilibrium under the torques applied as shown. The maximum shearing stress in the 6-in. segment is 10 ksi and the rotation of the end $C$ with respect to the section at $A$ is 0.010 rad counterclockwise looking left. Determine the torques $T_1$ and $T_2$.

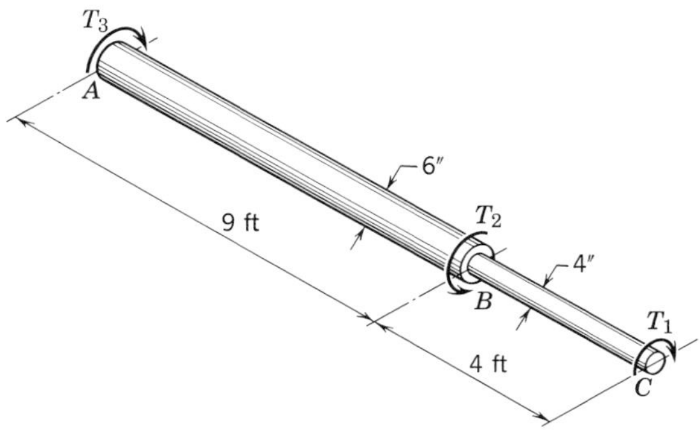


Fig. P4–30

The following problems[3] involve torsion of solids of revolution where the polar moment of inertia, $J$, is a function of $x$ (the axis of the solid). In each case, integration over the length of the solid is necessary. Assume that the taper is gradual enough that a conventional stress distribution exists on each section.

**4–31.*** The frustum of a cone with diameters $D_1$ and $D_2$ and length $L$ is subjected to a constant torque $T$. Determine the angle of twist of the solid in terms of $T$, $L$, $D_1$, $D_2$, and $G$ (see Fig. P4–31).

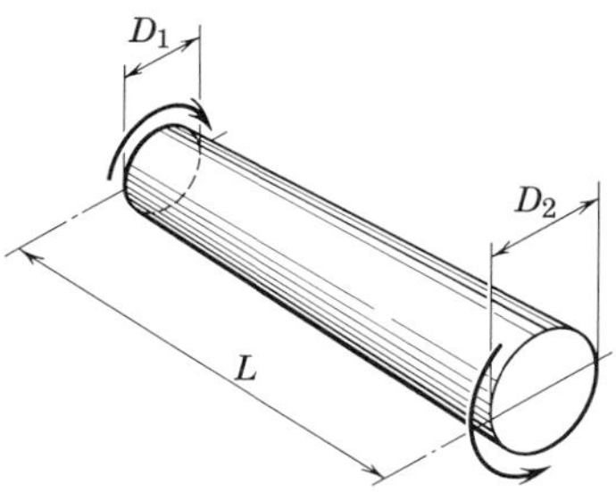


Fig. P4–31

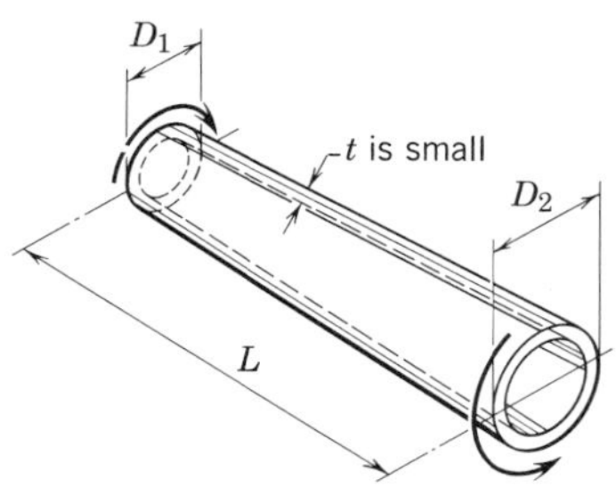


Fig. P4–32

**4–32.** For the hollow frustum of a cone shown in Fig. P4–32, determine the angle of twist in a length $L$ subjected to a constant torque $T$.

[3] Presented by Sidney Kravitz of Picatinny Arsenal in *Product Engineering* of 17 Aug., 1964, pp. 125, 126.

**4–33.*** Determine the angle of twist of a frustum of a cone with a cylindrical hole in it, as shown in Fig. P4–33, when subjected to a constant torque $T$.

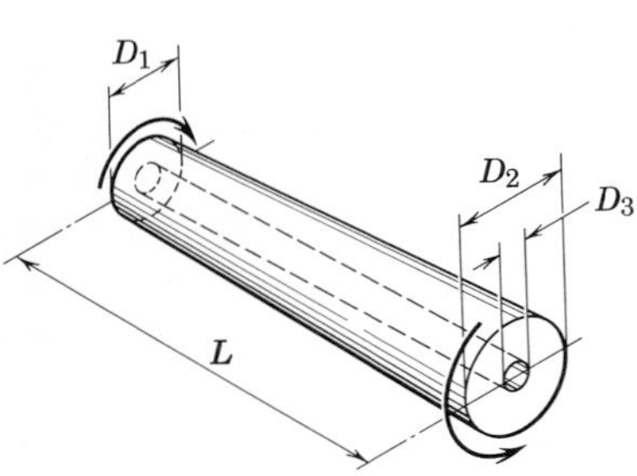


Fig. P4–33

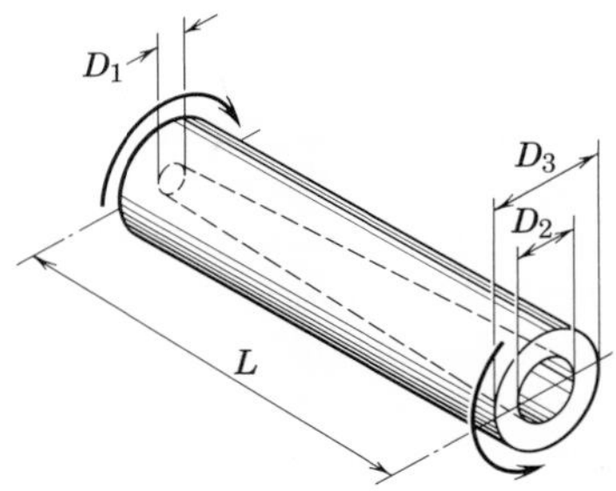


Fig. P4–34

**4–34.** The cylinder in Fig. P4–34 has a frustum of a cone removed. Determine the angle of twist in a length $L$ when the body is subjected to a constant torque $T$.

**4–3. Stresses on Oblique Planes.** As stated in Art. 4–1, the torsion formula can be used to evaluate the shearing stress on a transverse plane. It is necessary to ascertain if the transverse plane is a plane of maximum stress and if there are other significant stresses induced by torsion. For this study, the stresses at point $A$ in the shaft of Fig. 4–7*a* will be analyzed. Figure 4–7*b* shows a differential cube taken from the shaft at $A$ and the stresses acting on transverse and longitudinal planes. The stress $\tau_{xy}$ may be determined by means of the torsion formula, and $\tau_{yx} = \tau_{xy}$ (see Art. 1–4). Applying the equations of equilibrium to the free-body diagram of Fig. 4–7*c*, the following results are obtained:

$$\stackrel{+\nearrow}{\bigcirc} \qquad \Sigma F_t = 0;$$

$$\tau_{nt}\, dA + \tau_{yx}\, dA\, (\cos\theta)(\cos\theta) - \tau_{xy}\, dA\, (\sin\theta)(\sin\theta) = 0,$$

from which

$$\tau_{nt} = -\tau_{xy}(\cos^2\theta - \sin^2\theta) = -\tau_{xy}\cos 2\theta,$$

and

$$\stackrel{+\searrow}{\bigcirc} \qquad \Sigma F_n = 0;$$

$$\sigma_n\, dA + \tau_{yx}\, dA\, (\cos\theta)(\sin\theta) + \tau_{xy}\, dA\, (\sin\theta)(\cos\theta) = 0,$$

from which

$$\sigma_n = -\tau_{xy}(2)(\cos\theta)(\sin\theta) = -\tau_{xy}\sin 2\theta.$$

These results are shown in the graph of Fig. 4–8 from which it is apparent that the maximum shearing stress occurs on transverse and longitudinal (diametral) planes. The graph also shows that the maximum normal stresses occur on planes oriented at 45° with the axis of the bar and perpendicular to the surface of the bar, and that on one of these planes ($\theta = 45°$ for Fig. 4–7), the stress is compression and on the other ($\theta = 135°$), the stress is tension. Furthermore, *all these stresses have the same magnitude;* hence, the torsion formula will give the magnitude of the maximum normal and shearing stresses at a point in a circular shaft subjected to pure torsion.

Any of the stresses discussed in the preceding paragraph may be

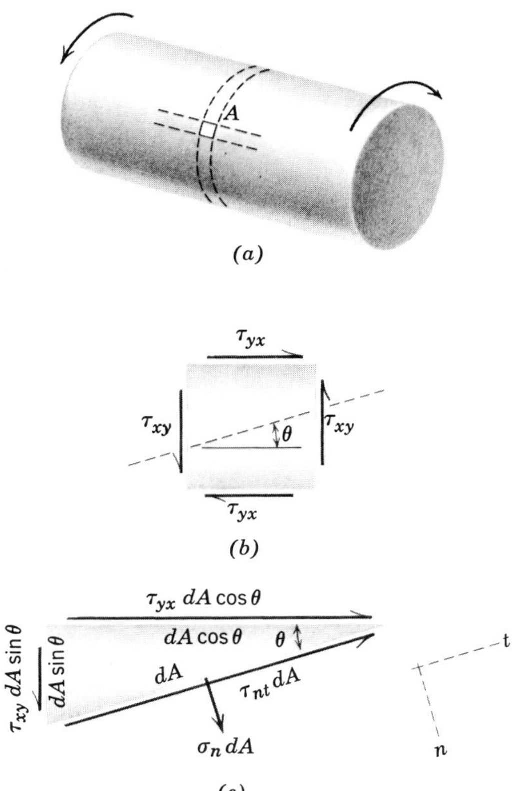


Fig. 4–7

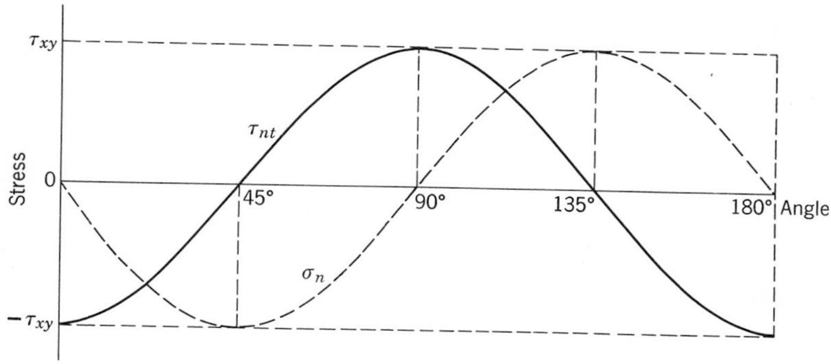


Fig. 4–8

significant in a particular problem. Compare, for example, the failures shown in Fig. 4–9. In Fig. 4–9*a*, the steel truck rear axle split longitudinally. One would also expect this type of failure to occur in a shaft of wood with the grain running longitudinally. In Fig. 4–9*b*, the compressive stress caused the thin-walled aluminum alloy tube to buckle along one 45° plane, while the tensile stress caused tearing on the other 45° plane. Buckling of thin-walled tubes (and other shapes) subjected to torsional loading is a matter of paramount concern to the designer. In Fig. 4–9*c*, the tensile stresses caused the gray cast iron to fail in tension—typical of any brittle material subjected to torsion; and in Fig. 4–9*d*, the low-carbon steel failed in shear on a plane which is almost transverse—a typical failure for ductile material. The reason the fracture in Fig. 4–9*d* did not occur on a transverse plane is that under the large plastic twisting deformation before rupture (note the spiral lines indicating elements which were originally parallel to the axis of the bar), longitudinal elements were subjected to axial tensile loading because the grips of the testing machine would not permit the bar to shorten as the elements were twisted into spirals. This axial tensile stress (not shown in Fig. 4–7) changes the plane of maximum shearing stress from a transverse to an oblique plane (resulting in a warped surface of rupture).[4]

For problems involving stresses on a given oblique plane, the methods of Art. 1–5, 1–6, and 1–7 are all available and there is no par-

[4] The tensile stress is not entirely due to the grips because the plastic deformation of the outer elements of the bar is considerably greater than that of the inner elements. This results in a spiral tensile stress in the outer elements and a similar compressive stress in the inner elements.

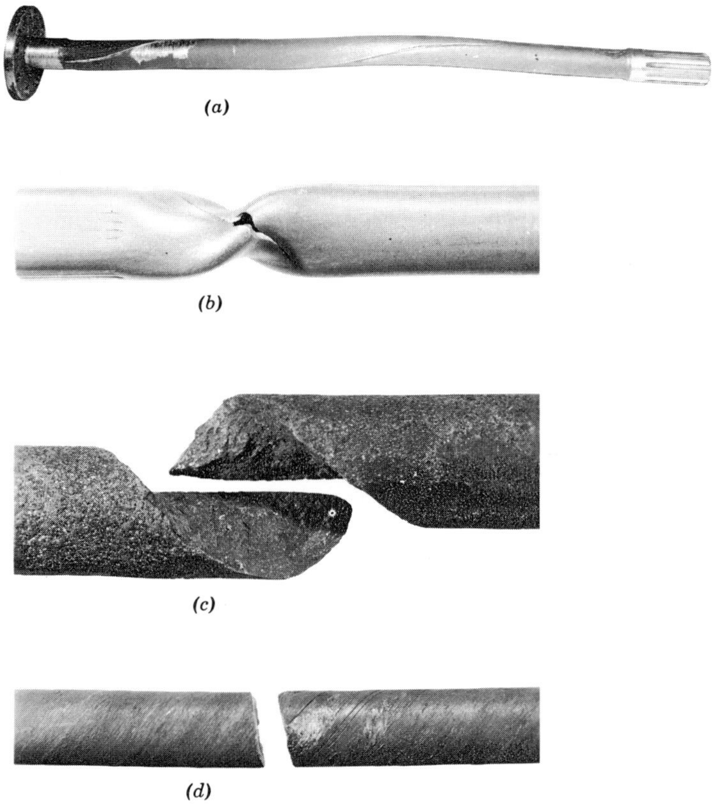

(a)

(b)

(c)

(d)

Fig. 4–9

ticular advantage for one method over the others. None of these methods is required for the principal stresses as they are equal in magnitude to the maximum shearing stress as given by the torsion formula.

## PROBLEMS

**4–35.*** For the shaft shown in Fig. P4–35:

(*a*) Determine the stresses which occur at point $A$ (at the surface of the shaft) on the plane $B$–$B$ (which is normal to the surface of the shaft at point $A$, and makes an angle of 40° with the axis of the shaft). Show these stresses on an enlarged sketch of the differential area representing $A$.

(*b*) Determine and show on a sketch of area $A$ the maximum normal stresses occurring at the point.

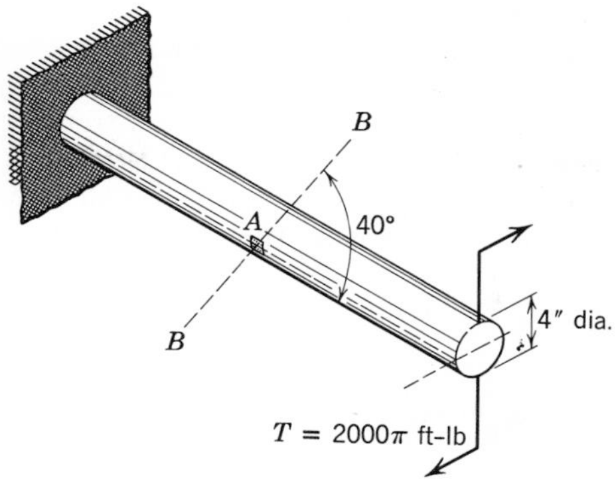


Fig. P4–35

**4–36.*** Determine the maximum torque that can be resisted by a hollow circular shaft having an inside diameter of 1 in. and an outside diameter of 2 in. without exceeding a normal stress of 10,000 psi $T$ or a shearing stress of 11,000 psi.

**4–37.** (*a*) A solid circular steel shaft 42 in. long is required to twist through 0.03 rad. If the shearing stress is not to exceed 9000 psi, determine the maximum permissible diameter for the shaft.

(*b*) For the shaft of part *a* determine the normal stress on plane *a–a* which is normal to the surface of the shaft at point *A* and has a slope of 3 to 4 with the axis of the shaft. See Fig. P4–37.

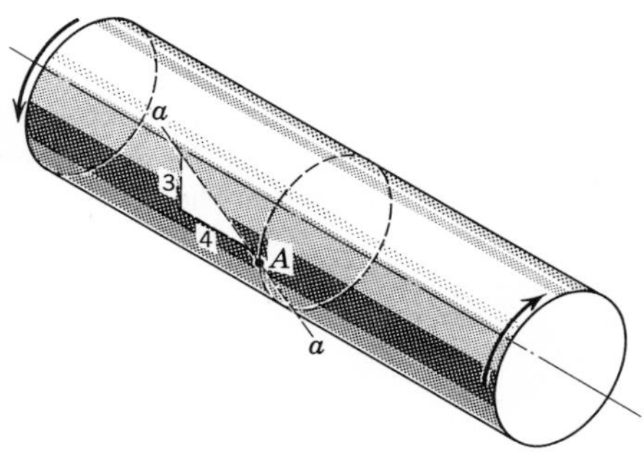


Fig. P4–37

**4–38.*** A hollow circular steel shaft, 3 in. outside diameter, 2 in. inside diameter, and 3 ft long is held at one end and subjected to a torque at the other end as shown in Fig. P4–38. Determine:

(*a*) The shearing stress at the inner surface on a transverse plane.

(*b*) The distance point *B* moves as the torque is applied.

(*c*) With the outside diameter unchanged, calculate the maximum inside diameter for an allowable compressive stress of 9000 psi.

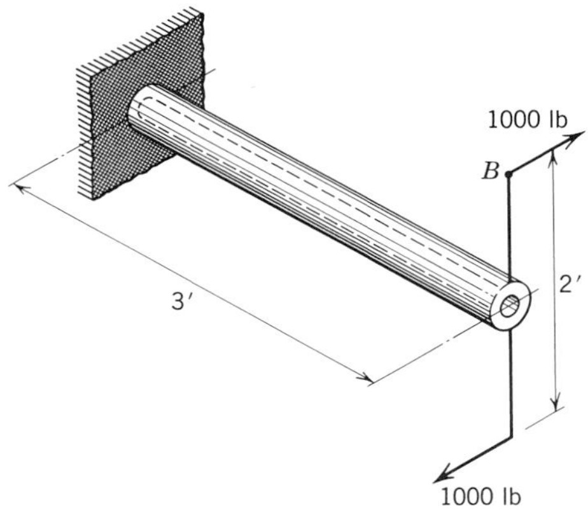


**Fig. P4–38**

**4–39.** A cylindrical tube is fabricated by butt welding $\frac{1}{4}$-in. steel plate along a spiral seam as shown in Fig. P4–39. For an allowable compressive stress in the shell of 12 ksi, determine:

(*a*) The maximum permissible torque, *T*, to which the tube may be subjected (note that the cross section may be treated as a thin annular area for computation of $J/c$).

(*b*) The factor of safety with respect to fracture for the weld if the ultimate strengths of the weld metal are 36 ksi shear and 60 ksi tension. The thickness of a butt weld is the same as that of the plate.

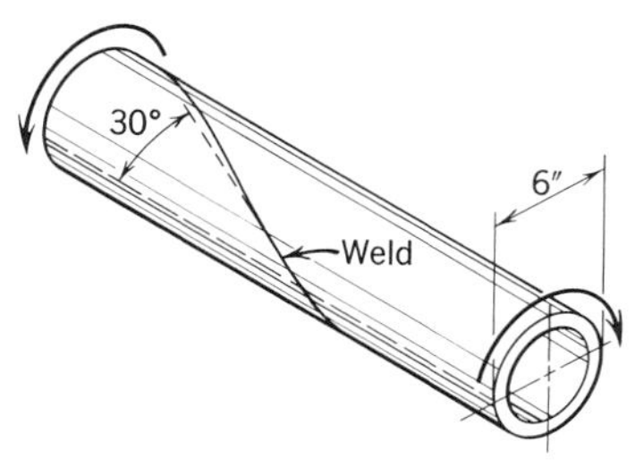


**Fig. P4–39**

**4–40.*** When the two torques of Fig. P4–40 are applied to the steel shaft as shown, point $A$ moves $0.54/\pi$ in. in the direction indicated by $T$. Determine:

(*a*) The torque $T$.

(*b*) The maximum tensile stress in the shaft assembly.

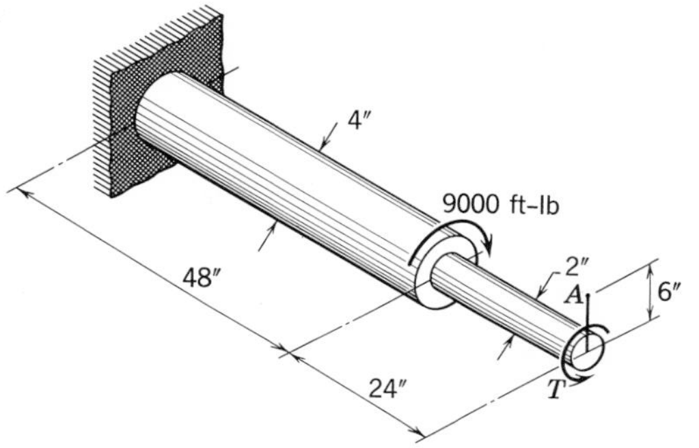


Fig. P4–40

**4–4. Combined Axial and Torsional Loads.** There are numerous situations where a shaft or other machine member is subjected to both an axial and a torsional load. Examples are the drill rod on a well or other drilling machine and the propeller shaft on a ship (bending stresses, to be studied later, will also probably be present in this case).

For stresses below the proportional limit of the material, the method of superposition (see Art. 2–5) applies. Thus each stress can be calculated separately and all stresses on a given plane superimposed. Once the stresses on a pair of mutually perpendicular planes through a point have been determined, the procedures of Art. 1–6 and 1–7 are available to calculate the principal stresses and the maximum shearing stress at the point. For the combination of axial and torsional loads, the axial stress is the same at all points in the body while the torsional stress is maximum at points on the outer surface of the shaft. Therefore, the investigation for critical stresses would normally be made for points on the surface of the shaft.

The following example illustrates the procedure for combined torsional and axial loads.

**Example 4–2:** A hollow circular shaft of outer diameter 4 in. and inner diameter 2 in. is loaded as shown in Fig. 4–10*a*.

(*a*) Determine the principal stresses and the maximum shearing stress at the point (or points) where the stress situation is most severe.

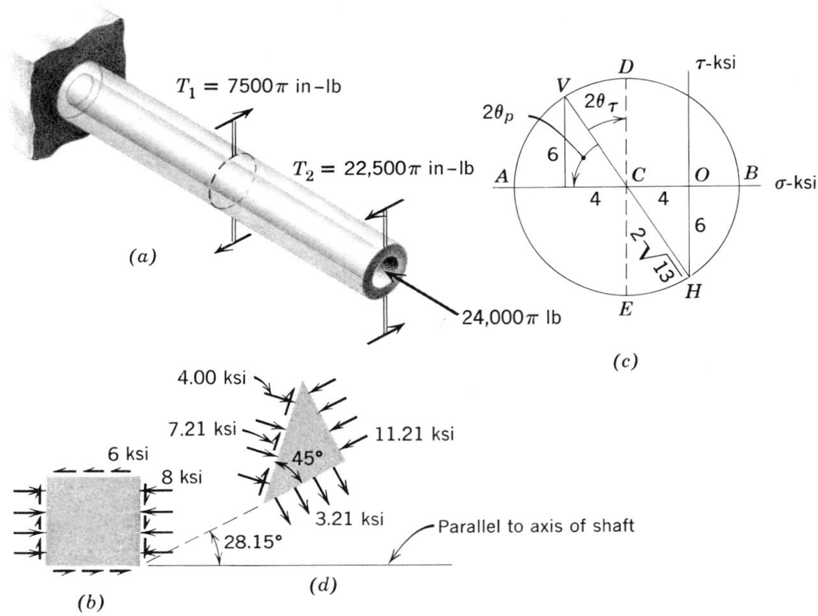


Fig. 4–10

(*b*) On a sketch show the approximate directions of these stresses.

*Solution:* The axial stress is constant throughout the shaft and is equal to

$$\sigma = \frac{P}{A} = \frac{24,000\pi}{\pi(2^2 - 1^2)} = 8000 \text{ psi } C.$$

The torsional load is highest to the right of the torque $T_1$ and the torsional stress is maximum on the outer surface of the shaft; therefore, the maximum torsional shearing stress is

$$\tau = \frac{Tc}{J} = \frac{22,500\pi(2)}{\pi(2^4 - 1^4)/2} = 6000 \text{ psi}.$$

These stresses are shown acting on mutually perpendicular planes through a point on the surface of the right portion of the shaft in Fig. 4–10*b*. Mohr's circle for the stresses at any point on the surface of the right portion of the shaft is shown in Fig. 4–10*c*.

(*a*) The principal stresses $OA$ and $OB$ from Mohr's circle are

$$\sigma_{p1} = OA = -4.00 - 2\sqrt{13} = -11.21 = \underline{11.21 \text{ ksi } C} \qquad \text{Ans.}$$

and

$$\sigma_{p2} = OB = -4.00 + 2\sqrt{13} = \underline{3.21 \text{ ksi } T.} \qquad \text{Ans.}$$

The maximum shearing stress $CD$ or $CE$ from Mohr's circle is

$$\tau_{max} = CD = 2\sqrt{13} = \underline{7.21 \text{ ksi.}} \qquad \text{Ans}$$

(*b*) From the Mohr's circle,

$$\tan 2\theta_p = 6/4 = 1.5,$$

which gives

$$2\theta_p = 56.30°⤸$$

and

$$\theta_p = 28.15°⤸.$$

The stresses are shown in their proper directions in Fig. 4–10*d*.

## PROBLEMS

**4–41.*** The solid circular shaft of Fig. P4–41*a* has a radius of 1 in. It is subjected to an axial tensile load of 10,000$\pi$ lb and a torque of 6000$\pi$ in-lb, as shown.

(*a*) Determine the principal stresses, the maximum shearing stress, and the planes on which these stresses act at point $A$ on the surface of the shaft.

(*b*) Determine the stresses on plane $B$–$B$, oriented as indicated in Fig. P4–41*b* at point $A$. Draw a complete stress block with answers shown on it.

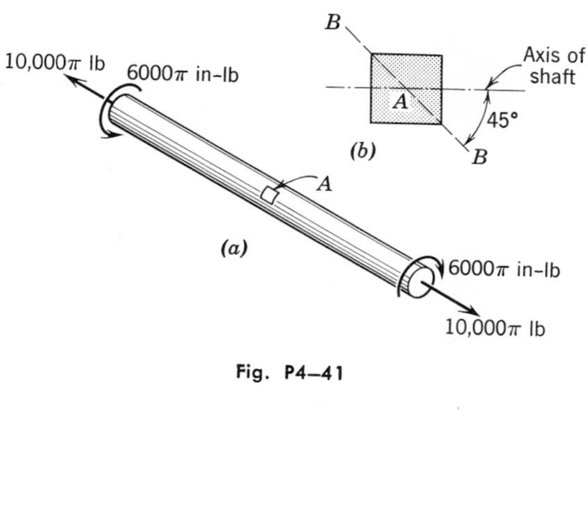


Fig. P4–41

**4–42.*** A shaft 4 in. in diameter is subjected to a torque of $4000\pi$ in-lb and an axial tension of $10{,}000\pi$ lb, see Fig. P4–42. Determine the principal stresses and the maximum shearing stress at point $A$, and show their magnitudes and approximate directions on a sketch.

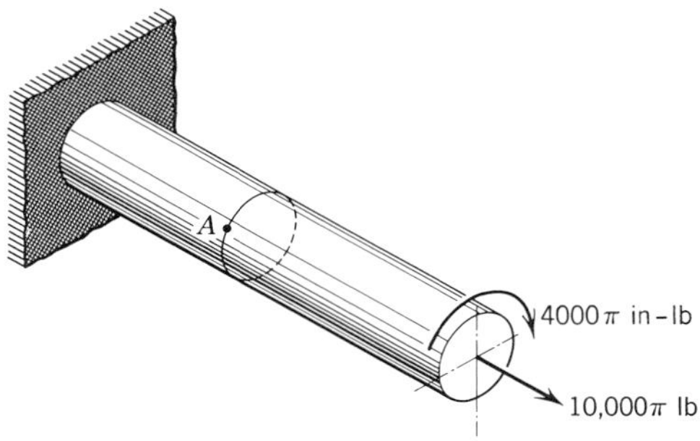


Fig. P4–42

**4–43.** ($a$) A solid circular shaft 4 in. in diameter is made of brass and steel segments and is loaded as shown in Fig. P4–43. Determine, and show on a sketch, the principal stresses and the maximum shearing stress at a point where the stresses are greatest. Use $6(10^6)$ psi for the modulus of rigidity of the brass.

($b$) Determine the angle of twist of section $C$ with respect to section $A$.

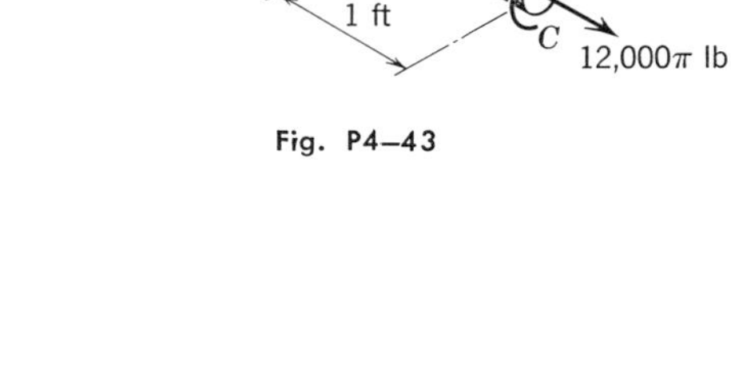


Fig. P4–43

**4–44.*** A 6-in. diameter shaft is loaded as shown in Fig. P4–44.

(*a*) At what point (or points) are the stresses the highest?

(*b*) Determine and show on a sketch the principal stresses and the maximum shearing stress at point *A*.

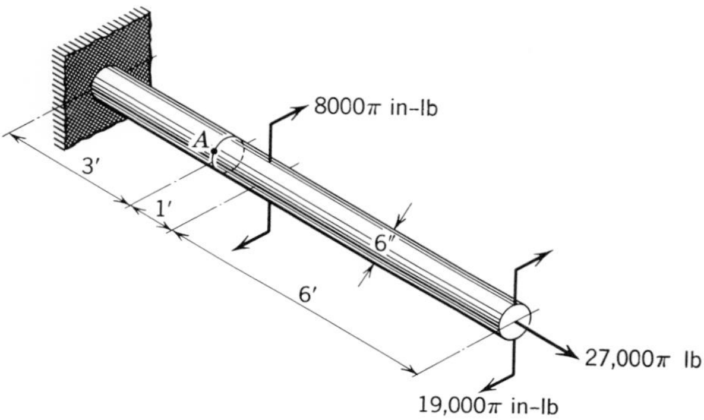


**Fig. P4–44**

**4–45.** A short length of 6-in. diameter steel shaft is subjected to a torque of $9000\pi$ ft-lb in addition to an axial tensile load, *P*, as indicated in Fig. P4–45. Determine the maximum allowable value of the axial tensile load if the shearing stress is not to exceed 10,000 psi and the tensile stress is not to exceed 15,000 psi.

**Fig. P4–45**

**4–46.*** A solid circular bronze shaft with diameters as shown in Fig. P4–46 is loaded and supported as indicated. The allowable shearing stress is 8 ksi and the allowable tensile stress is 12 ksi. With $T_2$ equal to $P$ in-kips and $T_1$ equal to $P/2$ in-kips, determine the maximum allowable value of $P$.

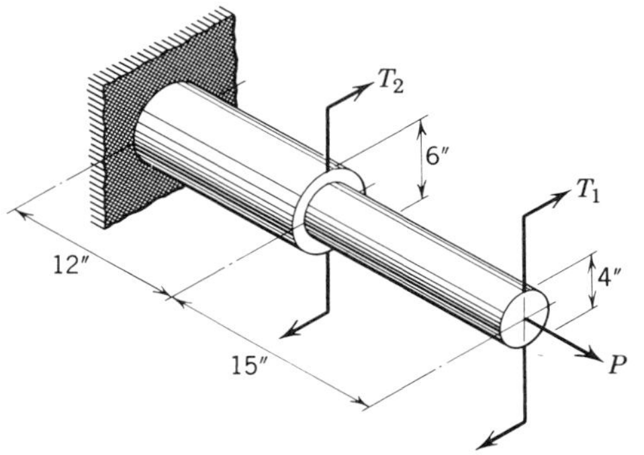


Fig. P4–46

**4–47.** For the shaft and allowable stresses of Prob. 4–46, with $P$ equal to 50.0 kips and $T_1$ equal to 80.0 in-kips, determine the maximum allowable value of $T_2$. *Note:* the stresses in the right portion of the shaft are less than the allowable values.

**4–48.*** A circular steel shaft with the left portion solid and the right portion hollow is loaded as shown in Fig. P4–48. The allowable shearing and normal stresses are 10.00 ksi and 18.00 ksi respectively. Determine the maximum allowable axial load $P$.

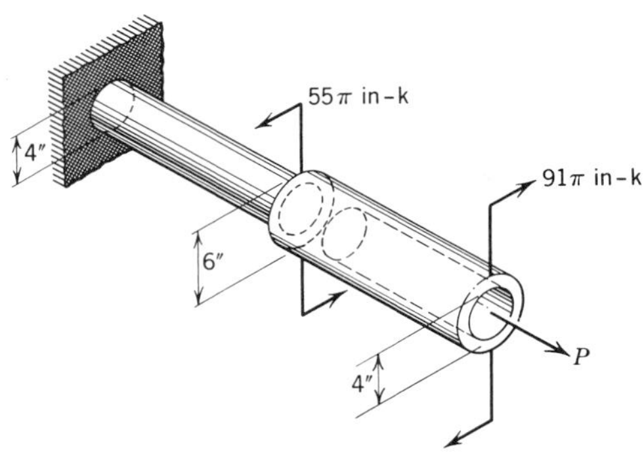


Fig. P4–48

**4–5. Power Transmission.** Almost anyone will recognize that one of the most common uses for the circular shaft is the transmission of power; therefore, no discussion of torsion would be adequate without including this topic. Power is the time rate of doing work, and the basic relationship for work done by a constant torque is $W_k = T\phi$ where $W_k$ is work and $\phi$ is the angular displacement of the shaft in radians. The time derivative of this expression gives

$$\frac{dW_k}{dt} = T\frac{d\phi}{dt} = T\omega,$$

where $dW_k/dt$ is power in foot-pounds per minute (or similar units), $T$ is a constant torque in foot-pounds, and $\omega$ is the angular velocity (of the shaft) in radians per minute. All units, of course, may be changed to any other consistent set of units. Since the angular velocity is usually given in revolutions per minute (rpm), the conversion of revolutions to radians will often be necessary. Also, power is usually given in units of horsepower, and the relation, one horsepower equals 33,000 ft-lb per min, will be found useful. An example of shaft design follows:

**Example 4–3:** A Diesel engine for a small commercial boat is to operate at 200 rpm and deliver 800 hp through a gear box with a ratio of 1 to 4 (increase) to the propeller shaft. Both the shaft from engine to gear box, and the propeller shaft are to be solid and made of heat treated alloy steel. Determine the minimum permissible diameters for the two shafts if the allowable stress is 20 ksi and the angle of twist in a 10-ft length of the propeller shaft is not to exceed 4°. Neglect power loss in the gear box and assume (incorrectly because of thrust stresses) that the propeller shaft is subjected to pure torsion.

*Solution:* The first step is the determination of the torques to which the shafts are to be subjected. By means of the expression, Power = $T\omega$, the torques are obtained as follows:

$$800(33{,}000) = T_1(200)(2\pi),$$

from which

$$T_1 = 66{,}000/\pi \text{ ft-lb},$$

which is the torque at the crank shaft of the engine. As the propeller shaft speed is four times that of the crank shaft and power loss in the gear box is to be neglected, the torque on the propeller shaft is one-fourth that on the crank shaft and is equal to $66{,}000/(4\pi)$ ft-lb. Using the torsion formula, the shaft sizes necessary to satisfy the stress specification are determined. For the main shaft,

$$\frac{J}{c} = \frac{T}{\tau} = \frac{66{,}000(12)}{\pi(20)(10^3)} = \frac{(\pi/2)c_1^4}{c_1},$$

$$c_1^3 = 8.02, \quad \text{and} \quad c_1 = 2.002 \text{ in.}$$

or the shaft from engine to gear box should be

$$\underline{4.00 \text{ in. in dia.}} \qquad \text{Ans.}$$

The torque on the propeller shaft is one-fourth that on the main shaft, and this is the only change in the expression for $c_1{}^3$; therefore,

$$c_2{}^3 = 8.02/4, \quad \text{and} \quad c_2 = 1.261 \text{ in.}$$

The size of propeller shaft necessary to satisfy the distortion specification will be determined as follows:

$$\gamma = \frac{c\theta}{L} = \frac{\tau}{G} = \frac{Tc}{JG},$$

from which

$$\frac{4(\pi/180)}{10(12)} = \frac{66{,}000(12)/(4\pi)}{(\pi c_3{}^4/2)(12)(10^6)},$$

$$c_3{}^4 = 5.75, \quad \text{and} \quad c_3 = 1.548 > 1.261;$$

therefore, the propeller shaft must be

$$\underline{3.10 \text{ in. in dia.}} \qquad \text{Ans.}$$

## PROBLEMS

**4–49.*** A 6-ft long hollow steel shaft has an outside diameter of 4 in. and an inside diameter of 2 in. The maximum shearing stress in the shaft is 8000 psi and the angular velocity is 120 rpm. Determine:

(*a*) The horsepower being transmitted by the shaft.

(*b*) The magnitude of the angle of twist in the shaft.

**4–50.*** A solid circular steel shaft is 3 in. in diameter and 4 ft long. The maximum permissible angle of twist is 0.024 rad. The working shearing stress is 10,000 psi. Determine the maximum horsepower that this shaft can deliver:

(*a*) When rotating at 225 rpm.

(*b*) When rotating at 180 rpm.

**4–51.*** (*a*) Determine the minimum permissible diameter of the solid circular structural steel (see App. A) shaft required to transmit 200 hp at a speed of 300 rpm, if the allowable shearing stress is 10,000 psi and the allowable angle of twist is 0.014 rad. The shaft is 2 ft long.

(*b*) If a shaft with the diameter determined in part *a* were used, at what speed should this horsepower be delivered if the stress is not to exceed 8000 psi?

**4–52.** A solid circular steel shaft 10 ft long transmits $30\pi^2$ hp at 180 rpm. The allowable shearing stress in the shaft is 8000 psi and the maximum allowable angle of twist is 0.040 rad. Determine the minimum allowable diameter of the shaft.

**Example 4–4:** A shaft of structural steel 4 in. in diameter is subjected to a pure torque of $7\pi$ ft-kips. Assume the steel is elastoplastic and that the yield point in shear is one-half the yield point in tension. Determine the maximum shearing stress in the shaft and the magnitude of the angle of twist in a 10-ft length.

*Solution:* Assume elastic action first; then

$$\tau = \frac{Tc}{J} = \frac{7\pi(12)(2)}{(\pi/2)(2^4)} = 21 \text{ ksi} > \frac{36}{2}.$$

Therefore, plastic action takes place, and the maximum stress is 18 ksi, unless strain hardening occurs. To check this, first determine how much of the cross section is behaving elastically by going back to the fundamental concept of resisting torque being the integral of the resisting torques of differential annular areas; thus

$$T = \int^{\text{Area}} \tau(2\pi\rho \, d\rho)(\rho).$$

For the part undergoing elastic action (of radius $r$),

$$\tau = \frac{\tau_r}{r}(\rho) = \frac{18}{r}(\rho),$$

and, for the rest, $\tau$ is constant at 18 ksi. Then, the resisting torque is given by

$$T = \frac{18}{r}(2\pi)\int_0^r \rho^3 \, d\rho + 18(2\pi)\int_r^2 \rho^2 \, d\rho,$$

$$7\pi(12) = 2\pi \frac{18}{r}\frac{r^4}{4} + 2\pi(18)\frac{1}{3}(2^3 - r^3),$$

from which $r^3 = 4$, and $r = 1.587$ which is the radius of the area undergoing elastic action.

From Hooke's law,

$$\gamma_r = \frac{\tau}{G} = \frac{18}{12{,}000} = 0.0015,$$

and, with Eq. (4–2),

$$\gamma_c = \frac{c\theta}{L} \quad \text{or} \quad \gamma_r = \frac{r\theta}{L}$$

from which

$$\gamma_c = \gamma_r \frac{c}{r} = 0.0015 \frac{2}{1.587} = 0.001{,}890.$$

This small value of strain indicates that strain hardening does not occur, for, as was stated in Art. 2–1, the plastic strain at the yield point for structural steel may be on the order of fifteen times the elastic strain before strain hardening occurs; therefore,

$$\tau_{max} = \underline{18 \text{ ksi.}} \qquad \text{Ans.}$$

The angle of twist may be obtaind from Eq. (4–2); thus,

$$\gamma = \frac{\tau}{G} = \frac{r\theta}{L} = \frac{18}{12(10^3)} = \frac{1.587(\theta)}{10(12)},$$

from which $\theta = \underline{0.1134 \text{ rad.}}$ Ans.

## PROBLEMS

**4–56.*** An aluminum alloy tube is to be used to transmit a torque in a control mechanism. The tube has an outside diameter of 1.375 in. and a wall thickness of $\frac{1}{16}$ in. Because of the tendency of thin sections to buckle, the maximum compressive stress is specified as 7500 psi. Determine:

(*a*) The maximum allowable torque which may be applied.

(*b*) The factor of safety with respect to failure by buckling for a torsional modulus of rupture of 32,000 psi.

**4–57.*** An elastoplastic solid circular steel shaft has a shearing yield point of 22 ksi. Determine:

(*a*) The shearing strain at the surface of the shaft when the shearing stress at three eighths of the radius from the center of the shaft reaches the yield point.

(*b*) The ratio of the torque carried by the shaft of part *a* to the torque carried by an identical shaft when the shearing stress at the outer surface reaches the yield point.

**4–58.*** A wrought aluminum alloy 6061-T6 tube has an outside diameter of 2.00 in. and an inside diameter of 1.84 in. The torsional modulus of rupture for this tubing is 54 per cent of the ultimate tensile strength. (See App. A.) Determine the approximate torque required to collapse the tube.

**4–59.** An elastoplastic hollow steel shaft has an outside diameter of 4 in. and an inside diameter of 2 in. The shearing yield point is 20 ksi. Determine:

(*a*) The maximum allowable torque that the shaft can transmit with an allowable shearing stress of 18 ksi.

(*b*) The percentage increase in the torque of part *a* if the shearing stress at the inside surface is allowed to reach the yield point.

**4–60.*** A 4-in. diameter solid circular shaft is made of a hypothetical material with a shearing stress-strain diagram in Fig. P4–60. It is proposed that the diagram be approximated by the two straight lines *OA* and *AB* for stresses less than 36 ksi. Determine:

(*a*) The torque which will be developed in the shaft when the maximum shearing stress is 32 ksi.

(*b*) The magnitude of the angle of twist per foot of shaft with this stress.

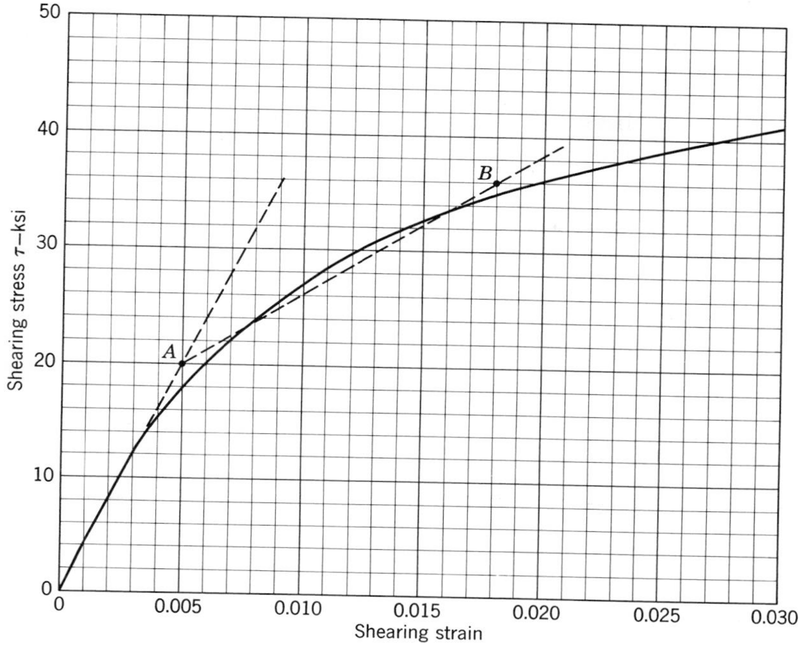


Fig. P4–60

**4–61.** Solve Prob. 4–60 for a hollow shaft with an inside diameter of 2 in. and an outside diameter of 4 in.

**4–7. Statically Indeterminate Members.** In all problems discussed in the preceding articles, the equations of equilibrium sufficed to determine the resisting torque at any section. Occasionally, torsionally loaded members are so constructed and loaded that the number of independent equilibrium equations is less than the number of unknowns, and the member is statically indeterminate. When this occurs, it becomes necessary to write the requisite number of distortion equations so that the total number of equations agrees with the number of unknowns to be determined.

The distortion equations will involve the angles of twist; and a simplified distortion diagram will often be of assistance in obtaining the correct equations. The following examples will serve to illustrate the application of the principles.

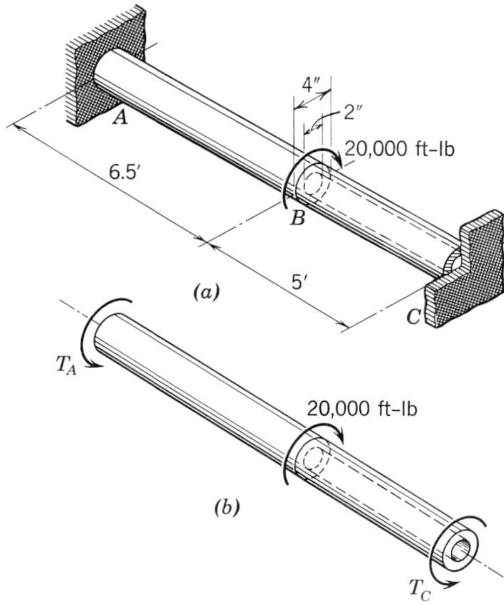


Fig. 4–12

**Example 4–5:** The circular shaft $AC$ in Fig. 4–12*a* is fixed to rigid walls at $A$ and $C$. The section $AB$ is solid annealed bronze and the hollow section $BC$ is made of aluminum alloy 2024-T4. There is no stress in the shaft before the 20,000-ft-lb torque, $T$, is applied. Determine the maximum shearing stress in each shaft after $T$ is applied.

*Solution:* Figure 4–12*b* is a free-body diagram of the shaft. The torques $T_A$ and $T_C$, produced by the supports, are unknown, and the summation of moments about the axis of the shaft gives

$$T_A + T_C = 20{,}000(12). \tag{a}$$

This is the only independent equation of equilibrium; thus, the problem is statically indeterminate. A second equation can be obtained from the deformation of the shaft, since each section must undergo the same angle of twist; thus,

$$\theta_{AB} = \theta_{BC}. \tag{b}$$

Equations (a) and (b) can both be expressed in terms of the shearing stresses in the two sections. When the torsion formula is applied to Eq. (a) it becomes

$$\frac{\tau_{AB}\pi(2^4)/2}{2} + \frac{\tau_{BC}\pi(2^4 - 1^4)/2}{2} = 240{,}000. \tag{c}$$

Since the angle of twist is

$$\theta = \frac{\gamma L}{c} = \frac{\tau L}{Gc},$$

Eq. (b) becomes

$$\frac{\tau_{AB}(6.5)12}{6.5(10^6)2} = \frac{\tau_{BC}(5)12}{4(10^6)2}$$

from which

$$\tau_{AB} = 5\tau_{BC}/4. \tag{d}$$

Equations (c) and (d) can be solved simultaneously to give

$$\tau_{AB} = \underline{10{,}900 \text{ psi},} \qquad \text{Ans.}$$

$$\tau_{BC} = \underline{8{,}730 \text{ psi}.} \qquad \text{Ans.}$$

Note: These maximum stresses, which occur at the surfaces of the shafts, do not exceed the proportional limits of the materials, therefore, only elastic behavior is involved and the results are valid.

**Example 4–6:** Segment *CD* of Fig. 4–13*a* is a solid shaft of cold-rolled bronze, and segment *EF* is a hollow shaft of aluminum alloy 2024-T4 with a core of low-carbon steel. The ends *C* and *F* are to be considered fixed to rigid walls and the steel core of *EF* is connected to the flange at *E* so that the aluminum and steel act together as a unit. The two flanges *D* and *E* are bolted together and the bolt clearance permits flange *D* to rotate through 0.03 rad before *EF* will carry any of the load. Determine the maximum shearing stress in any shaft material when the torque of 20 ft-kips is applied. See App. A for mechanical properties.

*Solution:* Figure 4–13*b* is a free-body diagram of the assembly. It is necessary to place two unknown torques $T_A$ and $T_S$ on the right end because the torques and stresses will need to be related by the torsion formula which is limited to cross sections of homogeneous material. Summing moments with respect to the axis of the shaft gives the following equation:

$$20(12) = T_B + T_A + T_S. \tag{e}$$

This is the only independent equilibrium equation for this problem and there are three unknown torques; hence, two distortion equations need to be written. A three-dimensional distortion diagram is not easy to draw and the cross-sectional drawing of Fig. 4–13*c* is adequate. The angles indicated represent the twist in segment *CD* at *D* and in segment *EF* at *E* (the coupling is considered rigid). The fact that segment *EF* is non-homogeneous does not invalidate the assumption of the plane section remaining plane and the diameter remaining straight; the strains are still proportional to the distance from the shaft axis. However, the stresses are not proportional to the radii throughout the entire cross section because *G* is not single valued.

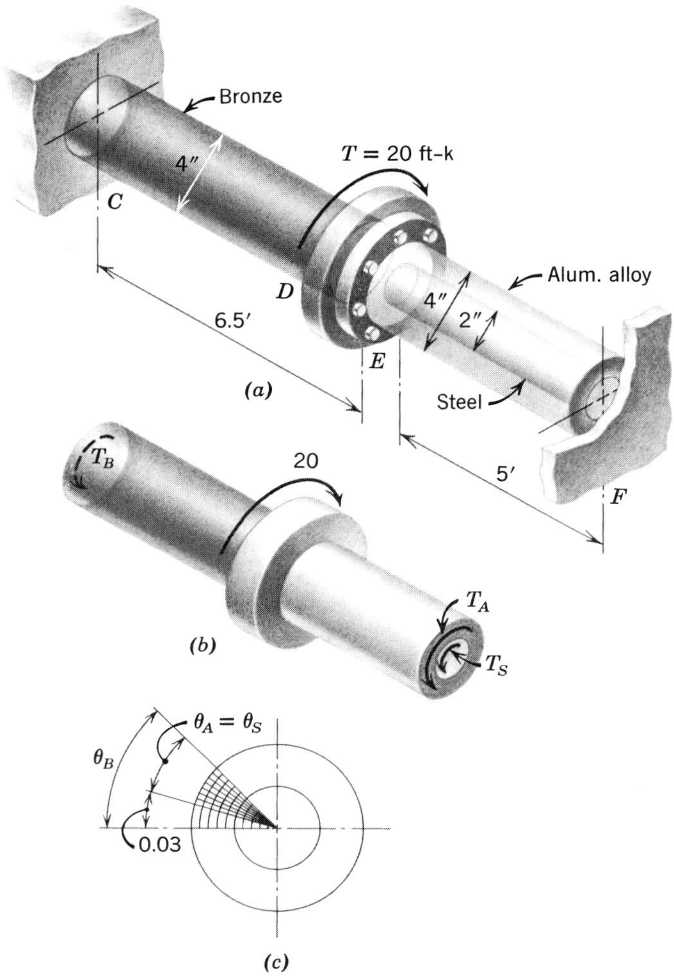


Fig. 4–13

From Fig. 4–13*c*, the distortion equations are

$$\theta_B = \theta_A + 0.03 \tag{f}$$

and

$$\theta_A = \theta_S. \tag{g}$$

The three equations (e), (f), and (g) can be written in terms of the same three unknowns (torque, angle, or stress) and solved simultaneously. Since the maximum stress is required, Eq. (f) and (g) will be written in terms of the maximum stress in each material; thus,

$$\theta = \frac{\gamma L}{c} = \frac{\tau L}{Gc};$$

(f) $$\frac{\tau_B(6.5)(12)}{6.5(10^3)(2)} = \frac{\tau_A(5)(12)}{4(10^3)(2)} + 0.03,$$

from which $$\tau_B = (5/4)\tau_A + 5, \qquad (\text{therefore } \tau_B > \tau_A) \qquad \text{(h)}$$

and (g) $$\frac{\tau_A(5)(12)}{4(10^3)(2)} = \frac{\tau_S(5)(12)}{11.6(10^3)(1)},$$

from which $$\tau_S = 1.45\tau_A = 1.45(4\tau_B/5 - 4). \qquad \text{(i)}$$

Writing Eq. (e) in terms of maximum stresses in each material and using the torsion formula, gives

$$20(12) = \frac{\tau_B(\pi/2)(2^4)}{2} + \frac{\tau_A(\pi/2)(2^4 - 1^4)}{2} + \frac{\tau_S(\pi/2)(1^4)}{1}$$

or $$960/\pi = 16\tau_B + 15\tau_A + 2\tau_S. \qquad \text{(j)}$$

Solving Eq. (h), (i), and (j) simultaneously gives the results

$$\tau_B = 12.44 \text{ ksi},$$

$$\tau_A = 5.95 \text{ ksi},$$

and $$\tau_S = 8.63 \text{ ksi}.$$

The maximum shearing stress is at the surface of the bronze and is

$$\underline{12.44 \text{ ksi.}}$$ Ans.

## PROBLEMS

**4–62.*** For the steel shaft of Fig. P4–62, determine the maximum shearing stress and the angle of rotation of the section where the given torque is applied.

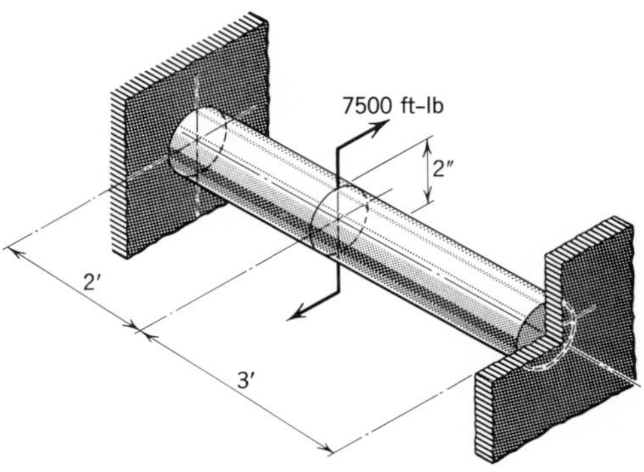


Fig. P4–62

**4–63.*** The steel shaft of Fig. P4–63 is rigidly secured at the ends. The right 10 ft of the shaft is hollow, having an inside diameter of 2 in. Determine the maximum shearing stress in the shaft.

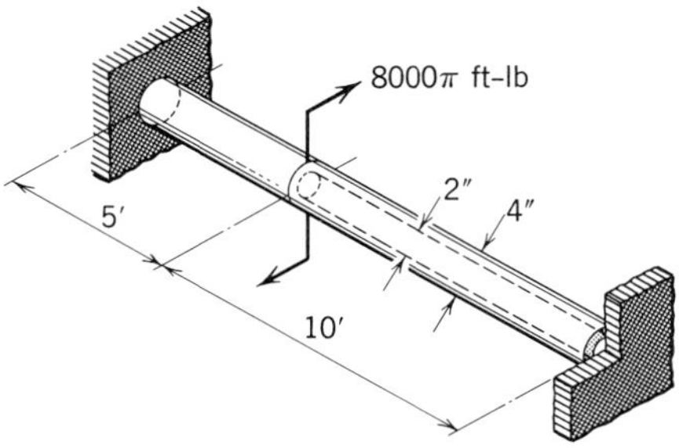


Fig. P4–63

**4–64.*** Shown in Fig. P4–64 is a hollow brass shaft ($G$ = 5000 ksi) with a 6-in. outside diameter and a 4-in. inside diameter. Inside and concentric with this brass shaft is a solid steel shaft with a 2-in. diameter. The two shafts are rigidly connected to a plate at the right end and to the wall at the left end. Determine the rotation of the plate $AB$ due to the couple shown.

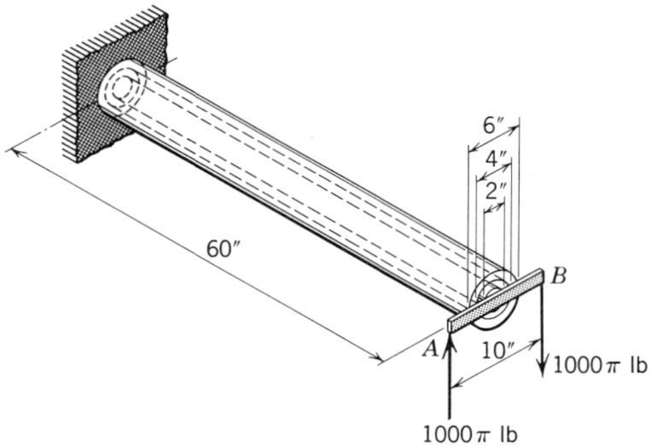


Fig. P4–64

**4–65.*** A 3-in. diameter by 30-in. steel shaft has hubs at the ends to which are fastened securely the ends of a brass tube, 6-in. outside diameter by $\frac{1}{2}$ in. thick as shown in Fig. P4–65. This composite shaft has a torque of $11{,}800\pi$ ft-lb applied at the ends. Determine the magnitude of the angle of twist in the 30-in. length and the maximum shearing stresses in the steel and the brass. The modulus of rigidity for the brass is 5000 ksi.

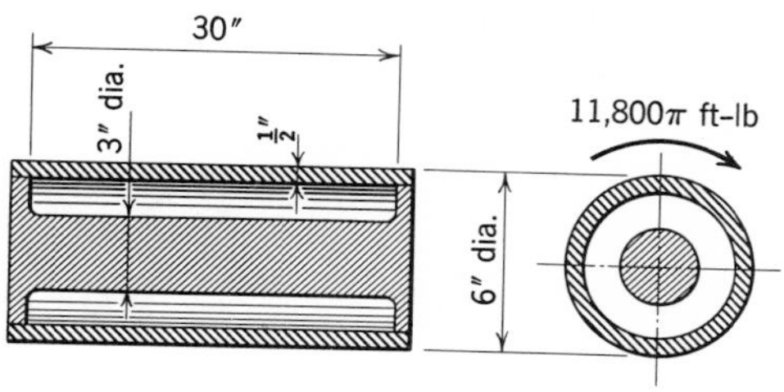


Fig. P4–65

**4–66.** Two 3.6-in. diameter solid circular shafts are rigidly connected and supported as shown in Fig. P4–66. An unknown torque $T$ is applied at the junction of the two shafts as indicated. Assume the shearing strengths of these materials to be 0.6 of their strengths in tension, see App. A. Determine the maximum allowable torque based on a factor of safety of 2 with respect to failure by slip.

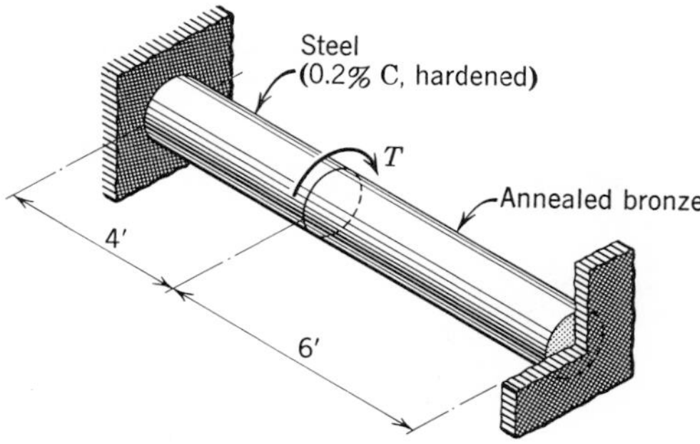


Fig. P4–66

**4–67.*** The aluminum alloy ($G = 4000$ ksi) shaft of Fig. P4–67 is rigidly fixed to the wall at $C$, but the flange at $A$ allows the left end of the shaft to rotate 0.012 rad before the bolts provide rigid support. Determine the maximum torque that can be applied at section $B$ if the shearing stress is not to exceed 7 ksi in either section of the shaft.

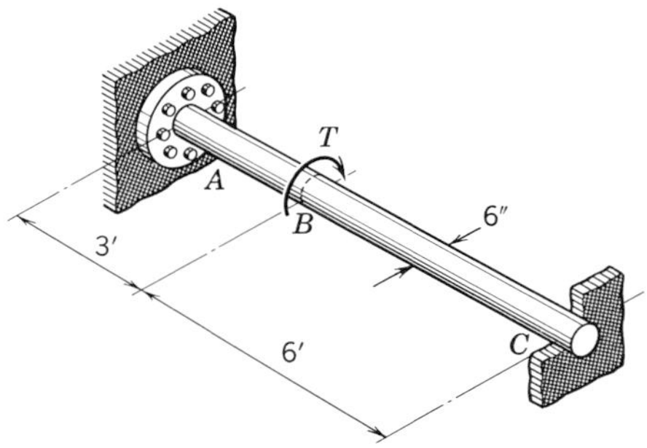


Fig. P4–67

**4–68.** A disk and two circular shafts are connected and supported between the two rigid walls of Fig. P4–68. What maximum permissible torque may be applied to the disk, which is securely attached to $A$ but rotates through 0.005 rad before engaging $B$?

| Shaft | Material | Diameter, in. | Length, in. | Modulus of Rigidity, ksi | Allowable Shearing Stress, ksi |
|---|---|---|---|---|---|
| $A$ | Bronze | 2 | 20 | 6,000 | 9 |
| $B$ | Steel | 3 | 30 | 12,000 | 16 |

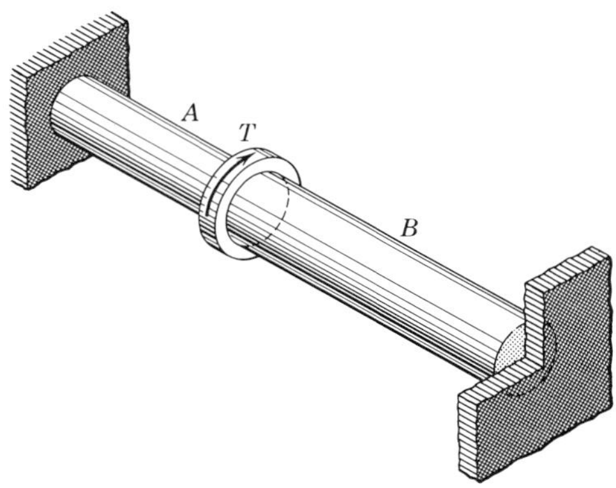


Fig. P4–68

**4–69.** A torque of 6000 ft-lb is applied to the steel shaft of Fig. P4–65 without the brass shell. The brass shell is then slipped in place and attached to the steel, after which the 6000-ft-lb torque is released. Determine the torque in the brass shell after release of the original torque.

**4–70.*** The shaft of Fig. P4–70 is made up of a 6-ft hollow steel section, *AB*, and a 4-ft solid section of aluminum alloy 2024-T4, *CD* (see App. A). The torque *T* of $130\pi$ in-kips is applied initially only to the steel section *AB*. Section *CD* is then connected and the torque, *T*, is released. When the torque is released, the connection slips 0.012 rad before the aluminum section *CD* takes any load. Determine the final maximum shearing stress in the aluminum alloy.

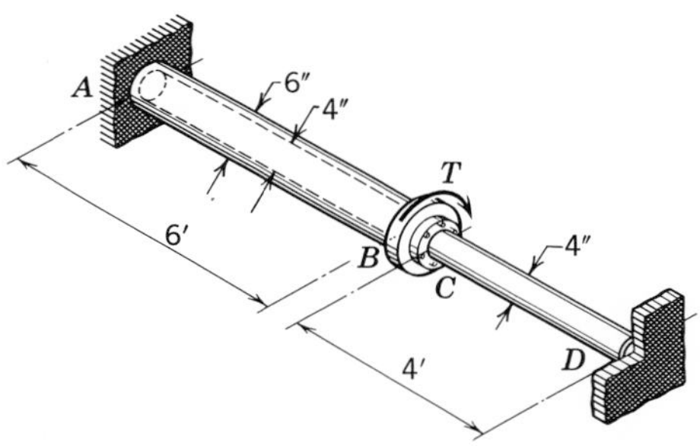


Fig. 4–70

**4–71.** A steel shaft 7 ft long extends through, and is attached to a hollow bronze shaft 4 ft long, as shown in Fig. P4–71. Both shafts are fixed at the wall. Use 6000 ksi for the modulus of rigidity of bronze. Determine:

(*a*) The maximum shearing stress in each material.

(*b*) The angle of twist of the right end.

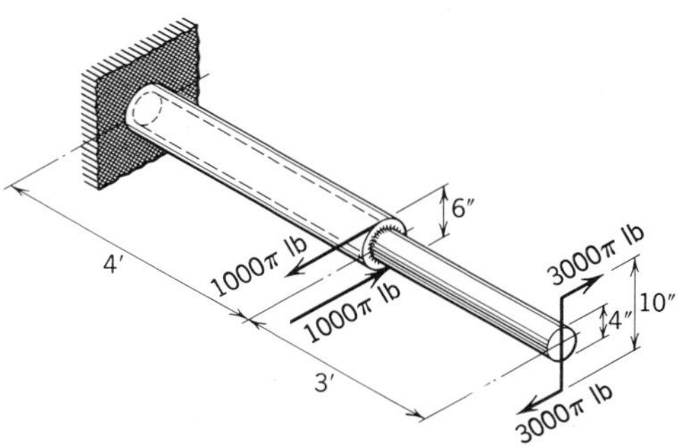


Fig. P4–71

**4–72.*** The 6-in. diameter steel shaft of Fig. P4–72 has a 4-in. diameter bronze core ($G = 6000$ ksi) inserted in 4 ft of the right end and securely bonded to the steel. When the torques of $11{,}000\pi$ ft-lb and $5000\pi$ ft-lb are applied as shown, determine:

(*a*) The maximum shearing stress in the steel.

(*b*) The rotation of the free end of the shaft.

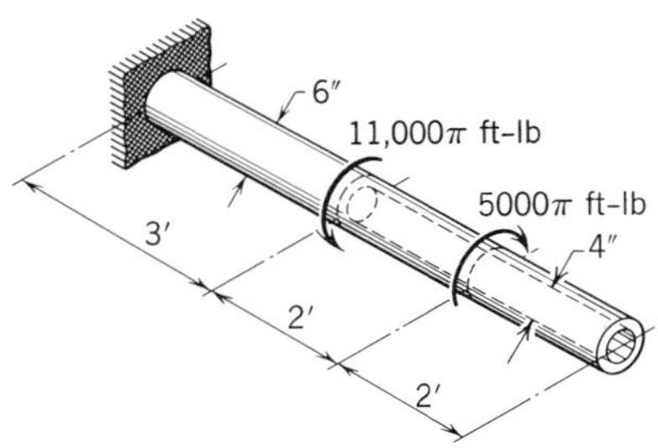


**Fig. P4–72**

**4–73.*** The inner surface of the aluminum alloy (2024-T4) collar $A$ and the outer surface of the steel shaft $B$ of Fig. P4–73 are smooth. Both the collar and the shaft are rigidly fixed to the wall at $D$. The 0.500-in. diameter pin $C$ fills a hole drilled completely through a diameter of the collar and shaft. The shearing deformation in the pin and the bearing deformation between the pin and shaft can be neglected. Calculate the maximum torque, $T$, which can be applied to the steel shaft as shown without exceeding an average shearing stress of 3200 psi on the cross-sectional area of the pin at $E$, the interface between the shaft and collar.

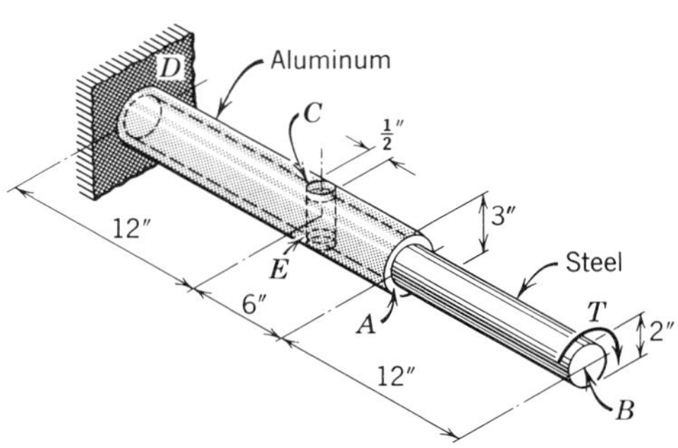


**Fig. P4–73**

**4–74.** In the assembly of Prob. 4–73, the hole in the collar is drilled oversize so that the pin moves 0.0003 in. before it makes contact at the outside surface of the aluminum. Determine:

(*a*) The maximum torque, $T$, as requested in Prob. 4–73.

(*b*) The angle of twist of the right end of shaft $B$ relative to the wall.

**4–75.*** The solid circular shafts $A$ (aluminum alloy, $G = 4000$ ksi) and $B$ (steel) are fixed at one end and connected by the linkage at $C$, see Fig. P4–75. The play in the connection permits arm $D$ to rotate through 0.006 rad before the slack is taken up. Neglect the elastic deformation of the linkage and compute the maximum permissible torque $T$ if the torsional shearing stresses are not to exceed 9 ksi in the aluminum alloy and 12 ksi in the steel.

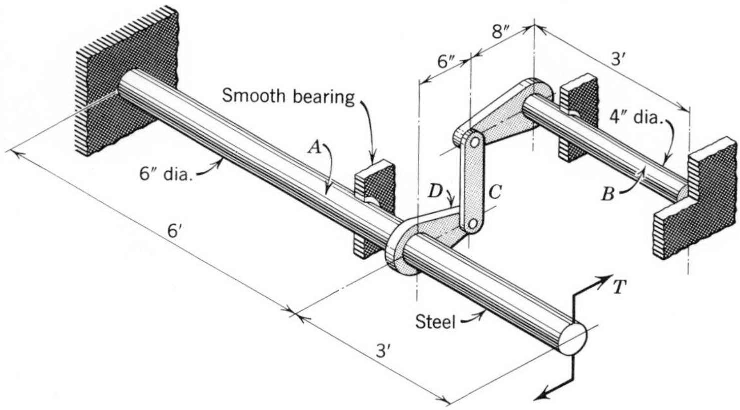


**Fig. P4–75**

**4–76.** The 4-in. diameter shaft of Fig. P4–76 is composed of a brass segment ($G = 6000$ ksi) and a steel segment. Determine the maximum permissible magnitude for the torque, $T$, applied at $C$ if the allowable shearing stresses are 5 ksi for the brass and 12 ksi for the steel.

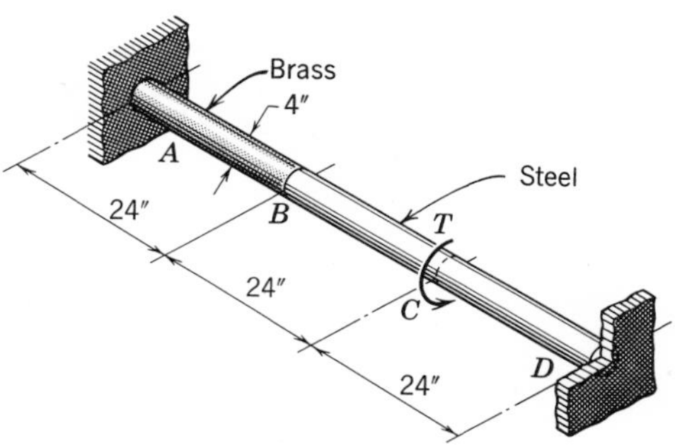


**Fig. P4–76**

**4–77.*** Solve Prob. 4–76 with a second torque of magnitude $2T$ applied at $B$ with the same sense as that of the torque shown.

**4–78.** The torsional assembly of Fig. P4–78 consists of an aluminum alloy [$G = 4(10^6)$ psi] segment $AB$ securely connected to a steel segment $BCD$ by means of a flange coupling with four bolts. The diameters of both segments are 4 in., the cross-sectional area of each bolt is 0.2 sq in., and the bolts are located 3 in. from the center of the shaft. If the cross shearing stress in the bolts is limited to 6000 psi, determine the maximum permissible magnitude for the torque $T$ applied at $C$.

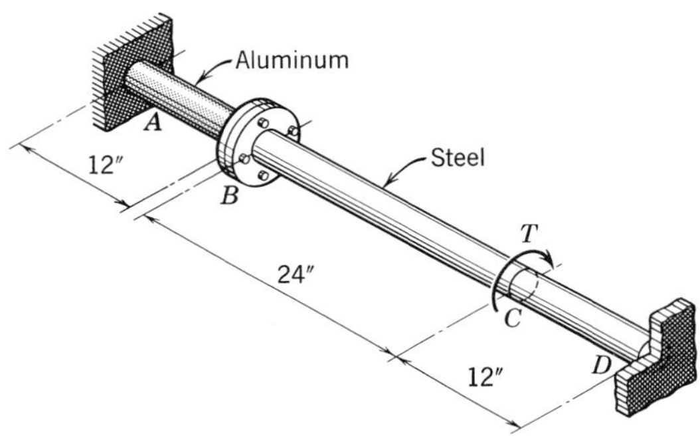


Fig. P4–78

**4–79.*** The circular shaft of Fig. P4–79 consists of a steel segment $ABC$ securely connected to a bronze ($G = 6000$ ksi) segment $CD$, with ends $A$ and $D$ fixed. Determine the maximum shearing stresses in the steel and the bronze when $T$ has a magnitude of $35\pi$ in-kips.

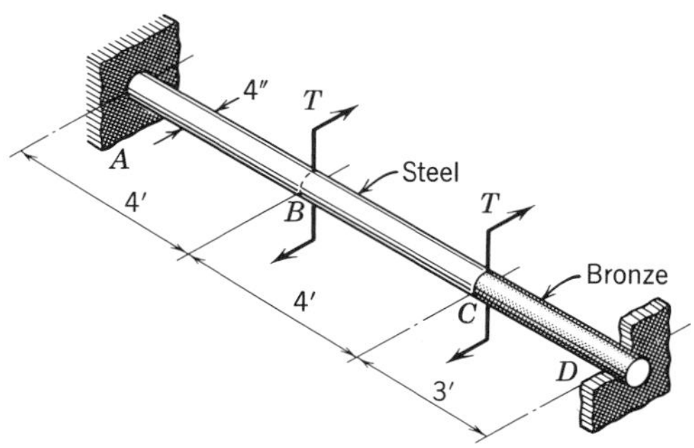


Fig. P4–79

**4–8. Non-circular Sections.** Prior to 1820, when A. Duleau published experimental results to the contrary, it was thought that the shearing stresses in any torsionally loaded member were proportional to the distance from its axis. Duleau proved experimentally that this is not true for rectangular cross sections. An examination of Fig. 4–14 will verify Duleau's conclusion. If the stresses in the rectangular bar were proportional to the distance from its axis, the maximum stress would occur at the corners. However, if there were a stress of any magnitude at the corner, as indicated in Fig. 4–14*a*, it could be resolved into the components indicated in Fig. 4–14*b*; and if these components existed, the two components shown dashed would also exist. These last components cannot exist, since the surfaces on which they are shown are free boundaries. Therefore, the shearing stresses at the corners of the rectangular bar must be zero.

The first correct analysis of the torsion of a prismatic bar of non-circular cross section was published by Saint Venant in 1855. The following discussion presents some aspects of this method.[6] Consider

[6] A more complete discussion of this theory is presented in various books such as *Mathematical Theory of Elasticity,* 2nd Ed., I. S. Sokolnikoff, McGraw-Hill, New York, 1956, pp. 109–134.

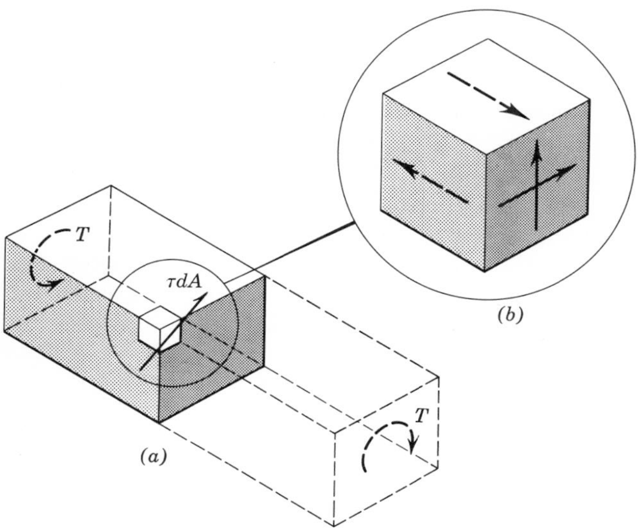


Fig. 4–14

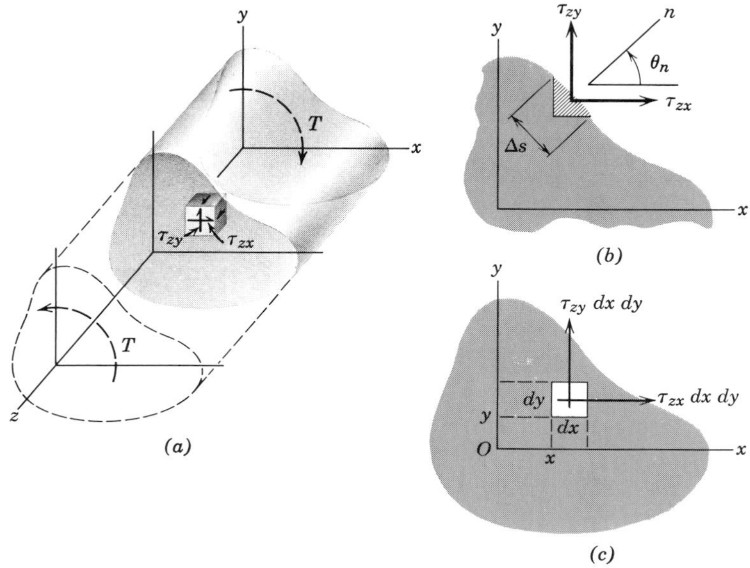


Fig. 4–15

the torsion of the prismatic bar in Fig. 4–15*a* where the $z$ axis passes through the centroid of each cross section. Assume that the only nonzero stresses are $\tau_{zx}$ and $\tau_{zy}$, since this is the simplest state of stress which will transmit a torque between adjacent cross sections. When body forces are neglected, the equations of equilibrium, Eq. (3–10), reduce to

$$\frac{\partial \tau_{zx}}{\partial z} = 0, \qquad \frac{\partial \tau_{zy}}{\partial z} = 0, \qquad \frac{\partial \tau_{zx}}{\partial x} + \frac{\partial \tau_{zy}}{\partial y} = 0.$$

The first two of these equations indicate that the stresses are independent of $z$. All three of the equations will be satisfied by stresses obtained from a stress function $\varphi(x, y)$ as

$$\tau_{zx} = \partial\varphi/\partial y, \qquad \tau_{zy} = -\partial\varphi/\partial x. \tag{4–3}$$

All of the compatibility equations [see Eq. (3–3)] are satisfied by the assumed stresses except the fourth and fifth ones which, with Eq. (4–3), become

$$\nabla^2(\tau_{zy}) = \nabla^2\left(-\frac{\partial\varphi}{\partial x}\right) = -\frac{\partial}{\partial x}(\nabla^2\varphi) = 0,$$

and similarly

$$\frac{\partial}{\partial y}(\nabla^2\varphi) = 0.$$

These equations are satisfied if

$$\nabla^2\varphi(x, y) = C, \tag{4–4}$$

where $C$ is a constant.

If the lateral surface of the bar is to be stress-free, the resultant shearing stress on any cross section at the boundary must be tangent to the boundary; that is, the component stress normal to the boundary (see Fig. 4–15*b*) must be zero. Thus

$$\tau_{zx}\cos\theta_n + \tau_{zy}\sin\theta_n = 0. \tag{a}$$

Since $\cos\theta_n$ is $dy/ds$ and $\sin\theta_n$ is $-dx/ds$, Eq. (a) becomes

$$\tau_{zx}\frac{dy}{ds} - \tau_{zy}\frac{dx}{ds} = 0.$$

When $\tau_{zx}$ and $\tau_{zy}$ are replaced by their equivalent expressions from Eq. (4–3), this last equation becomes

$$\frac{\partial\varphi}{\partial y}\frac{dy}{ds} - \left(-\frac{\partial\varphi}{\partial x}\right)\frac{dx}{ds} = \frac{d\varphi}{ds} = 0,$$

which indicates that $\varphi$ is constant along the boundary. For solid cross sections, the constant may be set equal to zero and

$$\varphi\Big|_{\text{boundary}} = 0. \tag{4–5}$$

The torque transmitted by the shearing stress is obtained by integrating the moments, about the $z$ axis, of the differential forces acting on the element $dx\,dy$, see Fig. 4–15*c*. The torque is

$$T = \iint_A (x\tau_{zy} - y\tau_{zx})\,dx\,dy. \tag{b}$$

When the expressions for the shearing stresses in Eq. (4–3) are substituted in Eq. (b), it can be integrated by parts to give

$$T = 2\iint_A \varphi\,dx\,dy \tag{4–6}$$

in which the integration is to be carried out over the area $A$.

The strain-deformation relations, Eq. (3–9), can be used to investigate the angle of twist of the bar as well as any possible warping of the cross sections. Such an analysis, however, is beyond the scope of

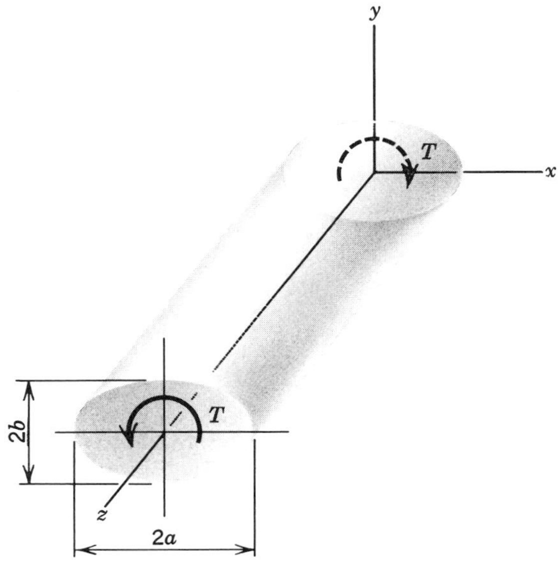


Fig. 4–16

this elementary discussion, but such a study will show that in general, *except for members with circular cross sections*, every section will warp (not remain plane) when the bar is twisted.

**Example 4–7:** The shaft in Fig. 4–16 has an elliptic cross section and is twisted by the torque $T$. Determine the magnitude and location of the maximum shearing stress in the shaft.

*Solution:* The equation of the ellipse is

$$\frac{x^2}{a^2} + \frac{y^2}{b^2} - 1 = 0$$

and Eq. (4–4) and (4–5) will both be satisfied if $\varphi$ is assumed to be

$$\varphi = K\left(\frac{x^2}{a^2} + \frac{y^2}{b^2} - 1\right).$$

The constant $K$ can be determined from Eq. (4–6), which, due to symmetry, may be written as

$$T = 8K \int_0^b \int_0^{(a/b)\sqrt{b^2-y^2}} \left(\frac{x^2}{a^2} + \frac{y^2}{b^2} - 1\right) dx\, dy.$$

When the integral of this equation is evaluated, the torque is found to be

$$T = -\pi abK$$

and consequently the stress function is

$$\varphi = -\frac{T}{\pi ab}\left(\frac{x^2}{a^2} + \frac{y^2}{b^2} - 1\right).$$

The shearing stresses are, from Eq. (4–3),

$$\tau_{zx} = \frac{\partial \varphi}{\partial y} = \frac{-2Ty}{\pi ab^3}$$

and

$$\tau_{zy} = -\frac{\partial \varphi}{\partial x} = \frac{2Tx}{\pi a^3 b}.$$

The maximum values of $\tau_{zx}$ and $\tau_{zy}$ occur at $y = \pm b$ and at $x = \pm a$ respectively, and are

$$(\tau_{zx})_{\max} = \frac{2T}{\pi ab^2} \quad \text{and} \quad (\tau_{zy})_{\max} = \frac{2T}{\pi a^2 b}.$$

Since $a$ is greater than $b$, the maximum shearing stress is $(\tau_{zx})_{\max}$ and it occurs at $y = \pm b$, that is

$$\tau_{\max} = \underline{2T/(\pi ab^2)}. \qquad \text{Ans.}$$

In example 4–7 the maximum shearing stress was found to occur at the points on the boundary nearest the axis of the bar. The following analogy, commonly called the membrane analogy, provides a convenient method for identifying regions of large shearing stress.

When a thin membrane (such as a soap film) is stretched over a plane opening having the shape of the cross section of the bar, a slight pressure difference, $p$, applied to it will cause the membrane to distend. The differential equation for small deflections, $w$, of the membrane is[7]

$$\nabla^2 w = p/S$$

where $S$ is the surface tension in the membrane, a constant. The deflection of the membrane is zero at the boundary and since the membrane deflection satisfies the same differential equation and boundary conditions as the stress function, it follows, from Eq. (4–3), that the slope of the membrane at any point is proportional to the shearing stress in the cross section at that point. Note that the shearing stress acts in a direction normal to the direction in which the slope is measured. Also, Eq. (4–6) indicates that the volume between the membrane and the plane of the opening is proportional to the applied torque in the shaft. Figure 4–17 indicates several sections through a membrane

[7] See *Foundations of Solid Mechanics,* Y. C. Fung, Prentice-Hall, Englewood Cliffs, N.J., 1965, pp. 170–172.

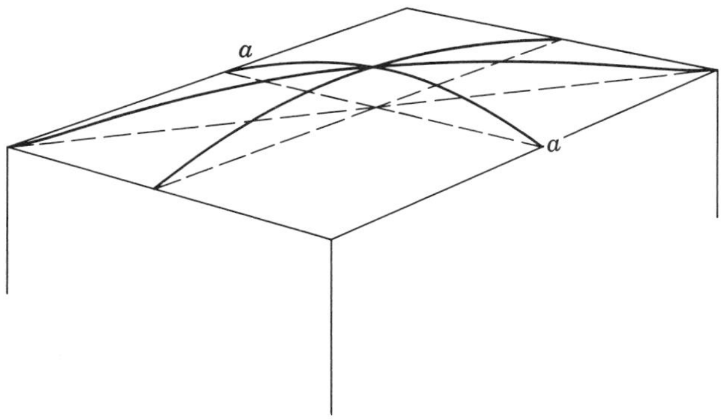


Fig. 4–17

for a rectangular cross section, and these sections indicate that the shearing stress is zero (where the slope of the membrane must be zero) at the corners and at the center and is a maximum at points $a$ and $a$. Thus, one observes that for a rectangular section, the maximum shearing stress is in the boundary material closest to the axis of the shaft rather than in the most distant material. In like manner, one can visualize the location of maximum torsional stresses in shaft keyways and how the addition of fillets in the re-entrant corners of the keyway reduces the magnitude of the stress.

The discussion of the techniques involved in the evaluation of stresses by means of the membrane analogy will be left to others.[8]

## PROBLEMS

**4–80.** Show that the results of Ex. 4–7 reduce to the elementary theory for a circular shaft when $a$ is made equal to $b$. This demonstrates that the theory developed in Art. 4–1 agrees exactly with the solution obtained using the equations of elasticity.

**4–81.*** The function

$$\varphi = \left(\frac{15\sqrt{3}\,T}{2a^5}\right)\left(x + \frac{a}{3}\right)\left(x + \sqrt{3}y - \frac{2a}{3}\right)\left(x - \sqrt{3}y - \frac{2a}{3}\right)$$

is proposed as the stress function for the torsion of the prismatic triangular bar whose cross section is shown in Fig. P4–81.

($a$) Given that the function satisfies Eq. (4–6), show that it meets the other requirements of the torsion stress function.

($b$) Develop expressions for the shearing stresses.

($c$) Determine the maximum shearing stress and state where it occurs.

[8] *Handbook of Experimental Stress Analysis,* M. Hetenyi, John Wiley and Sons, New York, 1950.

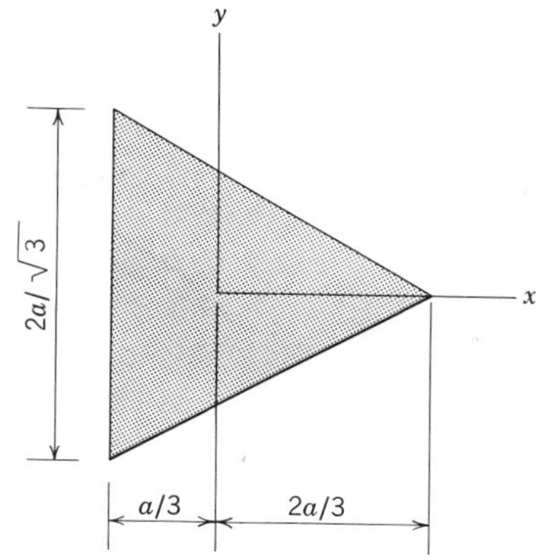


Fig. P4–81

**4–82.*** The function

$$\varphi = C\left[(x^2 + y^2) - 2ax + \frac{2ab^2x}{x^2 + y^2} - b^2\right]$$

is proposed as a stress function for the torsion of the prismatic bar whose cross section is shown in Fig. P4–82. Given that the function satisfies Eq. (4–4) show that it satisfies Eq. (4–5) and, hence, meets the other requirements of a torsion-stress function. Calculate the shearing stresses at points $A$, $B$, and $D$ and state at which point the shearing stress is largest.

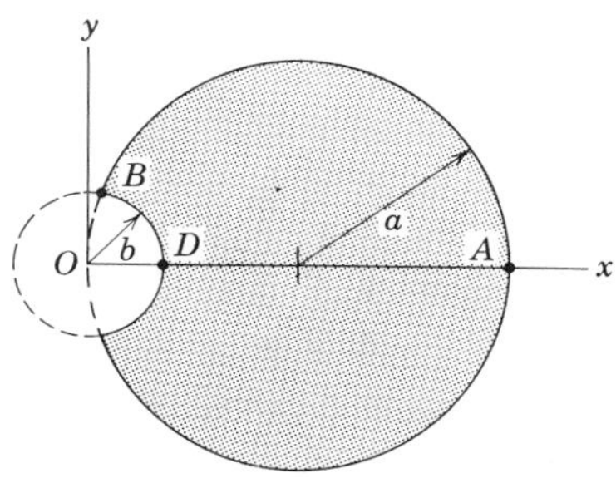


Fig. P4–82

**4–9. Shear Flow.** Although the elementary torsion theory presented in Art. 4–1 is limited to circular sections, one class of non-circular sections which can be readily handled by elementary methods is the thin-walled section such as the one illustrated in Fig. 4–18*a* which

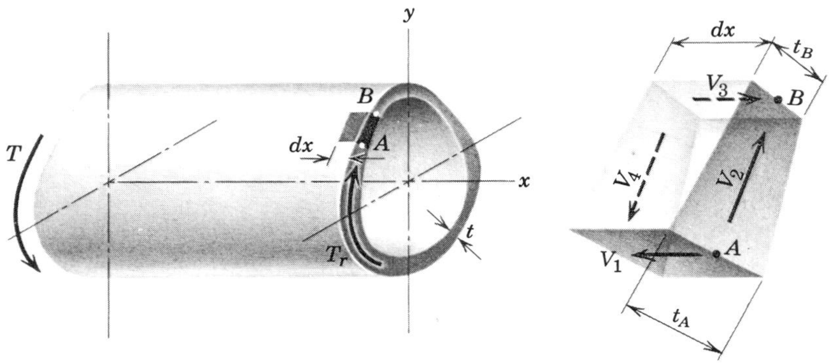


(a) (b)

Fig. 4–18

represents a non-circular section with a wall of variable thickness ($t$ varies).

A useful concept associated with the analysis of thin-walled sections is *shear flow*, $q$, defined as internal shearing force per unit of length of the thin section. Typical units for $q$ are pounds per inch. In terms of stress, $q$ equals $\tau(1)(t)$, where $\tau$ is the average shearing stress across the thickness $t$. It will now be demonstrated that the shear flow on a cross section is constant even though the thickness of the section wall varies. Figure 4–18$b$ shows a block cut from the member of Fig. 4–18$a$ between $A$ and $B$, and as the member is subjected to pure torsion, the shear forces $V_1 \cdots V_4$ alone (no normal forces) are necessary for equilibrium. Summing forces in the $x$ direction gives

$$V_1 = V_3,$$

or

$$q_1\,dx = q_3\,dx,$$

from which

$$q_1 = q_3,$$

and, as $q = \tau(t)$,

$$\tau_1 t_A = \tau_3 t_B. \quad \text{(a)}$$

The shearing stresses at point $A$ on the longitudinal and transverse planes have the same magnitude; likewise, the shearing stresses at point $B$ have the same magnitude on the two orthogonal planes; hence, Eq. (a) may be written

$$\tau_A t_A = \tau_B t_B,$$

or

$$q_A = q_B$$

which was to be proved.

To develop the expression for the resulting torque on the section, consider the force $dF$ acting through the center of a differential length of perimeter, $ds$, as shown in Fig. 4–19. The resisting torque, $T_r$, is the resultant of the forces $dF$; that is,

$$T_r = \int (dF)\rho = \int (q\,ds)\rho = q \int \rho\,ds.$$

This integral may be difficult to evaluate by formal calculus; however, the quantity $\rho\,ds$ is twice the area of the triangle shown shaded in Fig. 4–19, which makes the integral equal to twice the area, $A$, *enclosed by the median line.* The resulting expression is

$$T_r = q(2A)$$

or, in terms of stress,

$$\tau = T/(2At), \tag{4–7}$$

where $\tau$ is the *average* shearing stress across the thickness $t$ (and tangent to the perimeter), and is reasonably accurate when $t$ is relatively small. For example, in a round tube with a diameter to wall thickness ratio of 20, the stress as given by Eq. (4–7) is 5 per cent less than that given by the torsion formula. It must be emphasized that Eq. (4–7) applies only to "closed" sections, that is, sections which have a continuous periphery. If the member were slotted longitudinally (see, for example, Fig. 4–20), the resistance to torsion would be diminished considerably from that for the closed section.

It is sometimes necessary to evaluate the twist of a thin-walled member subjected to torsion. Since the ratio $\gamma/\rho$ for a section such as that of Fig. 4–19 is not a constant as it is for a circular section, Eq.

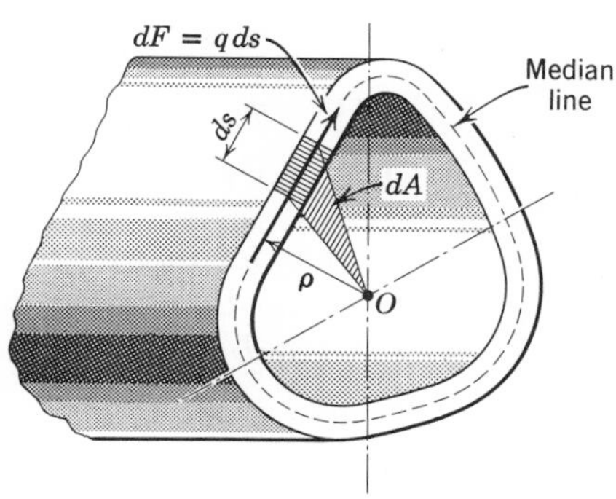


Fig. 4–19

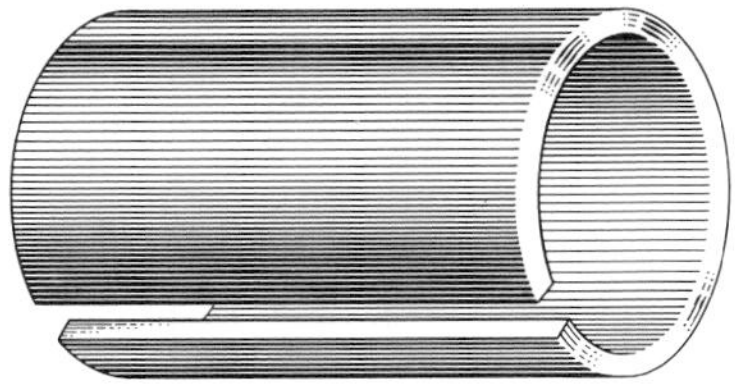

Fig. 4–20

(4–2), $\theta = L\gamma/\rho$, is not easily adapted to such a section. Instead, the principle of strain energy will be found most useful in the development of an expression for the angle of twist.

Assuming that energy is conserved, the work done by the torque in twisting a member through an angle, $\theta$, is equal to the strain energy stored in the member, and if the torque-twist relation is linear (which limits stresses to values within the proportional limit), the expression becomes

$$W_k = U = \frac{T\theta}{2} = \int^{\text{Vol.}} u\,dv, \tag{b}$$

where $u$ is the strain energy per unit volume of the material in the member. All elements of the member are subjected to pure shear (see Fig. 4–18); therefore, the expression for $u$ is $u = \tau^2/(2G)$, see Art. 2–2, and Eq. (b) becomes

$$U = \int^{\text{Vol.}} \frac{\tau^2}{2G}\,dv. \tag{c}$$

When $\tau$ in Eq. (c) is replaced by Eq. (4–7), and $dv$ by $t\,dx\,ds$ (see Fig. 4–18 and 4–19), the expression becomes

$$U = \int^{\text{Vol.}} \frac{T^2}{8A^2tG}\,dx\,ds. \tag{d}$$

In general, $t$ is a function of $s$, but for many practical problems, the integral may be replaced by a summation of $n$ elements, each having a constant thickness, $t_i$, and a length, $L$; thus

$$U = \sum_{i=1}^{n} \frac{T^2L}{8A^2G}\left(\frac{\Delta s_i}{t_i}\right). \tag{e}$$

Substituting Eq. (e) into Eq. (b) yields

$$\theta = \frac{TL}{4A^2G} \sum_{i=1}^{n} \frac{\Delta s_i}{t_i}, \tag{4–8}$$

where $\Delta s_i/t_i$ is the ratio of plate width (measured along the periphery of the section) to thickness. Note that $L$ and $A$ have been treated as constants; hence, the shape and size of the cross section must be maintained constant (possibly by diaphragms, bulkheads, or cross bracing), and all the elements must have the same length. The following example will illustrate the use of Eq. (4–7) and (4–8).

**Example 4–8:** A rectangular box section of aluminum alloy, 2024-T4, has outside dimensions 4 by 2 in. The plate thickness is 0.06 in. for the 2-in. sides and 0.10 in. for the 4-in. sides. Determine the maximum permissible torque to which the section may be subjected if a factor of safety of 2 with respect to failure by slip is specified. For this torque, determine the angle of twist in a 40-in. length of the member.

*Solution:* The allowable stress is 28/2 = 14 ksi, and this will occur in the thinnest plate; therefore, $q = 14(0.06) = 0.84$ kips/in. The torque is

$$T = 2qA = 2(0.84)(4 - 0.06)(2 - 0.10) = \underline{12.58 \text{ in-kips.}} \qquad \text{Ans.}$$

The angle of twist is

$$\theta = \frac{LT}{4GA^2} \sum \frac{s_i}{t_i} = \frac{40(12.58)}{4(4)(10^3)[3.94(1.90)]^2} \left( \frac{3.94(2)}{0.1} + \frac{1.90(2)}{0.06} \right)$$

$$= \underline{0.0797 \text{ rad.}} \qquad \text{Ans.}$$

## PROBLEMS

**Note: In all the following problems, assume that buckling of the shell is prevented and that the shape of the section is maintained by ribs, diaphragms, and stiffeners.**

**4–83.*** A 16-in. wide by 0.10-in. thick steel sheet is to be formed into a hollow section by bending through 360° and welding the long edges together (butt joint). The shape may be (1) circular, (2) square, or (3) a 6 by 2-in. rectangle. Assume a median length of 16 in. (no stretching of the sheet due to bending), and square corners for non-circular sections. The allowable shearing stress is 12 ksi. For each of the shapes listed, determine the magnitude of:

(*a*) The maximum permissible torque.
(*b*) The angle of twist in a 100-in. length.

**4–84.*** Solve Prob. 4–83 if the sheet is 24 in. wide by 0.125 in. thick and the shapes to be considered are (1) circular, (2) square, and (3) an equilateral triangle.

**4–85.** Figure P4–85 shows a cross section of the leading edge of an airplane wing. The length along the curve is 26 in. and the enclosed area is 82 sq. in. Sheet thicknesses are shown on the diagram and the material is 2024-T4 aluminum alloy. For a torque of 80 in-kips, determine the magnitude of:
(*a*) The maximum shearing stress.
(*b*) The angle of twist per foot of length.

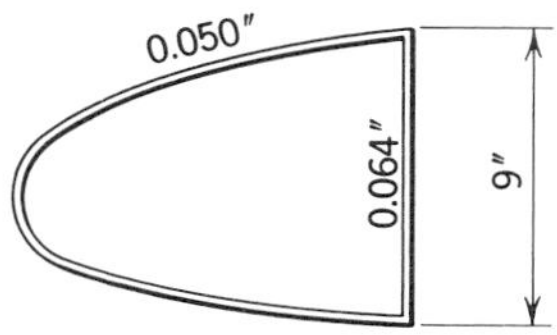


**Fig. P4–85**

**4–86.*** Figure P4–86 represents the cross section of an airplane fuselage made of 2024-T4 aluminum alloy. The sheet thicknesses are 0.050 in. from *A* to *B* and *C* to *D*, 0.040 in. from *B* to *C*, and 0.032 in. from *D* to *A*. For an allowable shearing stress of 12 ksi, determine the magnitude of:
(*a*) The maximum permissible torque.
(*b*) The angle of twist per foot of length.

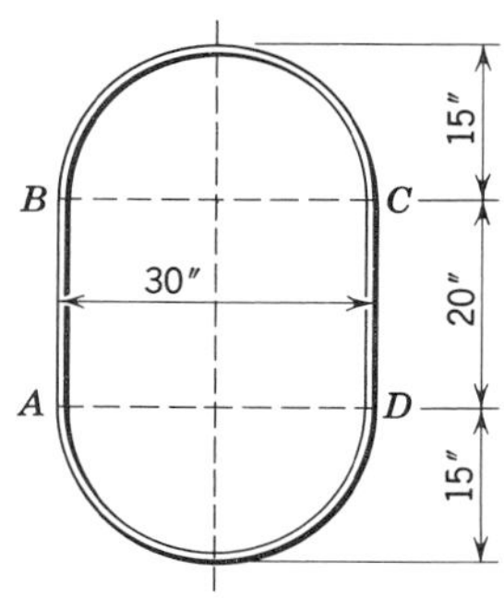


**Fig P4–86**

**4–87.** If an airplane fuselage with the cross section of Fig. P4–86 is subjected to a torque of 1000 in-kips, and the allowable shearing stress is 8 ksi, what is the minimum thickness of sheet (constant for the entire periphery) needed to resist the torque? For the given data, what is the magnitude of the angle of twist per foot of fuselage if the material is 2024-T4 aluminum alloy?

**4–88.*** Figure P4–88 shows the cross section of a torque box for an airplane wing. The top and bottom sheets are 40 in. long and the mean depth of the box is 11.2 in. Sheet thicknesses are shown on the figure and the material is 2024-T4 aluminum alloy. For a shear flow of 400 lb per in., determine the magnitude of:

(*a*) The torque applied to the box.
(*b*) The maximum shearing stress in each sheet.
(*c*) The angle of twist per foot of wing.

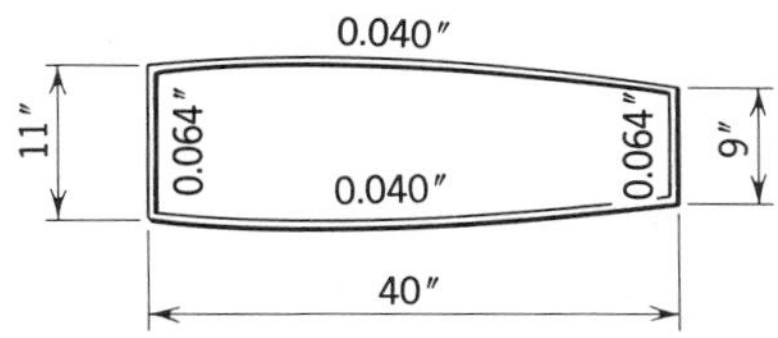


Fig. P4–88

**4–89.** Figure P4–89 shows the cross section of an airplane wing torque box. The lengths of the top and bottom sheets are 40.2 in. and the mean depth of the box is 14.6 in. Sheet thicknesses are shown on the figure and the material is 2024-T4 aluminum alloy. For an applied torque of 500 in-kips, determine the magnitude of:

(*a*) The maximum shearing stress in each sheet.
(*b*) The angle of twist per foot of wing.

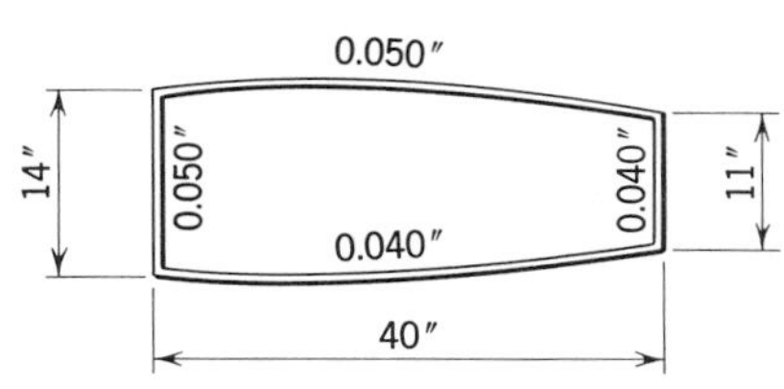


Fig. P4–89

**4–90.*** Determine the torque necessary to twist a 12-in. length of the torque box of Prob. 4–89 through a 0.12° angle. For this angle, what are the maximum shearing stresses in each sheet?

CHAPTER 5

# *Flexural Loading: Stresses*

**5–1. Introduction.** The beam, or flexural member, is frequently encountered in structures and machines, and its elementary stress analysis constitutes one of the more interesting facets of mechanics of materials. A beam is a member subjected to loads applied transverse to the long dimension, causing the member to bend. For example, Fig. 5–1 is a photograph of an aluminum alloy I-beam, $AB$, simply supported in a testing machine and loaded at the one-third points, and Fig. 5–2 depicts the shape (exaggerated) of the beam when loaded.

Before proceeding with a discussion of stress analysis for flexural members, it may be well to classify some of the various types of beams and loadings encountered in practice. Beams are frequently classified on the basis of the supports or reactions. A beam which is supported by pins, rollers, or smooth surfaces at the ends is called a *simple beam.* A simple support will develop a reaction normal to the beam but will not produce a couple. If either or both ends of a beam project beyond the supports, it is called a *simple beam with overhang.* A beam which has more than two simple supports is a *continuous beam.* Figure 5–3$a$, $b$, and $c$ shows respectively a simple beam, a beam with overhang, and a continuous beam. A *cantilever beam* is one in which one end is built into a wall or other support so that the built-in end cannot move transversely or rotate. The built-in end is said to be *fixed* if no rotation occurs and *restrained* if a limited amount of rotation occurs. The supports shown in Fig. 5–3$d$ and $e$ represent fixed ends unless otherwise stated. The beams in Fig. 5–3$d$, $e$, and $f$ are, in order, a cantilever beam, a beam fixed (or restrained) at the left end and simply supported near the other end, which has an overhang, and a beam fixed (or restrained) at both ends.

Cantilever beams and simple beams have only two reactions (two forces or one force and a couple), and these reactions can be obtained

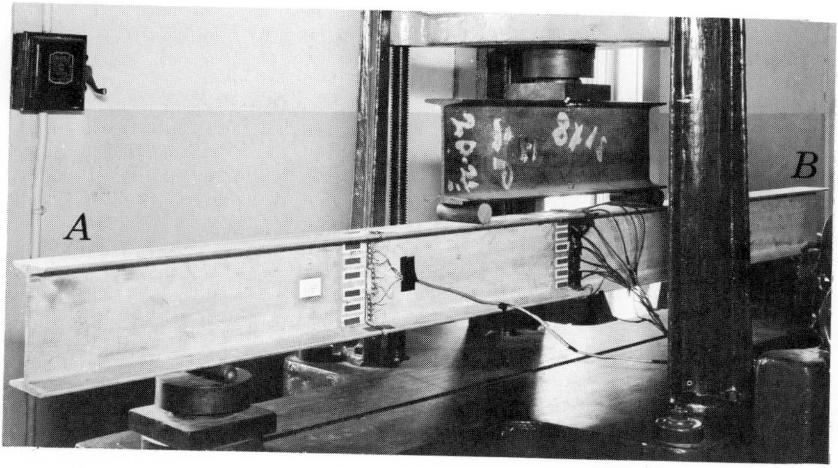


Fig 5–1. Set-up for measuring longitudinal strains in a beam.

from a free-body diagram of the beam by applying the equations of equilibrium. Such beams are said to be *statically determinate* since the reactions can be obtained from the equations of equilibrium. Continuous and other beams, with only transverse loads, which have more than two reaction components are called *statically indeterminate* since there are not enough equations of equilibrium to determine the reactions. Statically indeterminate beams are discussed in Chap. **7**.

The beams shown in Fig. **5–3** are all subjected to uniformly distributed loads and to concentrated loads, and, although shown as horizontal, they may have any orientation. Distributed loads will be shown on the side of the beam on which they are acting; that is, if drawn on the bottom of the beam, the load is pushing upward, and, if drawn on the right side of a vertical beam, the load is pushing to the left. The deflection curves shown beneath the beams are greatly exaggerated to assist in visualizing the shape of the loaded beam.

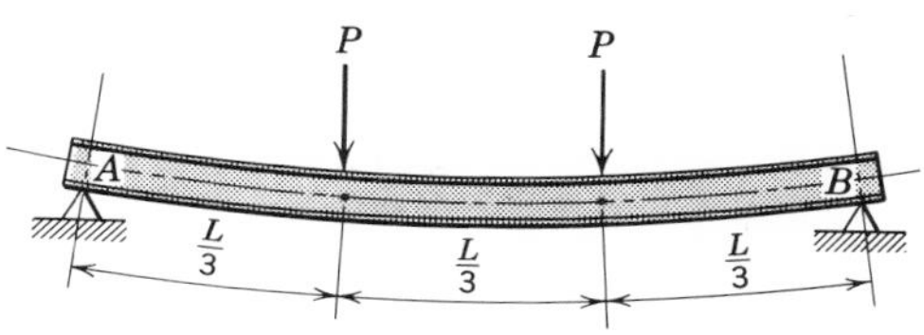


Fig. 5–2

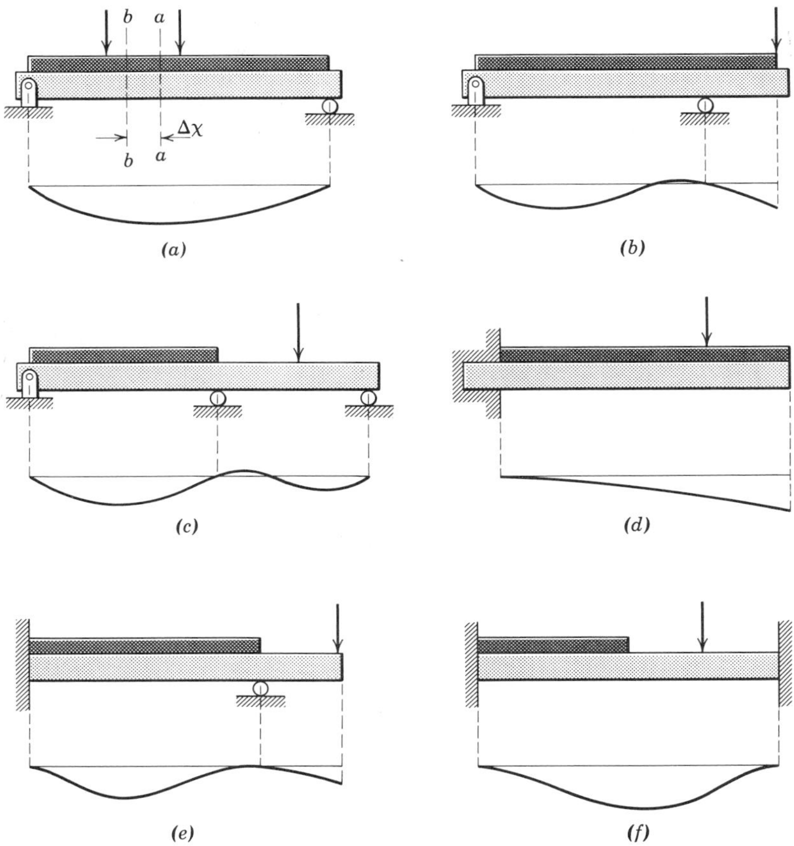


Fig. 5–3

**5–2. Flexural Stresses.** Examining the deflection curve of Fig. 5–3$a$, one observes that longitudinal elements of the beam near the bottom are stretched and those near the top are compressed, indicating the existence of both tensile and compressive stresses on transverse planes. These stresses are designated *fiber* or *flexural* stresses. A free-body diagram of the portion of the beam between the left end and plane $a$–$a$ is shown in Fig. 5–4. A study of this diagram reveals that a transverse force $V_r$ and a couple $M_r$ at the cut section and a force $R$ (a reaction) at the left support are needed to maintain equilibrium. The force $V_r$ is the resultant of the shearing stresses at the section (on plane $a$–$a$) and is called the resisting shear, and the couple $M_r$ is the

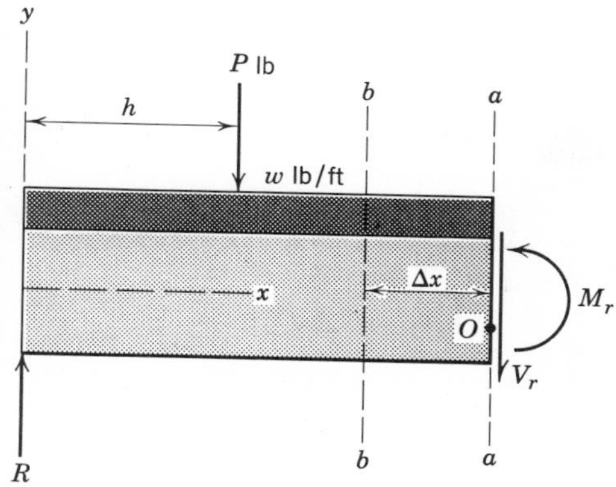


Fig. 5–4

resultant of the normal stresses at the section and is called the resisting moment. The magnitudes and senses of $V_r$ and $M_r$ may be obtained from the equations of equilibrium, $\Sigma F_y = 0$ and $\Sigma M_O = 0$ where $O$ is any axis perpendicular to plane $xy$ (the reaction, $R$, must be evaluated first from a free-body of the entire beam). For the present, the shearing stresses will be ignored while the normal stresses are studied. The magnitudes of the normal stresses can be computed if $M_r$ is known and also if the law of variation of the normal stresses on the plane $a$–$a$ is known. At this time, a brief résumé of the history of this problem is presented for a better appreciation of the problem.

The first recorded hypothesis on the normal stress distribution is that of Galileo Galilei (1564–1642) who, possibly observing that beams of stone when subjected to bending loads broke somewhat as indicated in Fig. 5–5, concluded that the material at point $A$ acted as a fulcrum and that the resultant of the tensile stresses was at the center of the cross section (as for axial loading).[1] This theory would lead to the stress distribution diagram of Fig. 5–6$a$ in which $F_T$ is the resultant of the tensile stresses. About fifty years after Galileo's observations, E. Mariotte (1620–1684), a French physicist, still retaining the concept of the fulcrum at the compression surface of the beam, but reasoning that the extensions of the longitudinal elements of the beam (fibers) would be proportional to the distance from the

[1] *Two New Sciences,* Galileo Galilei, 1638. Actually, Galileo's published work contains no record of test results; therefore, the manner in which he arrived at this conclusion is mere speculation.

Fig. 5–5. Failure of plaster of paris beam subjected to flexure.

fulcrum, suggested that the tensile stress distribution was as indicated in Fig. 5–6*b*. Mariotte later rejected the concept of the fulcrum and observed that part of the beam on the compression side was subjected to compressive stress having a triangular distribution; however, his expression for the ultimate load was still based on his original concept and was accepted by famous scientists such as Jacob Bernoulli, Leonard Euler, and others until C. A. Coulomb (1736–1806), a French military engineer, in a paper published in 1773 discarded the fulcrum

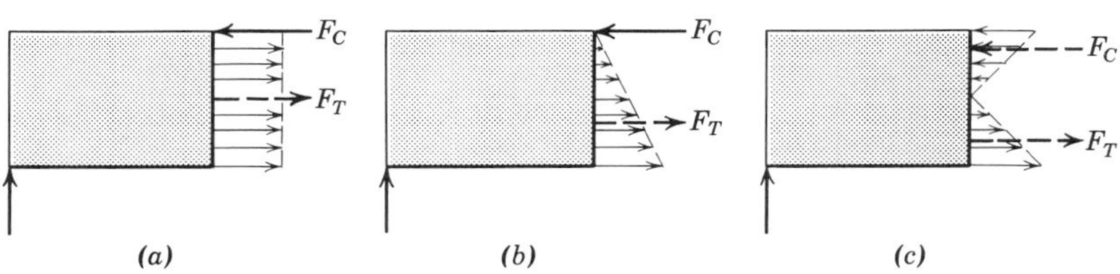


Fig. 5–6

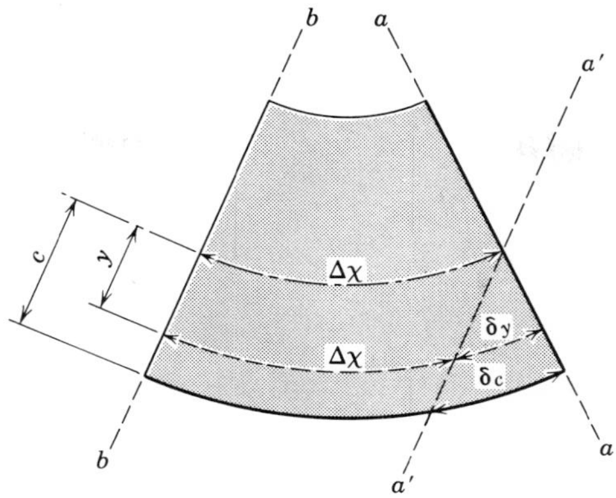


Fig. 5–7

concept and proposed the distribution of Fig. 5–6*c* in which both the tensile and compressive stresses have the same linear distribution.[2]

The correctness of the Coulomb theory can be demonstrated as follows: A segment of the beam of Fig. 5–4 between the planes *a–a* and *b–b* is shown in Fig. 5–7 with the distortion greatly exaggerated. The assumption is made that *a plane section before bending remains plane after bending.* For this to be strictly true, it is necessary that the beam be bent only with couples (no shear on transverse planes); the beam must be so proportioned that it will not buckle, and the loads applied so that no twisting occurs (this last limitation will be satisfied if the loads are applied in a plane of symmetry—a sufficient though not a necessary condition). One observes in Fig. 5–7 that at some distance $c$ above the bottom of the beam, the longitudinal elements (sometimes referred to as fibers) undergo no change in length. The curved surface formed by these elements is referred to as the *neutral surface*, and the intersection of this surface with any cross section is called the *neutral axis* of the section. All elements (fibers) on one side of the neutral surface are compressed, and those on the opposite side are elongated. The assumption is made that *all longitudinal elements have the same*

[2] Sixty years before Coulomb's paper, Parent (1666–1716), a French mathematician, proposed the correct triangular stress distribution. However, his work remained in obscurity, and the majority of engineers continued to use formulas based on Mariotte's theory. For a more complete discussion of the history of the flexure problem, the reader is referred to the *History of Strength of Materials,* S. P. Timoshenko, McGraw-Hill, New York, 1953, from which most of the foregoing discussion is taken.

*length.* This assumption imposes the restriction that the beam be initially straight and of constant cross section; however, in practice, considerable deviation from these last restrictions is often tolerated. If Fig. 5–7 is referred to,

$$\frac{\delta_y}{y} = \frac{\delta_c}{c},$$

from which

$$\delta_y = y\,\frac{\delta_c}{c}.$$

Since all elements had the same initial length, $\Delta x$, the strain of any element can be determined by dividing the deformation by the length of the element; the strain becomes

$$\epsilon = y\,\frac{\epsilon_c}{c},$$

where $\epsilon$ is the strain of a longitudinal element at a distance $y$ from the neutral surface and $\epsilon_c$ is the strain of a longitudinal element at the outer surface of the beam. On the basis of the general notation presented in Art. 1–6, both $\epsilon$ and $\epsilon_c$ should be designated $\epsilon_x$. However, for simplicity, the subscript $x$ is deleted on $\epsilon_x$ and $\sigma_x$ throughout the chapter, except where the general notation is necessary; for example, in Art. 5–11. The preceding equation indicates that the strain of any fiber is directly proportional to the distance of the fiber from the neutral surface. This variation can be demonstrated experimentally by means of strain gages attached to a beam, as shown in Fig. 5–1. The strains, as measured by gages on two different sections, are plotted against the vertical position of the gages on the beam in Fig. 5–8. Curve 1 represents strains on a section at the center of the beam where pure bending occurs (no transverse shear), and curve 2 shows strains at a section near one end of the beam where both fiber stresses and transverse shearing stresses exist. These curves are both straight lines within the limits of the accuracy of the measuring equipment.[3]

With the premise that the longitudinal strains are proportional to the distance from the neutral surface accepted, the assumption is now made that *Hooke's law applies* (which restricts stresses to magnitudes

---

[3] A more exact analysis using principles developed in the theory of elasticity indicates that curve 2 should be curved slightly. *Note:* other experiments indicate that a plane section of an initially curved beam will also remain plane after bending and that the deformations will still be proportional to the distance of the fiber from the neutral surface. The strain, however, will not be proportional to this distance since each deformation must be divided by a different original length.

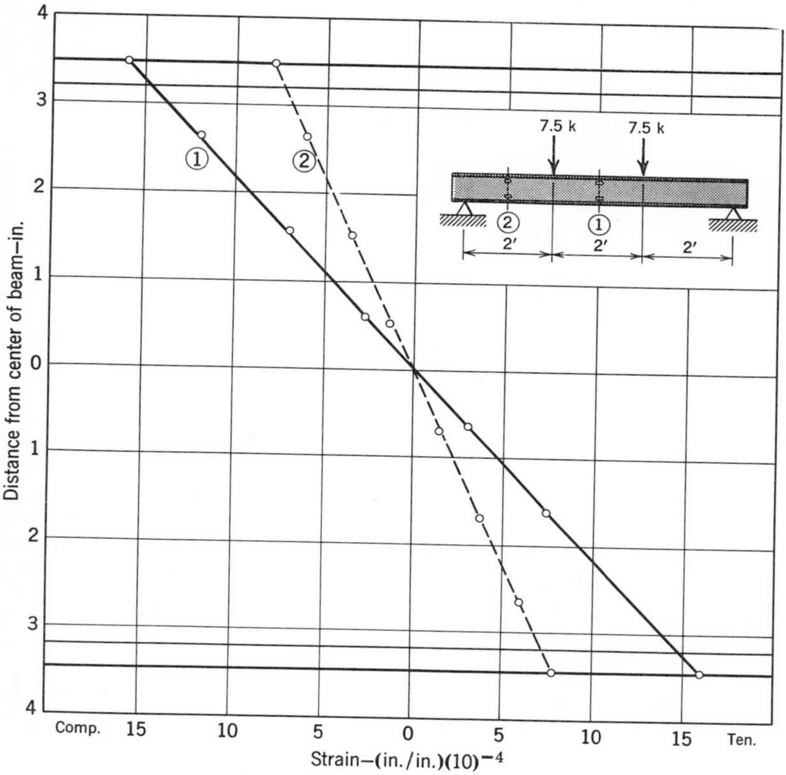


Fig. 5–8

within the proportional limit of the material). Then the equation becomes

$$\frac{\epsilon}{y} = \frac{\sigma}{Ey} = \frac{\sigma_c}{E_c c}.$$

The final assumption is made that *the modulus of elasticity has the same value in tension and compression at all points in the beam.* The attendant limitation is that the material be homogeneous. The final result is

$$\frac{\sigma}{y} = \frac{\sigma_c}{c} = \text{a constant,} \tag{a}$$

which verifies Coulomb's conclusion.

With the law of variation for the flexural stress known, Fig. 5–4 can

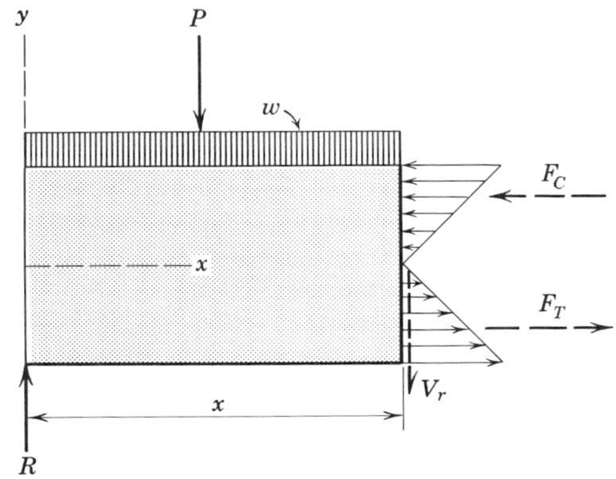


Fig. 5–9

now be redrawn as in Fig. 5–9. The vectors $F_C$ and $F_T$ are the resultants of the compressive and tensile flexural stresses respectively, and, because the sum of the forces in the $x$ direction must be zero, $F_C$ is equal to $F_T$; hence, they form a couple of magnitude $M_r$.

**5–3. Flexural Stresses by Equilibrium.** In order to evaluate flexural stresses, the neutral surface must first be located. Figure 5–10 shows the elevation of part of a beam and the beam cross section. The differential force, $dF$, is the normal force acting on the differential area $dA$ and equals $\sigma\, dA$. The equilibrium equation $\Sigma F_x = 0$ gives

$$\int^{A} \sigma\, dA = 0.$$

Substituting the expression for $\sigma$ from Eq. (a) in Art. 5–2, where $\sigma$ equals $(\sigma_c/c)y$, into this expression yields

$$\frac{\sigma_c}{c} \int^{A} y\, dA = \frac{\sigma_c}{c} A\bar{y} = 0.$$

Since $\sigma_c/c$ and $A$ are not zero, $\bar{y}$ must be zero, indicating that the *neutral axis coincides with the centroidal axis (c–c) of the cross section for flexural loading.* There are instances where the neutral axis (the line of zero longitudinal strain) moves away from the centroidal axis such as (1) a combined axial and flexural load (see Chap. 8), (2) a beam, with an unsymmetrical cross section, subjected to inelastic action, and (3) a curved beam.

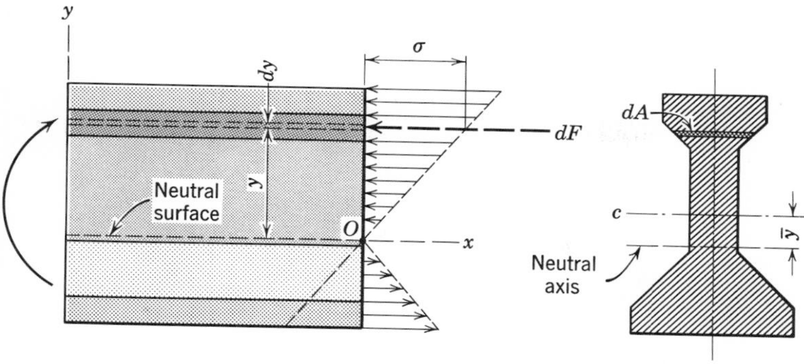


Fig. 5–10

The use of the concepts established thus far will now be demonstrated. *Note: Unless otherwise stated, as a simplifying assumption in all beam problems in this book, the weight of the beam itself is to be neglected.* Obviously this would not, as a rule, be done in professional practice.

**Example 5–1:** The beam of Fig. 5–3*a* has a rectangular cross section 4 in. wide by 12 in. deep and a span of 12 ft. It is subjected to a uniformly distributed load of 100 lb per ft and concentrated loads of 600 lb and 1200 lb at 2 ft and 7 ft respectively from the left support. Determine the maximum fiber stress at the center of the beam.

*Solution:* The reactions are evaluated from a free-body diagram of the entire beam and moment equations with respect to each of the two reactions. By this method, the reactions are evaluated independently, and summation of forces may be used to check the results. The left reaction, $R_L$, is 1600 lb, and the right reaction, $R_R$, is 1400 lb both upward. Figure 5–11*a* is a free-body diagram of the left half of the beam. The equation $\Sigma M_O = 0$ gives

$$\circlearrowleft + \qquad M_r + 600(4) + 100(6)(3) - 1600(6) = 0,$$

from which $M_r = 5400$ ft-lb or 5400(12) in-lb as shown.

The moment, $M_r$, is the resultant of the flexural stresses indicated in Fig. 5–11*b* and *c* in which $F_C$ and $F_T$ are the resultants of the compressive and tensile stresses respectively. The resultants $F_C$ and $F_T$ (which act through the centroids of the wedge-shaped stress distribution diagram of Fig. 5–11*c*) are located through the centroids of the triangular stress distribution diagrams of Fig. 5–11*b*. The magnitude of the couple $M_r$ equals $F_C(8)$ or $F_T(8)$, and $F_C = (\frac{1}{2})(\sigma)(6)(4) = 12(\sigma)$; therefore

$$M_r = 8(12)\sigma = 5400(12) \text{ in-lb},$$

from which $\sigma_{\max} = \underline{675 \text{ psi } C \text{ at top and } T \text{ at bottom.}}$ Ans.

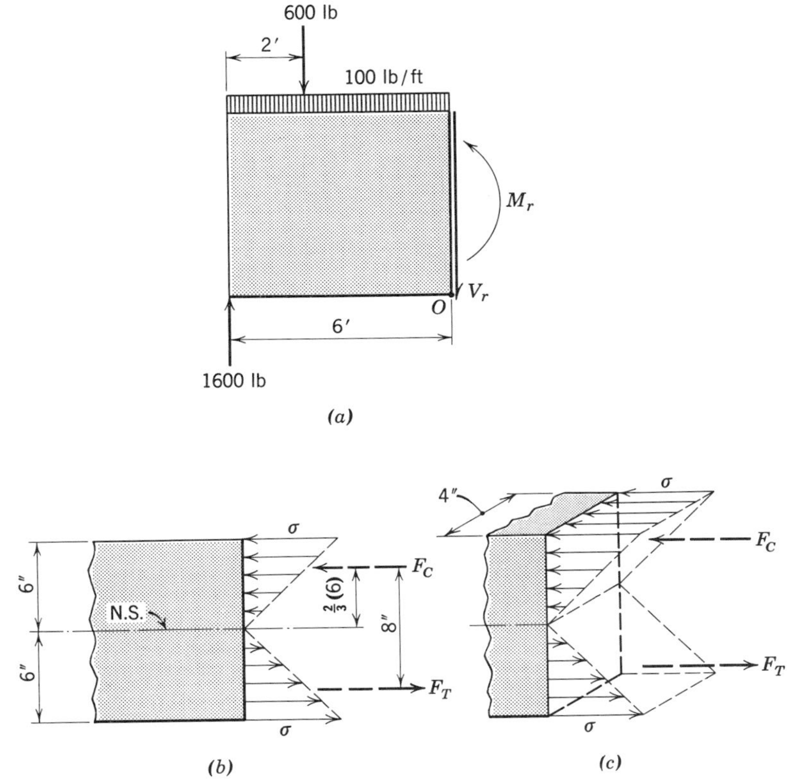


Fig. 5–11

**Example 5–2:** Solve Ex. 5–1 if the beam is composed of two 2 by 10-in. timbers[4] securely fastened together to form a symmetrical T-section with the flange on top.

*Solution:* The horizontal centroidal axis of the cross section is first determined to be 4 in. down from the top surface (see any text on statics). The fiber stress distribution diagrams are shown in Fig. 5–12. For convenience, the compression diagram is subdivided into three parts composed of two triangular distribution (wedge-shaped) diagrams and one uniform distribution diagram. This is done because the locations of the centroids of these diagrams are known. The values of the force components and their locations are shown in the figure. It will be noted that each component $F_1$, $F_2$, $F_3$, being the average stress multiplied by the area on which the stress acts, is equal to

[4] Structural timbers specified as rough sawn will be approximately full size. Cross-sectional dimensions of dressed timbers will be from ⅜ to ½ in. smaller than nominal. For convenience, all timbers in this book will be considered as full size.

the volume of the corresponding part of the three-dimensional diagram of Fig. 5–12*b*. The moment of the component forces with respect to $O$ is

$$M_r = F_1\left(\frac{10}{3}\right) + F_2\left(\frac{9}{3}\right) + F_3\left(\frac{4}{3}\right) + F_4\left(\frac{16}{3}\right)$$

which from the preceding example is equal to 5400(12); therefore

$$\left(\frac{5}{2}\right)\sigma\left(\frac{10}{3}\right) + 5\sigma\left(\frac{9}{3}\right) + \left(\frac{1}{2}\right)\sigma\left(\frac{4}{3}\right) + 8\sigma\left(\frac{16}{3}\right) = 5400(12),$$

from which $\sigma_{\max} = \underline{972 \text{ psi } T \text{ at bottom.}}$ Ans.

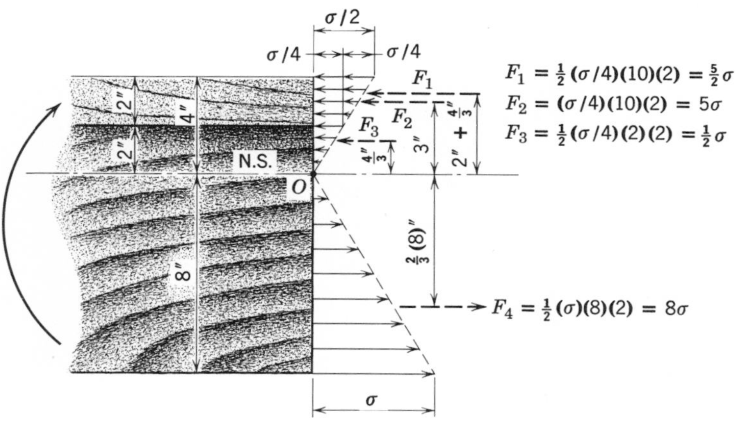


*(a)*

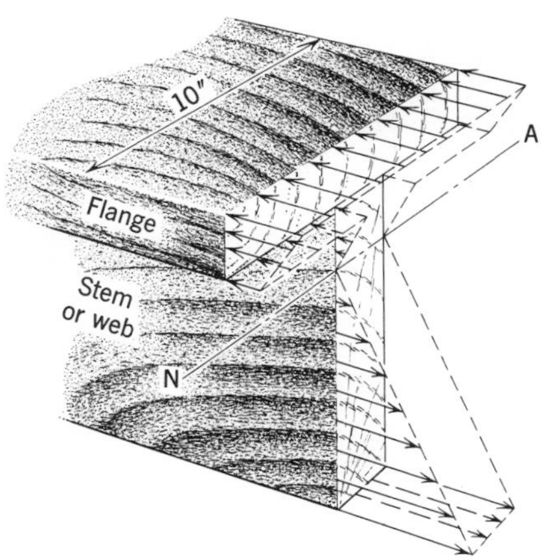


*(b)*

Fig. 5–12

## PROBLEMS

**5–1.*** A T-beam 7 in. in total depth has a 6-in. flange at the top. The thickness of both flange and stem is 2 in. The fiber stress at the top surface is 11,000 psi *T*. Determine the fiber stress at the bottom of the stem.

**5–2.*** On a cross section (see Fig. P5–2) of a beam the fiber stress at point *A* is 4000 psi *T*. Determine the maximum fiber stress on this cross section of the beam.

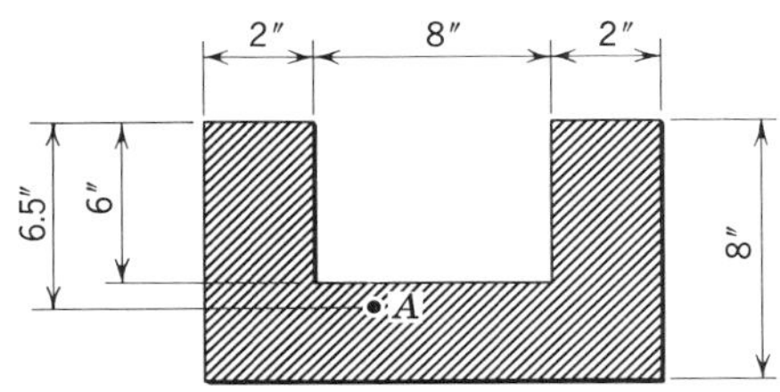


Fig. P5–2

**5–3.*** A beam is loaded so that the maximum fiber stress on the section of Fig. P5–3 is 9000 psi *C*. Determine the flexural force carried by the flange and its location on the cross section of the beam.

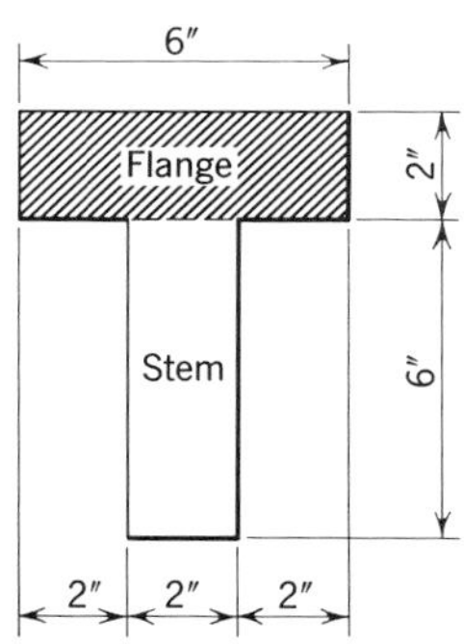


Fig. P5–3

**5–4.** If the maximum fiber stress on the beam cross section of Fig. P5–4 is 3500 psi *T*, determine the magnitude of the compressive force on the section.

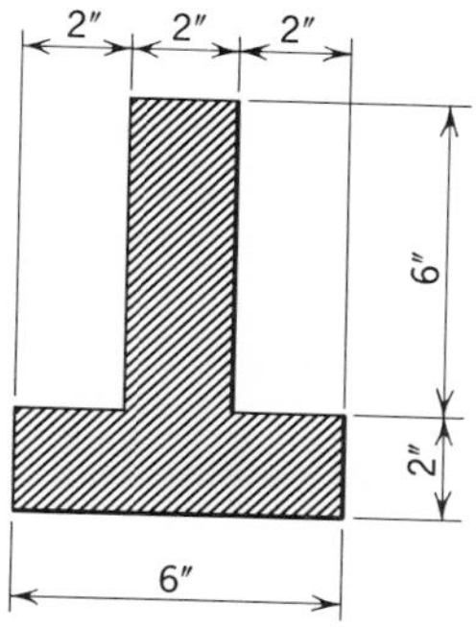


Fig. P5–4

**5–5.** The maximum fiber stress at a certain section in a rectangular beam 4 in. wide by 8 in. deep is 900 psi. Determine, by the free-body diagram method:

(*a*) The resisting moment in the beam at that section.

(*b*) The percentage decrease in this moment caused by removing the central (dotted portion) of the beam. See Fig. P5–5.

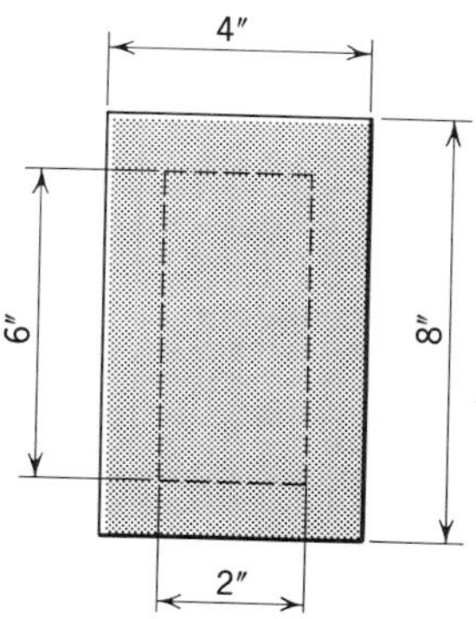


Fig. P5–5

**5–6.*** For the beam whose cross section is shown in Fig. P5–2 obtain an expression for the resisting moment by determining the magnitude and position of the components of the compressive and tensile forces. Express the result in terms of the maximum fiber stress.

**5–7.** A timber beam is made of three 2 by 6-in. planks fastened together to make an I-beam 6 in. wide by 10 in. deep. The allowable fiber stress is 1500 psi. Determine the maximum allowable bending moment to which the beam may be subjected. Use the free-body diagram method.

**5–8.*** A timber beam is made of three 2 by 4-in. members fastened together to make an I-beam 4 in. wide by 8 in. deep. Determine the maximum fiber stress on a section where the bending moment is 4000 ft-lb. Use the free-body diagram method.

**5–9.** An I-beam is fabricated by welding two 16 by 2-in. flange plates to a 28 by 1-in. web plate. The beam is loaded in the plane of symmetry parallel to the web. Determine, by the free-body diagram method, the percentage of the bending moment carried by the flanges.

**5–10.*** Solve Prob. 5–9 when the flange plates are 12 by 2-in. and the web plate is 20 by 1-in.

**5–4. The Flexure Formula.** The fundamental approach to the problem of flexural stress determination as outlined in Art. 5–3 becomes too time consuming for general use; therefore, a formula will be developed that has general applicability within the limitations noted in Art. 5–3. From Fig. 5–10, the moment of the differential force $dF$ with respect to $O$ is $y\,dF$, and the moment of all such forces on the cross section is

$$M_r = \int^A y\sigma\,dA.$$

Substitution of $\sigma = y\sigma_c/c$ from Eq. (a) of Art. 5–2 into this expression yields

$$M_r = \left(\frac{\sigma_c}{c}\right)\int^A y^2\,dA = \left(\frac{\sigma}{y}\right)\int^A y^2\,dA.$$

The integral is the familiar second moment of the cross-sectional area (commonly called "moment of inertia") *with respect to the centroidal axis*. When this integral is replaced by the symbol I, the flexure formula is obtained as

$$\sigma = \frac{My}{I}, \tag{5–1}$$

where $\sigma$ is the normal stress ($\sigma_x$ in general notation) at a distance $y$ from the neutral surface and on a transverse plane, and $M$ is the resisting moment of the section.

At any section of the beam, the fiber stress will be maximum at the surface farthest from the neutral axis, and Eq. (5–1) becomes

$$\sigma_{\max} = \frac{Mc}{I} = \frac{M}{Z}$$

where $Z = I/c$ is called the section modulus of the beam. Although the section modulus can be readily calculated for a given section, values of the modulus are often included in tables to simplify calculations. Observe that for a given area, $Z$ becomes larger as the shape is altered to concentrate more of the area as far as possible from the neutral axis. Commercial rolled shapes such as I- and H-beams and the various built-up sections are intended to optimize the area-section modulus relation. Note that in the problems involving reference to rolled shapes in App. C, the beam (unless otherwise noted) is so oriented that the maximum section modulus applies.

**Example 5–3:** Solve Ex. 5–2 by means of the flexure formula.

*Solution:* The second moment of the area can be obtained by integration or by composite areas,[5] and is 1600/3 in.[4] The maximum fiber stress is

$$\sigma_{\max} = \frac{5400(12)(8)}{1600/3} = \underline{972 \text{ psi } T \text{ at bottom.}} \qquad \text{Ans.}$$

## PROBLEMS

**5–11.** Solve Prob. 5–5 by means of the flexure formula.

**5–12.*** Solve Prob. 5–6 by means of the flexure formula.

**5–13.*** Solve Prob. 5–8 by means of the flexure formula.

**5–14.*** Solve Prob. 5–10 by means of the flexure formula.

**5–15.*** The beam loaded as shown in Fig. P5–15*a* has the cross section shown in Fig. P5–15*b*. At a point 4 ft from the left end and 2 in. above the neutral surface, determine the fiber stress.

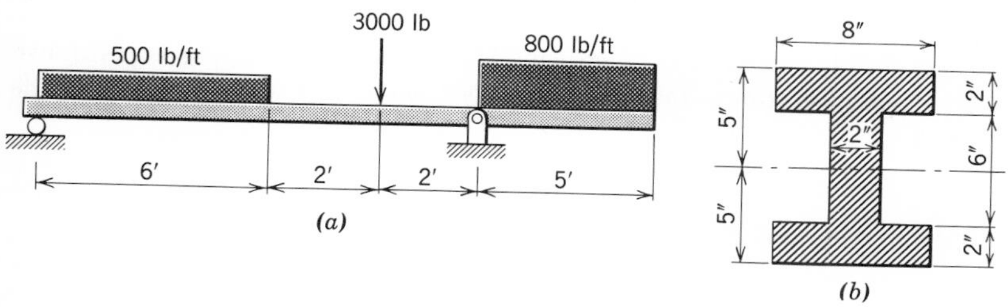


Fig. P5–15

---

[5] See App. B for formula; and *Engineering Mechanics*, vector Ed., A. Higdon and W. B. Stiles, Prentice-Hall, Englewood Cliffs, N.J., 1962, Chap. 6, for fundamentals.

**5–16.*** For the beam of Prob. 5–15, determine the fiber stress at a point 4 in. below the neutral surface on a section 4 ft from the right end.

**5–17.** For the beam of Prob. 5–15, determine the maximum fiber stress on a section 3 ft from the left end.

**5–18.*** Figure P5–18 shows the cross section of a beam loaded and supported as shown in Fig. P5–15. Determine the fiber stress on a section 9 ft from the left end:

(*a*) One inch above the neutral surface.
(*b*) One inch above the bottom of the section.

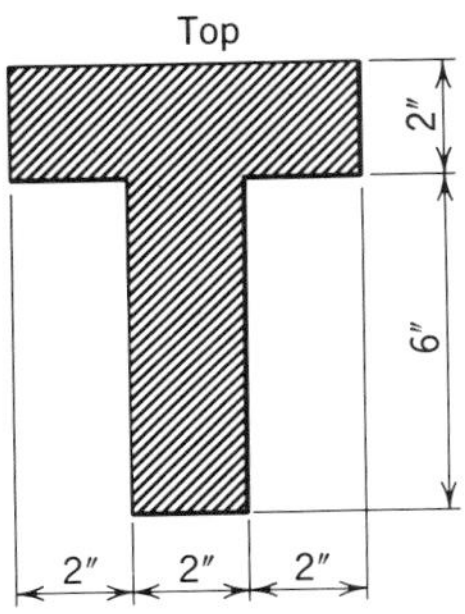


Fig. P5–18

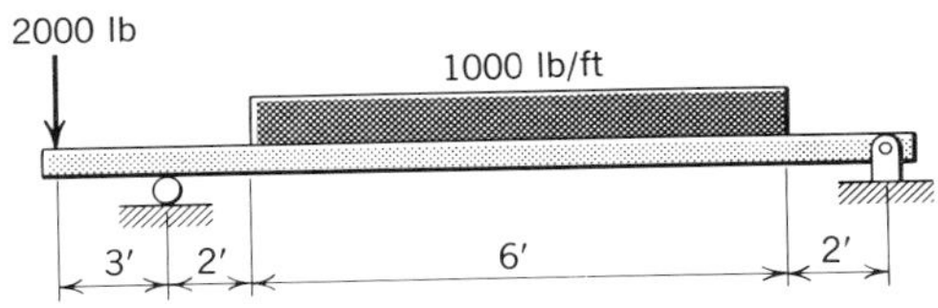


Fig. P5–19

**5–19.** A 7 I 15.3 steel beam (see App. C) is loaded and supported as shown in Fig. P5–19. Determine the maximum fiber stress on a section 7 ft from the right support.

**5–20.*** A 7 I 15.3 steel beam (see App. C) is loaded and supported as shown in Fig. P5–20. Determine the maximum fiber stress on a section 8 ft from the right end.

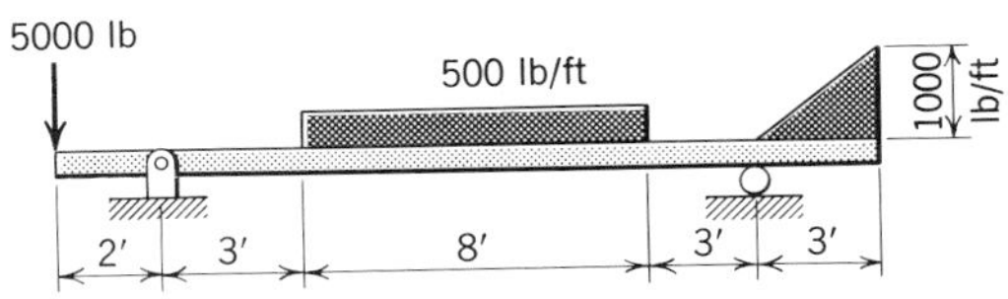


Fig. P5–20

**5–21.** An 8 WF 31 steel beam (see App. C) is loaded and supported as shown in Fig. P5–21. Determine the maximum fiber stress on a section 2 ft from the wall.

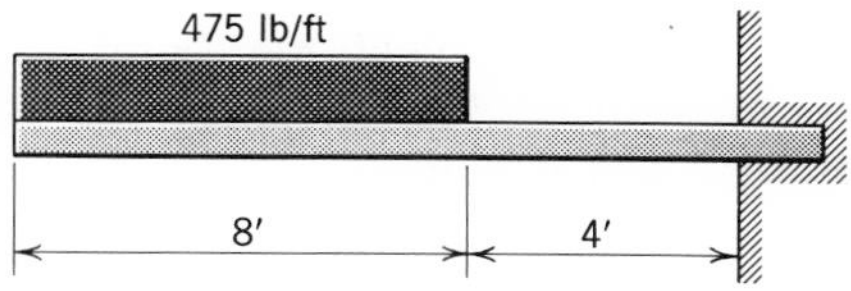


Fig. P5–21

**5–22.*** A 5 I 14.75 steel beam (see App. C) is loaded and supported as shown in Fig. P5–22. Determine the maximum fiber stress on a section 4 ft to the right of *B*.

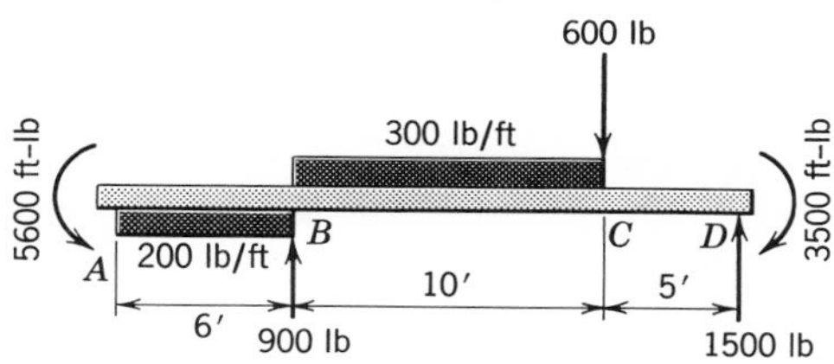


Fig. P5–22

**5–23.** For the beam shown in Fig. P5–23, the allowable fiber stresses on the section at the load are 6 ksi *T* and 10 ksi *C*. Determine the maximum allowable load *P*.

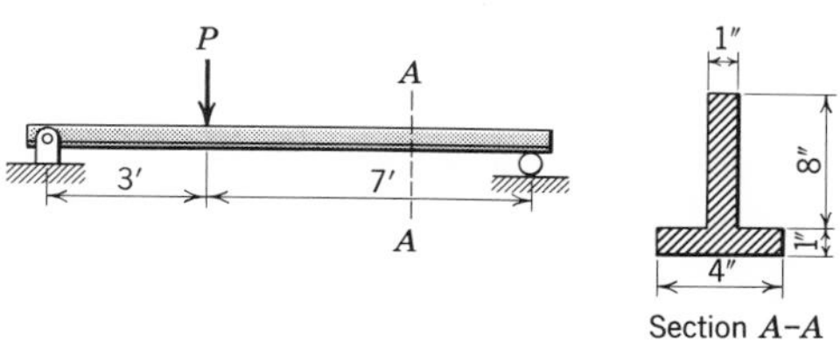


Fig. P5–23

**5–24.*** Solve Prob. 5–23 with an additional load, *P*, placed 3 ft to the left of the right support.

**5–5. Shear and Moment in Beams.** The procedures for determining flexural stresses outlined in Art. 5–3 and 5–4 are quite adequate if one wishes to compute the stresses at any specified section in a beam. However, if it is necessary to determine the maximum fiber stress in a given beam, it is desirable to have a convenient method for determining the maximum resisting moment.

In Art. 5–9 a relation will be developed between the resisting shear ($V_r$ of Fig. 5–4) and the longitudinal and transverse shearing stresses at the section. Since the maximum transverse shearing stress will occur at the section at which $V_r$ is maximum, a convenient method of determining such sections is likewise desirable.

When the equilibrium equation $\Sigma F_y = 0$ is applied to the free-body diagram of Fig. 5–4, the result can be written as

$$R - wx - P = V_r$$

or

$$V = V_r,$$

where $V$ is defined as the magnitude of the resultant of the external transverse forces acting on the part of a beam to either side of a section and is called the *transverse shear* or just the *shear* at the section.

As seen from the definitions of $V$ and $V_r$, the shear is equal in magnitude and opposite in sense to $V_r$. Since these shear forces are always equal in magnitude, they are frequently treated as though they were identical. For simplicity the symbol $V$ will be used henceforth to represent both the transverse shear and the resisting shear. The sign convention selected will have sufficient generality to apply to both quantities.

The resultant of the fiber stresses on any transverse section has been shown to be a couple (if only transverse loads are considered) and has been designated as $M_r$. When the equilibrium equation $\Sigma M_O = 0$ (where $O$ is any axis parallel to the neutral axis of the section) is applied to the free-body diagram of Fig. 5–4, the result can be written as

$$Rx - wx^2/2 - P(x - h) = M_r$$

or

$$M = M_r,$$

where $M$ is defined as the algebraic sum of the moments of the external forces, acting on the part of the beam to either side of the section, with respect to an axis in the section and is called the *bending moment* or just the *moment* at the section.

As seen from the definitions of $M$ and $M_r$, the bending moment is

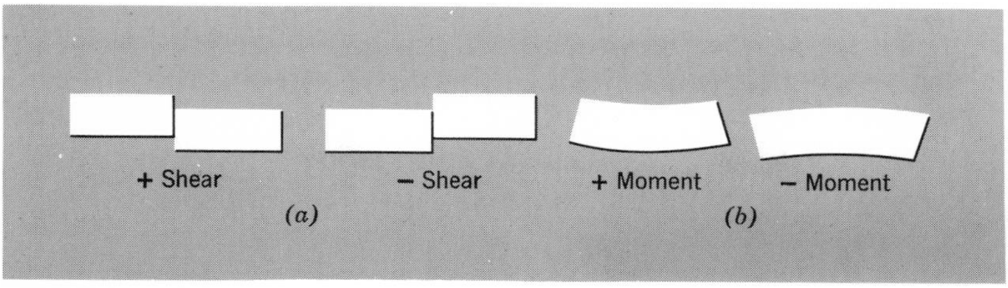


Fig. 5–13

equal in magnitude and opposite in sense to $M_r$. Since these moments are always equal in magnitude, they are frequently treated as though they were identical. For simplicity the symbol $M$ will be used henceforth to represent both the bending moment and the resisting moment. The sign convention selected will have sufficient generality to apply to both quantities.

The bending moment and the transverse shear are not normally shown on a free-body diagram. The usual procedure is to show each external force individually as indicated in Fig. 5–4. The variation of $V$ and $M$ along the beam can be shown conveniently by means of shear and moment diagrams as discussed in the following articles.

A sign convention is necessary for the correct interpretation of results obtained from equations or diagrams for shear and moment. The following convention will give consistent results regardless of whether one proceeds from left to right or from right to left. By definition, the shear at a section is positive when the portion of the beam to the left of the section (for a horizontal beam) tends to move upward with respect to the portion to the right of the section as shown in Fig. 5–13*a*. Also by definition, the bending moment in a horizontal beam is positive at sections for which the top of the beam is in compression and the bottom is in tension as shown in Fig. 5–13*b*. Observe that the signs of the terms in the preceding equations for $V$ and $M$ agree with this convention.

**5–6. Load, Shear, and Moment Relationships.** The mathematical relationships between loads and shears, and between shears and moments can be used to facilitate construction of shear and moment diagrams. The relationships can be developed from a free-body diagram of an elemental length of a beam as shown in Fig. 5–14 in which the upward

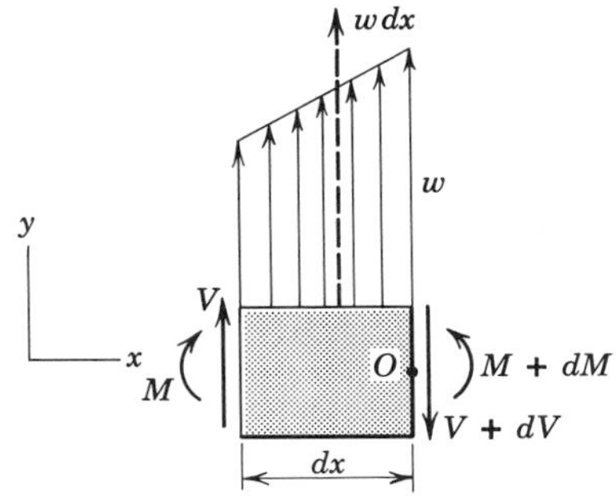


Fig. 5–14

direction is considered positive for the applied load, $w$, and the shears and moments are shown as positive according to the sign convention established in Art. 5–5. The element must be in equilibrium, and the equation $\Sigma F_y = 0$ gives

$$(\uparrow +) \qquad \Sigma F_y = V + w\,dx - (V + dV) = 0,$$

from which

$$dV = w\,dx \quad \text{or} \quad \frac{dV}{dx} = w.$$

The preceding equation indicates that at any section in the beam *the slope of the shear diagram is equal to the intensity of loading.* When $w$ is known as a function of $x$, the equation can be integrated between definite limits as follows:

$$\int_{V_1}^{V_2} dV = \int_{x_1}^{x_2} w\,dx = V_2 - V_1.$$

That is, *the change in shear between sections at $x_1$ and $x_2$ is equal to the area under the load diagram between the two sections.*

The equation $\Sigma M_O = 0$, applied to Fig. 5–14, gives

$$(\curvearrowleft +) \qquad \Sigma M_O = M + V\,dx + w\,dx(dx/2) - (M + dM) = 0,$$

from which

$$dM = V\,dx + w(dx)^2/2.$$

This expression reduces to

$$dM = V\,dx \quad \text{or} \quad V = \frac{dM}{dx},$$

when second-order differentials are considered as being negligible compared with first-order differentials. The preceding equation indicates that at any section in the beam *the slope of the moment diagram is equal to the shear.* The equation can be integrated between definite limits to give

$$\int_{M_1}^{M_2} dM = \int_{x_1}^{x_2} V\,dx = M_2 - M_1.$$

*Thus the change in moment between sections at $x_1$ and $x_2$ is equal to the area under the shear diagram between the two sections.*

Note that the equations in this article were derived with the $x$ axis positive to the right, the applied loads positive upward, and the shear and moment signs as indicated in Fig. 5–13. If one or more of these assumptions are changed, the algebraic signs in the equation may need to be altered.

The relations just developed will be useful in drawing shear and moment diagrams and in computing values of shear and moment at various sections along a beam as illustrated in the next article.

**5–7. Shear and Moment Diagrams.** As stated in Art. 5–5, shear and moment diagrams provide a convenient method of obtaining maximum values of shear and moment and other information which will be discussed later. *A shear diagram is a graph in which abscissas represent distances along the beam and ordinates represent the transverse shear at the corresponding sections. A moment diagram is a graph in which abscissas represent distances along the beam and ordinates represent the bending moment at the corresponding sections.*

Shear and moment diagrams can be drawn by calculating values of shear and moment at various sections along the beam and plotting enough points to obtain a smooth curve. Such a procedure is rather time consuming, and, although it may be desirable for graphical solutions of certain structural problems, other more rapid methods will be developed.

A convenient arrangement for constructing shear and moment diagrams is to draw a free-body diagram of the entire beam and construct shear and moment diagrams directly below. Two methods of procedure are presented in this article.

The first method consists of writing algebraic equations for the shear, $V$, and the moment, $M$, and constructing curves from the equations. This method has the disadvantage that unless the load is uniformly distributed or varies according to a known equation along the entire beam, no single elementary expression can be written for the shear, $V$, or the moment, $M$, which applies to the entire length of the beam. Instead, it is necessary to divide the beam into intervals bounded by the abrupt changes in the loading. An origin should be selected (different origins may be used for different intervals); positive directions should be shown for the coordinate axes, and the limits of the abscissa (usually $x$) should be indicated for each interval. Since $M$ and $V$ vary with $x$, they are functions of $x$, and equations for $M$ and $V$ can be obtained from free-body diagrams of portions of the beam as illustrated in Ex. 5–4 and 5–5.

Complete shear and moment diagrams should indicate values of shear and moment at each section where the load changes abruptly and at sections where they are maximum or minimum (negative maximum values). Sections where the shear and moment are zero should also be located.

The second method consists of drawing the shear diagram from the load diagram and the moment diagram from the shear diagram by means of the mathematical relationships developed in Art. 5–6. The latter method, though it may not produce a precise curve, is less time-consuming than the first and it does provide the information usually required.

When all loads and reactions are known, the shear and moment at the ends of the beam can be determined by inspection. Both shear and moment are zero at a free end of a beam. At a simply supported or pinned end, the shear must equal the end reaction and the moment must be zero. At a built-in or fixed-end beam, the reactions are the shear and moment values.

Once a starting point for the shear diagram is established, the diagram can be sketched by using the definition of shear and the fact that the slope of the shear diagram can be obtained from the load diagram. When positive directions are chosen as upward and to the right, a positive distributed load, one acting upward, will result in a positive slope on the shear diagram, and a negative load will give a negative slope. A concentrated force will produce an abrupt change in shear. The *change* in shear between any two sections is given by the area under the load diagram between the same two sections. The change of shear at a concentrated force is equal to the concentrated force.

The moment diagram is drawn from the shear diagram in the same manner. The slope at any point on the moment diagram is given by the shear at the corresponding point on the shear diagram, a positive shear representing a positive slope and a negative shear representing a negative slope, *when upward and to the right are positive.* The *change* in moment between any two sections is given by the area under the shear diagram between the corresponding sections. A couple applied to a beam will cause the moment to change abruptly by an amount equal to the moment of the couple. Examples 5–4 and 5–5 illustrate these two methods for construction of shear and moment diagrams.

**Example 5–4:** A beam is loaded and supported as shown in Fig. 5–15*a*.

(*a*) Write equations for the shear and bending moment for any section of the beam in the interval $BC$.

(*b*) Draw complete shear and moment diagrams for the beam.

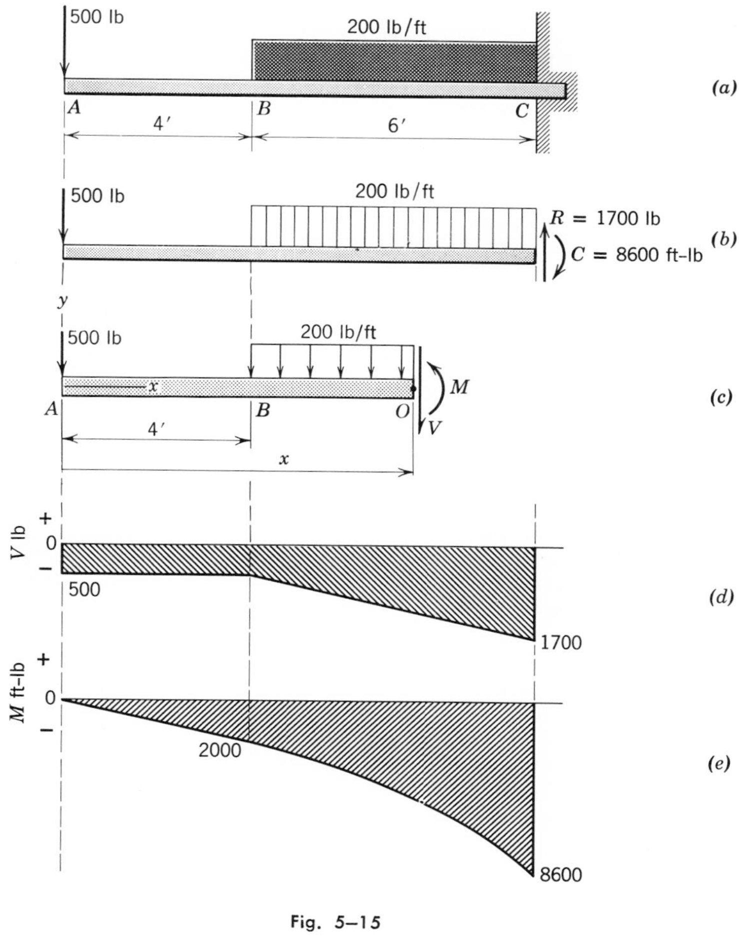


Fig. 5–15

*Solution:* A free-body diagram, or load diagram, for the beam is shown in Fig. 5–15*b*. The reactions at *C* are computed from the equations of equilibrium. It is not necessary to compute the reactions on a cantilever beam in order to write equations for the shear and bending moment or to draw shear and moment diagrams from the load diagram, but the reactions provide a convenient check.

(*a*) A free-body diagram of a portion of the beam from the left end to any section between *B* and *C* is shown in Fig. 5–15*c*. Note that the resisting shear, *V*, and the resisting moment, *M*, are shown as positive values. From the definition of *V* or from the equilibrium equation $\Sigma F_y = 0$,

$$V = -500 - 200(x - 4) = \underline{300 - 200x}. \qquad 4 \leq x \leq 10 \qquad \text{Ans.}$$

From the definition of *M*, or from the equilibrium equation $\Sigma M_0 = 0$,

$$\begin{aligned} M &= -500x - 200(x - 4)(x - 4)/2 \\ &= \underline{-100x^2 + 300x - 1600}. \qquad 4 \leq x \leq 10 \qquad \text{Ans.} \end{aligned}$$

(*b*) The equations for *V* and *M* in the interval *AB* can be written in the same manner as in part (*a*) after which the two equations for *V* can be plotted in the appropriate intervals to give the shear diagram in Fig. 5–15*d*. Likewise, the two equations for *M* can be plotted in the appropriate intervals to give the moment diagram in Fig. 5–15*e*.

The shear and moment diagrams can also be drawn, working directly from the load diagram without writing the shear and moment equations.

The shear diagram is drawn below the load diagram, Fig. 5–15*d*. The shear just to the right of the 500-lb load is −500 lb. The slope of the shear diagram is equal to the load; and, since no load is applied between *A* and *B*, the slope of the diagram is zero. From *B* to *C* the load is uniform; therefore the slope of the shear diagram is constant. The *change* of shear from *B* to *C* is equal to the applied load (1200 lb) from *B* to *C*, and the shear is −1700 lb at *C*. This shear is the same as the reaction at *C* which provides a check.

The moment diagram is drawn below the shear diagram as Fig. 5–15*e*. The moment is zero at the free end *A*. From *A* to *B* the shear is constant (−500); therefore the slope of the moment diagram is constant, and the moment diagram is a straight line. The shear increases (negatively) from −500 lb at *B* to −1700 lb at *C*; therefore, the slope of the moment diagram is negative from *B* to *C* and increases uniformly in magnitude. The change in moment from *A* to *B* is equal to the area under the shear diagram and is 500(4) = 2000 ft-lb. The area under the shear diagram from *B* to *C* is $(\frac{1}{2})(500 + 1700)(6) =$ 6600 ft-lb which represents the change in moment from *B* to C. Thus the moment at *C* is −2000 + (−6600) = −8600 ft-lb.

**Example 5–5:** A beam is loaded and supported as shown in Fig. 5–16*a*.

(*a*) Write equations for the shear and bending moment for any section of the beam in the interval *CD*.

(*b*) Draw complete shear and moment diagrams for the beam.

*Solution:* The reactions shown on the load diagram, Fig. 5–16*b*, are determined from the equations of equilibrium.

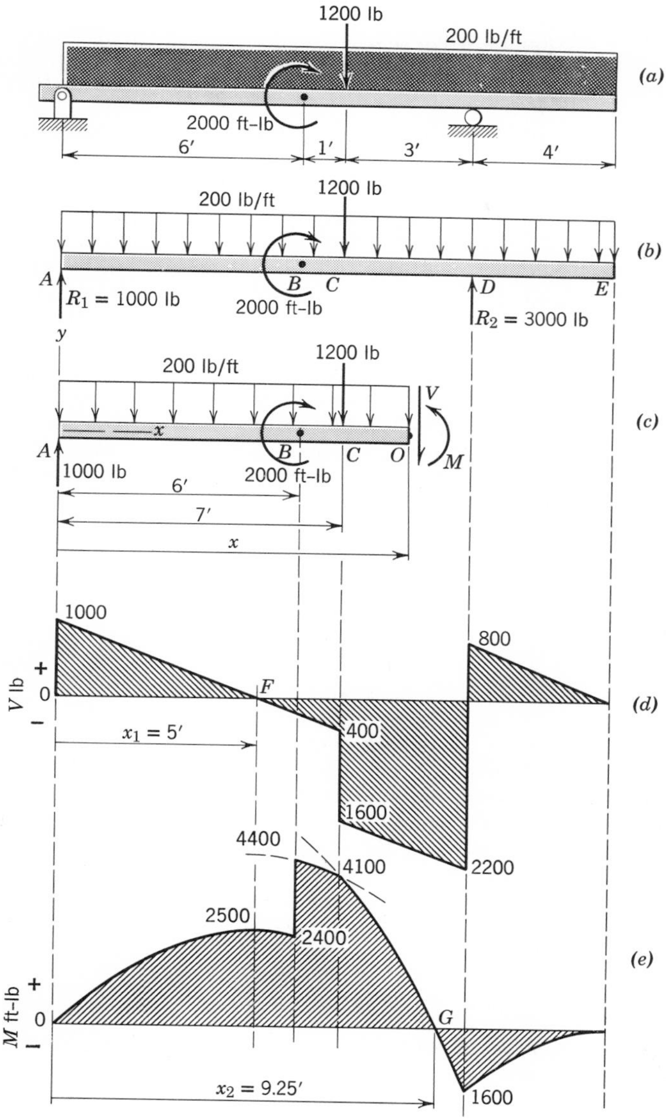

1200 lb
200 lb/ft
(a)
2000 ft-lb
6′
1′
3′
4′
200 lb/ft
1200 lb
(b)
A
$R_1$ = 1000 lb
B
C
2000 ft-lb
D
E
$R_2$ = 3000 lb
y
1200 lb
200 lb/ft
V
(c)
x
A
1000 lb
B
C
O
M
2000 ft-lb
6′
7′
x
1000
800
V lb
+
0
−
F
(d)
$x_1$ = 5′
400
1600
4400
4100
2200
2500
2400
(e)
M ft-lb
+
0
−
G
$x_2$ = 9.25′
1600


Fig. 5–16

(*a*) A free-body diagram of a portion of the beam from the left end to any section between $C$ and $D$ is shown in Fig. 5–16*c*. From the definition of $V$, or from the equilibrium equation $\Sigma F_y = 0$

$$V = 1000 - 200x - 1200 = \underline{-200x - 200.} \qquad 7 \le x \le 10 \qquad \text{Ans.}$$

From the definition of $M$, or from the equilibrium equation $\Sigma M_O = 0$

$$\begin{aligned} M &= 1000x - 200x(x/2) + 2000 - 1200(x - 7) \\ &= \underline{10{,}400 - 200x - 100x^2.} \qquad 7 \le x \le 10 \qquad \text{Ans.} \end{aligned}$$

(*b*) The equations for $V$ and $M$ in the other intervals can be written and the shear and moment diagrams obtained by plotting these equations. In this example, the second method is used and the shear diagram is drawn directly from the load diagram.

The shear just to the right of $A$ is 1000 lb and decreases at the rate of 200 lb each foot to $-400$ lb at $C$. The concentrated downward load at $C$ causes the shear to change suddenly from $-400$ lb to $-1600$ lb. The shear changes uniformly to $-2200$ lb just to the left of $D$. The reaction at $D$ changes the shear by 3000 lb (from $-2200$ lb to $+800$ lb). The shear decreases uniformly to zero at $E$. Note that the distributed load is uniform over the entire beam; consequently the slope of the shear diagram is constant. Points of zero shear, such as $F$ of Fig. 5–16*d*, are located from the geometry of the shear diagram. For example, the slope of the left portion of the shear diagram is 200 lb per ft. Therefore

$$\text{slope} = 200 = 1000/x_1,$$

and

$$x_1 = 5.00 \text{ ft.}$$

The moment is zero at $A$, and the slope of the moment diagram (equal to the shear) is 1000 ft-lb per ft. From $A$ to $C$ the shear and, hence, the slope of the moment diagram, decreases uniformly to zero at $F$ and to $-400$ at $C$. The abrupt change of shear at $C$ indicates a sudden change of slope of the moment diagram; thus, the two parts of the moment diagram at $C$ are not tangent. The slope of the moment diagram changes from $-1600$ at $C$ to $-2200$ just to the left of $D$. From $D$ to $E$ the slope changes from $+800$ to zero.

The change of moment from $A$ to $F$ is equal to the area under the shear diagram from $A$ to $F$ and is

$$\Delta M = 1000(5)/2 = 2500 \text{ ft-lb},$$

which is the moment at $F$ since the moment at $A$ is zero. From $F$ to $B$, $\Delta M = -100$ ft-lb and $M_B = +2400$ ft-lb. The 2000-ft-lb couple (point moment) is applied at $B$. Since the moment just to the left of $B$ is positive and the couple contributes an additional positive moment to sections to the right of $B$, the moment changes abruptly from $+2400$ ft-lb to $+4400$ ft-lb. Moments at $C$, $D$, and $E$ can be determined from the shear diagram areas which give the changes in moment from section to section. If the moment at $E$ is not equal to zero, it indicates that an error has occurred.

The point of inflection $G$, where the moment is zero, can be determined by setting the expression for the moment from part ($a$) equal to zero and solving for $x$. The result is

$$x_2 = -1 \pm \sqrt{105} = 9.247 = \underline{9.25 \text{ ft.}} \qquad \text{Ans.}$$

Note in the example that maximum and minimum moments may occur at sections where the shear curve passes through zero. In general, the shear curve may pass through zero at a number of points along the beam and each such crossing indicates a point of *possible* maximum moment (in engineering, the moment with the largest absolute value is the maximum moment). It should be emphasized that the shear curve does not indicate the presence of abrupt discontinuities in the moment curve; hence, the maximum moment may occur where a couple is applied to the beam; not where the shear passes through zero. All possibilities should be examined to determine the maximum moment.

Sections where the bending moment is zero, called points of inflection or contraflexure, can be located by equating the expression for $M$ to zero. The fiber stress is zero at such sections, and if a beam must be spliced, the splice should be located at or near a point of inflection if there is one.

## PROBLEMS

**5–25.** For the beam in Fig. P5–25, and using the coordinate axes shown:

($a$) Write equations for the shear and moment for any section between the supports.

($b$) Draw the shear diagram and indicate sections of possible maximum moment.

($c$) Without drawing a moment diagram, determine the maximum moment in the beam.

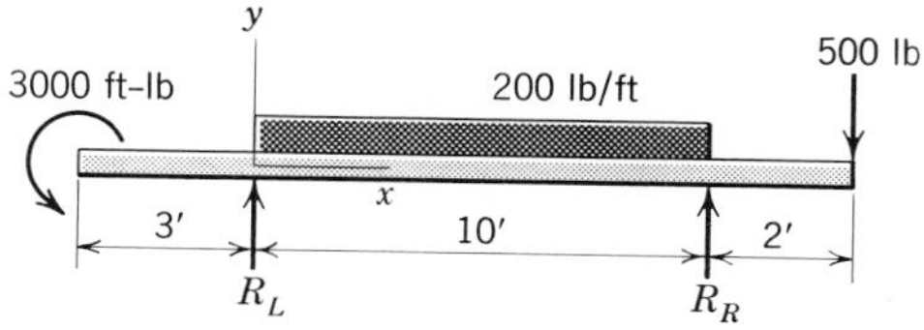


Fig. P5–25

**5–26.** For the beam of Fig. P5–26, and using the coordinate axes shown:

(*a*) Write the equations for the shear and moment for any section in the interval from *B* to *C*.

(*b*) Draw the shear diagram and indicate sections of possible maximum moment.

(*c*) Without drawing a moment diagram, determine the maximum moment in the beam.

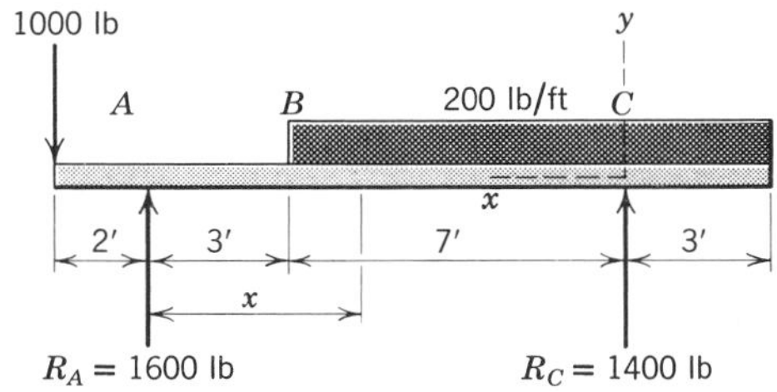


Fig. P5–26

**5–27.*** For the beam in Fig. P5–27:

(*a*) Write equations for the shear and moment for any section in the interval $0 \leq x \leq 6$.

(*b*) Draw complete shear and moment diagrams for the beam.

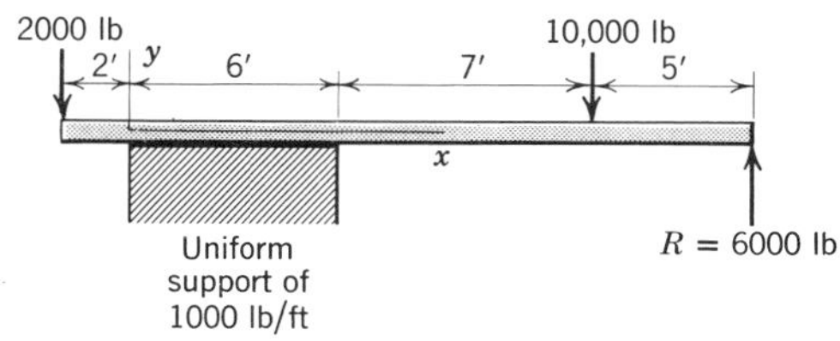


Fig. P5–27

**5–28.*** For the beam in Fig. P5–28:

(*a*) Write equations for the shear and moment for any section in the interval $0 \leq x \leq 12$.

(*b*) Draw complete shear and moment diagrams for the beam.

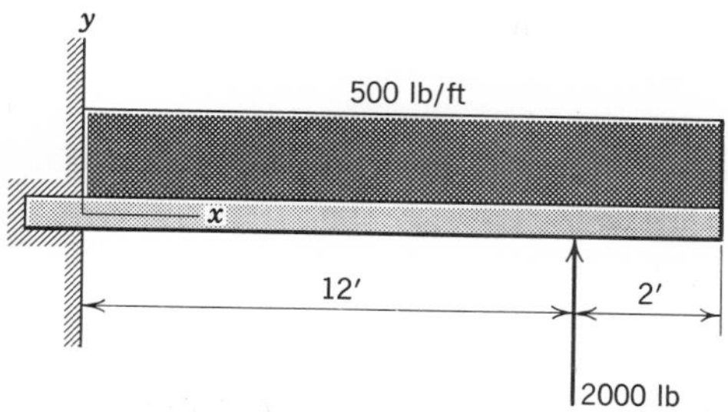


Fig. P5–28

**5–29.*** For the beam loaded and supported as shown in Fig. P5–29:

(*a*) Write equations for the shear and moment for any section in the interval $-4 \leq x \leq 0$.

(*b*) Draw complete shear and moment diagrams for the beam.

(*c*) Determine the minimum allowable section modulus for the beam for a working fiber stress of 20 ksi.

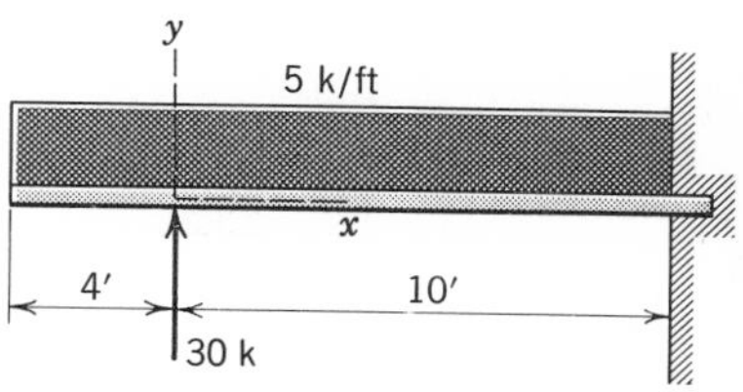


Fig. P5–29

**5–30.** For the beam of Fig. P5–30:

(*a*) Write equations for the shear and moment for any section in the center 5-ft interval.

(*b*) Draw complete shear and moment diagrams for the beam.

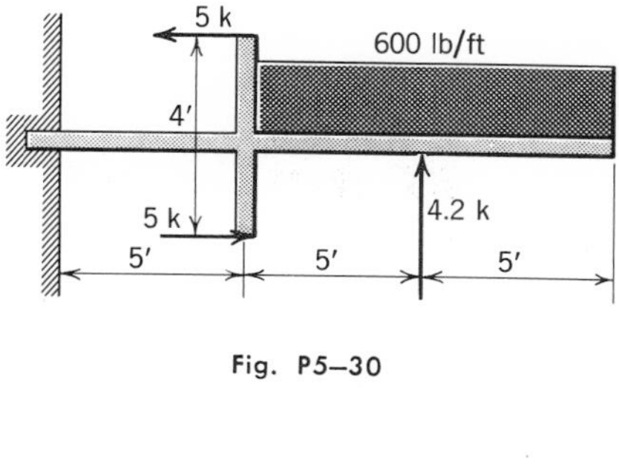


Fig. P5–30

**5–31.** For the beam of Fig. P5–31, and using the given coordinate axes:

(*a*) Write the equations for the shear and moment at any section of the beam in the interval $0 \le x \le 6$.

(*b*) Draw complete shear and moment diagrams for the beam.

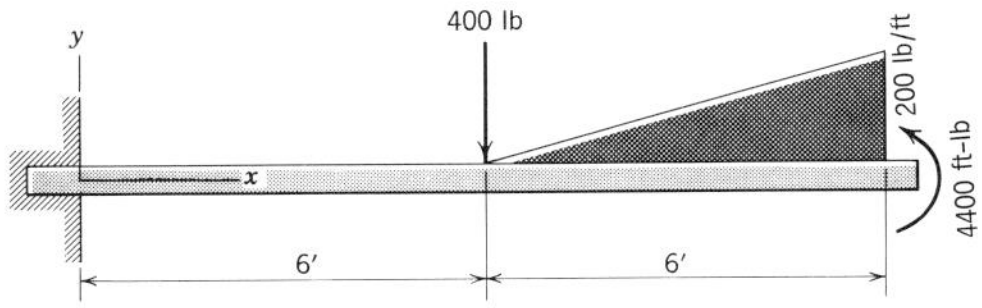


Fig. P5–31

**5–32.*** For the beam of Fig. P5–32:

(*a*) Write the equations for the shear and moment at any section of the beam.

(*b*) Draw complete shear and moment diagrams for the beam.

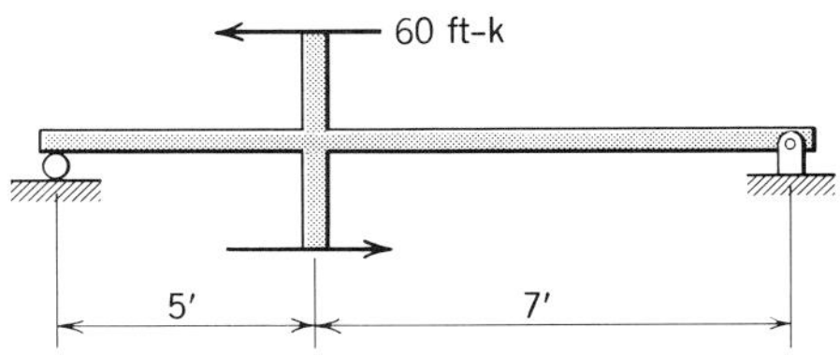


Fig. P5–32

**5–33.** For the beam of Fig. P5–33:

(*a*) Draw complete shear and moment diagrams.

(*b*) Write the equations for the shear and moment at any section in the interval $2 \le x \le 8$.

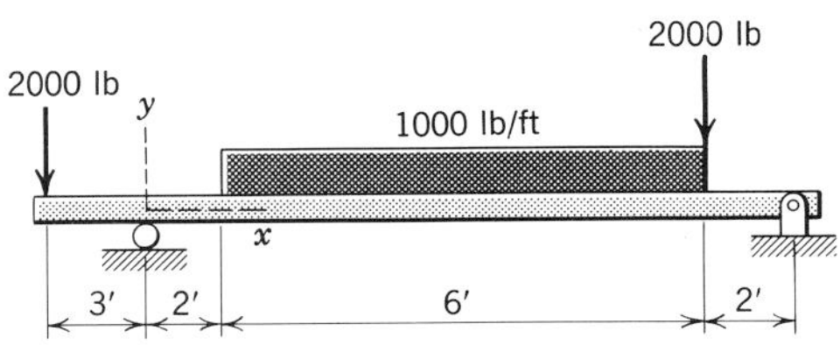


Fig. P5–33

**5–34.*** For the beam in Fig. P5–34, and using the coordinate axes shown:

(*a*) Write the equations for the shear and moment at any section of the beam in the interval $15 \leq x \leq 20$.

(*b*) Draw complete shear and moment diagrams for the beam.

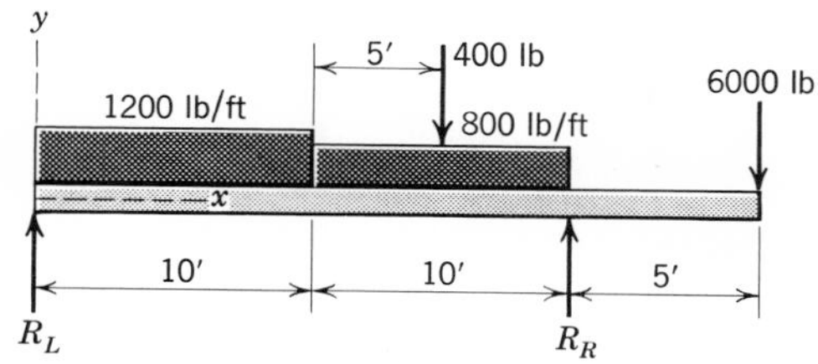


Fig. P5–34

**5–35.*** For the beam in Fig. P5–35:

(*a*) Write the equations for the shear and moment at any section of the beam.

(*b*) Draw complete shear and moment diagrams for the beam.

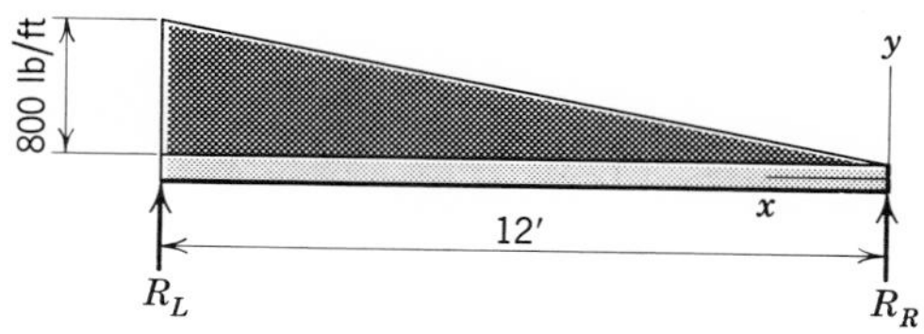


Fig. P5–35

**5–36.** For the beam loaded and supported as shown in Fig. P5–36:

(*a*) Write the equations for the shear and moment at any section of the beam in the uniformly loaded interval.

(*b*) Draw complete shear and moment diagrams for the beam.

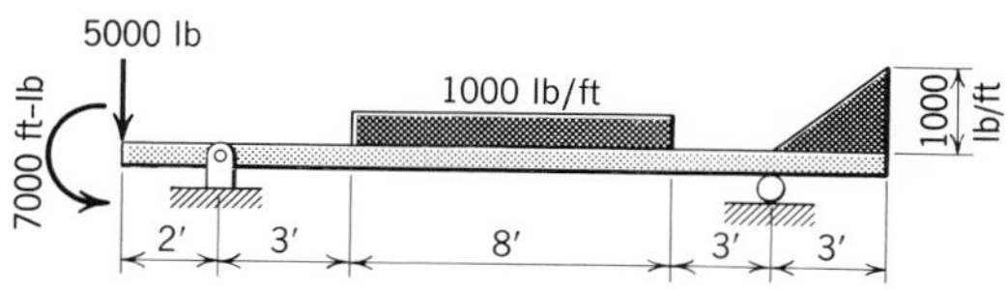


Fig. P5–36

**5–37.*** For the beam in Fig. P5–37:

(*a*) Write the equations for the shear and bending moment at any section of the beam in the interval $0 \leq x \leq 4$.

(*b*) Draw complete shear and moment diagrams for the beam.

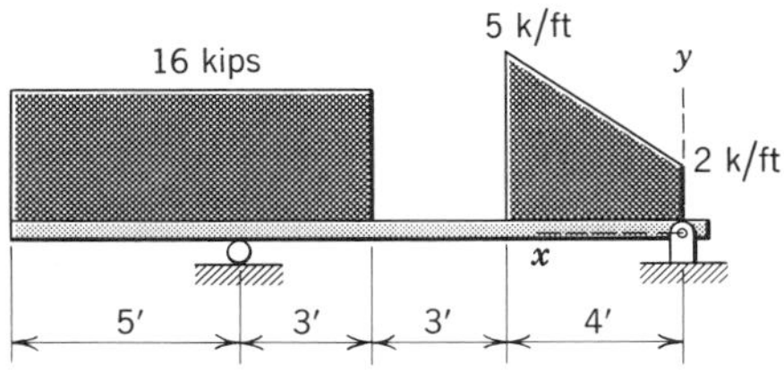


Fig. P5–37

**5–38.** (*a*) Draw complete shear and moment diagrams for the beam in Fig. P5–38.

(*b*) Write equations for the shear and moment at any section in the interval *BC*.

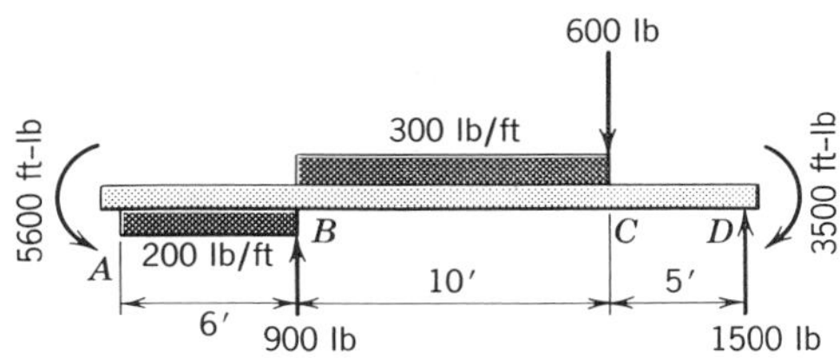


Fig. P5–38

**5–39.*** (*a*) Construct complete shear and moment diagrams for the beam of Fig. P5–39.

(*b*) Write the moment equation for the interval from *C* to *D* of Fig. P5–39.

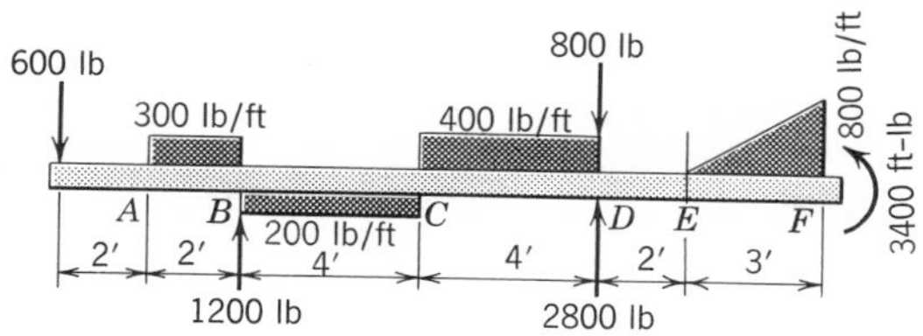


Fig. P5–39

**5–40.** Figure P5–40*a* shows a wooden airplane wing rib. For convenience, assume a chordwise load distribution as shown in Fig. P5–40*b*, and draw complete shear and moment diagrams for the rib.

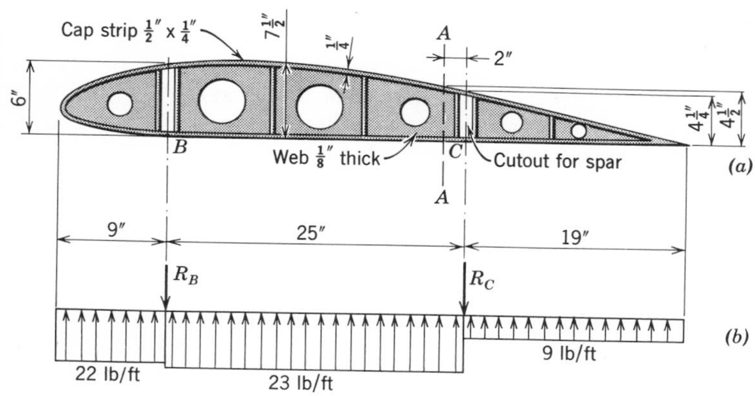


Fig. P5–40

**5–41.*** In Fig. P5–41 is shown one span of a continuous beam. The moment at support $A$ is $-wL^2/20$ ft-lb and that at support $B$ is $-wL^2/10$ ft-lb. Draw complete shear and moment diagrams for the portion of the beam between supports $A$ and $B$.

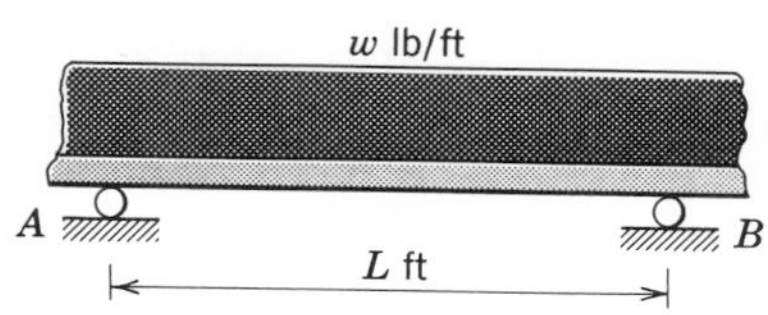


Fig. P5–41

**5–42.** The beam of Fig. P5–42 has a pin-connected joint at $C$. The moment and shear at $A$ are $+6000$ ft-lb and $+1200$ lb, respectively. Draw complete shear and moment diagrams for the length $ABCD$ of the beam, and write the moment equation for the interval $BCD$.

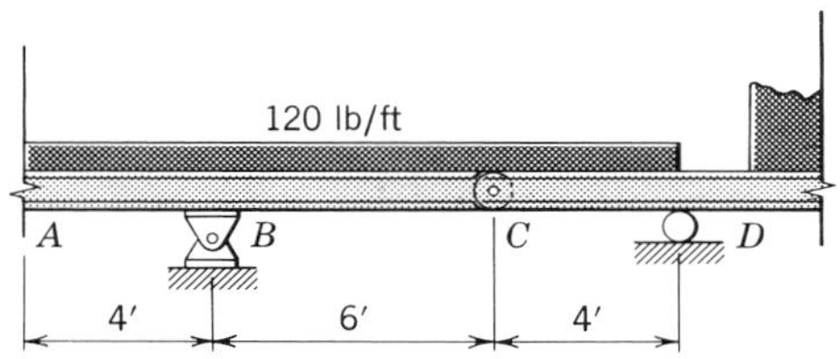


Fig. P5–42

**5–43.*** A barge which weighs 1000 lb per ft of length (not to be neglected) is loaded as shown in Fig. P5–43.

($a$) Determine the length $L$ if the barge is to float level (i.e., the water pressure must be uniform on the bottom of the barge).

($b$) Draw complete shear and moment diagrams for the barge.

($c$) Write the moment equation for the interval $BC$.

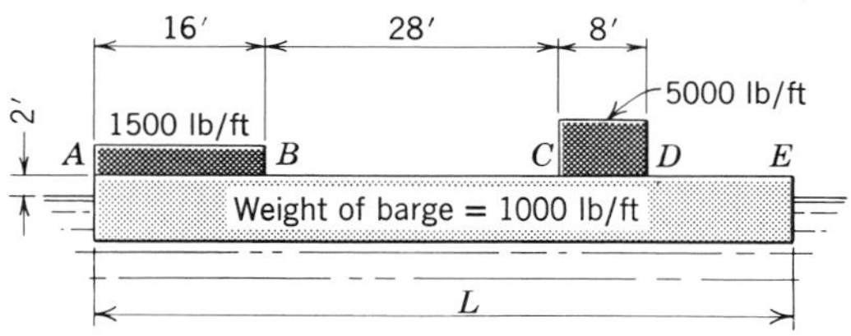


Fig. P5–43

**5–44.** ($a$) Draw complete shear and moment diagrams for the beam of Fig. P5–44.

($b$) Write the moment equation for the interval $EF$ of Fig. P5–44.

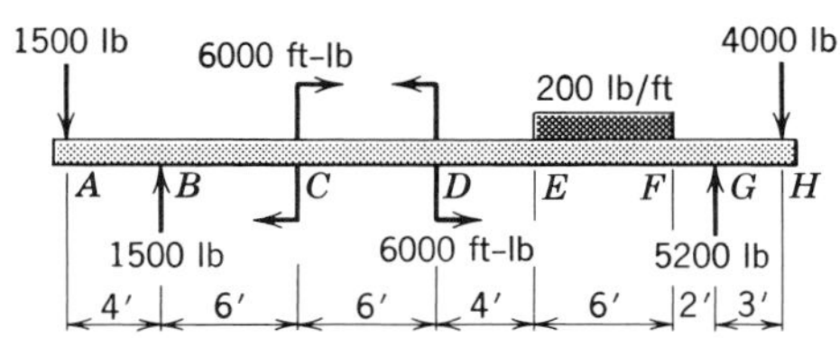


Fig. P5–44

**5–45.*** Draw complete shear and moment diagrams for the beam of Fig. P5–45, and write the moment equation for the interval $BC$.

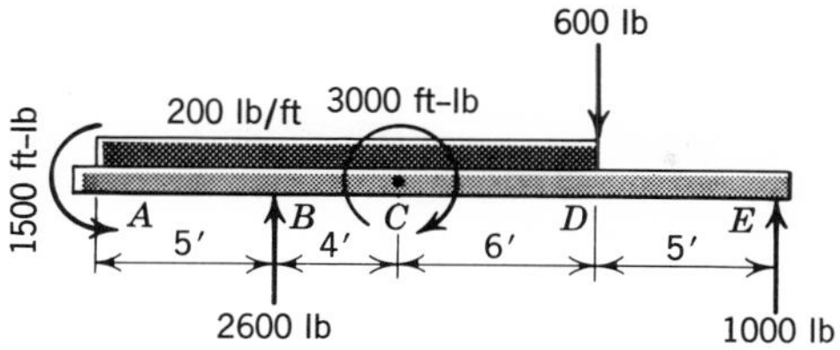


Fig. P5–45

**5–46.** Draw complete shear and moment diagrams for the beam of Fig. P5–46, and write the moment equation for the interval $BC$.

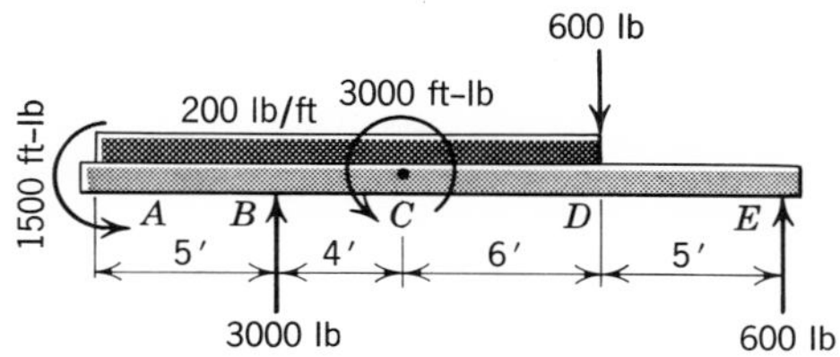


Fig. P5–46

**5–47.*** Draw complete shear and moment diagrams for the beam of Fig. P5–47, and write the moment equation for the interval $CD$.

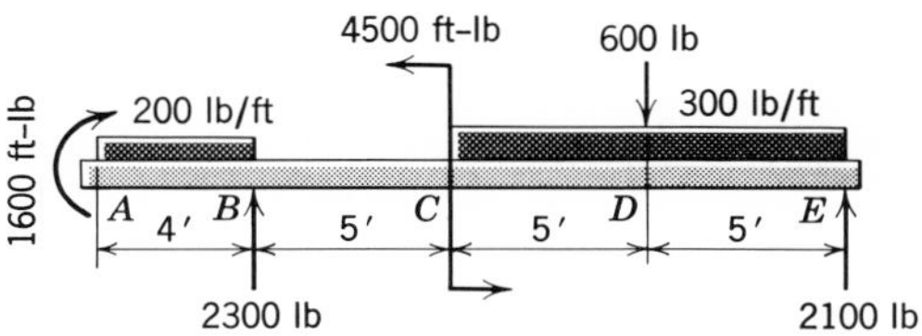


Fig. P5–47

**5–48.** Draw complete shear and moment diagrams for the beam of Fig. P5–48, and write the moment equation for the interval *DE*.

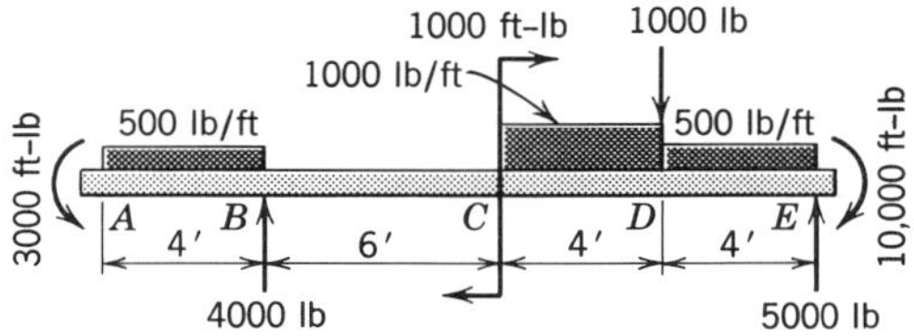


Fig. P5–48

**5–49.** (*a*) Draw complete shear and moment diagrams for the beam of Fig. P5–49.

(*b*) Write the moment equation for the interval *AB* of the beam of part (*a*).

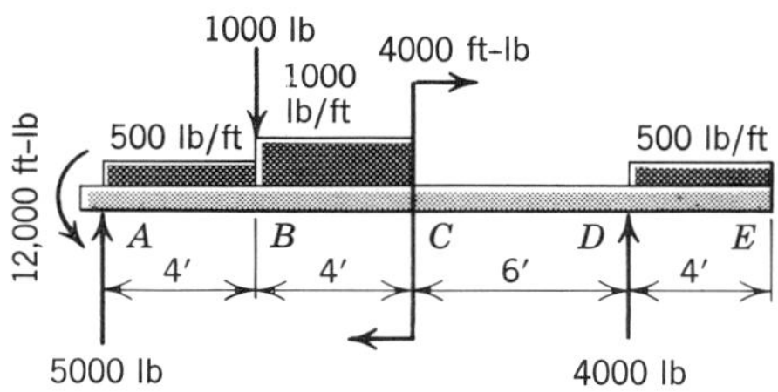


Fig. P5–49

**5–50.*** For the beam loaded as shown in Fig. P5–50, draw the complete shear and moment diagrams. Write the moment equation for the interval *EF*.

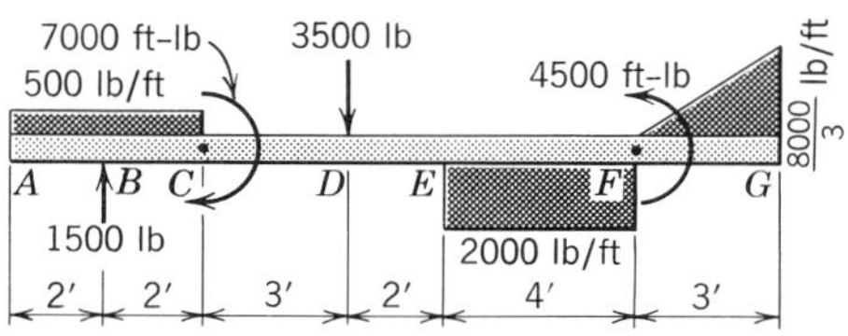


Fig. P5–50

**5–51.** Using a working fiber stress of 20,000 psi, select the lightest allowable steel I-beam which may be used for the beam in Fig. P5–31.

**5–52.** Using a working fiber stress of 18,000 psi, select the lightest allowable steel WF-beam which may be used for the beam in Fig. P5–36.

**5–53.** Select the lightest rolled steel I- or WF-beam which may be used for the beam in Fig. P5–44. Use a working fiber stress of 16,000 psi.

**5–54.** Select the lightest rolled steel WF- or I-beam which may be used for the beam in Prob. 5–47. Use a working fiber stress of 18,000 psi.

**5–55.*** If the beam in Prob. 5–42 has the cross section shown in Fig. P5–55, determine the maximum fiber stress between sections $A$ and $D$, and state where it occurs.

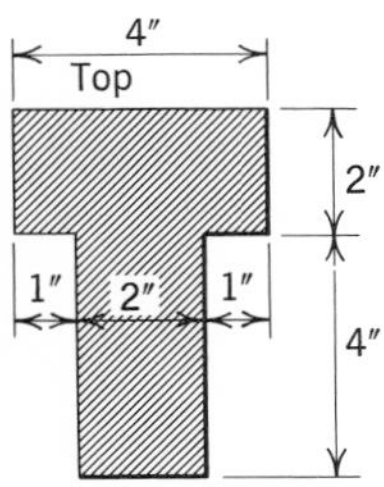


Fig. P5–55

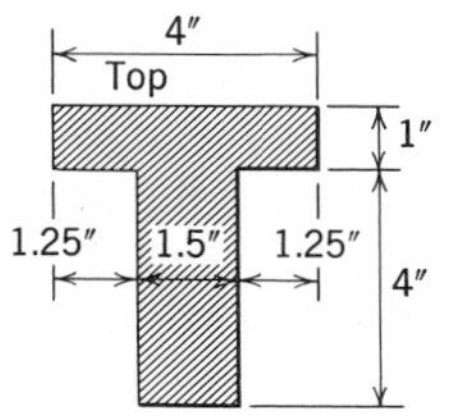


Fig. P5–56

**5–56.** The beam in Prob. 5–45 has the cross section shown in Fig. P5–56. Determine the maximum compressive fiber stress in the beam, and state where it occurs.

**5–57.*** The beam in Prob. 5–49 has the cross section shown in Fig. P5–57. Determine the maximum tensile fiber stress in the beam and state where it occurs.

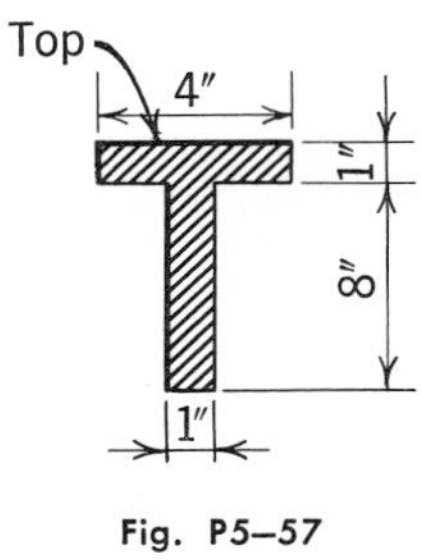


Fig. P5–57

**5–58.** The shear diagram for a beam, with zero moment on each end and with no couples applied between the ends, is shown in Fig. P5–58. Draw the free-body diagram and the moment diagram for the beam.

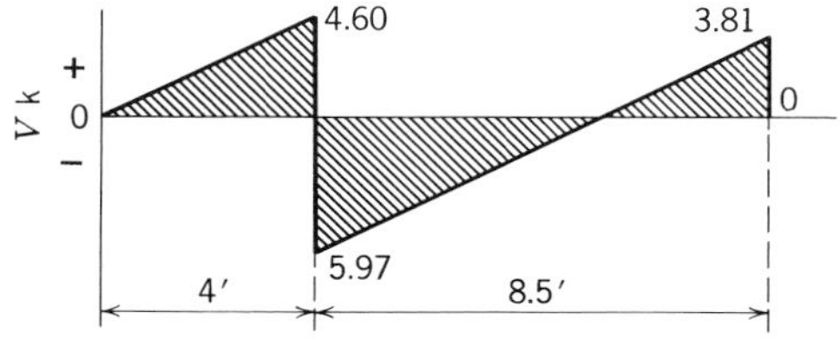


Fig. P5–58

**5–59.** The shear diagram for a beam fixed at the right end is shown in Fig. P5–59. There are no applied couples anywhere else along the beam. Draw the load and moment diagrams for the beam.

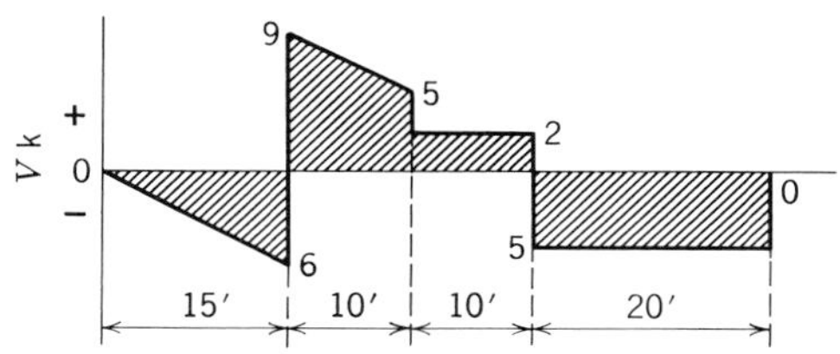


Fig. P5–59

**5–60.** The moment diagram for a beam is shown in Fig. P5–60. Construct the shear and load diagrams for the beam.

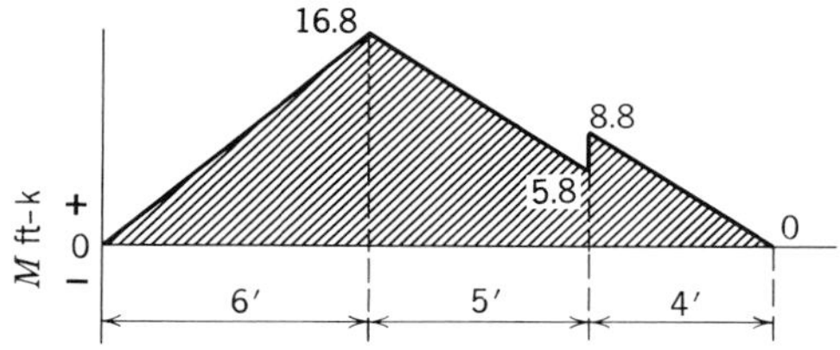


Fig. P5–60

**5–61.** Draw shear and load diagrams for the beam which has the moment diagram in Fig. P5–61.

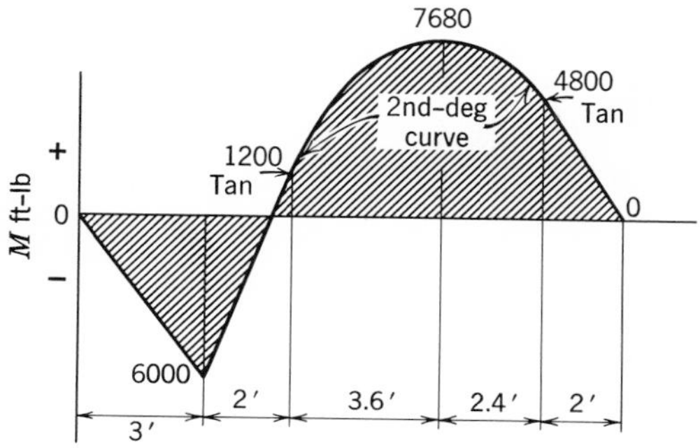


Fig. P5–61

**5–8. Shearing Stresses by Equilibrium.** In Art. 5–2, the discussion of shearing stresses in beams was by-passed while flexural stresses were studied. This procedure seems to be in keeping with the historical record on the study of beam stresses. From the time of Coulomb's paper which contained the correct theory of distribution of flexural stresses, approximately seventy years elapsed before D. J. Jourawski (1821–1891), a Russian engineer, while designing timber railroad bridges in 1844–1850, developed the elementary shear stress theory which is used today. Saint-Venant in 1856 developed a rigorous solution for beam shearing stresses, but the elementary solution of Jourawski, being much easier to apply and quite adequate, is the one in general use today by engineers and architects, and the one that will be presented here.

If one constructs a beam by stacking flat slabs one on top of another without fastening them together, and then loads this beam in a direction normal to the surface of the slabs, the resulting deformation will appear somewhat like that in Fig. 5–17*a*. The same type of deformation can be observed by taking a pack of cards and bending them, noting the relative motion of the ends of the cards with respect to each other. The fact that a solid beam does not exhibit this relative movement of longitudinal elements (see Fig. 5–17*b*, in which the beam is identical to that of Fig. 5–17*a* except that the layers are glued together) indicates the presence of shearing stresses on longitudinal

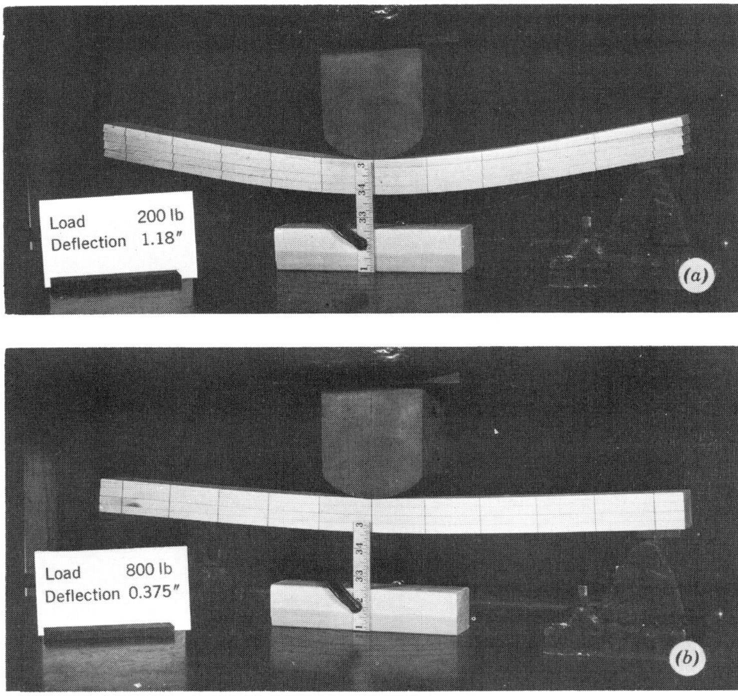


Fig. 5–17

planes. The evaluation of these shearing stresses will now be studied by means of the free-body diagram and equilibrium approach, as outlined in the following example.

**Example 5–6:** A 12-in. horizontal segment of the beam of Ex. 5–2 in Art. 5–3 is subjected to a moment of +5400 ft-lb at the left end and a constant shear of +900 lb. Determine the average shearing stress on a horizontal plane 4 in. above the bottom of the beam.

*Solution:* A free-body diagram of the segment is shown in Fig. 5–18*a*, where *AB* designates the plane on which the shearing stress is to be evaluated. The moment, $M_2$, obtained from a moment equilibrium equation, is +6300 ft-lb. A free-body diagram of the portion of the beam segment below plane *AB* is shown in Fig. 5–18*b*. The force $V_H$ is the resultant of the shearing stresses on plane *AB*. The forces $F_1$ and $F_2$ are the resultants of the normal stresses on the ends of the segment and are equal to the average fiber stress (which oc-

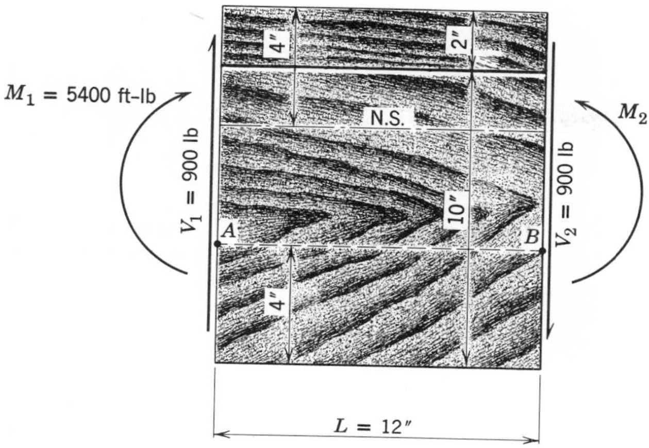


*(a)*

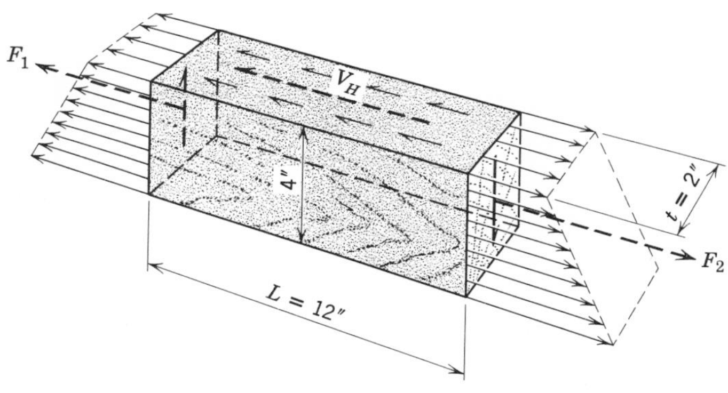


*(b)*

Fig. 5–18

curs 6 in. below the neutral surface) on the end area multiplied by the area; that is,

$$F_1 = \frac{M_1 y}{I} A = \frac{5400(12)(6)}{1600/3} (2)(4) = 108(54) \text{ lb},$$

and

$$F_2 = \frac{M_2 y}{I} A = \left(\frac{M_2}{M_1}\right) F_1 = \frac{63}{54} (108)(54) = 108(63) \text{ lb}.$$

Summing forces in the horizontal direction on Fig. 5–18*b* gives

$$V_H = F_2 - F_1 = 108(63) - 108(54) = 972 \text{ lb.}$$

The average shearing stress on plane *AB* is

$$\tau = \frac{V_H}{A_H} = \frac{972}{12(2)} = \underline{40.5 \text{ psi.}} \qquad \text{Ans.}$$

In this example, the shearing stress on plane *AB* does not vary with the length; therefore, since the shearing stresses on orthogonal planes have the same magnitude, the shearing stresses on the transverse (vertical) planes at *A* and *B* are also 40.5 psi in the directions shown.

## PROBLEMS

**5–62.*** A 12-in. section of a 6-in. deep by 3-in. wide timber beam is shown in Fig. P5–62*a*, and a free-body diagram of the bottom 2 in. of the beam is indicated in Fig. P5–62*b*.

(*a*) Determine values for $M_2$, $F_1$, and $F_2$.

(*b*) By applying an equilibrium equation to Fig. P5–62*b*, determine the resultant of the shearing stresses indicated on the area *ABCD*.

(*c*) Compute the average shearing stress on the area *ABCD*.

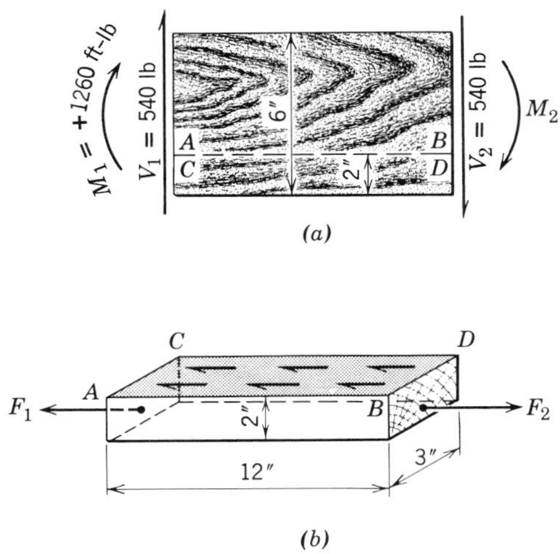


Fig. P5–62

**5–63.*** A timber T-beam is made up of two 2 by 6-in. planks glued together at *ABCD* as shown in Fig. P5–63*a*. This glued joint must offer shearing resistance as shown in Fig. P5–63*b*. The vertical shear at sections 1–1 and 2–2 is +1360 lb. The bending moment is +1020 ft-lb at section 1–1 and +2380 ft-lb at section 2–2.

(*a*) After computing values for $F_1$ and $F_2$, apply an equilibrium equation to Fig. P5–63*b*, and determine the total resistance offered by the glue.

(*b*) Determine the average shearing stress at the joint.

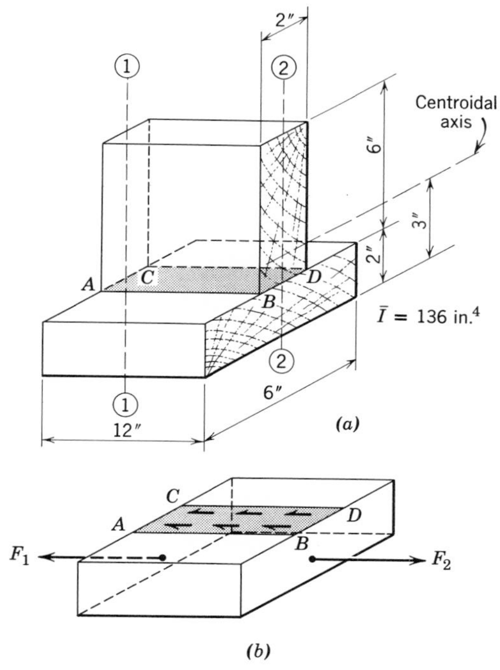


Fig. P5–63

**5–64.*** If the beam of Prob. 5–63 is made up of one 2 by 8-in. plank and two 2 by 2-in. sticks as shown in Fig. P5–64, determine:

(*a*) The total resistance offered by the glue on the surface *ABCD*.

(*b*) The average shearing stress at the joint.

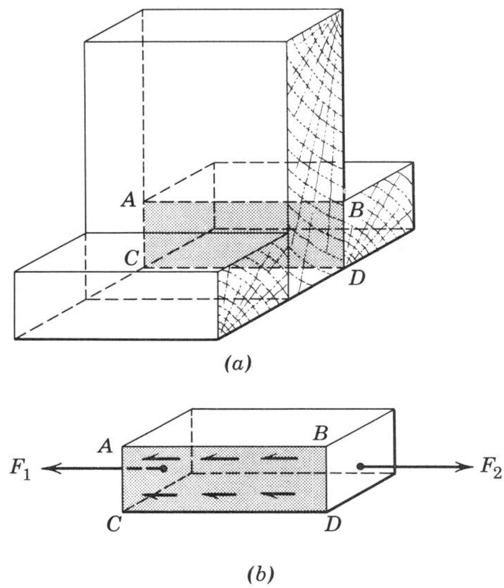


Fig. P5–64

**5–65.** A timber beam is to be fabricated of 2-in. planks to form the cross section of Fig. P5–65. A 9-in. length of the beam for which the shear is constant will be subjected to bending moments of +10.8 ft-kips and +16 ft-kips at the ends. Which method of fabrication, (*a*) or (*b*) of Fig. P5–65, will result in the least average shearing stress in the glued joint? Compute these stresses by the free-body diagram method.

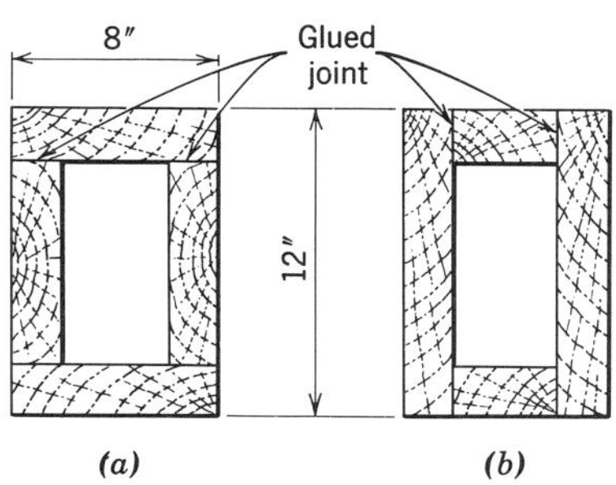


Fig. P5–65

**5–66.** A steel 18 WF 96 beam is to have 9 by ½-in. cover plates welded to the top and bottom flanges (detail in Fig. P5–66). A 16-in. length of the beam is subjected to bending moments of +5400 in-kips and +3000 in-kips at the ends, and the shear is constant. The fillet weld has an allowable strength of 2400 lb per in. Using the free-body diagram method, determine the number of fillet welds, each 2 in. long, needed on each side of the cover plate.

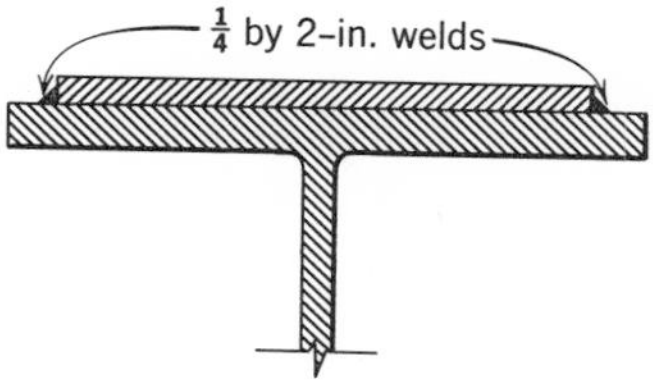


Fig. P5–66

## 5–9. The Shearing Stress Formula.

The shearing stress as determined in Ex. 5–6 is reasonably accurate only if the width of the section on plane $AB$ is not large compared to the depth of the beam and the transverse shear in the beam is constant throughout the length of the segment $AB$. If the transverse shear varies, the length, $L$, must be reduced to a short length $\Delta x$. The method outlined in the example, though excellent for emphasizing the concept of horizontal shear, is too laborious for general use; therefore, a formula will be developed which will be generally applicable within limits imposed by the assumptions. For this development refer to Fig. 5–19, which is a free-body diagram similar to that of Fig. 5–18$b$. The force $dF_2$ is the nor-

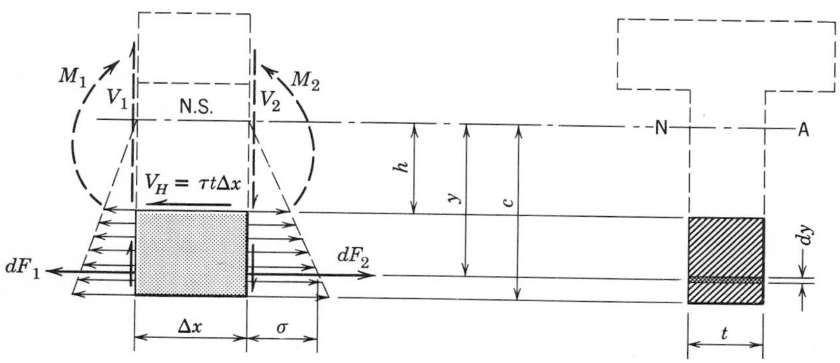


Fig. 5–19

mal force acting on a differential area, $dA$, and is equal to $\sigma\, dA$. The resultant of these differential forces is $F_2$ (not shown), and $F_2 = \int \sigma\, dA$ integrated over the shaded area of the cross section, where $\sigma$ is the fiber stress at a distance $y$ from the neutral surface and is given by the expression $\sigma = My/I$. When the two expressions are combined, the force $F_2$ becomes

$$F_2 = \frac{M_2}{I} \int y\, dA = \frac{M_2}{I} \int_h^c ty\, dy.$$

Similarly, the resultant force on the left side of the element is

$$F_1 = \frac{M_1}{I} \int_h^c ty\, dy.$$

The summation of forces in the horizontal direction on Fig. 5–19 yields

$$V_H = F_2 - F_1 = \frac{M_2}{I} \int_h^c ty\, dy - \frac{M_1}{I} \int_h^c ty\, dy = \frac{\Delta M}{I} \int_h^c ty\, dy.$$

The average shearing stress is $V_H$ divided by the area, from which

$$\tau = \lim_{\Delta x \to 0} \frac{\Delta M}{\Delta x} \left(\frac{1}{It}\right) \int_h^c ty\, dy = \frac{dM}{dx} \left(\frac{1}{It}\right) \int_h^c ty\, dy, \qquad \text{(a)}$$

and from Art. 5–6, $dM/dx$ equals $V$, the shear at the beam section where the stress is to be evaluated. Also, the integral is the first moment of that portion of the cross-sectional area between the transverse line where the stress is to be evaluated and the extreme fiber of the beam. The integral is designated $Q$, and, when $V$ and $Q$ are substituted into Eq. (a), the formula for horizontal (or longitudinal) shearing stress becomes

$$\tau = \frac{VQ}{It}. \qquad \text{(5–2)}$$

Because the flexure formula was used in the derivation, Eq. (5–2) is subject to the same assumptions and limitations. Although the stress given by Eq. (5–2) is associated with a particular point in a beam, it is averaged across the thickness, $t$, and hence is accurate only if $t$ is not too great. For a rectangular section having a depth twice the width, the maximum stress as computed by Saint-Venant's more rigorous method is about 3 per cent greater than that given by Eq. (5–2). If the beam is square, the error is about 12 per cent; and, if the width is four times the depth, the error is almost 100 per cent, from which one must conclude that, if Eq. (5–2) were applied to a point in the flange of the beam of Ex. 5–6, the result would be worthless. Furthermore, if Eq. (5–2) is applied to sections where the sides of the beam are not parallel, such as a triangular section, the average stress is subject to additional error because the transverse variation of stress is greater when the sides are not parallel.

As noted previously, at each point in a beam, the horizontal (longitudinal) and vertical (transverse) shearing stresses have the same magnitude; hence, Eq. (5–2) gives the vertical shearing stress at a point in a beam (averaged across the width). The variation of the transverse shearing stress in a beam will be demonstrated for the T-shaped beam of Ex. 5–6. Note that in Eq. (5–2) $V$ and $I$ are constant for any section, and only $Q$ and $t$ vary for different points in the section. The transverse shearing stress at any point in the stem of the T-section a distance $y_o$ from the neutral axis is, from Fig. 5–20*a* and Eq. (5–2),

$$\tau = \frac{V}{It_1}\int_{y_o}^{8} yt_1\,dy = \frac{V}{2I}(8^2 - y_o^2). \qquad (-2 \le y_o \le 8)$$

Note that $y_o$ is shown as positive downward. An expression for the average shearing stress in the flange can be written in a similar manner and is

$$\tau = \frac{V}{2I}(4^2 - y_o^2). \qquad (-4 \le y_o \le -2)$$

These are parabolic equations for the theoretical stress distribution, and the results are shown by Fig. 5–20*b*. The diagram has a discon-

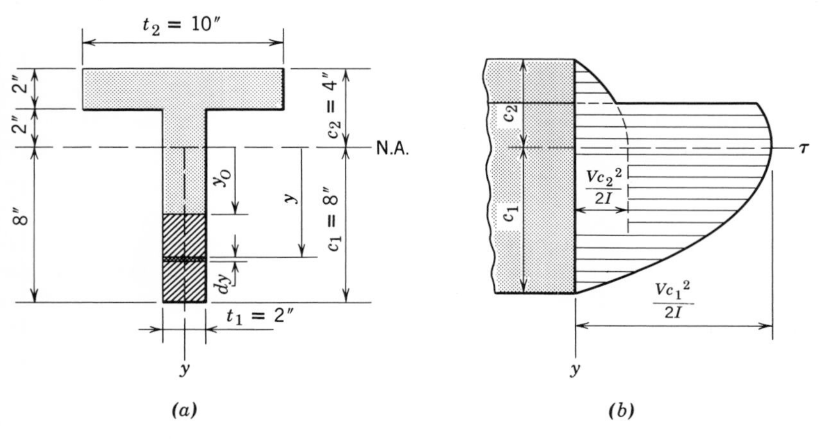


Fig. 5–20

tinuity at the junction of the flange and stem because the thickness of the section changes abruptly. As pointed out previously, the distribution in the flange is fictitious, since the stress at the bottom of the flange must be zero (a free surface). Also, because of the abrupt change in section, a stress concentration will exist, and the maximum shearing stress at the junction of flange and stem will be more than the average stress (see Art. 1–8). From Fig. 5–20*b* and Eq. (5–2), one may conclude that in general the maximum[6] longitudinal and transverse shearing stress occurs at the neutral surface at a section where the transverse shear is maximum. There may be some exceptions such as a beam with a cross section in the form of a triangle or a Greek cross with $Q/t$ at the neutral surface less than the value some distance from the neutral surface. The following example illustrates the use of the shearing stress formula.

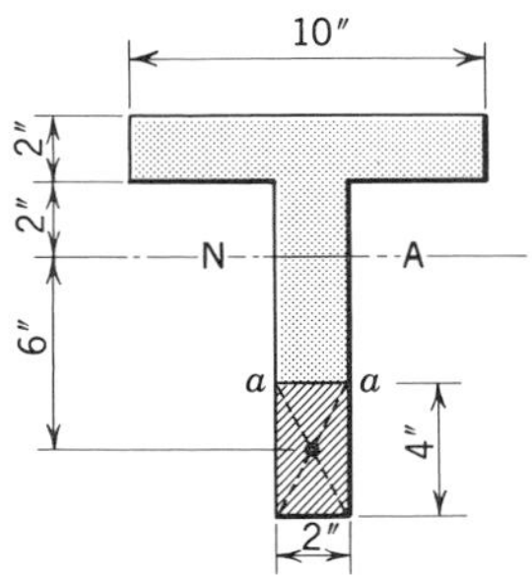


Fig. 5–21

**Example 5–7:** For the beam segment of Ex. 5–6, determine:
(*a*) The vertical shearing stress 4 in. from the bottom.
(*b*) The maximum vertical shearing stress.

*Solution:* (a) The moment of inertia taken from Ex. 5–2 is 1600/3 in.⁴ With reference to Fig. 5–21 the first moment, $Q$, is the shaded area multiplied by the distance from the centroid of the area to the neutral axis, which is 4(2)(6) in.³

Then
$$\tau = \frac{VQ}{It} = \frac{900(4)(2)(6)}{(1600/3)(2)} = \underline{40.5 \text{ psi.}} \qquad \text{Ans.}$$

(*b*) The maximum stress will occur at the neutral surface, and it is

$$\tau_{\max} = \frac{900(8)(2)(4)}{(1600/3)(2)} = \underline{54.0 \text{ psi.}} \qquad \text{Ans.}$$

Note that, in the computation of the value for $Q$, it is immaterial whether one takes the area above or below a transverse line such as *a–a* in Fig. 5–21. For example, in part *b*, $Q$ for the area above the neutral axis is $Q_{\text{flange}} + Q_{\text{web}}$ equals $10(2)(2 + 1) + 2(2)(1)$ equals $60 + 4$ which is the same as that used in part *b* above.

---

[6] In this book the term *maximum,* as applied to longitudinal and transverse shearing stress, will mean the average stress across the thickness, $t$, at a point where such average has the maximum value.

## PROBLEMS

**5–67.*** A beam simply supported at its ends is 10 ft long and carries a uniformly distributed load of 480 lb per ft over its entire length. The beam is 3 in. wide by 8 in. deep. Determine:

(*a*) The horizontal shearing stress in the beam at a point 2 ft from the left support and 5 in. above the bottom of the beam.

(*b*) The maximum vertical shearing stress in the beam.

**5–68.*** The total vertical shear at a certain section of a beam is 3200 lb. The beam has the cross section shown in Fig. P5–68. Determine:

(*a*) The maximum horizontal shearing stress, and indicate where it occurs within the cross section.

(*b*) The vertical shearing stress 3 in. below the top.

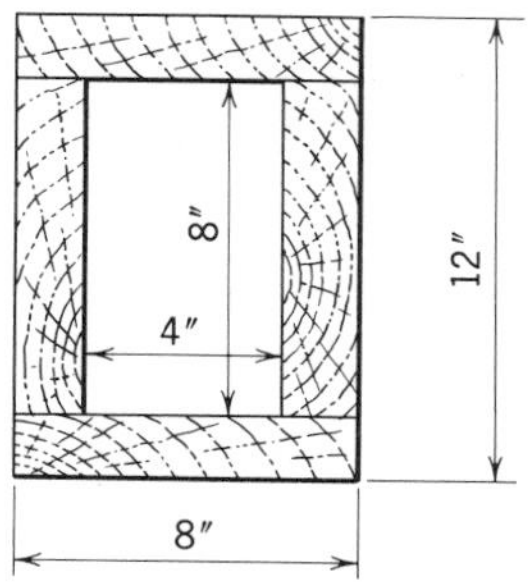


Fig. P5–68

**5–69.*** A beam 18 ft long is simply supported at its ends and has a cross section as shown in Fig. P5–69. The beam carries a uniformly distributed load of 360 lb per ft over its entire length. Determine:

(*a*) The vertical shearing stress at a point 2 ft from the right end and 3 in. below the top surface of the beam.

(*b*) The maximum horizontal and vertical shearing stresses, and state where each occurs.

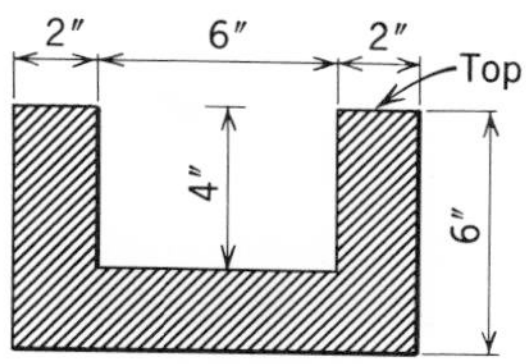


Fig. P5–69

**5–70.** A beam 15 ft long has the cross section shown in Fig. P5–70. It is simply supported at the ends and carries a uniformly distributed load of 400 lb per ft over its entire length. At a point 3 ft from the left end and 1 in. below the neutral surface, determine:

(*a*) The fiber stress.
(*b*) The horizontal shearing stress.
(*c*) The vertical shearing stress.

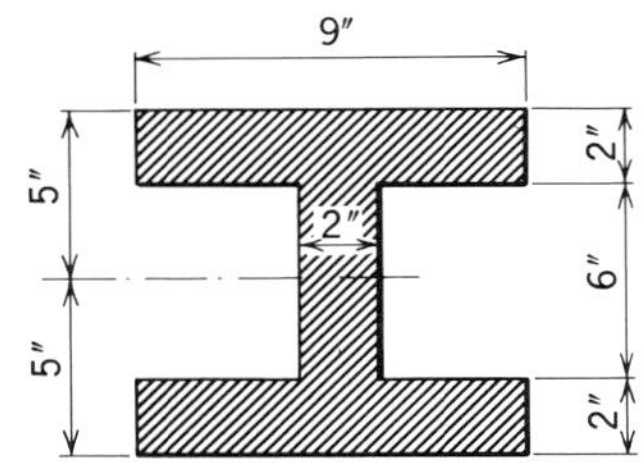


Fig. P5–70

**5–71.*** The beam of Fig. P5–71*a* has the cross section of Fig. P5–71*b*. The left reaction is 950 lb upward on the beam. Determine:

(*a*) The maximum fiber stress in the beam.
(*b*) The maximum horizontal shearing stress.

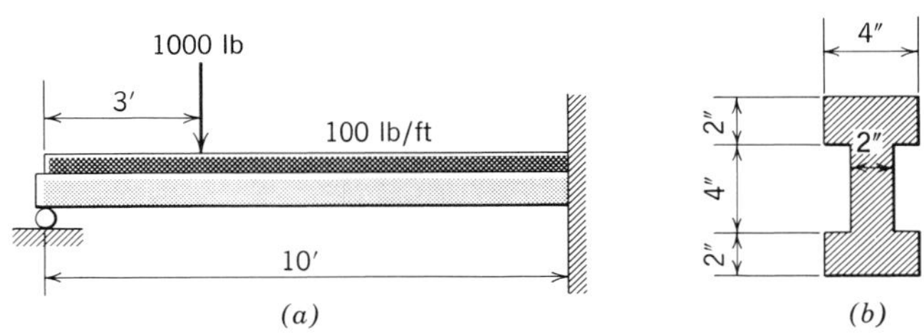


Fig. P5–71

**5–72.** The beam of Fig. P5–72*a* has the cross section shown in Fig. P5–72*b*. At section *a–a*, the shear is +1200 lb, and the bending moment is −7000 ft-lb. Determine the fiber stress and the vertical shearing stress 2 in. below the neutral surface at section *b–b*.

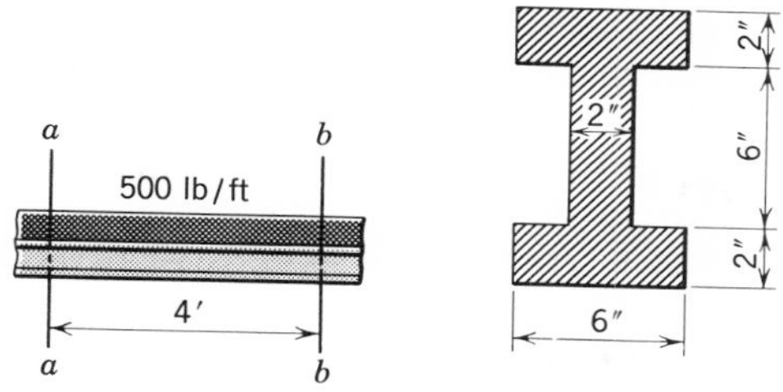


(a) (b)

Fig. P5–72

**5–73.*** Figure P5–73 shows the elevation and cross section of part of a beam. At section *b–b*, the vertical shear is −3400 lb, and the bending moment is +7000 ft-lb. Determine:

(*a*) The vertical shearing stress at point *A* in the section at *a–a*.

(*b*) The maximum compressive fiber stress at section *a–a*.

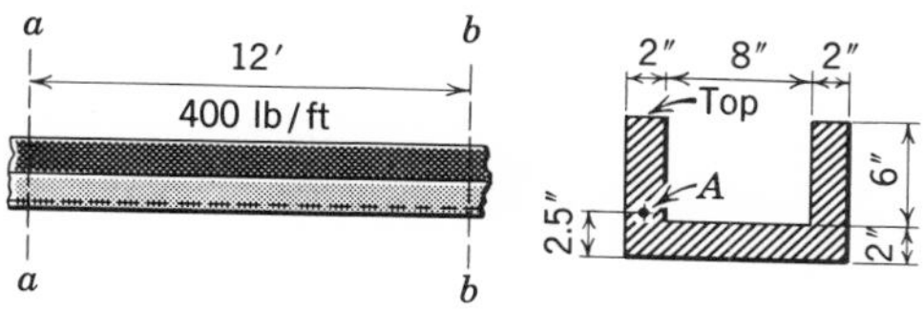


Fig. P5–73

**5–74.** The beam in Fig. P5–74 supports the indicated loads and has the cross section shown. Determine:

(*a*) The maximum compressive fiber stress.

(*b*) The maximum horizontal shearing stress.

(*c*) The vertical shearing stress 9 ft from the left end at a point 2 in. above the bottom of the beam.

(*d*) The flexural stress at a point 3 ft from the right end and 2 in. above the bottom of the beam.

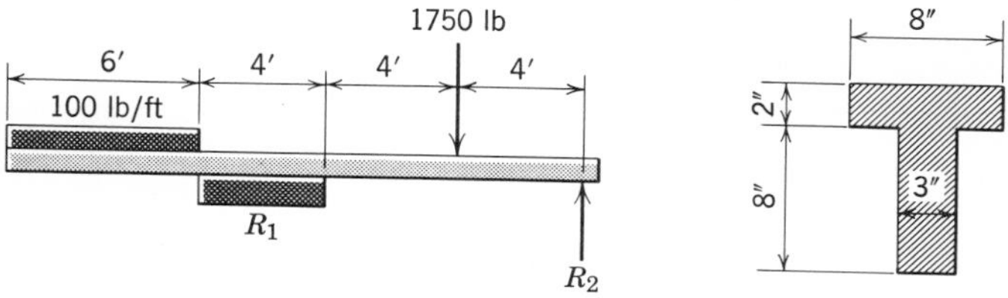


Fig. P5–74

**5–75.*** A 6-in. wide by 8-in. deep timber beam 12 ft long is simply supported at its ends and carries a concentrated load $P$ at mid-span. It is specified that the fiber stress in the beam shall not exceed 1250 psi and that the horizontal shearing stress shall not exceed 90 psi. Determine the maximum allowable value of the load $P$.

**5–76.** A beam simply supported at its ends is 6 in. wide by 12 in. deep by 9 ft long and carries a uniformly distributed load over its entire length. The allowable fiber stress in the beam is 1800 psi in tension or compression, and the allowable horizontal shearing stress is 120 psi. Determine the maximum allowable total load on the beam.

**5–77.*** The cross section of the timber beam in Fig. P5–77 is a rectangle 6 in. wide by 12 in. deep. The allowable fiber stress is 1200 psi, and the allowable vertical shearing stress is 90 psi. Determine the maximum allowable value of $w$.

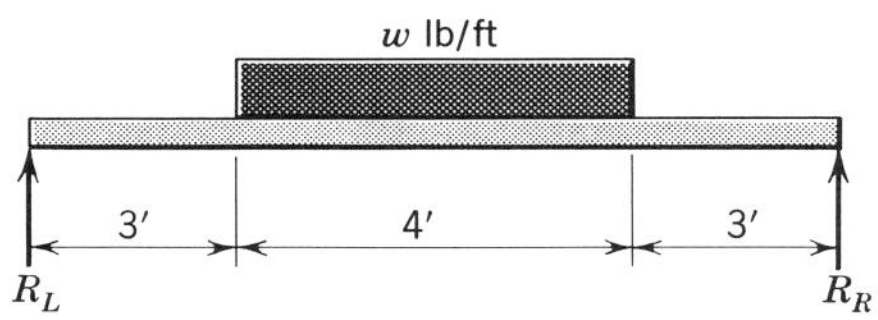


Fig. P5–77

**5–78.** The simple beam of Fig. P5–78*a* has the cross section shown in Fig. P5–78*b*. Determine the maximum permissible value for the load, $w$ (lb per ft), if the compressive flexural stress is not to exceed 900 psi, and the horizontal shearing stress is not to exceed 100 psi.

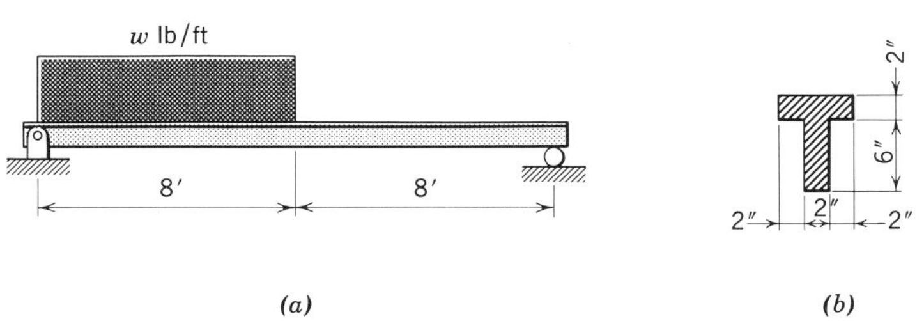


Fig. P5–78

**5–79.*** The beam of Fig. P5–79 is composed of 3 pieces of timber glued together as shown. The maximum horizontal shearing stresses are not to exceed 100 psi and 120 psi in the glue joint and the wood respectively. The tensile and compressive fiber stresses are not to exceed 1800 psi anywhere in the beam. What is the maximum permissible value of $P$?

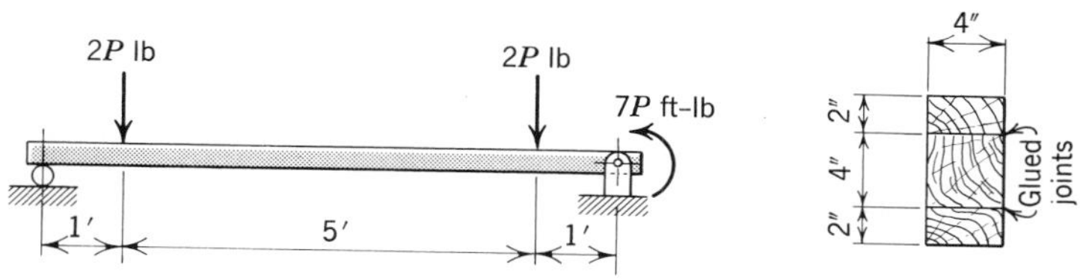


Fig. P5–79

**5–80.** A T-beam 12 ft long is simply supported at the ends and has a cross section as shown in Fig. P5–80. It is specified that the tensile fiber stress shall not exceed 1500 psi and that the horizontal shearing stress shall not exceed 120 psi. Determine the maximum concentrated load that may be placed 8 ft from the right end.

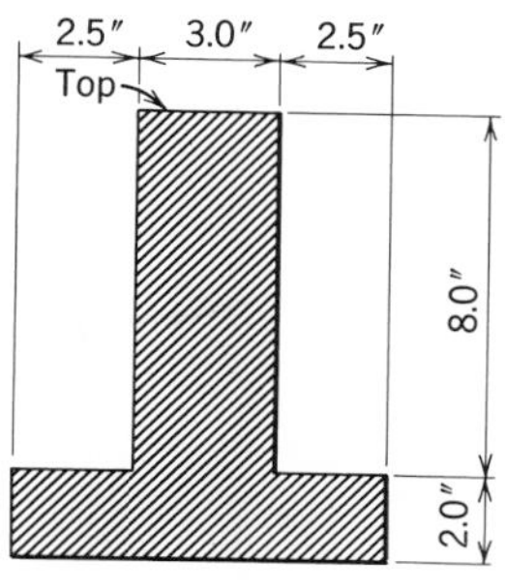


Fig. P5–80

**5–81.*** A certain beam with restrained ends has a span, $L$, of 10 ft and the cross section shown in Fig. P5–78$b$; it is subjected to a loading which produces the following maximum values: positive moment $wL^2/25$, negative moment $wL^2/10$, and shear $5wL/9$, where $L$ is in feet and $w$ is in pounds per foot. The working stresses are: fiber, 5000 psi $T$, 15,000 psi $C$; and horizontal shear, 2000 psi. Determine the maximum allowable uniformly distributed load, $w$, for the beam.

**5–82.** The beam in Fig. P5–82*a* has a cross section as shown in Fig. P5–82*b*. Determine:

(*a*) The maximum horizontal shearing stress in the beam.
(*b*) The maximum fiber stress in the beam.

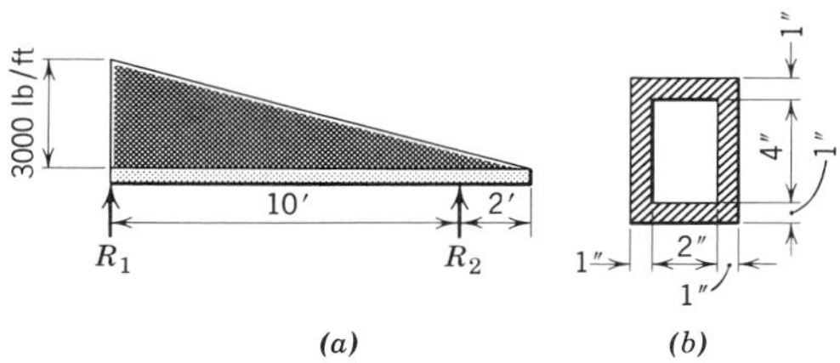


Fig. P5–82

**5–83.*** A horizontal wooden beam with the cross section of Fig. P5–83 is subjected to maximum moments of +600 ft-lb and −800 ft-lb, and a maximum shear of 400 lb—all in a vertical plane. Determine:

(*a*) The maximum horizontal shear stress.
(*b*) The maximum *tensile* fiber stress.

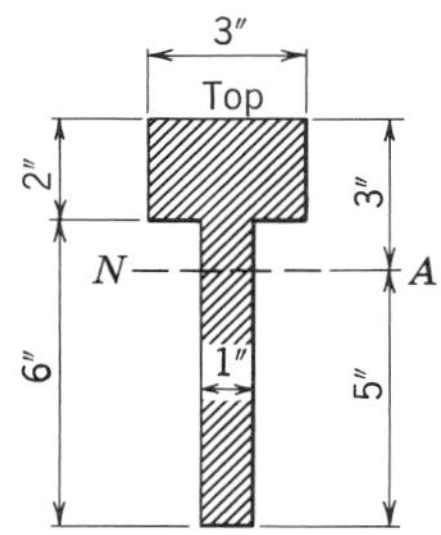


Fig. P5–83

**5–84.** The cross section of Fig. P5–84*a* is to be used for the beam of Fig. P5–84*b*. Determine:

(*a*) The maximum fiber stress, and state where it occurs.

(*b*) The maximum horizontal shearing stress, and state where it occurs.

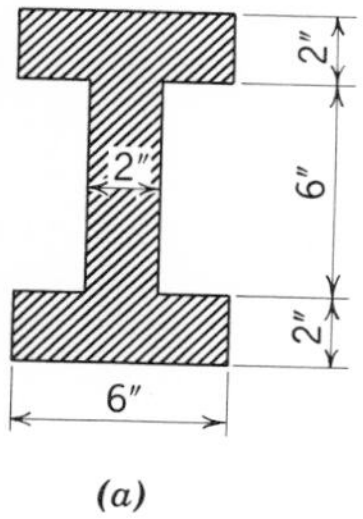


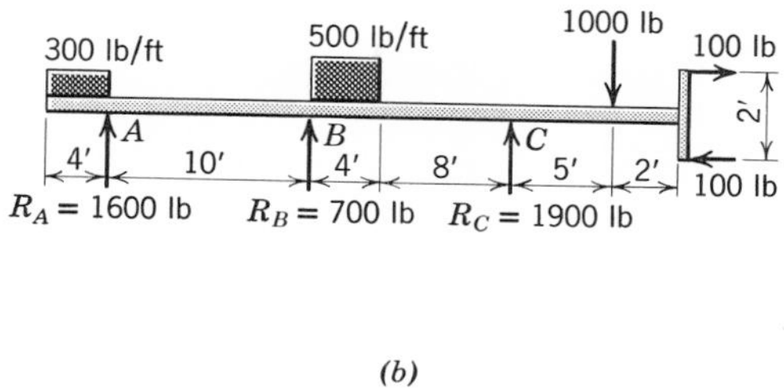


Fig. P5–84

**5–85.*** A horizontal beam having the cross section of Fig. P5–85 is subjected to vertical loads which produce the following maximum values: positive moment, 12,000 ft-lb; negative moment, 10,000 ft-lb; and shear, −2000 lb.

(*a*) Determine the maximum compressive fiber stress.

(*b*) Determine the maximum horizontal shearing stress.

(*c*) Determine the fiber stress at a point 2 in. above the bottom of the beam at a section where the bending moment is +6000 ft-lb.

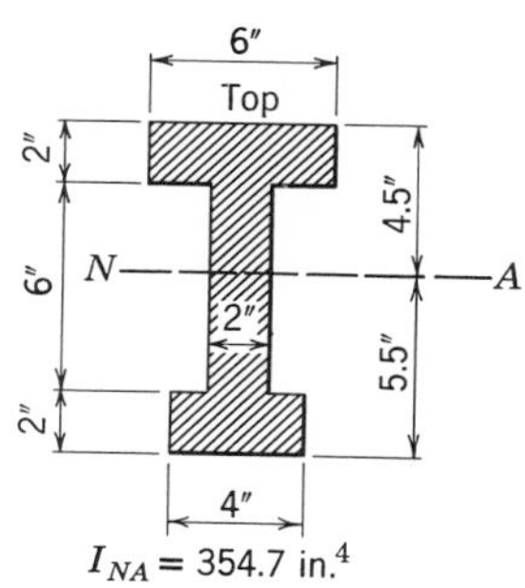


Fig. P5–85

**5–86.** For the beam of Fig. P5–86, determine the maximum fiber and horizontal shearing stresses, and state where they occur.

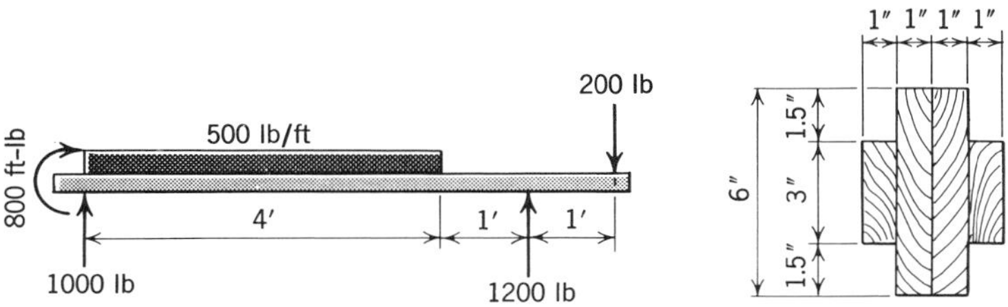


Fig. P5–86

**5–87.*** In Fig. P5–87 are shown one span of a continuous beam and shear and moment diagrams for this span. The beam cross section is given in Fig. P5–80. If the allowable fiber stress is 1600 psi $C$ and the allowable horizontal shearing stress is 120 psi, determine the maximum permissible value for the uniformly distributed load, $w$.

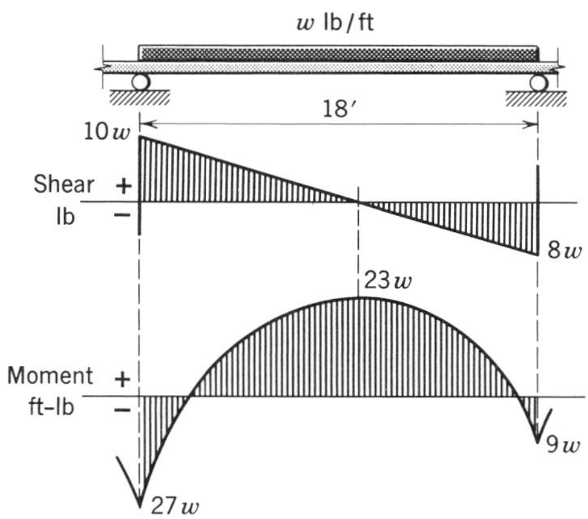


Fig. P5–87

**5–88.** The beam of Fig. P5–87 with a load $w$ of 200 lb per ft is to be a rectangular timber stringer with the depth twice the width. The fiber and horizontal shearing stresses are not to exceed 1200 psi and 100 psi, respectively. Determine the minimum permissible dimensions of the cross section.

**5–10. Principal Stresses for Flexure.** Methods for finding the fiber stress at any point in a beam were presented in Art. 5–3 and 5–4. Procedures for locating the critical sections of a beam (maximum $M$ and $V$) were developed in Art. 5–6 and 5–7. Methods for determining the transverse and longitudinal shearing stresses at any point in a beam were presented in Art. 5–8 and 5–9. However, the discussion of stresses in beams is incomplete without careful consideration of the principal and maximum shearing stresses at carefully selected points on the sections of maximum shear and bending moment.

The fiber stress is maximum at the top or bottom edge of a section, and the transverse and longitudinal shearing stresses are zero at these points. Consequently, the maximum tensile and compressive fiber stresses on a given section are also the principal stresses at these points, and the corresponding maximum shearing stress is equal to one-half the fiber stress $[\tau_{max} = (\sigma_p - 0)/2]$. The longitudinal and transverse shearing stresses are normally maximum for a given section at the neutral axis, where the fiber stress is zero. Thus, the evidence presented so far is limited to the extremes in a cross section. One might well wonder what the stress situation might be between these points. Unfortunately, the magnitude of the principal stresses throughout a cross section cannot be expressed for all sections as a simple function of position. However, in order to provide some insight into the nature of the problem, two typical sections will be discussed in the following paragraphs.

For a cantilever beam of rectangular cross section subjected to a concentrated load, the theoretical principal stress variation is indicated for two sections in Fig. 5–22. One observes from this figure that, at a distance from the load of one-fourth the depth of the beam, the maximum normal stress does not occur at the surface. However, at a distance of one-half the depth, the maximum fiber stress is the maximum normal stress. For either of these sections, Saint-Venant's principle (see Art. 1–8) indicates that the flexure (and transverse shearing stress) formula is inapplicable. Since such a small increase in bending moment is required to overcome the effect of the transverse shearing stress, the conclusion may be drawn that, for a rectangular cross section, in regions where the flexure formula applies, the maximum fiber stress is the maximum normal stress. Although, for the rectangular cross section, the maximum shearing stress will usually be one-half the maximum normal stress (at a surface of the beam), for materials having a longitudinal plane of weakness (for example, the usual timber beam), the longitudinal shearing stress may frequently be the significant stress, hence the emphasis on this stress.

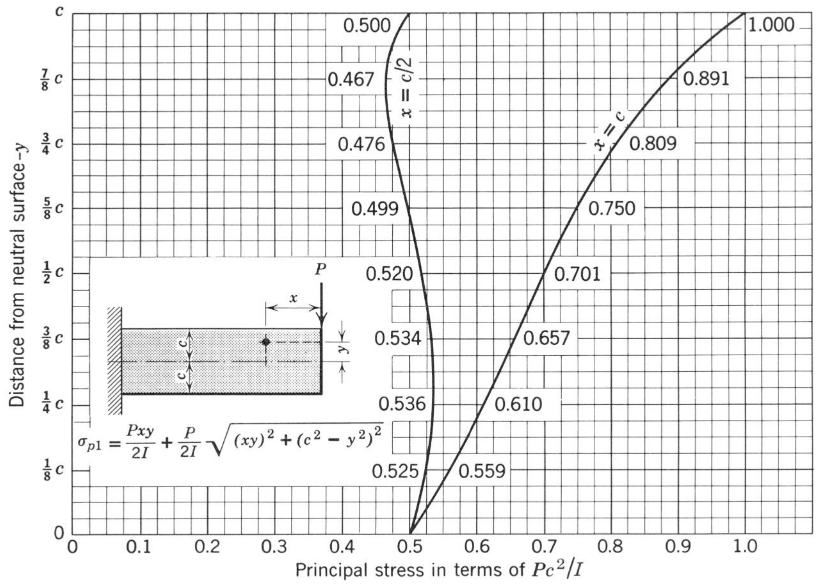


Fig. 5–22

The other section to be discussed is the deep wide flange section subjected to a combination of large shear and large bending moment. For this combination, the high fiber and transverse shearing stresses occurring simultaneously at the junction of flange and web sometimes yields a principal stress greater than the fiber stress at the surface of the flange (see Ex. 5–8). In general, at any point in a beam, a combination of large $M$, $V$, $Q$, and $y$ and a small $t$ should suggest a check on the principal stresses at such a point. Otherwise, the maximum flexural (fiber) stress will very likely be the maximum normal stress, and the maximum shearing stress will probably occur at the same point.

A knowledge of the directions of the principal stresses may aid in the prediction of the directions of cracks in a brittle material (concrete, for example) and thus may aid in the design of reinforcement to carry the tensile stresses. Curves drawn with their tangents at each point in the directions of the principal stresses are called *stress trajectories*. Since there are two principal stresses at each point, there are two stress trajectories passing through each point. These curves will be perpendicular since the principal stresses are orthogonal; one set of curves will represent the maximum stresses, whereas the other

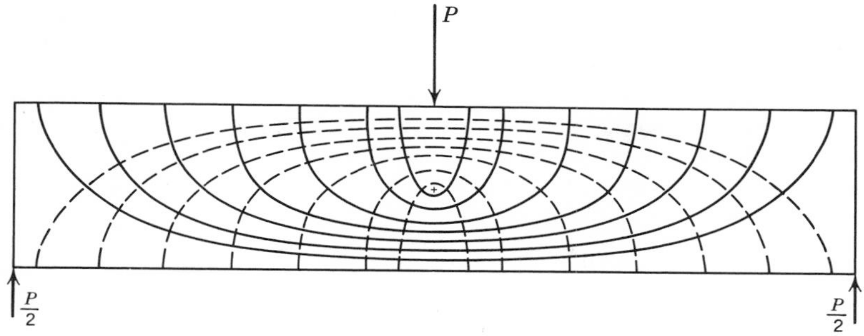


Fig. 5–23

represents the minimum stresses. The trajectories for a simply supported rectangular beam carrying a concentrated load at the midpoint are shown in Fig. 5–23 with dashed lines representing the directions of the compressive stresses and solid lines showing tensile stress directions. In the vicinities of the load and reactions there are stress concentrations, and the trajectories become much more complicated. Figure 5–23 neglects all stress concentrations.

In order to determine the principal stresses and the maximum shearing stresses at a particular point in a beam, it is necessary to calculate the fiber stresses and transverse (or longitudinal) shearing stresses at the point. With the stresses on orthogonal planes through the point known, the methods of Art. 1–6 or 1–7 can be used to calculate the maximum stresses at the point. The following example illustrates the procedure and provides an example in which the principal stresses at some point are greater than the maximum fiber stresses.

**Example 5–8:** A 24 WF 100 cantilever beam, supported at the left end, carries a uniformly distributed load of 11.60 kips per ft on a span of 8.00 ft. Determine the maximum normal and shearing stresses in the beam.

*Solution:* For a cantilever beam with a uniformly distributed load, both the maximum bending moment and the maximum transverse shear occur on the section at the support. In this case they are

$$M = \frac{wL^2}{2} = \frac{(11.6)(8^2)(12)}{2} = 4454 \text{ in-kips}$$

and

$$V = wL = (11.6)(8) = 92.8 \text{ kips.}$$

The upper half of the cross section of the beam at the wall is shown in Fig. 5–24*a*. The distribution of fiber stress for this half of the section is shown in

Fig. 5–24*b*, and the distribution of the average transverse shearing stress is as shown in Fig. 5–24*c*. The vertical stresses, $\sigma_y$, due to the pressure of the load on the top of the beam are considered negligible.

Values will be calculated for three points, namely, at the neutral axis, in the web at the junction of the web and the top flange, and at the top surface. Both the fillets and the stress concentrations at the junction of the web and flange will be neglected. At the neutral axis, the fiber stress is zero, and the transverse shearing stress is

$$\tau = \frac{VQ}{It} = \frac{92.8[12(0.775)(11.61) + 11.225(0.468)(5.61)]}{2987(0.468)} = 9.12 \text{ ksi}.$$

In the web at the junction with the top flange, the fiber stress is

$$\sigma = \frac{My}{I} = \frac{4454(11.225)}{2987} = 16.72 \text{ ksi } T,$$

and the transverse shearing stress is

$$\tau = \frac{VQ}{It} = \frac{92.8(12)(0.775)(11.61)}{2987(0.468)} = 7.16 \text{ ksi}.$$

At the top surface, the transverse shearing stress is zero, and the fiber stress is

$$\sigma = \frac{My}{I} = \frac{M}{Z} = \frac{4454}{248.9} = 17.88 \text{ ksi } T.$$

The principal stresses and maximum shearing stresses for each of the three selected points are shown in Fig. 5–24*d*. The calculations, from the equations of Art. 1–6, for the point at the junction of web and flange are

$$\sigma_p = \frac{16.72}{2} \pm \sqrt{\left(\frac{16.72}{2}\right)^2 + 7.16^2}$$

$$= 8.36 \pm 11.01,$$

which gives

$$\sigma_{p1} = 19.37 \text{ ksi } T, \quad \sigma_{p2} = 2.65 \text{ ksi } C, \quad \text{and} \quad \tau_{\max} = 11.01 \text{ ksi}.$$

The angle is not particularly important in this case, but it can be obtained from the equation

$$\tan 2\theta_p = -\frac{7.16}{8.36} = -0.857.$$

The result is

$$2\theta_p = -40.6°$$

or

$$\theta_p = 20.3° \circlearrowright.$$

The other stresses of Fig. 5–24*d* are obtained by inspection. It should be noted that the maximum tensile stress of 19.37 ksi is 8.32 per cent above the maxi-

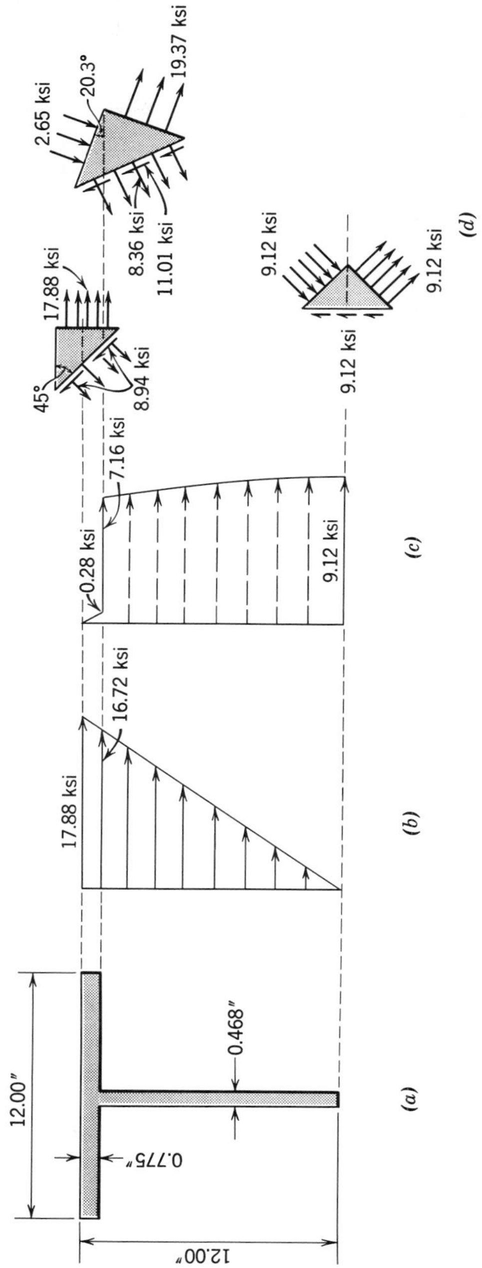

12.00″
0.775″
0.468″
12.00″
(a)
17.88 ksi
16.72 ksi
(b)
0.28 ksi
7.16 ksi
9.12 ksi
(c)
45°
17.88 ksi
8.94 ksi
9.12 ksi
2.65 ksi
20.3°
19.37 ksi
8.36 ksi
11.01 ksi
9.12 ksi
9.12 ksi
(d)


Fig. 5–24

mum tensile fiber stress of 17.88 ksi and that the maximum shearing stress of 11.01 ksi is 20.7 per cent above the maximum transverse shearing stress of 9.12 ksi.

## PROBLEMS

**In the following problems neglect both the fillets and the stress concentrations at the junction of the web and flange.**

**5–89.*** A 5-ft steel beam with the cross section shown in Fig. P5–89*a* is loaded and supported as shown in Fig. P5–89*b*. At point *A*, which is just to the right of the 15,000-lb load and just above the flange, determine and show on a sketch the principal stresses and the maximum shearing stress.

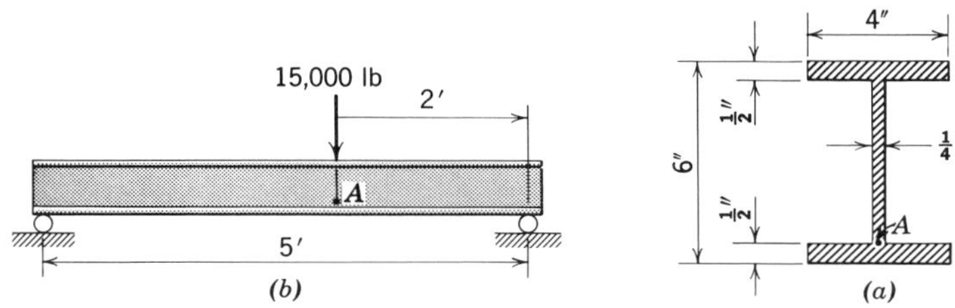


Fig. P5–89

**5–90.*** At the point *A* (just below the flange) on the cantilever beam shown in Fig. P5–90, determine and show on a sketch the principal and the maximum shearing stresses.

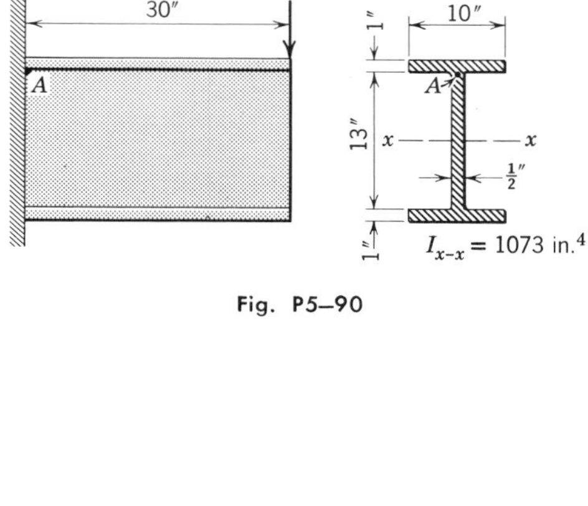


Fig. P5–90

**5–91.*** A hollow rectangular timber beam with the cross section in Fig. P5–91*a* is loaded as shown in Fig. P5–91*b*. Determine the principal and maximum shearing stresses at point *A*. Show these stresses in their approximate directions on a sketch.

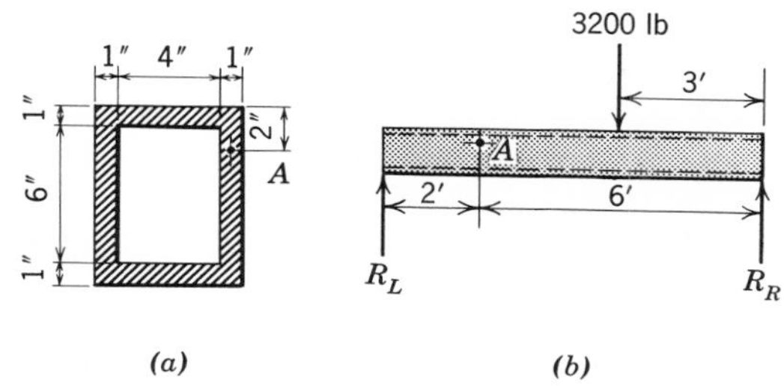


Fig. P5–91

**5–92.** The simple beam in Fig. P5–92*a* has the cross section shown in Fig. P5–92*b*. If the allowable stresses are 10,000 psi shear and 16,000 psi tension and compression, determine the maximum allowable load, *P*.

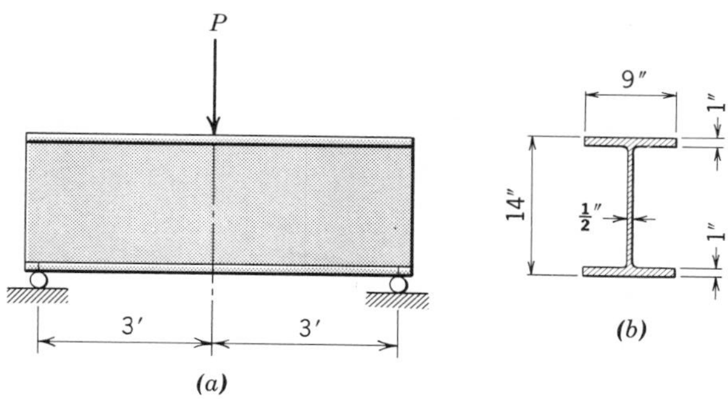


Fig. P5–92

**5–93.*** A 24 WF 130 cantilever beam with a span of 8 ft carries a uniformly distributed load of 12.0 kips per ft. Determine the maximum principal and shearing stresses in the beam.

**5–94.** For the beam of Prob. 5–93, calculate the principal stresses and the maximum shearing stress at a point 8 in. above the neutral axis on a section 1 ft from the support. Show these stresses in their correct directions on a sketch.

**5–95.*** A 14 WF 68 continuous beam has an overhanging end 4 ft long. The allowable normal and shearing stresses are 18.00 ksi and 12.00 ksi respectively. Determine the maximum allowable uniformly distributed load per foot for the overhang.

**5–96.** For the beam of Prob. 5–95, with a uniformly distributed load of 6.00 kips per ft, determine the maximum principal and shearing stresses, and show them on a sketch in their correct directions.

**5–11. Stresses in Beams by Theory of Elasticity.** In Art. 5–2 and 5–4, the flexure formula was developed by assuming that plane sections remain plane and that the beam is subjected to couples only. The formula for transverse shearing stresses was developed in Art. 5–9 for use with the flexure formula when transverse shearing forces are present. If shearing stresses exist, plane sections cannot remain plane, as Fig. 5–25 indicates, because the element $b$ at the left section will undergo the maximum shearing strain while elements $a$ and $c$ will experience no shearing strain. The result is a warped surface, as indicated. If the segment is in a region of constant shear, the warping of section $def$ will match that of $abc$; consequently, the ends of any longitudinal element will be displaced the same amount, and the normal strains will be proportional to the distance from the neutral surface just as though the section had remained plane. In a region of varying shear, the warping of section $def$ will not be the same as that

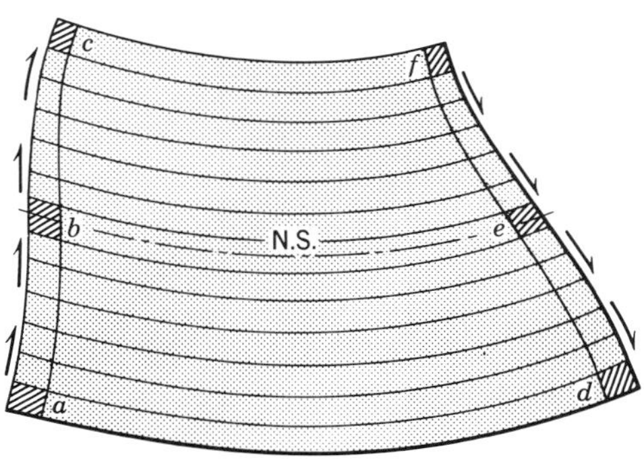


Fig. 5–25

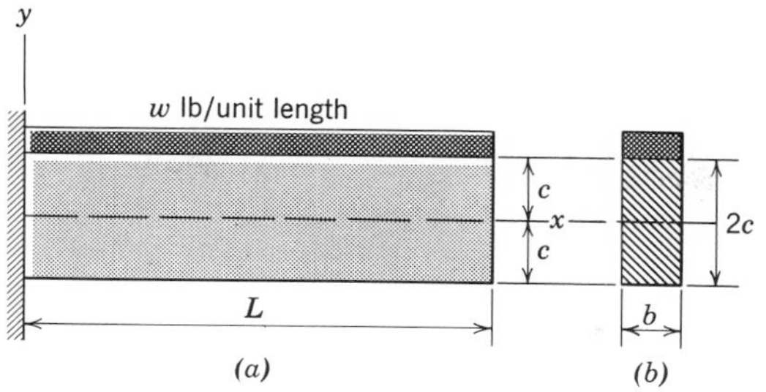


Fig. 5–26

of $abc$, and the normal strains will not be proportional to the distance from the neutral surface.

Although for most engineering applications the effects of cross-section warping can be neglected, the flexure and shearing stress formulas of Eq. (5–1) and (5–2) do not completely describe the state of stress for sections with distributed loading. In such sections, the transverse shearing force varies, and there must also be normal stresses in the vertical direction which are not considered in the elementary theory. The following example demonstrates how the Airy stress function (see Art. 3–7) can be used to study the stresses resulting from a uniform load acting on a rectangular cantilever beam. This analysis assumes a state of plane stress.

**Example 5–9**: The cantilever beam in Fig. 5–26 supports a uniformly distributed load of $w$ lb per unit length. The function

$$\varphi = K[5y^3(L - x)^2 + 30c^2xy(L - x) + 5c^2x^2(3y - 2c) + y^3(2c^2 - y^2)]$$

is suggested as a possible Airy stress function for the beam.

($a$) Show that $\varphi$ is biharmonic and thus satisfies the conditions of compatibility and equilibrium.

($b$) Determine the value of $K$ to satisfy the boundary condition on the surface $y = c$. Analyze this solution with regard to the other boundary requirements.

($c$) Compare the results of this solution with those obtained from the elementary theory.

*Solution:* ($a$) The biharmonic equation is, see Eq. (3–8),

$$\nabla^4\varphi = \frac{\partial^4\varphi}{\partial x^4} + 2\frac{\partial^4\varphi}{\partial x^2\,\partial y^2} + \frac{\partial^4\varphi}{\partial y^4} = 0.$$

These terms, for the given function, are

$$\frac{\partial^4\varphi}{\partial x^4} = 0,\ 2\frac{\partial^4\varphi}{\partial x^2\,\partial y^2} = 2K(60y) = 120Ky,\ \frac{\partial^4\varphi}{\partial y^4} = -K(120y),$$

therefore, the function is biharmonic and satisfies the conditions of compatibility and equilibrium.

(*b*) The stresses are determined from Eq. (3–7). Thus

$$\sigma_x = \partial^2\varphi/\partial y^2 = K[30y(L-x)^2 + 4y(3c^2 - 5y^2)], \tag{a}$$

$$\sigma_y = \partial^2\varphi/\partial x^2 = 10K(y^3 - 3c^2y - 2c^3), \tag{b}$$

$$\tau_{xy} = -\partial^2\varphi/(\partial x\,\partial y) = -30K(c^2 - y^2)(L - x). \tag{c}$$

The normal stress on the surface $y = c$ is $\sigma_y = -w/b$, therefore

$$-w/b = \sigma_y\Big]_{y=c} = 10K(c^3 - 3c^3 - 2c^3)$$

from which

$$K = \frac{w}{40bc^3} = \frac{w}{60I}$$

since $I = 2bc^3/3$.

In addition to the boundary condition for the load on the beam, the other boundary conditions are

$$\tau_{xy}\Big]_{y=\pm c} = 0, \quad \tau_{xy}\Big]_{x=L} = 0, \quad \sigma_y\Big]_{y=-c} = 0, \quad \sigma_x\Big]_{x=L} = 0.$$

Equation (c) indicates that the shearing stress satisfies the first two equations. The normal stress along the bottom of the beam is, from Eq. (b),

$$\sigma_y\Big]_{y=-c} = 10K(-c^3 + 3c^3 - 2c^3) = 0,$$

and thus, the third boundary equation is satisfied. The value of $\sigma_x$ on the free end of the beam is, from Eq. (a),

$$\sigma_x\Big]_{x=L} = 4Ky(3c^2 - 5y^2)$$

which is zero for $y = \pm c\sqrt{0.6}$ and zero. However, the resultant normal force due to the stresses acting on the surface $x = L$ is

$$R = \int_{-c}^{+c} \sigma_x b\,dy = 4Kb\int_{-c}^{+c}(3c^2y - 5y^3)\,dy = 4Kb\left[\frac{3c^2y^2}{2} - \frac{5y^4}{4}\right]_{-c}^{+c} = 0,$$

and the resultant moment of these same stresses about the center of the beam at $x = L$, is

$$M_O = \int_{-c}^{+c} y\sigma_x b\,dy = 4Kb\int_{-c}^{+c}(3c^2y^2 - 5y^4)\,dy = 4Kb[c^2y^3 - y^5]_{-c}^{+c} = 0.$$

Since the normal stresses ($\sigma_x$) on the beam at $x = L$ give rise to neither resultant axial force nor moment, it follows from Saint Venant's principle that, at sections somewhat removed from the end, the stress distribution is the correct one.

(*c*) The shear and moment in the beam are given by

$$V = w(L - x) \qquad \text{and } M = -w(L - x)^2/2.$$

The flexure and shearing stresses can be obtained from Eq. (5–1) and (5–2) but the signs must be reversed to provide agreement with the sign convention used with the elasticity theory. The stresses from the elementary theory are

$$\sigma_x = \frac{-My}{I} = \frac{w(L-x)^2 y}{2I}, \qquad \tau_{xy} = \frac{-VQ}{Ib} = \frac{-w(L-x)(c^2-y^2)}{2I}. \tag{d}$$

The elementary theory provides no information about $\sigma_y$.

When the value for $K$ is substituted in Eq. (a), (b), and (c) they become

$$\begin{aligned} \sigma_x &= \frac{w}{2I}(L-x)^2 y + \frac{w}{15I}(3c^2 - 5y^2)y, \\ \sigma_y &= \frac{w}{6I}(y^3 - 3c^2 y - 2c^3), \\ \tau_{xy} &= -\frac{w}{2I}(L-x)(c^2-y^2). \end{aligned} \tag{e}$$

When Eq. (d) are compared with Eq. (e), it is seen that $\tau_{xy}$ is the same for both solutions. The first term of the expression for $\sigma_x$ from Eq. (e) is the same as that from Eq. (d), and the second term may be considered to be a correction to the elementary theory. For most beams, the length is several times the depth, in which case, the maximum value of the correction term is very much less than the maximum fiber stress and can be neglected. (The correction is less than 0.2 per cent if the length is six times the depth.) The maximum value of $\sigma_y$ occurs at the top of the beam and it is also much less than the maximum value of $\sigma_x$ for beams whose length is several times the depth. Figures 5–27*a* and *b* show the variation of $\sigma_x$ with $y$ at the end of the beam ($x = L$) and at a section $2c$ from the end ($x = L - 2c$). The third curve, Fig. 5–27*c*, indicates the variation of $\sigma_y$ with $y$. Note that even for very short beams ($L = 2c$) the correction for the maximum fiber stress is less than seven per cent, and the maximum value of $\sigma_y$ is about 33 per cent of the maximum fiber stress. Thus, the elementary theory gives results which are in reasonably close agreement with results from the more exact theory of elasticity.

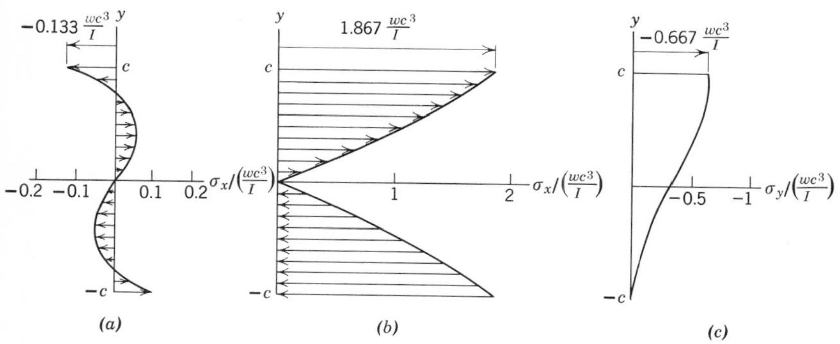


Fig. 5–27

## PROBLEMS

**5–97.*** The stress function $\varphi = axy + bxy^3$ is proposed for a cantilever beam of length $L$, depth $2c$, and thickness $t$. The origin is at the midpoint of the free end. The beam supports a transverse concentrated load $P$ at the free end. Determine the coefficients $a$ and $b$ to satisfy the boundary requirements. In what way, if any, does this equation fail to satisfy the boundary requirements? In what respects, if any, does this solution differ from the elementary solution?

**5–98.** The stress function

$$\varphi = \frac{\gamma x^3}{24c^3}(y^3 - 3c^2y - 2c^3) - \frac{\gamma x}{40c^3}(y^5 - 2c^2y^3 + c^4y)$$

is proposed for the cantilever beam subjected to a linearly varying load as shown in Fig. P5–98.

(*a*) Determine the equations for the stresses and show that all the boundary conditions are satisfied except that $\tau_{xy}$ is not zero on the surface $x = 0$.

(*b*) Show that the stress $\tau_{xy}$ on the surface $x = 0$ gives zero resultant force.

(*c*) Use Eq. (5–2) to obtain the expression for the shearing stress according to the simple theory. Identify the shearing stress correction term in the equation obtained in part (*a*).

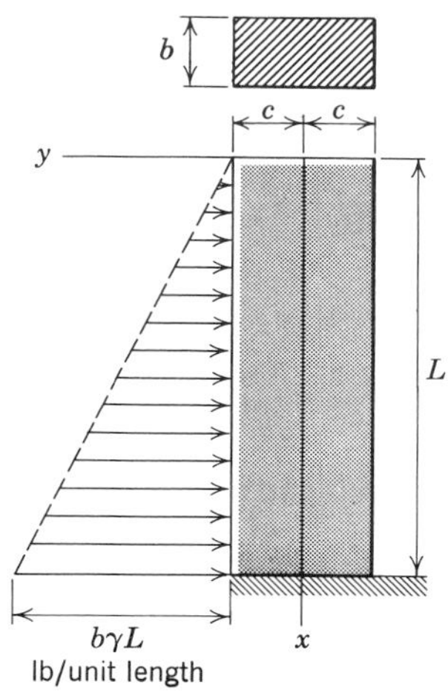


Fig. P5–98

**5–99.*** The Airy stress function for a beam with linearly varying depth and constant thickness subjected to a uniformly distributed load, as shown in Fig. P5–99, is

$$\varphi = \frac{w}{2(1 - \pi/4)}\left[-x^2 + xy + (x^2 + y^2)\left(\frac{\pi}{4} - \arctan\frac{y}{x}\right)\right].$$

(*a*) Determine the equation for $\sigma_x$ and evaluate this stress at points *A*, *B*, and *C*.

(*b*) Using the moment of inertia of section *ABC*, compare the stresses of part (*a*) with those given by the flexure formula.

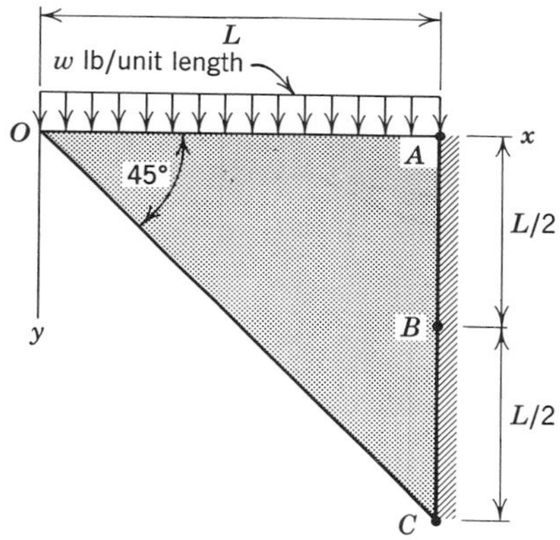


Fig. P5–99

**5–100.** A circular cantilever beam has a length *L* and a radius *R* and supports a downward load *P* at the free end. A set of axes is selected with the origin at the fixed end, the *x* axis downward, and the *z* axis along the axis of the beam. The following stresses are proposed to provide an exact solution.

$$\sigma_z = -\left(\frac{P}{I}\right)(L - z)x,$$

$$\tau_{zx} = \frac{(3 + 2\mu)P}{8(1 + \mu)I}\left(R^2 - x^2 - \frac{1 - 2\mu}{3 + 2\mu}y^2\right),$$

$$\tau_{zy} = -\frac{1 + 2\mu}{4(1 + \mu)I}Pxy.$$

All other stresses are zero.

(*a*) Show that these stresses satisfy the equations of equilibrium and compatibility for three dimensional stresses; see Eq. (3–10) and (3–13).

(*b*) Are the boundary conditions satisfied? Explain.

(*c*) Illustrate the variation of $\tau_{zx}$ along a horizontal diameter and compare this stress with that obtained from the elementary theory. Assume $\mu = 0.25$.

**5–12. Inelastic Bending.** A large proportion of structural and machine designs are based on elastic analysis, for which the flexure formula is applicable. However, for many designs, particularly where the weight of the structure is an important consideration (aircraft design, for example), the limitation of stresses within the proportional limit results in uneconomical or inefficient designs; therefore this limitation is sometimes discarded, necessitating either a modification of the flexure formula or the development of a new expression or technique.

The basic approach to plastic analysis is the same as that outlined in Art. 5–3 and 5–4. The plane section is assumed to remain plane, resulting in a linear distribution of longitudinal strain. However, the fiber stress distribution instead of being linear may be somewhat as shown in Fig. 5–28*a* which shows the distribution for a beam of symmetrical cross section made of a material having the stress-strain diagram of Fig. 5–28*c*. The resisting moment of the section is obtained from the expression

$$M_r = \int (\sigma)(dA)(y), \tag{a}$$

and, if the law of variation of the stress is known, $\sigma$ as some function of $y$ can be substituted in Eq. (a). If the resulting expression is integrable, the resisting moment can be evaluated.

The oldest approach to the problem is to simply replace the distribution $AOB$ of Fig. 5–28*a* by a linear distribution $DOE$ (shown dotted) with a maximum value of $R$, where $R$ is the *modulus of rupture in bending*. The value $R$ is selected to give the same resisting moment as that obtained from the actual stress distribution $AOB$. For the

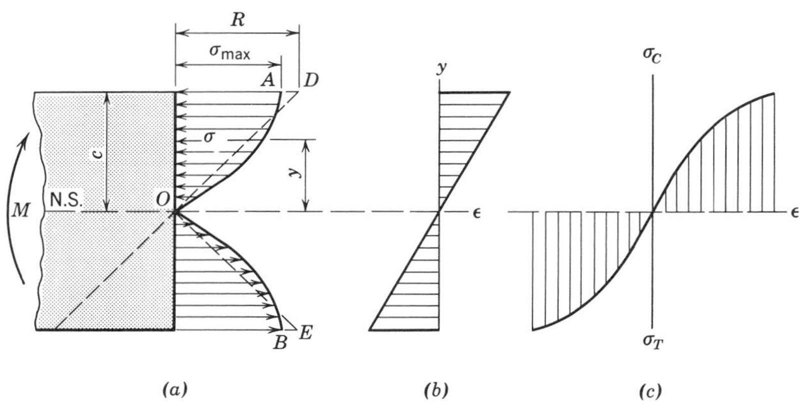


Fig. 5–28

assumed linear distribution *DOE*, the flexure formula is valid, and $R = Mc/I$. The value of $R$ is obtained by loading a beam to rupture and substituting the magnitude of the bending moment at rupture into the flexure formula. The modulus of rupture is, of course, useful only for predicting the rupture load, is valid only for a particular material and cross section, and in general may be dependent on some shape parameter (as, for example, the diameter to thickness ratio for tubing).

Of wider usefulness is the evaluation of Eq. (a) when the law of variation of fiber stress (*AOB*) is known or can be approximated. If the material is elastoplastic (see Art. 2–1), the large amount of strain accompanying the yield point will permit plastic action to take place as indicated in Fig. 5–29 which depicts the stress distribution for fully elastic action in (*a*), partially plastic action in (*b*), and essentially complete plastic action in (*c*) (there must always be a slight amount of elastic action if the strain at the neutral surface is zero). The action in Fig. 5–29*c* is idealized in (*d*); this is the assumed distribution used in plastic analysis of beams of structural steel.

To give some idea of the difference between plastic and elastic analysis applied to a rectangular section of elastoplastic material of width $b$ and depth $d$, the resisting moments for both cases will be computed. The moment necessary to produce the yield point stress, $\sigma_{yp}$, in the outer fibers, defined as the *elastic moment*, is (see method of Art. 5–3)

$$M_e = (\sigma_{yp}) \frac{1}{2} (b) \frac{d}{2} \frac{2d}{3} = \sigma_{yp} \frac{bd^2}{6}.$$

For yielding to occur throughout the section (as indicated in the idealized diagram of Fig. 5–29*d*), the yield point stress acts on the entire section, and the location of the resultant force on one-half the sec-

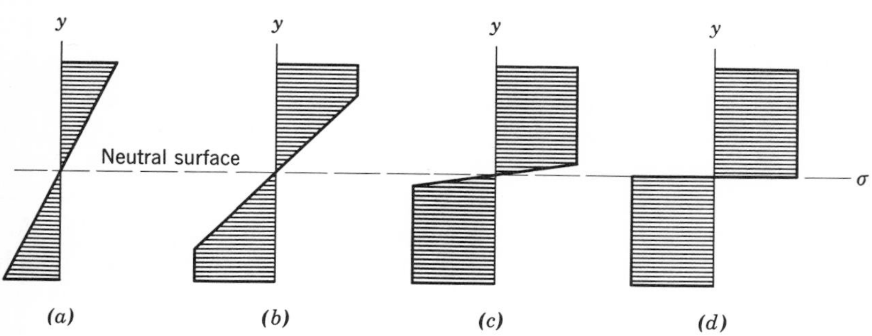


Fig. 5–29

tion is (½) $(d/2)$, rather than (⅔) $(d/2)$, from the neutral surface; therefore the resisting moment, defined as the *plastic moment*, is

$$M_p = (\sigma_{yp})(b)\frac{d}{2}\frac{d}{2} = \sigma_{yp}\frac{bd^2}{4},$$

from which the moment necessary to cause plastic action throughout the section is 50 per cent greater than that necessary to produce yielding in only the outer fibers. The loads on the beam would bear this same ratio; therefore, a factor of safety based on load would be one and one-half times that based on stress.

The difference between $M_p$ and $M_e$ is not as large for rolled sections. For example, the moments for a 36 WF 160 beam (neglecting fillets) are

$$M_e = (I/c)\sigma_{yp} = 541\sigma_{yp}$$

and $$M_p = (\sigma_{yp})(12)(1.02)(36 - 1.02) + (\sigma_{yp})(0.653)(18 - 1.02)^2(1/2)(2)$$

$$= 616\sigma_{yp},$$

from which $$M_p = 1.14M_e.$$

If the material does not have a yield point, and if a stress-strain diagram is available, the diagram can be used to obtain a stress distribution diagram for the beam since the strain variation in the beam is linear. Sometimes the equation of the resulting stress distribution diagram may be written. If not, or if the resulting Eq. (a) is not integrable, the beam depth can be divided into layers of finite thickness, and the stress on the end area of each layer can be obtained from the stress distribution diagram. As a simplifying technique, the stress-strain diagram may be approximated by a series of straight lines, resulting in a simplified stress distribution diagram and a reduction in the amount of work involved.

In the analysis of members subjected only to flexure, one must be conscious of the fact that, *when the action is plastic, the neutral axis does not necessarily coincide with a centroidal axis* as it does when the action is elastic. If the section is symmetrical and the stress-strain diagrams for tension and compression are identical, the neutral axis for plastic action coincides with the centroidal axis. However, if the section is unsymmetrical (for example a T-section), or if the stress-strain relations for tension and compression differ appreciably (for example cast iron), the neutral axis shifts away from the fibers which first undergo plastic action, and it is necessary to locate the axis before the resisting moment can be evaluated. The neutral axis is located by equating $\int \sigma(dA)$ over the cross section to zero and solving for the location of the axis where $\sigma$ is zero. Sometimes it may be possible to express the integral in terms of one unknown, in which case the solution may not be difficult. However, if the stress-strain relation for the material does not approach the ideal elastoplastic diagram, the solution will probably involve a number of trials.

The following example illustrates the concepts of plastic design as applied to a beam of symmetrical section.

**Example 5–10:** A beam is to be fabricated of three 6 by 1-in. plates welded together to form a symmetrical I-section. Determine the maximum elastic and plastic bending moments that the beam will sustain if the material is:

(*a*) Elastoplastic steel with a proportional limit (equal to the yield point) of 40 ksi.

(*b*) An aluminum alloy with the stress-strain diagram of Fig. 5–30*a*. Assume the diagram is the same in tension and compression, and limit the strain to 0.010.

*Solution:* (*a*) For the elastic moment, the flexure formula $M = \sigma I/c$ applies, and $I$ is found to be 166 in.$^4$ Then

$$M_e = 40(166/4) = \underline{1660 \text{ in-kips.}} \qquad \text{Ans.}$$

For this material, the ideal stress distribution of Fig. 5–29*d* may be assumed when computing the plastic moment, which is

$$M_p = \left[40(6)(1)\frac{7}{2} + 40(3)(1)\frac{3}{2}\right][2] = \underline{2040 \text{ in-kips.}} \qquad \text{Ans.}$$

To check the degree of approximation involved in assuming the ideal stress distribution for this section, the maximum strain before strain hardening takes place will be assumed to be sixteen times the elastic strain (a reasonable value for structural steel). When this strain is reached in the outer fibers, the maximum elastic strain occurs at $(1/16)(4)$ or $1/4$ in. above and below the neutral surface; which means that the actual stress distribution diagram is like that of Fig. 5–29*c*, the distance from the neutral surface to the point where the stress is constant being $1/4$ in. This distribution makes the moment of 2040 in-kips computed above too large by the amount

$$\Delta M_r = \frac{40}{2}\frac{1}{4}(1)\frac{1}{3}\frac{1}{4}(2) = 0.833 \text{ in-kip}$$

or 0.04 per cent.

(*b*) The elastic moment is given by the flexure formula, where $\sigma$ is the proportional limit which from Fig. 5–30*a* is 30 ksi. Then

$$M_e = 30(166/4) = \underline{1245 \text{ in-kips.}} \qquad \text{Ans.}$$

The plastic moment will first be determined with the aid of the stress distribution diagram of Fig. 5–30*c* which was obtained from Fig. 5–30*a* by assuming the maximum strain in the beam to be 0.010, and, since the strains are proportional to the distance from the neutral surface of the beam, the stress for any point in the cross section may be read from the diagram of Fig. 5–30*a*. With the assumption that the curve in Fig. 5–30*c* is a series of chords connecting the points at 1-in. vertical intervals, the following solution is obtained:

$$M_p = 2\left[6(1)(48)\frac{7}{2} + (50 - 48)\frac{1}{2}(6)(1)\left(3 + \frac{2}{3}\right)\right.$$
$$+ 43.3(1)(1)\frac{5}{2} + (48 - 43.3)\frac{1}{2}\left(2 + \frac{2}{3}\right)$$
$$+ 26.5(1)(1)\frac{3}{2} + (43.3 - 26.5)\frac{1}{2}\left(1 + \frac{2}{3}\right)$$
$$\left. + \frac{26.5}{2}\frac{2}{3}\right]$$
$$= \underline{2414 \text{ in-kips.}} \qquad \text{Ans.}$$

This result will be checked by assuming the curve below the flange to be a parabola. The flange moment will be the same as above and equals 2060 in-kips; then

$$M_p = 2060 + \frac{2}{3}(48)(3)(1)\frac{5}{8}(3)(2) = \underline{2420 \text{ in-kips.}} \qquad \text{Ans.}$$

One more technique will be applied to this problem. The stress-strain diagram of Fig. 5–30*a* will be approximated by the dotted lines shown, and the resulting stress distribution diagram for the beam will be as in Fig. 5–30*d*. With the use of this diagram, the plastic moment becomes

$$M_p = 2\left[49(6)(1)\left(\frac{7}{2}\right) + 40\left(\frac{3}{2}\right)(1)\left(\frac{3}{2} + \frac{3}{4}\right)\right.$$
$$+ (49 - 40)\left(\frac{1}{2}\right)\left(\frac{3}{2}\right)(1)\left(\frac{3}{2} + 1\right)$$
$$\left. + 40\left(\frac{1}{2}\right)\left(\frac{3}{2}\right)(1)\left(\frac{2}{3}\right)\left(\frac{3}{2}\right)\right]$$
$$= \underline{2422 \text{ in-kips.}} \qquad \text{Ans.}$$

The last three results are so close together that any of the three techniques are adequate for this problem.

From the preceding discussion and examples, one may observe that, if the load capacity of a beam is based on the plastic moment rather than the elastic moment, a considerable saving in material may be realized by using the plastic design procedure (with the same factor of safety). Furthermore, in the design of statically indeterminate beams and frames, the plastic method of analysis is considerably less time-consuming than the elastic analysis. This is not to say that the elastic method of analysis should be discarded as outmoded, for there are many designs which must be based on elastic action, and for some situations, particularly where repeated loading is involved, the application of plastic analysis may be dangerous.

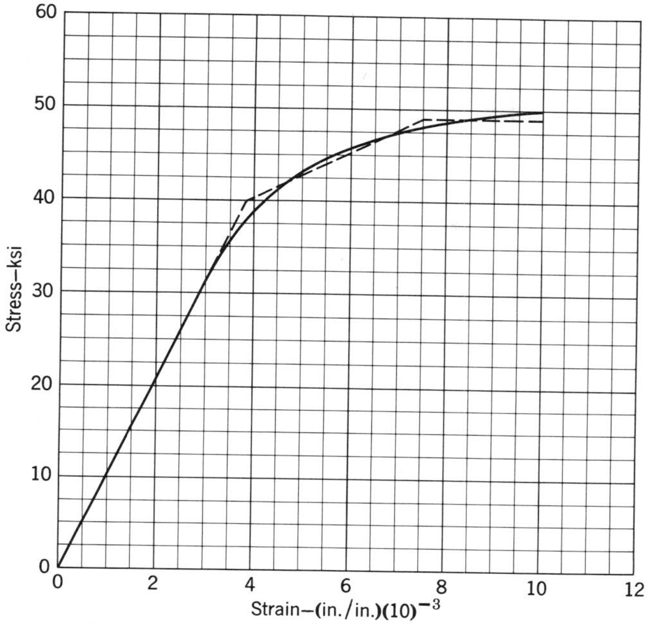

60
50
40
30
20
10
0
Stress–ksi
0
2
4
6
8
10
12
Strain–(in./in.)(10)$^{-3}$


(a)

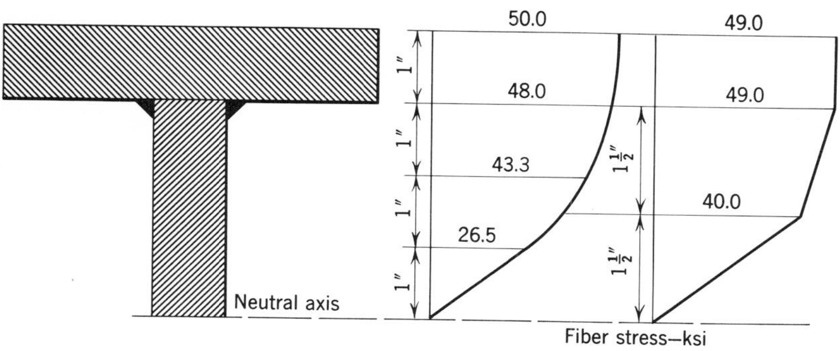

Neutral axis
50.0
48.0
43.3
26.5
1″
1″
1″
1″
49.0
49.0
40.0
1½″
1½″
Fiber stress–ksi


(b) (c) (d)

Fig. 5–30

## PROBLEMS

**Note: Neglect fillets and assume the material to be elastoplastic unless otherwise specified in the following problems.**

**5–101.*** Determine the plastic moment for a 16 WF 64 steel beam having a proportional limit of 36 ksi.

**5–102.*** Determine the plastic moment for a 21 WF 112 steel beam having a proportional limit of 40 ksi.

**5–103.** A 10 WF 21 structural steel beam (yield point, 36 ksi) is selected for the beam of Prob. 5–37. Determine the factor of safety:

(*a*) Based on the plastic moment.

(*b*) Based on the elastic strength.

**5–104.*** An 8 WF 31 structural steel beam (yield point, 36 ksi) is selected for the beam of Prob. 5–32. Determine the factor of safety:

(*a*) Based on the plastic moment.

(*b*) Based on the elastic strength.

**5–105.** Determine the ratio of the plastic moment to the elastic moment for a beam of elastoplastic material having a circular cross section.

**5–106.*** For a beam of elastoplastic material (with a yield point of 36 ksi) having the cross section of Prob. 5–80:

(*a*) Locate the neutral axis when the action is plastic over the entire section.

(*b*) Determine the plastic moment.

(*c*) Determine the ratio of plastic to elastic moment.

**5–107.** For the beam of Prob. 5–106:

(*a*) Locate the neutral axis when the stress in the outer fibers of the flange reaches the yield point.

(*b*) Determine the resisting moment when the condition in part *a* is reached.

**5–108.*** A beam is fabricated of two 6 by 1-in. plates and two 3 by 1-in. plates welded together to form a box section 3 in. wide by 8 in. deep. The material is aluminum alloy having the stress-strain diagram of Fig. 5–30*a*. Determine:

(*a*) The elastic moment.

(*b*) The plastic moment, using the stress distribution diagram of Fig. 5–30*d*.

**5–13. Shear Center.** One of the assumptions made in the development of the flexure formula was that the loads were applied in a plane of symmetry. When this assumption is not satisfied, the beam will, in general, twist about a longitudinal axis. It is possible, however, to place the loads in such positions that the beam will not twist. When any load is applied in such a position, the line of action of the load will pass through the *shear center* defined as the point in the plane of the cross section through which the resultant of the transverse shearing stresses due to flexure (no torsion) will pass for any orientation of transverse loads. The shear center of any section can be located by considering the shearing stresses which tend to twist the beam.

A common example of a beam loaded in a plane other than the plane of symmetry is a channel section loaded as a cantilever beam as in Fig. 5–31*a*. In order to simplify the analysis, it is assumed that the flanges have a constant thickness which is small compared with the width and depth of the section. The flange thickness need not be the same as the web thickness. The shearing stresses on the flanges at section *a–a* can be determined from a free-body diagram of the small block *B* shown enlarged in Fig. 5–31*b*. The forces $T$ and $T + dT$ at the ends of the block are

$$T = \sigma A = \frac{My}{I} A = \frac{Mh}{2I} (zt_1) \nwarrow$$

and

$$T + dT = \frac{(M + dM)h}{2I} (zt_1). \searrow$$

The force $F_1$ is the difference of these two forces and is

$$F_1 = [(h\, dM)/(2I)](zt_1). \nwarrow$$

The shearing stress on the longitudinal plane is

$$\tau = \frac{F_1}{\text{Area}} = \frac{[(h\, dM)/(2I)](zt_1)}{t_1\, dx} = \frac{Vhz}{2I}, \tag{a}$$

where $dM/dx$ is the transverse shear as shown in Art. 5–6. This same shearing stress acts in the $z$ direction on the transverse plane a distance $z$ from the edge of the flange, see Art. 1–4. Equation (a) shows that the horizontal shearing stress in the flange varies linearly from zero at the outer edge to $Vhb/(2I)$ at the web. The resultant shearing force on the top flange is

$$F_2 = (\tau_{\text{avg.}})\ \text{Area} = \left[\frac{Vhb}{4I}\right] (bt_1) = \frac{Vhb^2t_1}{4I},$$

as shown in Fig. 5–31*c*. A similar analysis indicates that the force $F_3$ on the lower flange is equal in magnitude and opposite in direction to $F_2$. These two forces constitute a clockwise twisting moment which must be balanced by the resultant force in the web and the applied load if the beam is not to twist.

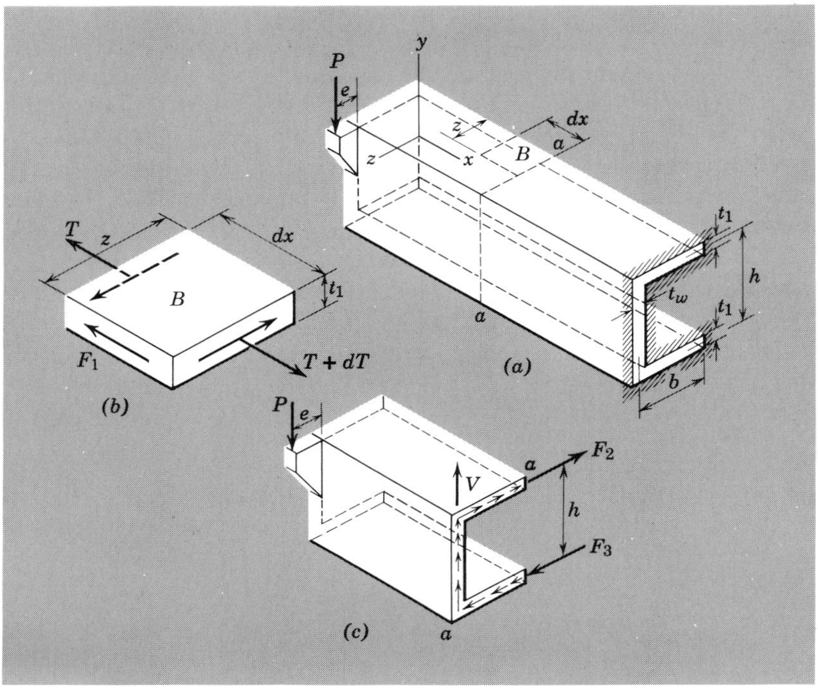


Fig. 5–31

The resultant of the shearing forces in the web is $V$ and must be equal to the applied load, $P$, for equilibrium. It is assumed that the vertical shearing force transmitted by the flanges is small enough to neglect (see Art. 5–9). The forces $V$ and $P$ will provide a counterclockwise couple to balance $F_2$ and $F_3$ if $P$ is located a distance $e$ from the center of the web as indicated in Fig. 5–31$c$. When the moments of these two couples are equated, the result is

$$Ve = hF_2 = \frac{Vh^2b^2t_1}{4I},$$

from which

$$e = \frac{h^2b^2t_1}{4I}.$$

If the load $P$ of Fig. 5–31$a$ were applied in the $z$ direction, the forces corresponding to $F_2$ and $F_3$ (entirely different magnitudes) would be equal, and both would be directed to oppose the load $P$. There would also be two shearing forces in the web (parallel to the web) of equal magnitude and oppositely directed; hence a moment equation would establish the fact that the shear center lies on the $z$ axis of symmetry. If there is to be no twisting of the beam (buckling is assumed not to occur), any load must pass through the intersection of this axis of symmetry and the line of action of force $P$ of Fig. 5–31$a$. If the load is inclined, it may be resolved into components parallel to these axes, and the stresses may be determined using the principle of superposition.

## PROBLEMS

**5–109.*** A channel is 15 in. deep over-all, and has flanges which are 3.4 in. wide and 0.650 in. thick. The web is 0.400 in. thick. Locate the shear center with respect to the outside edge of the web.

**5–110.** An eccentric H-section is made by welding two 1 by 10-in. steel flanges to a 1 by 12-in. steel web as indicated in Fig. P5–110. The section is used as a 5-ft cantilever beam and carries a concentrated load of 100 kips at the end, parallel to the web.

($a$) Determine the magnitude of the resultant shearing force in the top flange, and locate the center of shear of the section. Assume that all of the section is effective in resisting fiber stresses but that only the web resists vertical shearing stresses.

($b$) Would the solution be any different if the web were assumed to be effective only for shearing stresses (but not for fiber stress)? Locate the shear center for this assumption if different from the initial assumption.

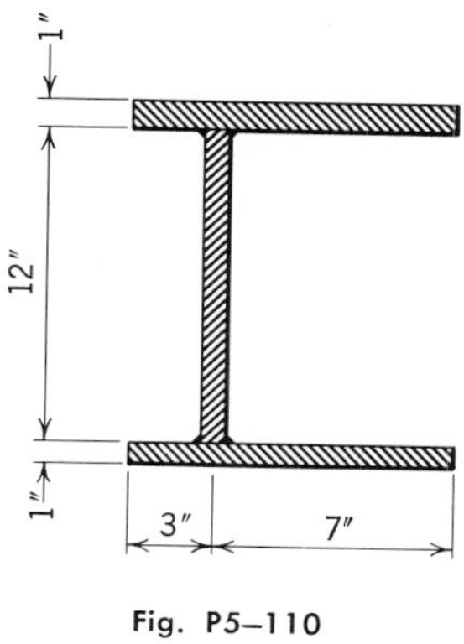


Fig. P5–110

**5–111.*** Two ½-in. thick plates are welded to the flanges of the channel in Prob. 5–109, as shown in Fig. P5–111. Determine the width of these plates which will cause the shear center to be located at the center of the web.

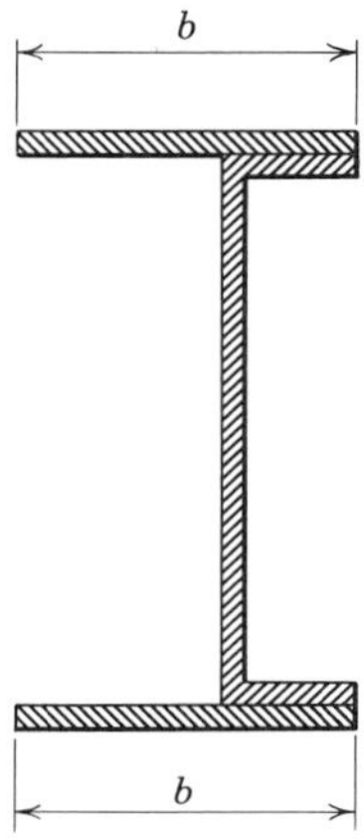


Fig. P5–111

**5–112.** A thin-walled cylindrical tube is cut longitudinally to make a semi-cylinder which is used as a cantilever beam. The load acts parallel to the cut section. Demonstrate that the shear center is $4R/\pi$ from the cut section. Use the element shown in Fig. P5–112.

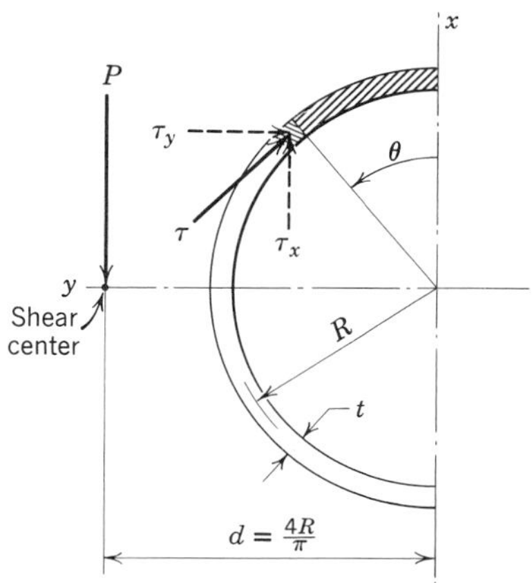


Fig. P5–112

**5–14. Beams of Two Materials.** In Art. 5–2, the assumption was made that a plane section before bending remained plane after bending. With this assumption, it was demonstrated that the strain of any fiber of an initially straight beam is directly proportional to the distance of the fiber from the neutral surface. Consequently, the fiber stress is directly proportional to the distance of the fiber from the neutral surface if Hooke's law applies and if the modulus of elasticity does not vary. For beams of more than one material, the assumption regarding the plane section is still valid; however, the modulus of elasticity is, in general, different for unlike materials, and it is necessary to start with the accepted law of strain distribution (it is assumed that the different materials are securely fastened together to act as a unit) and develop a stress distribution pattern.

The beam of Fig. 5–32, composed of a central portion of material $A$ and two outer layers of material $B$, will serve as a model for the development of the stress relationship. As long as neither material is subjected to stresses above the proportional limit, Hooke's law applies, and the strain relation

$$\epsilon_b = \epsilon_a \frac{b}{a}$$

becomes

$$\frac{\sigma_b}{E_B} = \frac{\sigma_a}{E_A}\left(\frac{b}{a}\right)$$

or

$$\sigma_b = \sigma_a \left(\frac{E_B}{E_A}\right)\left(\frac{b}{a}\right).$$

From this relation it is evident that, at the junction between the two materials where distances $a$ and $b$ are equal, there is an abrupt change

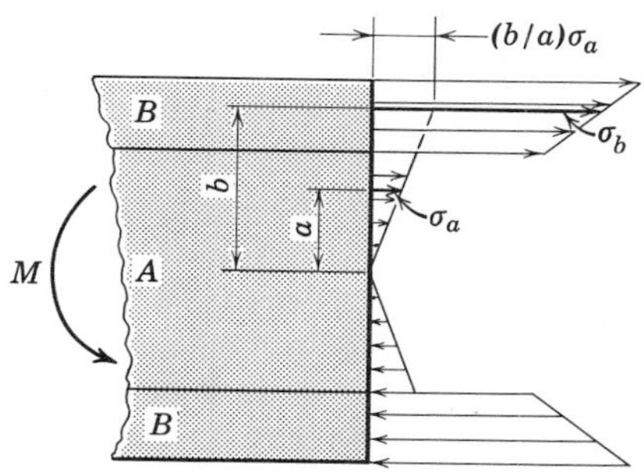


Fig. 5–32

in stress determined by the ratio, $n = E_B/E_A$, of the two moduli. The normal force on a differential end area of element $B$ is given by the expression

$$dF_B = \sigma_b\, dA = n\,\frac{b}{a}\,\sigma_a\, dA = \frac{\sigma_a}{a}\, nbt\, db,$$

where $t$ is the width of the beam at a distance $b$ from the neutral surface. The last expression may be interpreted as $[(\sigma_a/a)(b)](nt)\, db$, where $(\sigma_a/a)(b)$ is the stress at any distance $b$ in a homogeneous material $A$ and where $(nt)$ is the width of the beam (at distance $b$) of material $A$. Therefore, if material $B$ were replaced by $n$ times as much material $A$ at the same distance from the neutral surface, the fiber stress in the equivalent material would be proportional to the distance from the neutral surface, and the substitute beam would provide exactly the same resistance to bending as the actual beam. For this reason, beams of two materials can readily be solved by using an equivalent or transformed section and proceeding as for one material. A *transformed cross section* is obtained by replacing either material by an equivalent amount of the other material as determined by the ratio, $n$, of their elastic moduli. This method is not limited to two materials; however, the use of more than two materials for one beam would be unusual.

The equilibrium method of Art. 5–3 could be used directly without a transformed section. However, it is perhaps easier to obtain the transformed section and then use the flexure formula; this method will be illustrated in the following example.

**Example 5–11:** A timber beam 4 in. wide by 8 in. deep has an aluminum alloy plate with a net section $1\frac{2}{3}$ in. wide by $\frac{1}{2}$ in. deep securely fastened to its bottom face. The working stresses in the timber and aluminum alloy are 1200 psi and 16,000 psi, respectively. The moduli of elasticity for the timber and aluminum alloy are 1250 ksi and 10,000 ksi, respectively. Determine the maximum allowable resisting moment for the beam.

*Solution:* The ratio of the moduli of the aluminum and timber is

$$n = 10{,}000/1250 = 8.$$

The actual cross section (timber $A$ and aluminum $B$) and the transformed timber cross section are shown in Fig. 5–33$a$ and $b$. The neutral axis of the transformed section is located by the principle of moments as

$$\bar{y} = \frac{\Sigma M}{\Sigma A} = \frac{(40/3)(1/4) + 32(9/2)}{(40/3) + 32} = 3.25 \text{ in.}$$

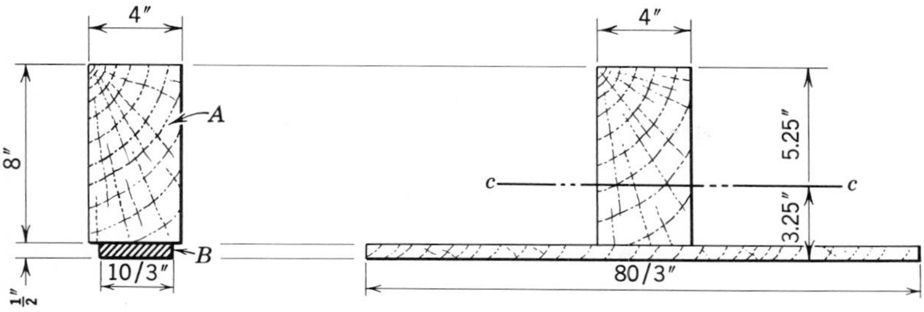


*(a)* *(b)*

Fig. 5–33

above the bottom of the aluminum plate. The moment of inertia of the cross-sectional area with respect to the neutral axis is

$$I_c = \frac{1}{12}\left(\frac{80}{3}\right)\left(\frac{1}{2}\right)^3 + \frac{40}{3}(3)^2 + \frac{1}{12}(4)(8)^3 + 32\left(\frac{5}{4}\right)^2 = 341 \text{ in.}^4.$$

The stress relation

$$\sigma_b = \sigma_a\left(\frac{E_B}{E_A}\right)\left(\frac{b}{a}\right)$$

will be used to determine which working stress controls the capacity of the section. From Fig. 5–33*b* it is seen that the distance to the outermost timber fiber is 5.25 in. and that the distance to the outermost aluminum fiber is 3.25 in. With 1200 psi for $\sigma_a$ the maximum stress in the aluminum is

$$\sigma_b = 1200(8)(3.25/5.25) = 5940 \text{ psi}$$

which is well below the allowable stress. Therefore the stress in the timber is the controlling stress. From the flexure formula

$$M = \frac{\sigma I}{c} = \frac{1200(341)}{5.25} = \underline{77{,}900 \text{ in-lb.}} \qquad \text{Ans.}$$

## PROBLEMS

**5–113.*** A 4-in. wide by 6-in. deep timber cantilever beam 7 ft long is reinforced by two ½-in. wide by 6-in. deep structural aluminum plates cemented securely to the sides of the timber beam. The flexural moduli of elasticity of this timber and aluminum are 1000 ksi and 10,000 ksi, respectively. Determine the maximum tensile fiber stresses in the two materials when a static load of 1000 lb is placed on the free end of the beam. The 6-in. dimension is vertical.

**5–114.*** A composite beam 4½ in. wide by 12 in. deep and 14 ft long is made by fastening two timber planks, 2 in. by 12 in., to the sides of a steel plate, ½ in. by 12 in., as shown in Fig. P5–114. $E_{\text{wood}} = 1200$ ksi. The composite beam, simply supported at the ends, acts as a single unit in supporting a uniformly distributed load of 1080 lb per ft. Determine the maximum fiber stresses in the steel and wood.

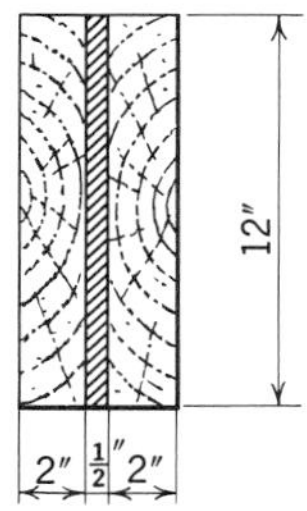


Fig. P5–114

**5–115.** A timber beam 4 in. wide by 10 in. deep has a 0.4-in. deep by 4-in. wide steel plate cemented securely to its bottom face. The modulus of elasticity of the timber is $1.5(10^6)$ psi. Determine the maximum fiber stress in the timber when that in the steel is 10,000 psi $T$.

**5–116.*** A 4 by 12-in. timber beam has a ½ by 3-in. steel plate cemented to its lower face. The flexural modulus of elasticity of the timber is $1.5(10^6)$ psi. Determine the maximum fiber stress in the steel when the maximum fiber stress in the timber is 960 psi $C$.

**5–117.** A cantilever beam 6 ft long carries a load of 4000 lb on the free end. The beam consists of a wooden section 4 in. wide and 8 in. deep to which are cemented 4-in. wide by ½-in. deep steel plates top and bottom. The modulus of elasticity for this wood is $1.8(10^6)$ psi. Determine the maximum fiber stress in the wood and in the steel.

**5–118.*** A 4-in. wide by 12-in. deep timber beam 15 ft long is reinforced by a 3.2-in. wide by ½-in. deep steel plate fastened securely to its lower face. The beam is simply supported and carries a uniformly distributed load of 720 lb per ft. The modulus of elasticity of the timber is $1.5(10^6)$ psi. Determine the maximum fiber stresses in the timber and the steel.

**5–119.** A 6-in. wide by 10-in. deep timber beam has a ½-in. deep by 4-in. wide steel plate fastened securely to its lower face. The modulus of elasticity of the timber is $1.5(10^6)$ psi.

(*a*) Determine the maximum fiber stress in the steel when that in the timber is 1000 psi $C$.

(*b*) Determine the resisting moment of the cross section for a maximum fiber stress of 1000 psi in the timber.

**5–120.*** A timber beam 8 in. wide by 12 in. deep has a 7-in. wide by ½-in. deep steel plate cemented to its bottom face. The flexural modulus of elasticity of the timber is 1500 ksi.

(*a*) Determine the maximum fiber stress in the steel when that in the timber is 1000 psi *C*.

(*b*) Determine the load *P* that must be placed on this beam at the center of a 20-ft simply supported span to give these stresses.

**5–121.** A timber beam 6 in. wide by 12 in. deep has a 6 by ½-in. steel plate securely fastened to its bottom face. It is to be used on a simple span of 16 ft carrying a uniformly distributed load. If the allowable fiber stress is 16,000 psi in the steel and 1200 psi in the wood, determine the maximum allowable load in pounds per foot. The flexural modulus of elasticity of the wood is 1500 ksi.

CHAPTER 6

# *Flexural Loading: Deflections*

**6–1. Introduction.** The important relations between applied load and stress (fiber and shear) developed in a beam were presented in Chap. 5. However, a beam design is frequently not complete until the amount of deflection has been determined for the specified load. Failure to control beam deflections within proper limits in building construction is frequently reflected by the development of cracks in plastered walls and ceilings. Beams in many machines must deflect just the right amount for gears or other parts to make proper contact. In innumerable instances the requirements for a beam involve a given load-carrying capacity with a specified maximum deflection.

The deflection of a beam depends upon the stiffness of the material and the dimensions of the beam as well as upon the applied loads and supports. Four common methods for calculating beam deflections due to fiber stresses are presented here. They are: (1) the double integration method, (2) the area-moment method, (3) the superposition method, and (4) an energy method. The material is presented so that methods 1 and 2 can be studied independently of each other. Deflections due to shearing stresses are also discussed.

**6–2. The Radius of Curvature.** When a straight beam is loaded and the action is elastic, *the longitudinal centroidal axis of the beam becomes a curve which is defined as the elastic curve.* In regions of constant bending moment, the elastic curve is an arc of a circle of radius $\rho$ as indicated in Fig. 6–1, in which the portion $AB$ of a beam is bent only with couples. Therefore, the plane sections $A$ and $B$ remain plane (see Art. 5–2), and the deformation (elongation and compression) of the fibers is proportional to the distance from the neutral surface

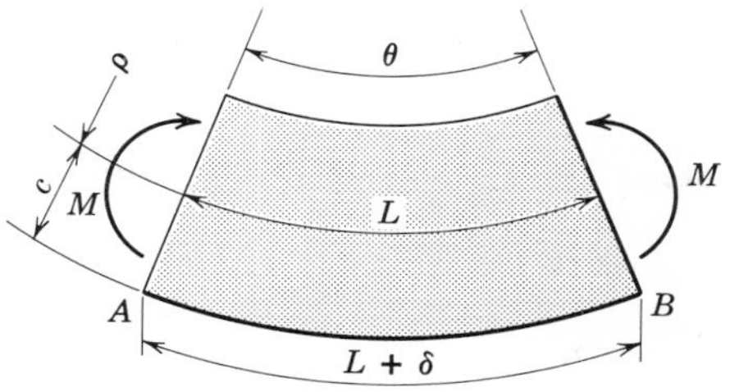


Fig. 6-1

which is unchanged in length. From Fig. 6–1,

$$\theta = \frac{L}{\rho} = \frac{L+\delta}{\rho+c},$$

from which

$$\frac{c}{\rho} = \frac{\delta}{L} = \epsilon = \frac{\sigma}{E} = \frac{Mc}{EI}.$$

Therefore

$$\frac{1}{\rho} = \frac{M}{EI}, \tag{6–1}$$

which relates the radius of curvature of the neutral surface of the beam to the bending moment, $M$; the stiffness of the material, $E$; and the moment of inertia of the cross-sectional area, $I$. The quantity $EI$ is sometimes referred to as the *flexural stiffness.*

## PROBLEMS

**6–1.*** The beam shown in Fig. P6–1*a* is made of two steel bars each 1 by 4 in. The bars are to be securely fastened together as shown in either Fig. P6–1*b* or *c*. Determine:

(*a*) The radius of curvature and maximum fiber stress for configuration *b*.
(*b*) The radius of curvature and maximum fiber stress for configuration *c*.

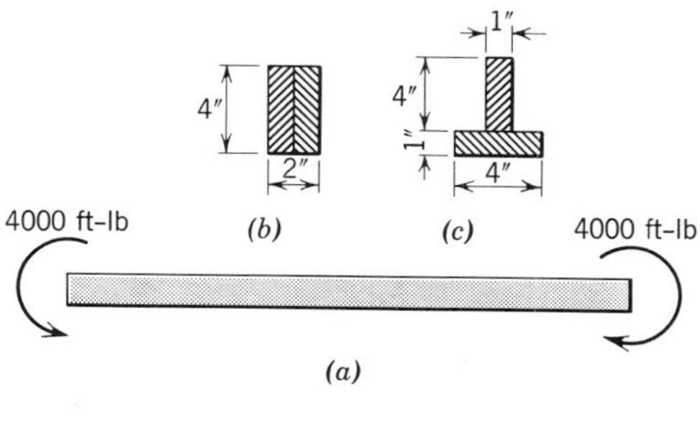


Fig. P6-1

**6–2.*** The 12-ft beam of Fig. P6–2 is a 5 I 14.75 steel beam.
(*a*) Determine the radius of curvature at any point between the supports.
(*b*) What is the shape of the elastic curve?

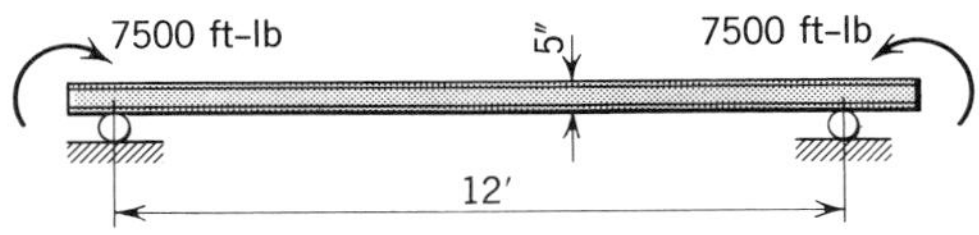


Fig. P6-2

**6–3.*** The boards for a concrete form are to be bent to a circular curve of 15-ft radius. What maximum thickness can be used if the stress is not to exceed 2000 psi? The timber is saturated, and Young's modulus may be taken as 800,000 psi.

**6–4.*** Determine the radius of curvature of the aluminum cantilever beam of Fig. P6–4 and from it the deflection of the left end of the beam when loaded as shown. Use $10^7$ psi for Young's modulus.

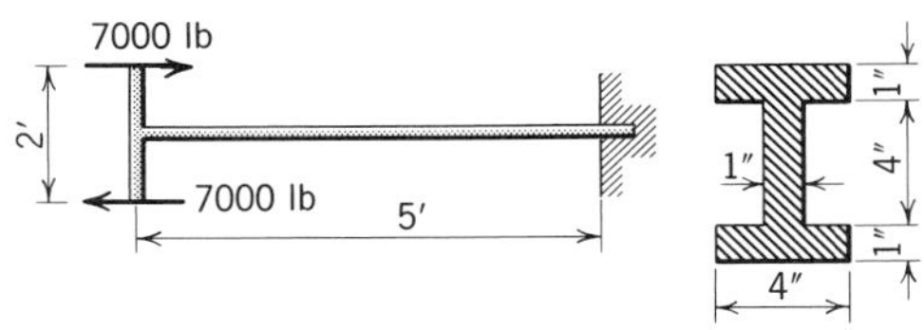


Fig. P6-4

**6–3. The Differential Equation of the Elastic Curve.** The expression for the curvature of the elastic curve derived in Art. 6–2 is useful only when the bending moment is constant for the interval of the beam involved. For most beams the bending moment is a function of the position along the beam and a more general expression is required.

The curvature from the calculus (see any standard calculus textbook) is

$$\frac{1}{\rho} = \frac{d^2y/dx^2}{[1 + (dy/dx)^2]^{3/2}}.$$

For actual beams the slope, $dy/dx$, is very small, and its square can be neglected in comparison to unity. With this approximation,

$$\frac{1}{\rho} = \frac{d^2y}{dx^2},$$

and Eq. (6–1) becomes

$$EI\frac{d^2y}{dx^2} = M_x, \qquad (6\text{–}2)$$

which is the differential equation for the elastic curve of a beam where the subscript $x$ on $M$ is used as a reminder that $M$ is a function of $x$.

The differential equation of the elastic curve can also be obtained from the geometry of the bent beam as shown in Fig. 6–2 where it is evident that $dy/dx = \tan\theta = \theta$ (approximately) for small angles and that $d^2y/dx^2 = d\theta/dx$. Again from Fig. 6–2

$$d\theta = \frac{dL}{\rho} = \frac{dx}{\rho} \qquad \text{(approximately) for small angles.}$$

Therefore

$$\frac{d^2y}{dx^2} = \frac{d\theta}{dx} = \frac{1}{\rho} = \frac{M_x}{EI},$$

or

$$EI\frac{d^2y}{dx^2} = M_x. \qquad (6\text{–}2)$$

The sign convention for bending moments established in Art. 5–5 will be used for Eq. (6–2). Both $E$ and $I$ are always positive; therefore the signs of the bending moment and the second derivative must be consistent. With the coordinate axes selected as shown in Fig. 6–3, the slope changes from positive to negative in the interval from $A$ to $B$; therefore the second derivative is negative, which agrees with the

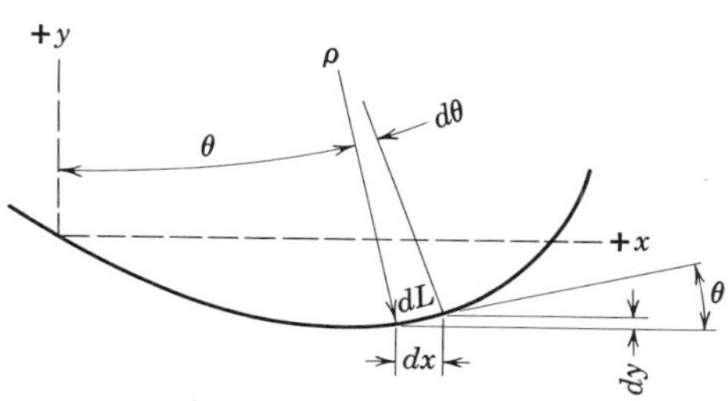


Fig. 6-2

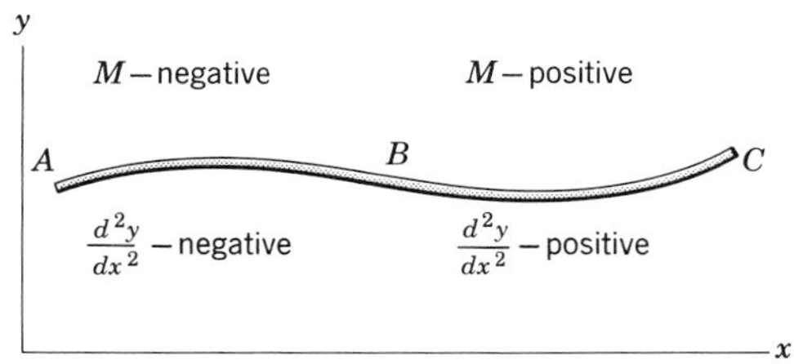


Fig. 6-3

sign convention of Art. 5–5. For the interval $BC$, both $d^2y/dx^2$ and $M_x$ are seen to be positive.

Careful study of Fig. 6–3 reveals that the signs of the bending moment and the second derivative are also consistent when the origin is selected at the right with $x$ positive to the left and $y$ positive upward. Unfortunately, the signs are inconsistent when $y$ is positive downward. In this book, for horizontal beams, $y$ will always be selected positive upward for beam deflection problems.

Before proceeding with the solution of Eq. (6–2), it is desirable to correlate the successive derivatives of the deflection, $y$, of the elastic curve with the physical quantities that they represent in beam action. They are:

$$\text{Deflection} = y,$$

$$\text{Slope} = \frac{dy}{dx},$$

$$\text{Moment} = EI\frac{d^2y}{dx^2} \text{ (from Eq. 6–2)},$$

$$\text{Shear} = \frac{dM}{dx} \text{ (from Art. 5–6)} = EI\frac{d^3y}{dx^3} \text{ (for } EI \text{ constant)},$$

$$\text{Load} = \frac{dV}{dx} \text{ (from Art. 5–6)} = EI\frac{d^4y}{dx^4} \text{ (for } EI \text{ constant)},$$

where the signs are as given in Art. 5–5 and 5–6.

In Art. 5–7 a method based on these differential relations was presented for starting from the load diagram and drawing first the shear diagram and then the moment diagram. This method can readily be extended to construction of the slope diagram from the relation

$$M = EI\frac{d\theta}{dx},$$

from which

$$\int_{\theta_A}^{\theta_B} d\theta = \int_{x_A}^{x_B} \frac{M}{EI}\,dx.$$

This relation shows that except for a factor $EI$ the area under the moment diagram between any two points along the beam gives the change in slope between the same two points. Likewise the area under the slope diagram between two points along a beam gives the change in deflection between these points. These relations have been used to construct the complete series of diagrams shown in Fig. 6–4 for a simply supported beam with a concentrated load at the center of the span. The geometry of the beam was used to locate the points of zero slope and deflection, required as starting points for the construction. More commonly used methods for calculating beam deflections will be developed in succeeding articles.

Before proceeding with specific methods for calculating beam deflec-

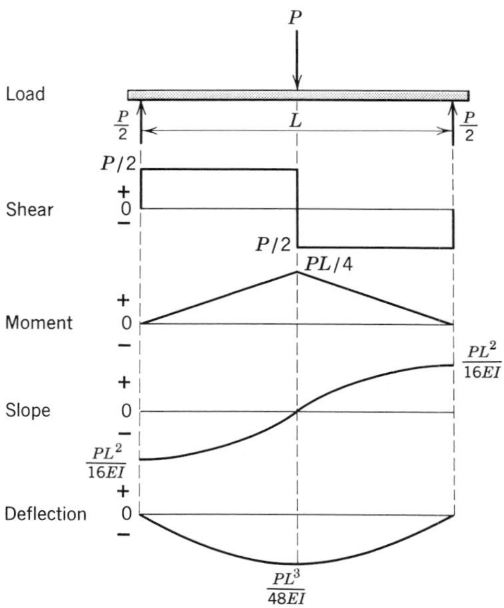


Fig. 6-4

tions, it is advisable to consider the assumptions used in the development of the basic relation, Eq. (6–2). All of the limitations that apply to the flexure formula apply to the calculation of deflections because the flexure formula was used in the derivation of Eq. (6–2). It is further assumed that:

1. The square of the slope of the beam is negligible compared to unity.
2. The beam deflection due to shearing stresses is negligible (a plane section is assumed to remain plane). See Art. 6–12 for further information.
3. The values of $E$ and $I$ remain constant for any interval along the beam. In case either of them varies and can be expressed as a function of the distance $x$ along the beam, a solution of Eq. (6–2) may be possible which takes this variation into account.

**6–4. Deflections by Integration.** Whenever the assumptions of the previous paragraph are essentially correct and the bending moment can be readily expressed as an integrable function of $x$, Eq. (6–2) can be solved for the deflection, $y$, of the elastic curve of a beam at any point, $x$, along the beam. The constants of integration can be evaluated from the applicable boundary or matching conditions.

*A boundary condition is defined as a known set of values for $x$ and $y$, or $x$ and $dy/dx$, at a specific location along the beam.* One boundary condition (one set of values) can be used to determine one and only one constant of integration.

Many beams are subjected to abrupt changes in loading along the beam, such as concentrated loads, reactions, or even distinct changes in the amount of a uniformly distributed load. Since the expressions for the bending moment on the left and right of any abrupt change in load are different functions of $x$, it is impossible to write a single equation for the bending moment in terms of ordinary algebraic functions which is valid for the entire length of the beam. This can be resolved by writing separate bending moment equations for each interval of the beam. Although the intervals are bounded by abrupt changes in load, the beam is continuous at such locations and, consequently, the slope and the deflection at the junction of adjacent intervals must match. *A matching condition is defined as the equality of slope or deflection, as determined at the junction of two intervals from the elastic curve equations for both intervals.* One matching condition (for example, at $x$ equals $L/3$, $y$ from the left equation equals $y$ from the right equation) can be used to determine one and only one constant of integration.

The procedure for obtaining beam deflections where matching conditions are required is lengthy and tedious. A method is presented in the next article in which *singularity functions* are used to write a single equation for the bending moment which is valid for the entire length of the beam; this eliminates the need for matching conditions and, accordingly, reduces the labor involved.

Calculating the deflection of a beam by the double integration method involves four definite steps, and the following sequence for these steps is strongly recommended.

1. Select the interval or intervals of the beam to be used; next, place a set of coordinate axes on the beam *with the origin at one end of an interval,* and then indicate the range of values of $x$ in each interval. For example, two adjacent intervals might be

$$0 \leq x \leq L \quad \text{and} \quad L \leq x \leq 3L.$$

2. List the available boundary and matching (where two or more adjacent intervals are used) conditions for each interval selected. Remember that two conditions are required to evaluate the two constants of integration for each interval used.

3. Express the bending moment as a function of $x$ for each interval selected, and equate it to $EI\, d^2y/dx^2$.

4. Solve the differential equation or equations from item 3, and evaluate all constants of integration. Check the resulting equations for dimensional homogeneity. Calculate the deflection at specific points where required.

A roller or pin at any point in a beam, see Fig. 6–5$a$ and $b$, represents a simple support at which the beam cannot deflect (unless other-

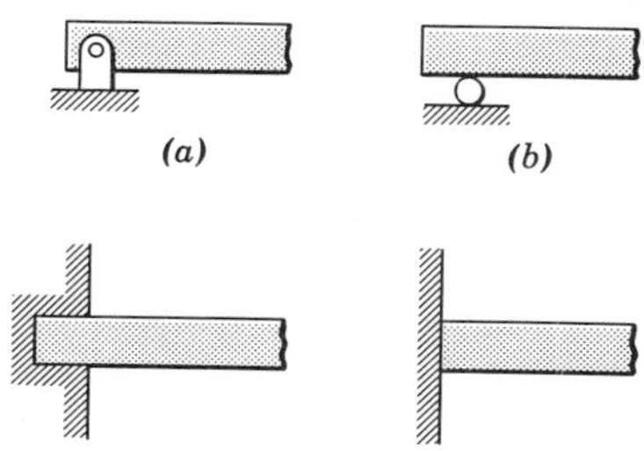

(a) (b) (c) (d)

Fig. 6-5

wise stated in the problem) but can rotate. At a fixed end as represented by Fig. 6–5*c* and *d*, the beam can neither deflect nor rotate unless otherwise stated.

The following examples illustrate the use of the double integration method for calculating beam deflections.

**Example 6–1**: For the beam loaded and supported as shown in Fig. 6–6*a* locate the point of maximum deflection between the supports.

*Solution:* From a free-body diagram of the beam and the equations of equilibrium, the left reaction is found to be $7wL/12$ upward. As indicated in Fig. 6–6*b*, the origin of coordinates is selected at the left support, and the interval to be used is $0 \leq x \leq L$.

The two required boundary conditions are $y = 0$ when $x = 0$ and $y = 0$ when $x = L$. From the free-body diagram of a portion of the beam as shown in Fig. 6–6*b*, the expression for the bending moment at any point in the selected interval is seen to be

$$EI\frac{d^2y}{dx^2} = M_x = \frac{7wL}{12}x - \frac{wL^2}{12} - wx\left(\frac{x}{2}\right).$$

Successive integration gives

$$EI\frac{dy}{dx} = \frac{7wL}{24}x^2 - \frac{wL^2}{12}x - \frac{w}{6}x^3 + C_1$$

and

$$EIy = \frac{7wL}{72}x^3 - \frac{wL^2}{24}x^2 - \frac{w}{24}x^4 + C_1x + C_2.$$

Substitution of the boundary condition $y = 0$ when $x = 0$ yields $C_2 = 0$. Substitution of the remaining boundary condition $y = 0$ when $x = L$ gives

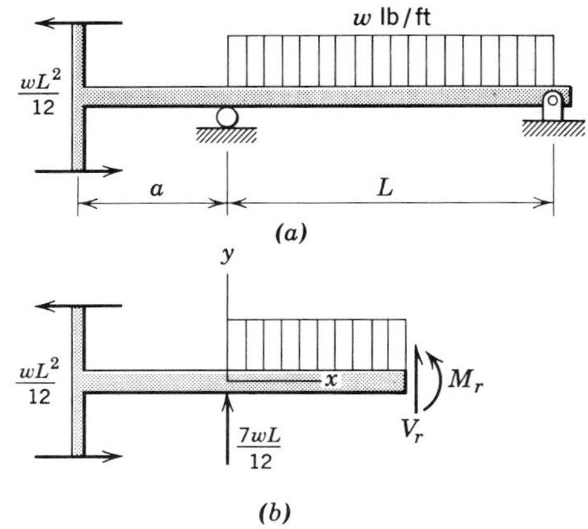


Fig. 6-6

$C_1 = -wL^3/72$, and therefore the elastic curve equation is

$$EIy = \frac{7wL}{72}x^3 - \frac{wL^2}{24}x^2 - \frac{w}{24}x^4 - \frac{wL^3}{72}x.$$

The maximum deflection occurs where the slope, $dy/dx$, is zero or

$$0 = \frac{7wL}{24}x^2 - \frac{wL^2}{12}x - \frac{w}{6}x^3 - \frac{wL^3}{72}.$$

The solution of this cubic equation in $x$ gives the point of maximum deflection at

$$x = \underline{0.541L \text{ to the right of the left support.}} \qquad \text{Ans.}$$

Although not requested here the amount of the maximum deflection can readily be obtained by substituting $0.541L$ for $x$ in the elastic curve equation. The result is

$$y = \frac{-7.88wL^4}{10^3EI} = \frac{7.88wL^4}{10^3EI} \text{ downward.}$$

**Example 6–2:** For the beam loaded and supported as shown in Fig. 6–7*a* determine the deflection of the right end.

*Solution:* From the free-body diagram of Fig. 6–7*b* the equations of equilibrium give the shear at the wall as $wL/3$ upward and the moment at the wall

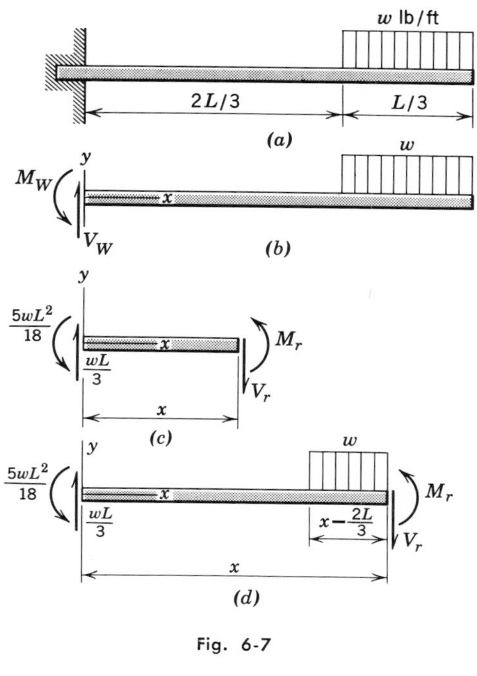


Fig. 6-7

as $5wL^2/18$ counterclockwise. With the sign conventions established in Art. 5–5 these results are

$$V_W = +\frac{wL}{3}, \qquad M_W = -\frac{5wL^2}{18}.$$

In Fig. 6–7*b* there are two intervals to be considered, namely, the loaded and the unloaded portions of the beam. The loaded portion of the beam must be used because it contains the point where the deflection is required. However, a quick check reveals the absence of boundary conditions in this interval. It therefore becomes necessary to use both intervals as well as matching and boundary conditions; hence, the origin of coordinates is selected, as shown. For clarity, the two intervals are written side by side with available boundary conditions, matching conditions, and equations listed under the appropriate interval. The intervals are

| $0 \le x \le 2L/3$ | $2L/3 \le x \le L.$ |
|---|---|

The available boundary conditions are

$$\frac{dy}{dx} = 0 \quad \text{when } x = 0$$

and

$$y = 0 \quad \text{when } x = 0.$$

The available matching conditions are

$\dfrac{dy}{dx}$ from the left equation $= \dfrac{dy}{dx}$ from the right equation when $x = 2L/3$

and

$y$ from the left equation $= y$ from the right equation when $x = 2L/3$.

Four conditions (two boundary and two matching) are sufficient for the evaluation of the four constants of integration (two in each of the two elastic curve differential equations); therefore the problem can be solved in this manner.

From the free-body diagram of an unloaded portion of the beam in Fig. 6–7*c*, the expression for the bending moment at any point in the left interval is seen to be

$$EI\frac{d^2y}{dx^2} = M_x = \frac{wL}{3}x - \frac{5wL^2}{18}.$$

From the free-body diagram of Fig. 6–7*d* where the beam is cut in the loaded interval, the bending moment at any point in the right interval is found to be

$$EI\frac{d^2y}{dx^2} = M_x = \frac{wL}{3}x - \frac{5wL^2}{18} - w\left(x - \frac{2L}{3}\right)\left(\frac{x - 2L/3}{2}\right).$$

Integration of each differential equation gives

$$EI\frac{dy}{dx} = \frac{wL}{6}x^2 - \frac{5wL^2}{18}x + C_1$$

$$EI\frac{dy}{dx} = \frac{wL}{6}x^2 - \frac{5wL^2}{18}x - \frac{w}{6}\left(x - \frac{2L}{3}\right)^3 + C_3.$$

Substitution of the boundary condition

$$\frac{dy}{dx} = 0 \text{ when } x = 0$$

gives $C_1 = 0.$

Substitution of the matching condition

$$\frac{dy}{dx} \text{ from left Eq} = \frac{dy}{dx} \text{ from right Eq when } x = \frac{2L}{3}$$

gives $C_1 = C_3 = 0.$

Integration of the resulting differential equations gives

$$EIy = \frac{wL}{18}x^3 - \frac{5wL^2}{36}x^2 + C_2$$

$$EIy = \frac{wL}{18}x^3 - \frac{5wL^2}{36}x^2 - \frac{w}{24}\left(x - \frac{2L}{3}\right)^4 + C_4.$$

Use of the remaining boundary condition yields $C_2 = 0$, and use of the final matching condition yields $C_2 = C_4 = 0$.

The deflection of the right end of the beam can now be obtained from the elastic curve equation for the right interval by replacing $x$ by its value $L$. The result is

$$EIy = \frac{wL}{18}L^3 - \frac{5wL^2}{36}L^2 - \frac{w}{24}\left(L - \frac{2L}{3}\right)^4,$$

from which

$$y = -\frac{163}{1944}\frac{wL^4}{EI} = \underline{\frac{163}{1944}\frac{wL^4}{EI} \text{ downward.}} \qquad \text{Ans.}$$

## PROBLEMS

**6–5.** Determine the deflection of the left end of the beam of Prob. 6–4 by the double integration method.

**6–6.*** (*a*) Derive the equation of the elastic curve of the beam of Prob. 6–2 by the double integration method in terms of $E$, $I$, $L$, and $M$.

(*b*) Explain any discrepancy between the answers to Prob. 6–2*a* and Prob. 6–6*a*.

**6–7.*** For the beam loaded and supported as shown in Fig. P6–7, derive the equation for the elastic curve in terms of $E$, $I$, $w$, and $L$. Use the designated axes.

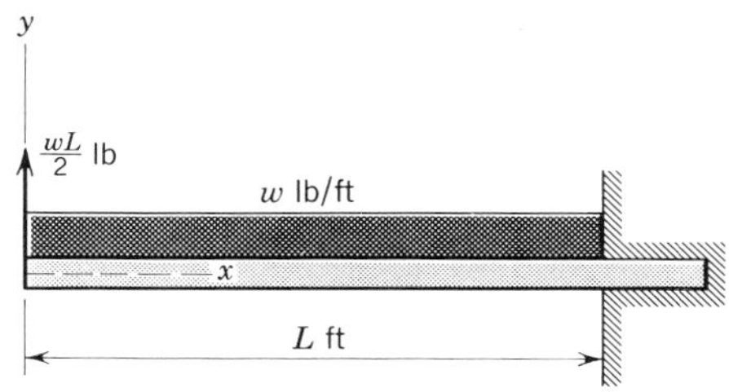


Fig. P6-7

**6–8.*** A beam of length $L$ with a uniformly distributed load of $w$ lb per unit length over the entire span is simply supported at the left end and restrained at the right end. The ends remain on the same horizontal line, and a load cell shows that the left reaction is one fifth of the total load. Determine the equation of the elastic curve of the beam in terms of $E$, $I$, $w$, and $L$.

**6–9.*** For the beam in Fig. P6–9 determine the equation of the elastic curve for the portion of the beam between the supports in terms of $E$, $I$, $w$, and $L$.

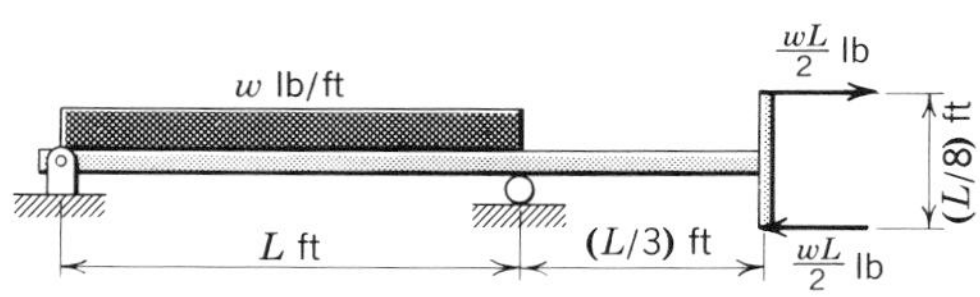


Fig. P6-9

**6–10.** For the beam loaded and supported as shown in Fig. P6–10 determine the deflection at the point of application of the load $P/2$ in terms of $P$, $L$, $E$, and $I$.

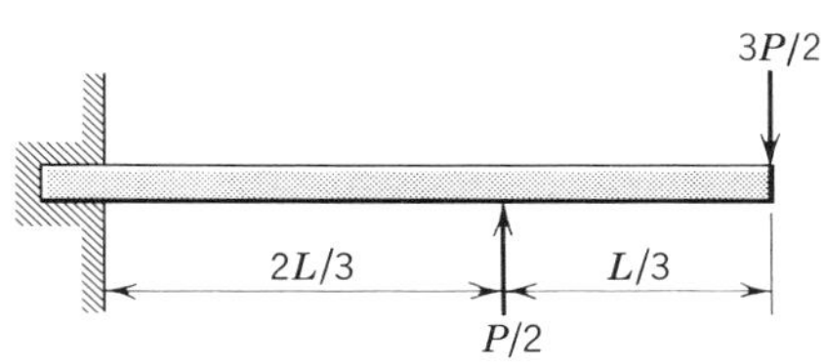


Fig. P6-10

**6–11.** Determine, in terms of $w$, $L$, $E$, and $I$, the elastic curve equation for the interval between the supports of the beam loaded and supported as shown in Fig. P6–11.

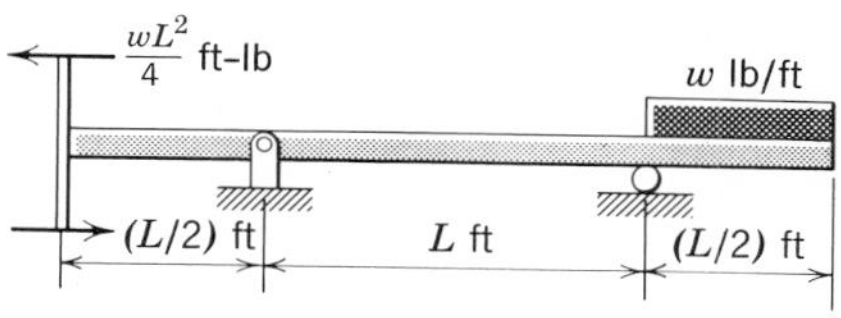


Fig. P6-11

**6–12.*** Develop, in terms of $w$, $L$, $E$, $I$, $x$, and $y$, the elastic curve equation for span $BC$ of the beam in Fig. P6–12. Use the indicated origin.

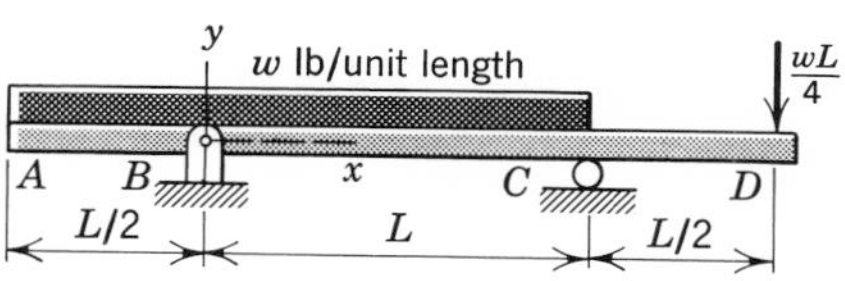


Fig. P6-12

**6–13.** A 15 I 42.9 steel beam 40 ft long is simply supported at its ends and carries a concentrated load $P$ at mid-span. Determine the value of $P$ which will produce deflections of 0.9 in. at points 10 ft from the supports. The web of the cross section is vertical.

**6–14.*** Determine, in terms of $w$, $L$, $E$, and $I$, the deflection of the free end of the cantilever beam loaded as shown in Fig. P6–14.

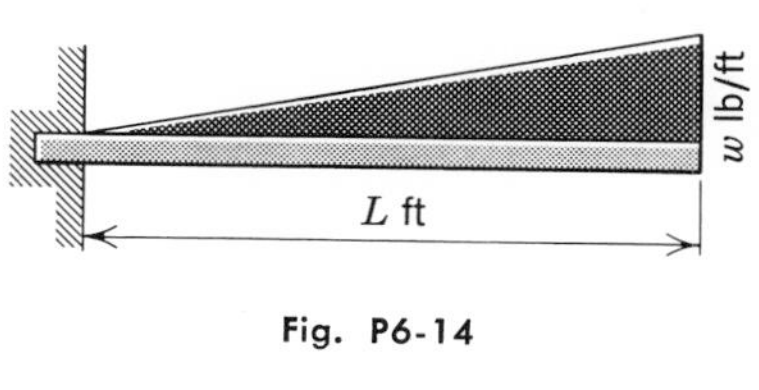


Fig. P6-14

**6–15.** A 4 by 10-in. timber having a modulus of elasticity of 1500 ksi is used as a post as shown in Fig. P6–15. Compute the maximum horizontal movement of the top of the post in inches.

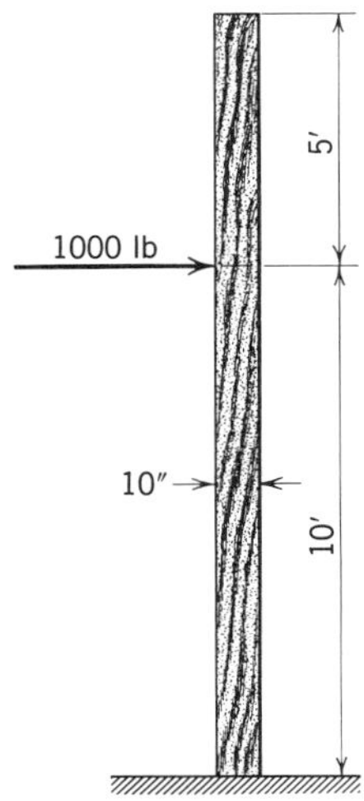


Fig. P6-15

**6–16.*** A timber beam 6 in. wide by 12 in. deep is simply supported at the ends of a span of 18 ft. It carries a concentrated load of 2000 lb at the center of the span. The modulus of elasticity of the timber is 1200 ksi. A pointer 5 ft long is attached to the right end as shown in Fig. P6–16.

(*a*) Determine the movement of the right end of the pointer in inches.

(*b*) Determine the maximum deflection of the beam.

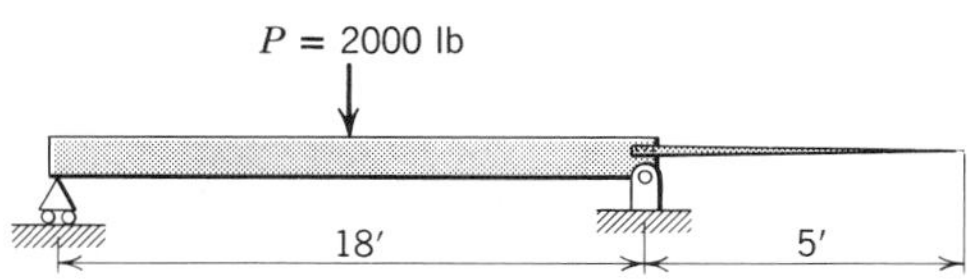


Fig. P6-16

**6–17.** A beam of length $L$ ft is fixed at the left end and carries a uniformly distributed load of $w$ lb per ft. The right end is subjected to a moment of $+3wL^2/8$.

(*a*) Derive the equation of the elastic curve.

(*b*) Determine the deflection of the right end when $I = 6$ in.$^4$, $E = 30(10^6)$ psi, $L = 10$ ft, and $w = 100$ lb per ft.

**6–18.*** For the cantilever beam loaded as shown in Fig. P6–18, determine the deflection of the free end in terms of $w$, $L$, $E$, and $I$.

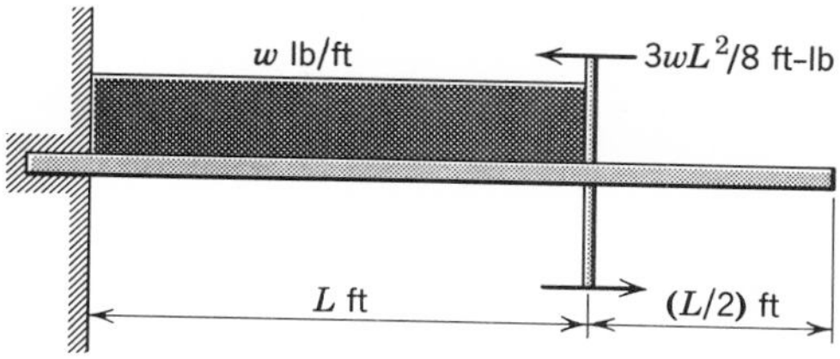


**Fig. P6-18**

**6–19.*** For the continuous beam in Fig. P6–19 the bending moment at the second support is $-wL^2/6$ and the vertical shear just to the right of the second support is $+7wL/12$. Determine the equation of the elastic curve for the portion of the beam between the second and third supports.

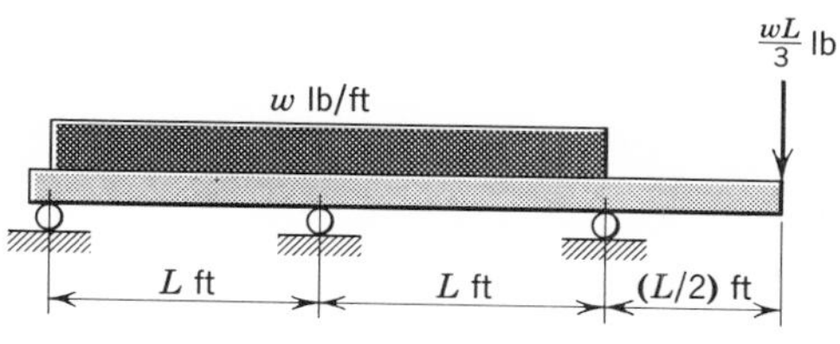


**Fig. P6-19**

**6–20.** One span of a continuous beam carries a uniformly distributed load of $w$ lb per unit length. The bending moment at the left end of the span is $-wL^2/12$, and the vertical shear on the right side of the left support is $+wL/4$, where $L$ is the span length. Determine the equation of the elastic curve of this span of the beam.

**6–21.*** The moment at support $C$ of the beam in Fig. P6–21 is $+wL^2/8$, and the shear to the right of $C$ is $-5wL/8$, where $w$ is the load per unit length on the left overhang. Determine the maximum deflection in terms of $w$, $L$, $E$, and $I$ in the span $CD$. *Note:* The reaction at $D$ is upward on the beam.

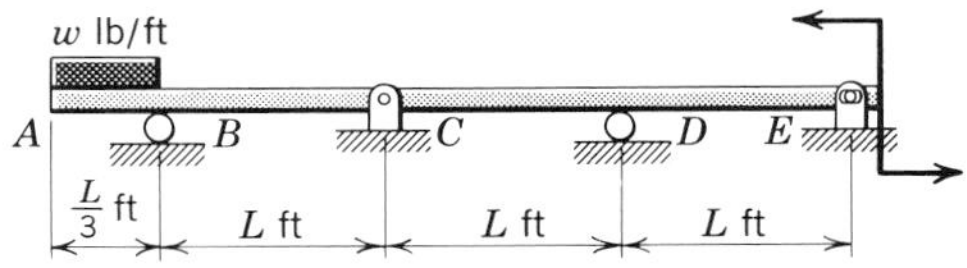


Fig. P6-21

**6–22.** Determine, for the beam of Fig. P6–22, the distance from the left reaction to the point of maximum deflection between the supports.

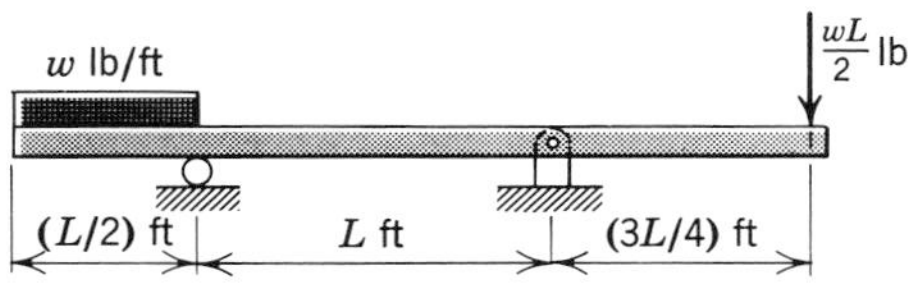


Fig. P6-22

**6–23.*** For the beam loaded and supported as shown in Fig. P6–23, determine the maximum deflection between the supports in terms of $P$, $L$, $E$, and $I$.

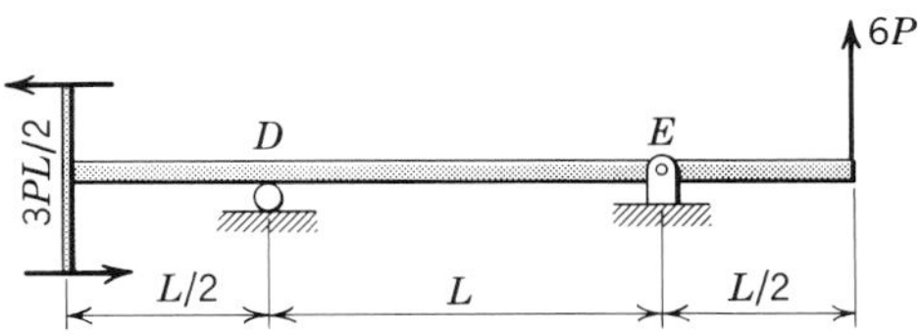


Fig. P6-23

**6–24.** Determine the maximum deflection in the left half of the beam loaded and supported as shown in Fig. P6–24, in terms of $M$, $L$, $E$, and $I$. The deflection at the center of the beam is zero due to symmetry.

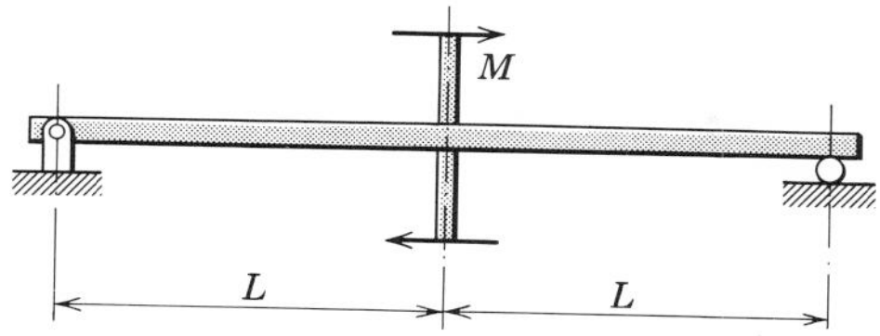


Fig. P6-24

**6–25.** Determine the maximum deflection in the interval $AB$ of the beam loaded and supported as shown in Fig. P6–25, in terms of $P$, $L$, $E$, and $I$.

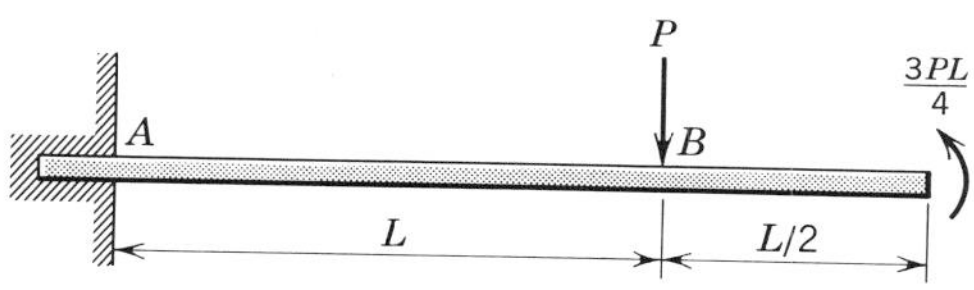


Fig. P6-25

**6–26.*** Determine, in terms of $P$, $L$, $E$, and $I$, the maximum deflection in the beam of Fig. P6–26.

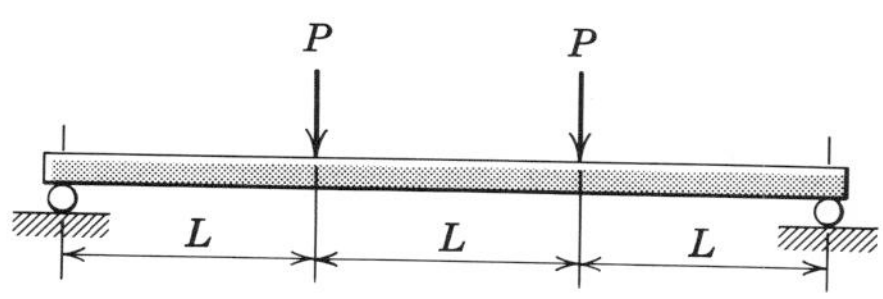


Fig. P6-26

**6–27.** Develop in terms of $w$, $L$, $E$, $I$, $x$, and $y$, the elastic curve equation for segment $BC$ of the beam of Fig. P6–27.

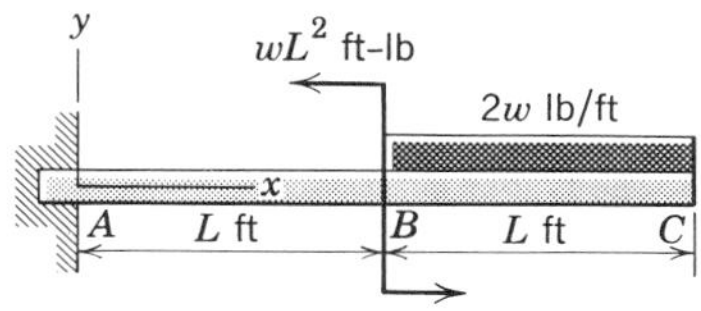


Fig. P6-27

**6–28.*** For the beam in Fig. P6–28, determine the deflection of the beam at $x = L$ in terms of $w$, $L$, $E$, and $I$.

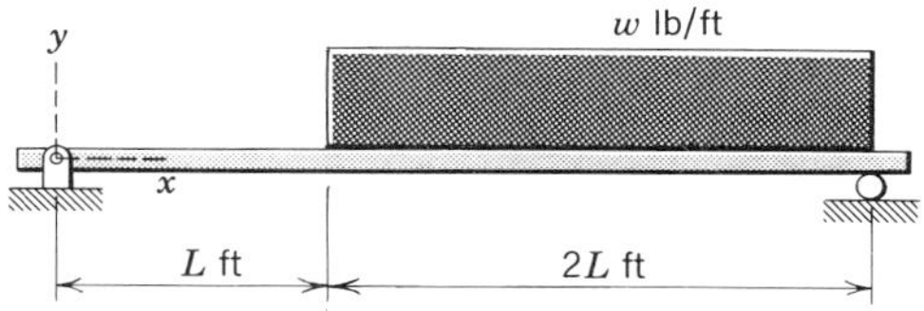


Fig. P6-28

**6–29.** For the beam loaded and supported as shown in Fig. P6–29, determine in terms of $P$, $L$, $E$, and $I$:

(*a*) The elastic curve equations.
(*b*) The location of the point of maximum deflection.
(*c*) The maximum deflection.

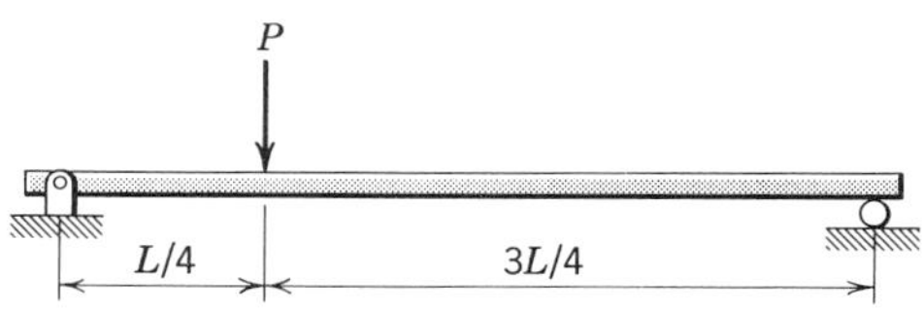


Fig. P6-29

**6–30.** A cantilever beam carries two concentrated loads as shown in Fig. P6–30. Determine, in terms of $P$, $L$, $E$, and $I$:

(*a*) The location of the point of maximum deflection.

(*b*) The deflection of the beam at the load $P$.

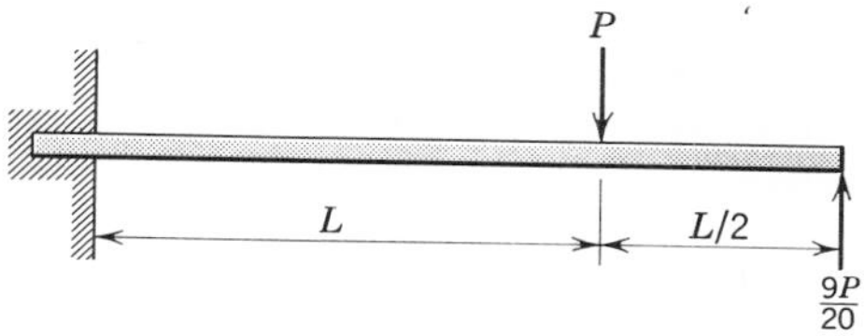


Fig. P6-30

**6–31.*** For the beam in Fig. P6–31:

(*a*) Derive the elastic curve equation for the loaded portion in terms of $E$, $I$, $w$, and $a$.

(*b*) Determine the maximum deflection when $I = 81$ in.$^4$, $E = 12(10^6)$ psi, $w = 1000$ lb per ft, $a = 3$ ft.

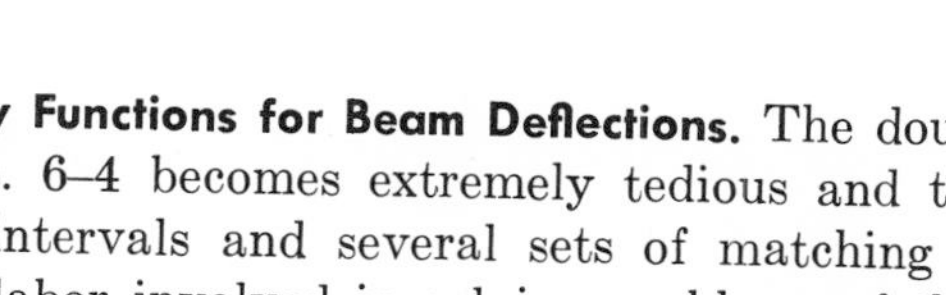


Fig. P6-31

**6–5. Singularity Functions for Beam Deflections.** The double integration method of Art. 6–4 becomes extremely tedious and time consuming when several intervals and several sets of matching conditions are required. The labor involved in solving problems of this type can be diminished by making use of *singularity functions* following the method originally developed in 1862 by the German mathematician A. Clebsch (1833–1872).[1]

---

[1] For a rather complete history on the Clebsch method and the numerous extensions thereof, see "Clebsch's Method for Beam Deflections," Walter D. Pilkey, *Journal of Engineering Education,* Jan., 1964, p. 170.

Singularity functions are closely related to the unit step function used by Heaviside[2] to analyze the transient response of electrical circuits. Singularity functions will be used here for writing one bending moment equation which applies in all intervals along a beam; thus eliminating the need for matching conditions.

A singularity function of $x$ is written as $\langle x - x_0 \rangle^n$, where $n$ is any *integer* (positive or negative) including zero, and $x_0$ is a constant equal to the value of $x$ at the initial boundary of a specific interval along a beam. Selected properties of singularity functions required for beam-

[2] O. Heaviside (1850–1925), a British physicist.

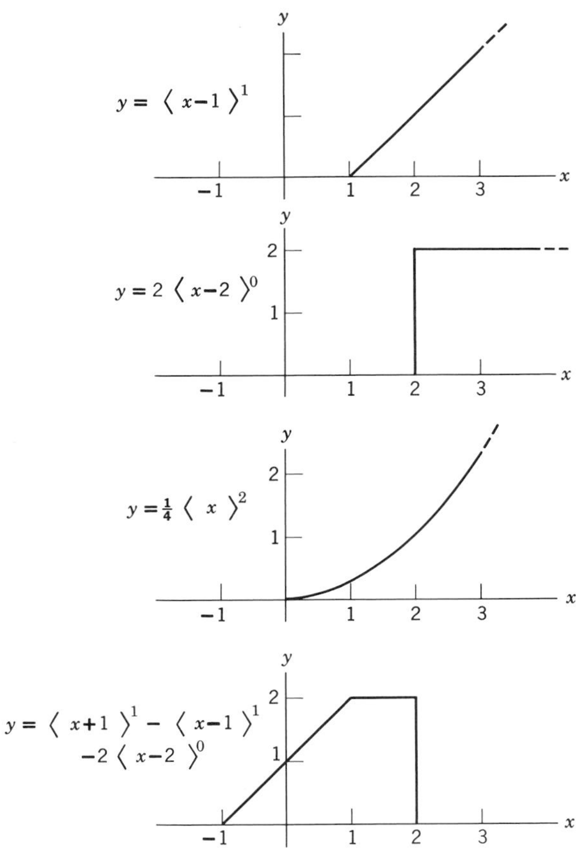


Fig. 6-8

deflection problems are listed here for emphasis and ready reference.

$$\langle x - x_0\rangle^n = \begin{cases} (x - x_0)^n & \text{when } n > 0 \text{ and } x \geq x_0. \\ 0 & \text{when } n > 0 \text{ and } x < x_0. \end{cases}$$

$$\langle x - x_0\rangle^0 = \begin{cases} 1 & \text{when } x \geq x_0. \\ 0 & \text{when } x < x_0. \end{cases}$$

$$\int \langle x - x_0\rangle^n\, dx = \frac{1}{n+1}\langle x - x_0\rangle^{n+1} + C \quad \text{when } n \geq 0.$$

$$\frac{d}{dx}\langle x - x_0\rangle^n = n\langle x - x_0\rangle^{n-1} \quad \text{when } n \geq 1.$$

Several examples of singularity functions are shown in Fig. 6–8.

By making use of these properties of singularity functions, one is able to write a single equation for the bending moment for a beam and obtain the correct value of the moment in any interval along the beam. To amplify this statement consider the beam of Fig. 6–9. By the methods of Art. 5–7 and 6–4, the moment equations at the four designated sections are:

$$\begin{aligned} M_1 &= R_L x, & 0 \leq x \leq x_1 \\ M_2 &= R_L x - P(x - x_1), & x_1 \leq x \leq x_2 \\ M_3 &= R_L x - P(x - x_1) + M_A, & x_2 \leq x \leq x_3 \\ M_4 &= R_L x - P(x - x_1) + M_A - (w/2)(x - x_3)^2. & x_3 \leq x \leq L \end{aligned}$$

These four moment equations can be combined into a single equation by means of singularity functions to give

$$M_x = R_L x - P\langle x - x_1\rangle^1 + M_A\langle x - x_2\rangle^0 - \frac{w}{2}\langle x - x_3\rangle^2. \quad 0 \leq x \leq L$$

Distributed beam loadings which are sectionally continuous (the distributed load cannot be represented by a single function of $x$ for all values of $x$) require special consideration. Moment equations for such

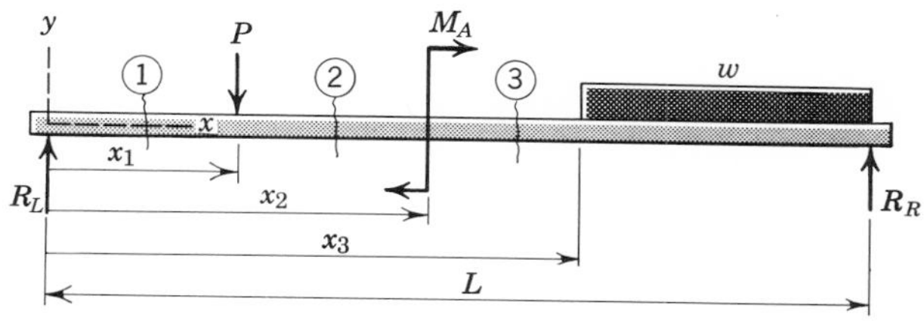


Fig. 6-9

loads can be readily obtained by superposition, as illustrated in the following example. Further applications of singularity functions to the determination of beam deflections are given in Examples 6–4 and 6–5.

**Example 6–3:** Use singularity functions to write a single equation for the bending moment at any section of the beam of Fig. 6–10*a*.

*Solution:* The loading on the beam of Fig. 6–10*a* can be considered as a combination of the loadings shown in Fig. 6–10*b*, *c*, *d*, and *e*; where downward-acting loads are shown on top of the beam and upward-acting loads on the bottom. The moment of the linearly varying load can be obtained from the geometry of the load diagram. The moment equation is

$$M_x = R_L x - \frac{w_1}{6(x_2 - x_1)} \langle x - x_1 \rangle^3 + \frac{w_1}{6(x_2 - x_1)} \langle x - x_2 \rangle^3 + \frac{w_1 - w_2}{2} \langle x - x_2 \rangle^2 + \frac{w_2}{2} \langle x - x_3 \rangle^2. \quad \text{Ans.}$$

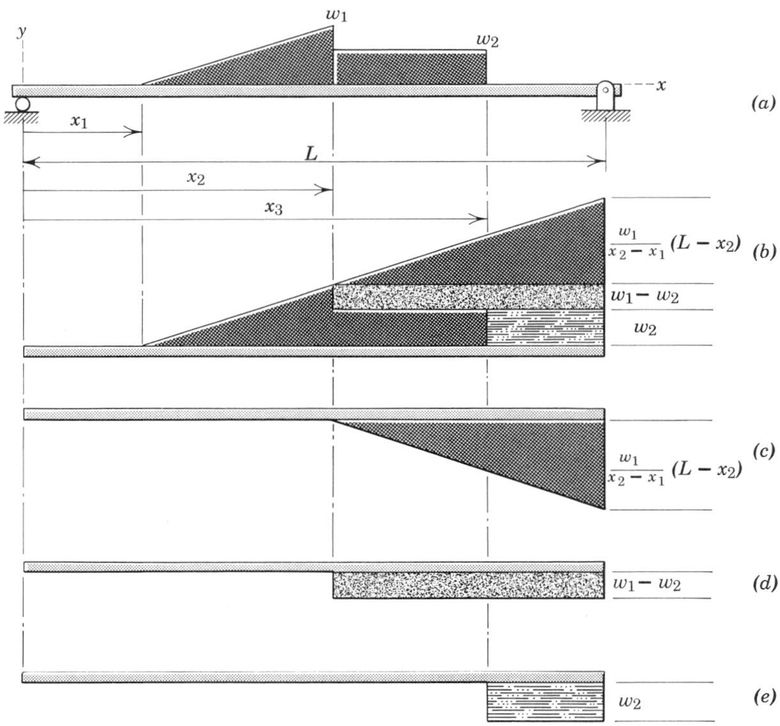


Fig. 6-10

**Example 6–4:** Solve Ex. 6–2, Art. 6–4, by use of singularity functions.

*Solution:* The boundary conditions, in the interval $0 \le x \le L$, are $dy/dx = 0$ when $x = 0$ and $y = 0$ when $x = 0$ with the axes as shown in Fig. 6–7*b*. From the free-body diagram of Fig. 6–7*d*, the expression for the bending moment is

$$EI\,\frac{d^2y}{dx^2} = -\frac{5wL^2}{18} + \frac{wLx}{3} - \frac{w}{2}\left\langle x - \frac{2L}{3}\right\rangle^2.$$

The first integration gives

$$EI\,\frac{dy}{dx} = -\frac{5wL^2x}{18} + \frac{wLx^2}{6} - \frac{w}{6}\left\langle x - \frac{2L}{3}\right\rangle^3 + C_1.$$

Substituting the boundary condition $dy/dx = 0$ when $x = 0$, and noting that for $x < 2L/3$, the term in the brackets is zero, gives $C_1 = 0$. Integrating again gives

$$EIy = -\frac{5wL^2}{36}x^2 + \frac{wL}{18}x^3 - \frac{w}{24}\left\langle x - \frac{2L}{3}\right\rangle^4 + C_2.$$

The boundary condition $y = 0$ when $x = 0$ gives $C_2 = 0$. The deflection at the right end is obtained by substituting $x = L$ in the above elastic curve equation. The result is

$$y = -\frac{163}{1944}\frac{wL^4}{EI} = \underline{\frac{163}{1944}\frac{wL^4}{EI}\text{ downward.}} \qquad \text{Ans.}$$

It will be observed that the result agrees with that of Ex. 6–2.

**Example 6–5:** Determine the deflection at the left end of the beam of Fig. 6–11*a*.

*Solution:* The equations of equilibrium, applied to the free-body diagram of Fig. 6–11*b*, give the reaction $R_L$ as $3wL/8$ upward. By use of singularity functions and the superposed loads of Fig. 6–11*c*, the bending-moment equation becomes

$$EI\,\frac{d^2y}{dx^2} = -\frac{w}{2}(x + L)^2 + \frac{w}{2}\left\langle x + \frac{L}{2}\right\rangle^2 + \frac{3wL}{8}\langle x\rangle^1 + \frac{wL^2}{2}\left\langle x - \frac{L}{2}\right\rangle^0.$$

The boundary conditions are $y = 0$ when $x = 0$, and $y = 0$ when $x = L$. Two integrations of the moment equation give

$$EI\,\frac{dy}{dx} = -\frac{w}{6}(x + L)^3 + \frac{w}{6}\left\langle x + \frac{L}{2}\right\rangle^3 + \frac{3wL}{16}\langle x\rangle^2 + \frac{wL^2}{2}\left\langle x - \frac{L}{2}\right\rangle^1 + C_1$$

and

$$EIy = -\frac{w}{24}(x + L)^4 + \frac{w}{24}\left\langle x + \frac{L}{2}\right\rangle^4 + \frac{wL}{16}\langle x\rangle^3 + \frac{wL^2}{4}\left\langle x - \frac{L}{2}\right\rangle^2 + C_1x + C_2.$$

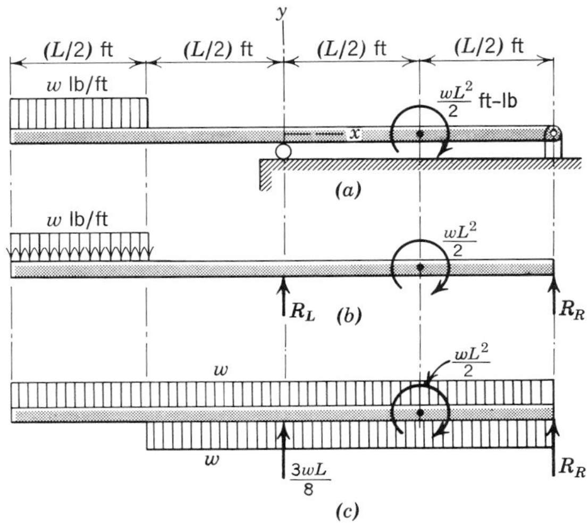


Fig. 6-11

The first boundary condition, $y = 0$ when $x = 0$, gives

$$0 = -\frac{wL^4}{24} + \frac{wL^4}{384} + 0 + 0 + 0 + C_2,$$

from which

$$C_2 = +\frac{5}{128}\,wL^4.$$

The second boundary condition, $y = 0$ when $x = L$, gives

$$0 = -\frac{16wL^4}{24} + \frac{81wL^4}{384} + \frac{wL^4}{16} + \frac{wL^4}{16} + C_1L + \frac{5wL^4}{128},$$

from which

$$C_1 = +\frac{7}{24}\,wL^3.$$

The deflection at $x = -L$ is found, by substitution, to be

$$y = -\frac{97}{384}\frac{wL^4}{EI} = \underline{\frac{97}{384}\frac{wL^4}{EI}\text{ downward.}} \qquad \text{Ans.}$$

## PROBLEMS

**6–32.*** Use singularity functions to solve Prob. 6–27.

**6–33.*** Use singularity functions to solve Prob. 6–28.

**6–34.** Use singularity functions to solve Prob. 6–29.

**6–35.** Use singularity functions to solve Prob. 6–30.

**6–36.*** Use singularity functions to solve Prob. 6–31.

**6–37.** Determine, in terms of $w$, $E$, $I$, and $L$, the deflection at the free end of the beam of Fig. P6–37.

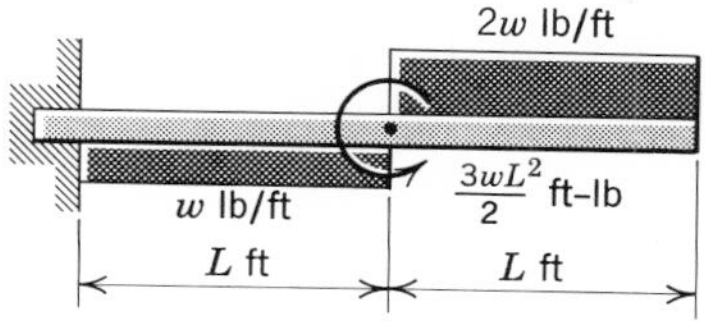


Fig. P6-37

**6–38.*** For the beam loaded and supported as shown in Fig. P6–38, determine the deflection at the mid-point in terms of $w$, $L$, $E$, and $I$.

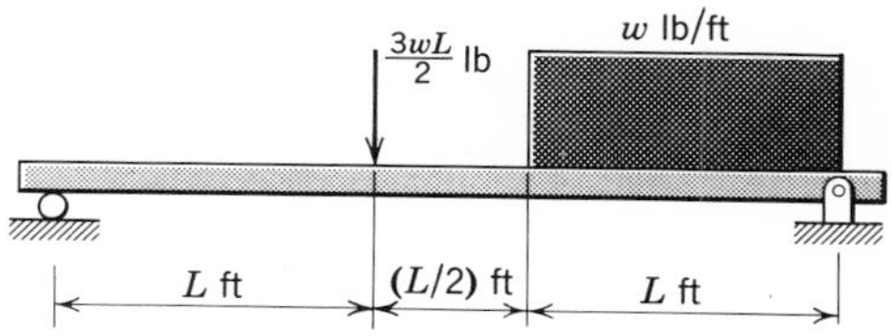


Fig. P6-38

**6–39.** A steel rod 1.00 in. in diameter and 20 in. long is used as a cantilever beam and is loaded as shown in Fig. P6–39. Determine the deflection of the free end of the beam.

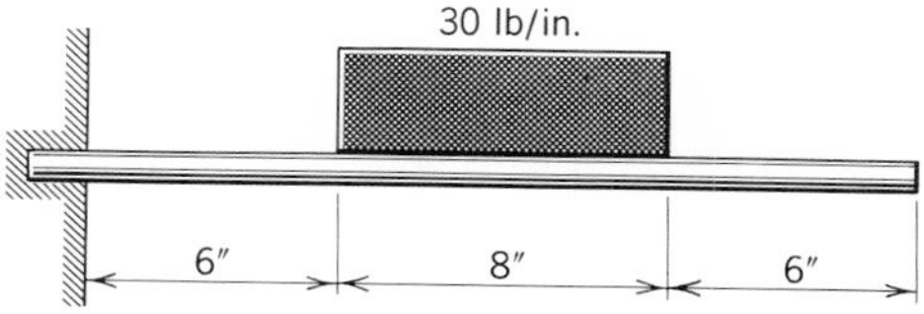


Fig. P6-39

**6–40.*** For the beam in Fig. P6–40 determine the deflection of the left end with $E = 10(10^6)$ psi and $I = 100$ in.$^4$.

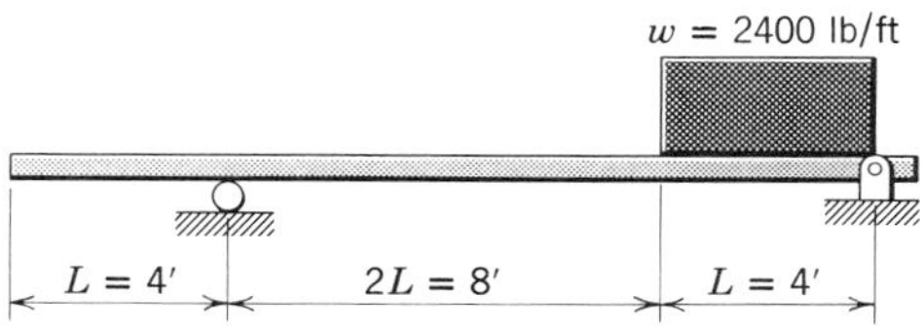


Fig. P6-40

**6–41.** For the beam loaded and supported as shown in Fig. P6–41, determine the deflection of the left end.

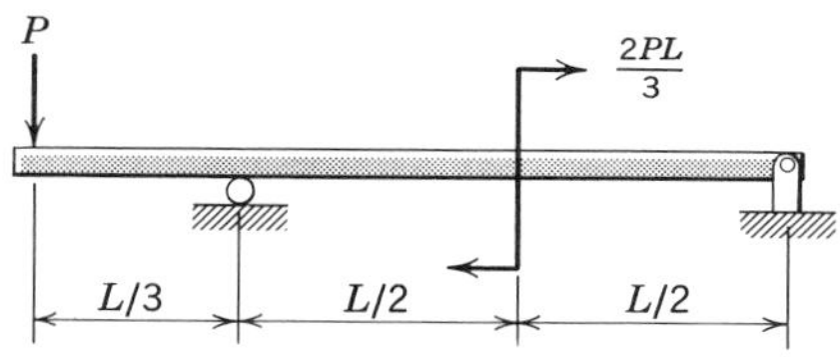


Fig. P6-41

**6–42.*** For the beam of Fig. P6–42, develop the equation of the elastic curve between the supports, and determine the deflection at mid-span in terms of $E$, $I$, $w$, and $a$.

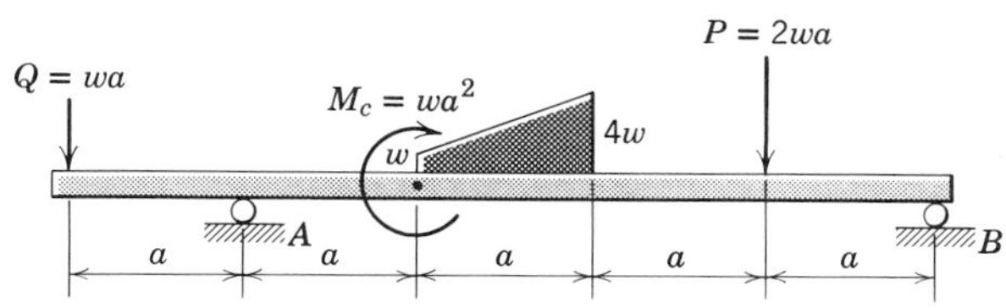


Fig. P6-42

**6–43.** For the beam of Fig. 6–10$a$, Ex. 6–3, determine the deflection at $x = x_2$ with $w_1 = 2w_2$, $x_1 = a$, $x_2 = 4a$, $x_3 = 5a$, and $L = 6a$. Express the result in terms of $E$, $I$, $w_2$, and $a$.

**6–6. The Development of the Area-Moment Method.** The area-moment method of determining the deflection at any specified point along a beam is a semigraphical method utilizing the relations between successive derivatives of the deflection, $y$, and the moment diagram. For problems involving several changes in loading, the area-moment method is usually much faster than the double integration method; consequently, it is widely used in practice. The method is based upon two propositions or theorems derived as follows.

From Fig. 6–2 it is seen that for any arc length $dL$ along the elastic curve of a loaded beam

$$d\theta = dL/\rho,$$

and from Eq. (6–1) this becomes

$$d\theta = \frac{M}{EI}\, dL.$$

For actual beams the curvature is so small that $dL$ can be replaced by the distance $dx$ along the unloaded beam, and the relation becomes

$$d\theta = \frac{M}{EI}\, dx.$$

Reference to Fig. 6–12 (which consists of a loaded beam, its moment diagram, and a sketch, greatly exaggerated, of the elastic curve of the

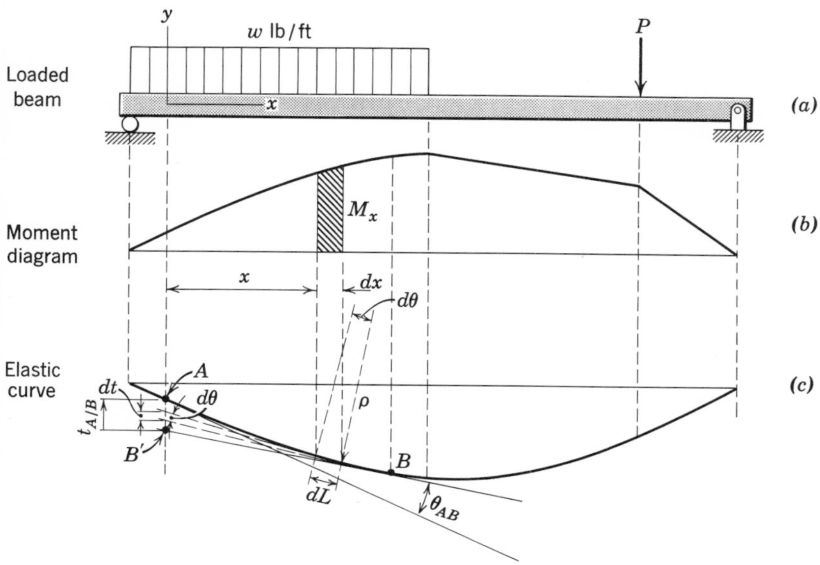


Fig. 6-12

deflected beam) shows that $d\theta$ is also the angle between tangents to the elastic curve at two points a distance $dL$ apart. When any two points $A$ and $B$ along the elastic curve are selected as shown in Fig. 6–12c and the relation for $d\theta$ is integrated between these limits, the result is

$$\int_{\theta_A}^{\theta_B} d\theta = \theta_B - \theta_A = \int_{x_A}^{x_B} \frac{M}{EI}\, dx.$$

The integral on the left, designated $\theta_{AB}$, is the angle between tangents to the elastic curve at $A$ and $B$. The theorem can be written as

$$\theta_{AB} = \int_{x_A}^{x_B} \frac{M\, dx}{EI}. \tag{6–3}$$

From Fig. 6–12b, the integral is clearly equal to $1/(EI)$ times the area under the moment diagram when $E$ and $I$ are constants (as they are in many beams). In case either $E$ or $I$ vary along the beam, it is advisable to construct an $M/(EI)$ diagram instead of a moment diagram. The first area-moment theorem can be expressed in words as follows: *The angle between the tangents to the elastic curve of a loaded beam at any two points A and B along the elastic curve equals the area under the M/(EI) diagram of the beam between ordinates at A and B.*

The correct interpretation of numerical results frequently depends upon a reliable sign convention, and the following convention is recommended. Areas under positive bending moments are considered positive. A positive area means that the angle is positive or counterclockwise when measured from the tangent to the point on the left to the other tangent, as shown in Fig. 6–13.

In order to obtain the second area-moment theorem, let $B'$ in Fig. 6–12c be the intersection of the tangent to the elastic curve at $B$ with a vertical from point $A$ on the elastic curve. Since the deflections of a beam are assumed to be small, it can be seen from Fig. 6–12c that

$$dt = x\, d\theta = \frac{xM}{EI}\, dx.$$

Integration gives

$$\int_0^{t_{A/B}} dt = t_{A/B} = \int_{x_A}^{x_B} \frac{Mx}{EI}\, dx. \tag{6–4}$$

A study of Fig. 6–12c shows that the left integral is the vertical (actually perpendicular to the beam) distance, $t_{A/B}$, of any point $A$ on the elastic curve from a tangent drawn to any other point $B$ on the elastic curve. This distance is frequently called the *tangential deviation* to distinguish it from the beam deflection $y$. The integral on the right, as

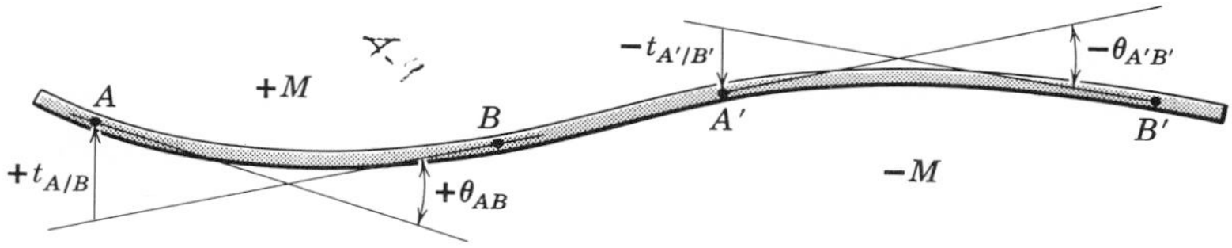


Fig. 6-13

seen from Fig. 6–12*b*, is the first moment, with respect to an axis through $A$ parallel to the $y$ axis, of the area under the $M/(EI)$ diagram between ordinates at $A$ and $B$.

The second area-moment theorem, Eq. (6–4), can be stated in words as follows. *The vertical distance (tangential deviation) of any point $A$ on the elastic curve of a beam from a tangent drawn at any other point $B$ on the elastic curve equals the first moment, with respect to an axis at $A$, of the area under the $M/(EI)$ diagram between ordinates at $A$ and $B$.*

A sign convention is required for those instances where the shape of the elastic curve is not immediately evident from the given loads. The following sign convention is recommended. The first moment of the area under a positive $M/(EI)$ diagram is considered positive, and from the second area-moment theorem a positive first moment gives a positive tangential deviation. A positive tangential deviation means that point $A$ on the beam (the moment center) is above the tangent drawn from the other point $B$. Examples of the sign conventions for both the first and second area-moment theorems are shown in Fig. 6–13.

The effectiveness of the area-moment method for determining beam deflections depends, to a considerable extent, upon the ease of calculating areas and first moments of areas under the $M/(EI)$ diagrams. Naturally these values can be obtained by integration of the general expressions from Eq. (6–3) and (6–4). In general, however, this procedure offers little or no advantage over the double integration method of Art. 6–3 and 6–4. The device of drawing a separate moment diagram for each load and reaction on a beam, as explained in the next article, greatly simplifies the finding of areas and moments of areas because in this way nearly all areas are triangular, rectangular, or parabolic.

**6–7. Moment Diagrams by Parts.** In Chap. 5 a composite or complete moment diagram was drawn which included the contribution or effect of all loads and reactions acting on a beam. In the application of the area-moment method for obtaining deflections, it is sometimes easier and more practical to draw a series of diagrams showing the moment due to each load or reaction on a separate sketch. For this purpose, each moment diagram is drawn from the load to an arbitrarily selected section called the reference section. It is preferable to select the reference section at one end of the beam or at one end of a uniformly distributed load. With a reference section at one end, start with the load or reaction at the opposite end of the beam, and draw the bending moment diagram immediately below the beam as if no other loads or reactions were acting on the beam except at reference section. Proceed toward the right (or left) drawing the moment diagram for each load and reaction below the previous one. The composite moment diagram can be obtained, where necessary, by superposition (addition) of the component diagrams. The actual or total moment on the beam at any section is the algebraic sum of the moments from the separate diagrams except at an interior reference section where the sum of the moments on either side of the reference section gives the total moment. The following example illustrates the procedure for drawing moment diagrams by parts.

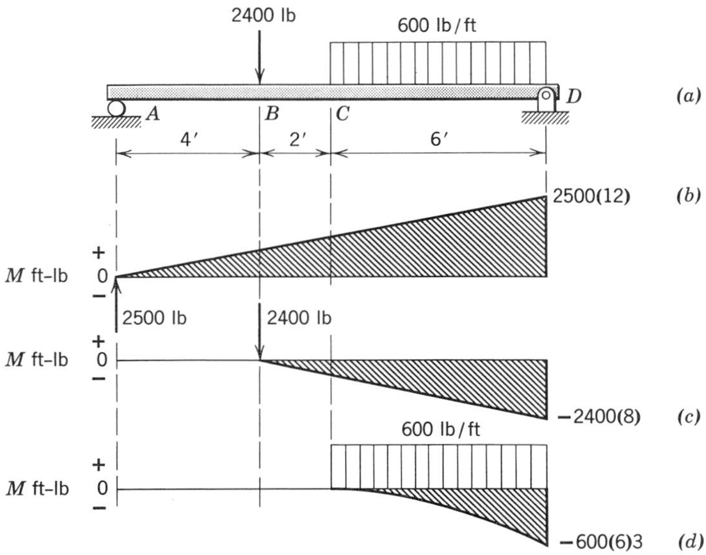


Fig. 6-14

**Example 6–6:** For the beam in Fig. 6–14*a* construct the moment diagram by parts.

*Solution:* From a free-body diagram of the beam the equations of equilibrium give the left reaction as 2500 lb upward and the right reaction as 3500 lb upward. The reference section can be selected anywhere along the beam, and each different reference section yields a different series of component moment diagrams. Three different solutions of the example are given to emphasize the importance of a careful selection of the reference section.

(*a*) The reference section is selected at the right end of the beam. The bending moment due to the 2500-lb left reaction is positive and equal to 2500 times the distance from the left end as shown in Fig. 6–14*b*. The bending moment due to the 2400-lb concentrated load is negative and equal to 2400 times the distance from the load as shown in Fig. 6–14*c*. The bending moment due to the distributed load is negative and equal to 300 times the square of the distance from $C$ (the left edge of the uniform load) as shown in Fig. 6–14*d*. Since the diagrams were drawn from left to right, the right reaction makes no contribution to the moment. It is in the reference section.

(*b*) The reference section is selected at $C$, the left edge of the distributed load. The bending moment due to the 3500-lb right reaction is positive, varies

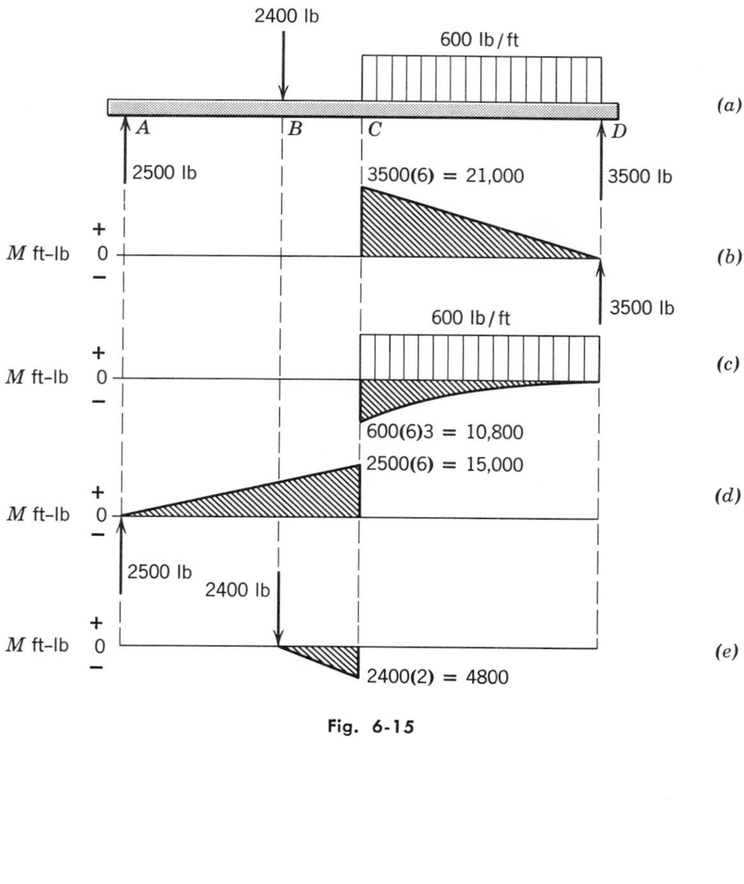


Fig. 6-15

directly as the distance from the right end, and is shown in Fig. 6–15*b*. The bending moment due to the distributed load is negative, equal to 300 times the square of the distance from the right end, and shown in Fig. 6–15*c*. Likewise, the moments due to the left reaction and the concentrated load are shown in Fig. 6–15*d* and *e*. The total bending moment at the reference section $C$ is the sum of the individual moments just to the left or right of the section, namely, +10,200 ft-lb.

(*c*) The reference section is selected at the left end. The bending moment due to the 3500-lb right reaction is shown in Fig. 6–16*b*. The bending moment due to the distributed load is shown in Fig. 6–16*c*, and the area is subdivided into a triangle, a rectangle, and a parabolic area. This shows the disadvantage of selecting the reference section away from one end of a distributed load. The bending moment due to the concentrated load is shown in Fig. 6–16*d*.

Study of the three solutions reveals that the amount of numerical work involved in finding areas and first moments is greatly altered by the choice of reference section to be used. In solution (*a*) there are three component areas, whereas solution (*b*) involves four component areas, and solution (*c*) is the least desirable with five component areas, three of which are due to the distributed load.

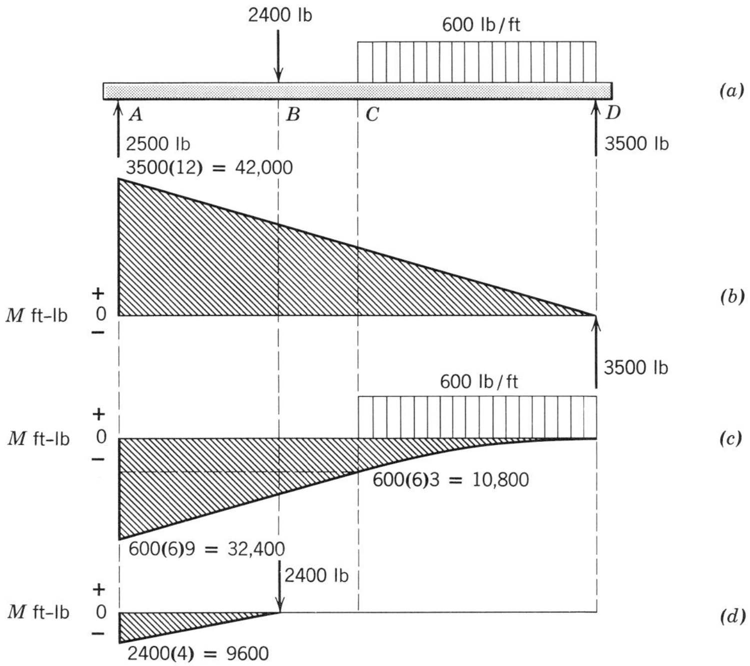


Fig. 6-16

## PROBLEMS

**6–44.** For the beam of Fig. P6–44, construct the moment diagram by parts.

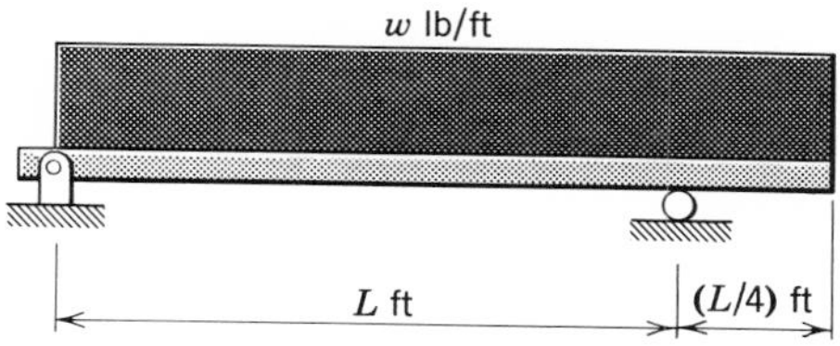


Fig. P6-44

**6–45.** Construct the moment diagram by parts for the beam of Fig. P6–31.

**6–46.** For the beam of Fig. P6–18, construct the moment diagram by parts.

**6–47.** Construct the moment diagram by parts for the beam of Fig. P6–11.

**6–48.** For the beam of Fig. P6–27, construct the moment diagram by parts.

**6–8. Deflections by Area Moments.** The deflection of a specific point along the elastic curve of a beam can normally be determined by one or more applications of the second area-moment theorem. The amount of time and effort required for a given problem depends to a considerable extent on the judgment used in selecting the moment center for the second theorem and in drawing the moment diagram by parts. The ability to construct the correct moment diagram and sketch the elastic curve is essential to the application of the method.

A few general suggestions are presented here to facilitate the application of the method.

(1) It is frequently easier to use a composite moment diagram instead of a moment diagram by parts for those cases where the beam carries only one load.

(2) It will be found that selection of the reference section of a cantilever beam at the fixed end for construction of the moment diagram by parts usually gives the minimum number of areas.

(3) For overhanging beams it is advisable to avoid taking a tangent to the elastic curve at the overhanging end.

(4) It is sometimes advantageous to solve the problem with symbols and substitute numerical values in the last step.

The list of steps given below for the solution of deflection problems by the area-moment method may prove helpful. There is no point in memorizing these steps as they are not a substitute for an understanding of the principles involved. They are:

(1) Sketch the loaded beam, the moment [or $M/(EI)$] diagrams (either by parts or composite diagram depending on the complexity of the problem), and the elastic curve (if the shape is not known, it may be assumed).

(2) Visualize which tangent lines may be most helpful, and draw such lines on the elastic curve. When a point of zero slope is known, either from symmetry or from supports, a tangent drawn at this point will frequently be useful.

(3) By application of the second area-moment theorem, determine the tangential deviation at the point where the beam deflection is desired and at any other points required.

(4) From geometry, determine the perpendicular distance from the unloaded beam to the tangent line at the point where the beam deflection is desired, and, using the results of step 3, solve for the required deflection.

For essential data on properties of curves see App. B.

Examples 6–7 and 6–8 illustrate the method.

**Example 6–7:** Determine the deflection of the free end of the cantilever beam of Fig. 6–17*a* in terms of $w$, $L$, $E$, and $I$.

*Solution:* Since $E$ and $I$ are constant, a bending moment diagram is used instead of an $M/(EI)$ diagram. The moment diagram is shown in Fig. 6–17*b*, and the area under it has been divided into rectangular, triangular, and parabolic parts for ease in calculating areas and moments. The elastic curve is shown in Fig. 6–17*c* with the deflections greatly exaggerated. Points $A$ and $B$ are selected at the ends of the beam because the beam has a horizontal tangent at $B$ and the deflection at $A$ is required. The vertical distance to $A$ from the tangent at $B$, $t_{A/B}$ in Fig. 6–17*c*, equals the deflection of the free end of the beam, $y_A$. For this reason, the second area-moment theorem can be used directly to obtain the required deflection. The area of each of the three portions of the area under the moment diagram is shown in Fig. 6–17*b* along with the distance from the centroid of each part to the moment axis at the free end $A$. The second area-moment theorem gives

$$EIt_{A/B} = -\frac{wL^3}{6}\left(\frac{5L}{4}\right) - \frac{wL^3}{2}(2L) - \frac{wL^3}{2}\left(\frac{13L}{6}\right),$$

from which

$$y_A = t_{A/B} = -\frac{55wL^4}{24EI} = \underline{\frac{55wL^4}{24EI}\text{ downward.}} \qquad \text{Ans.}$$

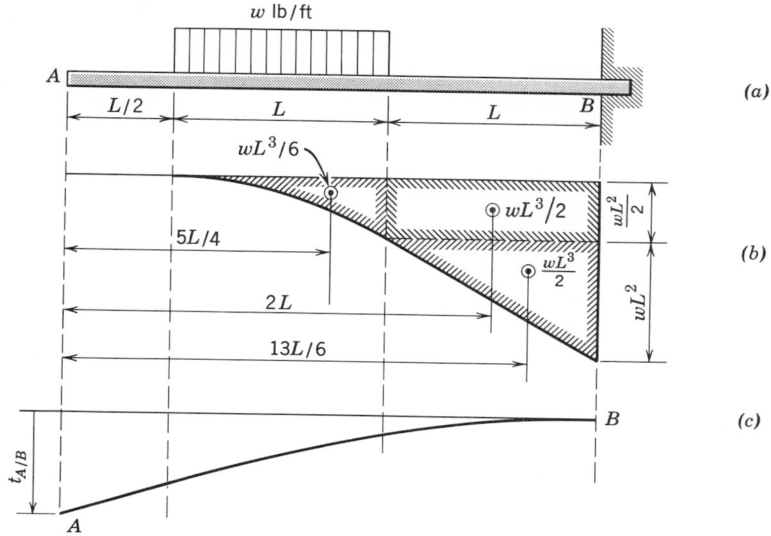


Fig. 6-17

**Example 6–8:** Determine the deflection of the left end of the beam loaded and supported as shown in Fig. 6–18*a*, in terms of $P$, $L$, $E$, and $I$.

*Solution:* The equations of equilibrium yield the left reaction equal to $4P/3$ upward and the right reaction equal to $P/3$ downward. Since $E$ and $I$ are constant, a moment diagram (instead of an $M/(EI)$ diagram) is drawn, and, with only one applied load, a composite moment diagram as shown in Fig. 6–18*b* is used.

The elastic curve with a tangent drawn at $C$ is shown in Fig. 6–18*c*. The tangent line could have been drawn at $B$ for an equally satisfactory solution. The second area-moment theorem gives the tangential deviation at $B$ as

$$EIt_{B/C} = -\frac{3PL^2}{2}(L) = -\frac{3PL^3}{2},$$

where the negative sign indicates that the moment center $B$ is below the tangent drawn from $C$. Another application of the second area-moment theorem gives the tangential deviation at $A$ as

$$EIt_{A/C} = -\frac{3PL^2}{2}(2L) - \frac{PL^2}{2}\left(\frac{2L}{3}\right) = -\frac{10PL^3}{3},$$

where the sign indicates $A$ is below the tangent from $C$. Finally

$$EIy_A = EI\left(t_{A/C} - \frac{4}{3}t_{B/C}\right) = -\frac{10PL^3}{3} - \frac{4}{3}\left(-\frac{3PL^3}{2}\right) = -\frac{4}{3}PL^3,$$

and

$$y_A = \underline{\frac{4}{3}\frac{PL^3}{EI}\text{ downward.}}$$ Ans.

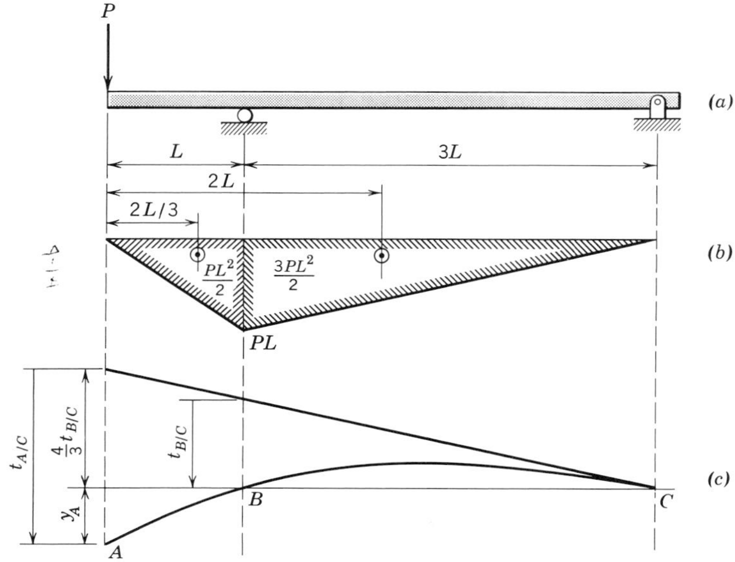


Fig. 6-18

## PROBLEMS

**6–49.*** Solve Prob. 6–30*b* by area moments.

**6–50.*** For the beam of Fig. P6–25, determine the deflection at the right end in terms of *E*, *I*, *P*, and *L*.

**6–51.** For the 10 I 25.4 steel beam in Fig. P6–51 determine the deflection under the load with $P = 5.0$ kips.

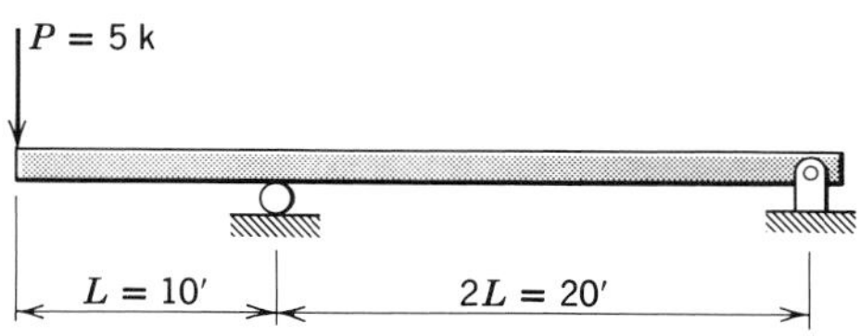


Fig. P6-51

**6–52.*** For the beam loaded and supported as shown in Fig. P6–52, determine the deflection at the right end in terms of $P$, $L$, $E$, and $I$.

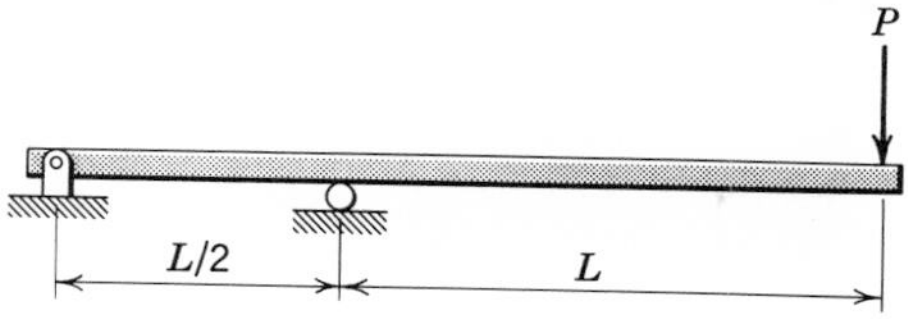


Fig. P6-52

**6–53.** For the beam of Fig. P6–27, determine the deflection at the right end in terms of $w$, $E$, $I$, and $L$.

**6–54.*** For the beam loaded and supported as shown in Fig. P6–54, determine the deflection at the right end in terms of $w$, $L$, $E$, and $I$.

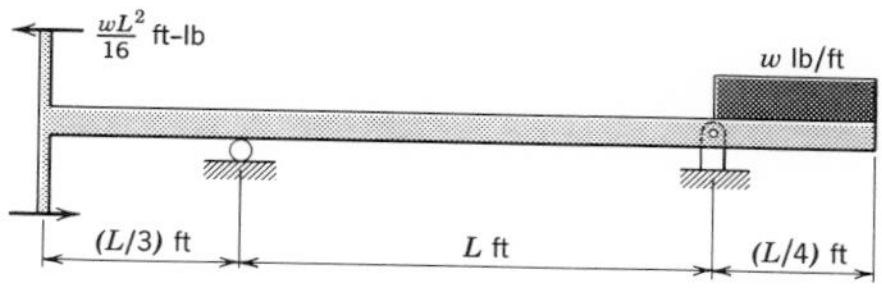


Fig. P6-54

**6–55.** For the beam loaded and supported as shown in Fig. P6–55, determine in terms of $P$, $L$, $E$, and $I$:

(a) The deflection at the left load.
(b) The deflection at the center of the span.

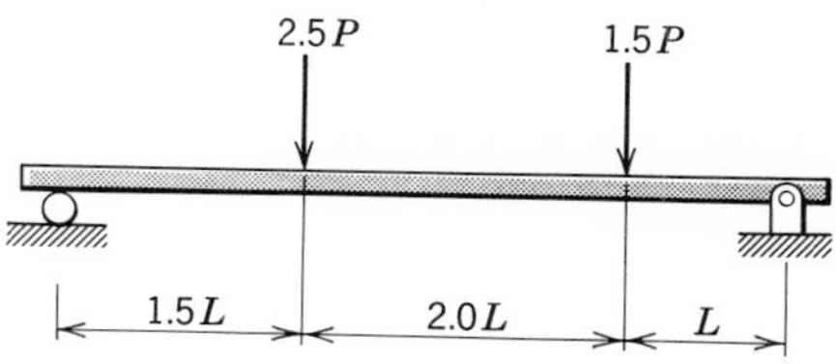


Fig. P6-55

**6–56.*** A 10 I 25.4 steel beam is shown in Fig. P6–56. When the deflection at the free end is 0.40 in. downward, determine:

(*a*) The load *P*.

(*b*) The maximum fiber stress in the beam.

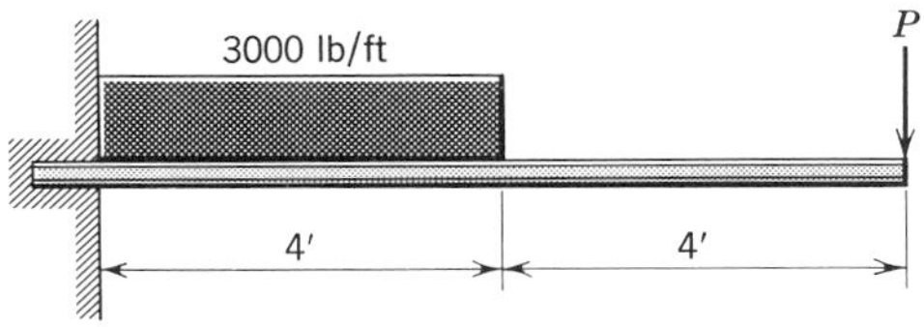


**Fig. P6-56**

**6–57.** Select the lightest allowable steel *WF* section for the beam of Fig. P6–57 with $L = 6$ ft, $w = 800$ lb per ft, an allowable fiber stress of 18,000 psi, and an allowable deflection at the free end of 0.72 in.

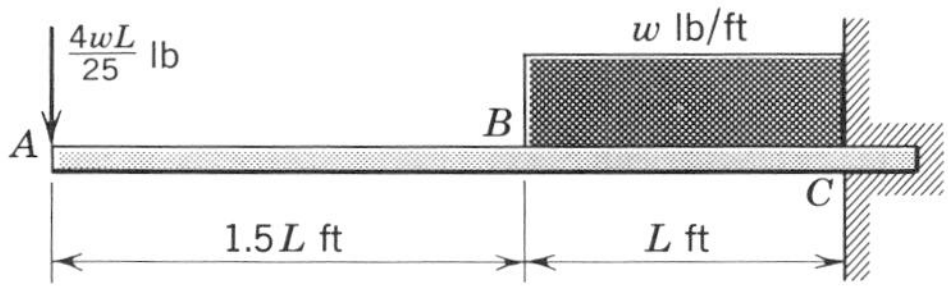


**Fig. P6-57**

**6–58.*** A 12 I 31.8 steel beam is loaded and supported as shown in Fig. P6–58. Determine the maximum allowable value of *P* for an allowable fiber stress of 24,000 psi and a maximum allowable deflection at mid-span of 0.60 in.

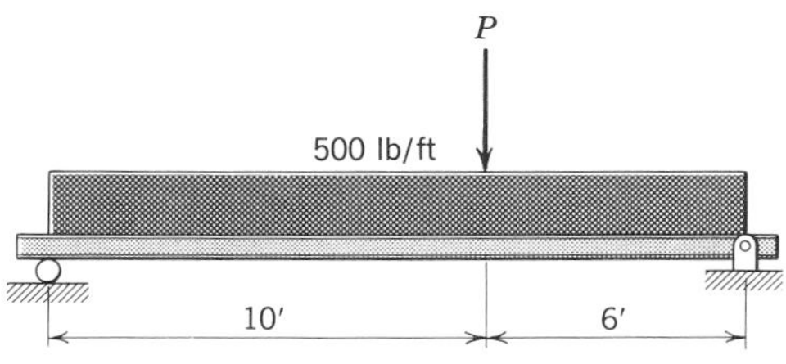


**Fig. P6-58**

**6–59.** A 6-in. wide by 12-in. deep timber cantilever beam is 10 ft long. It carries a uniformly distributed load of $w$ lb per ft over its entire length and a concentrated load of 500 lb 4 ft from the free end. The modulus of elasticity of the timber is 1200 ksi; the maximum allowable deflection is 0.5 in.; the allowable fiber stress is 1200 psi; and the allowable horizontal shearing stress is 100 psi. Determine the maximum allowable distributed load, $w$.

**6–60.** Solve Prob. 6–40 by area moments.

**6–61.** For the beam loaded and supported as shown in Fig. P6–61, determine the deflection at the concentrated load in terms of $w$, $L$, $E$, and $I$.

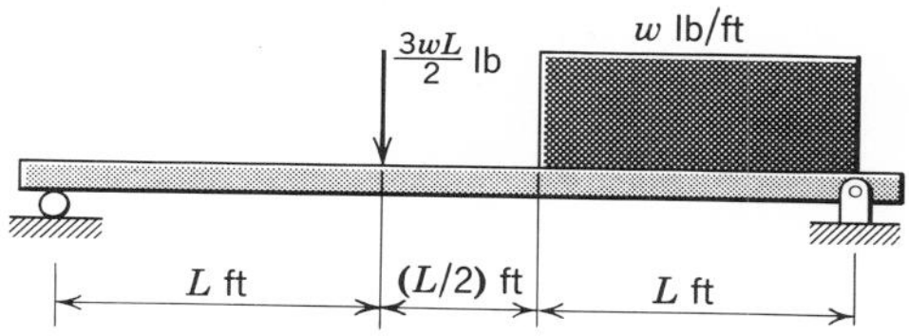


Fig. P6-61

**6–62.** For the beam loaded and supported as shown in Fig. P6–62, determine the deflection at the left end in terms of $w$, $L$, $E$, and $I$.

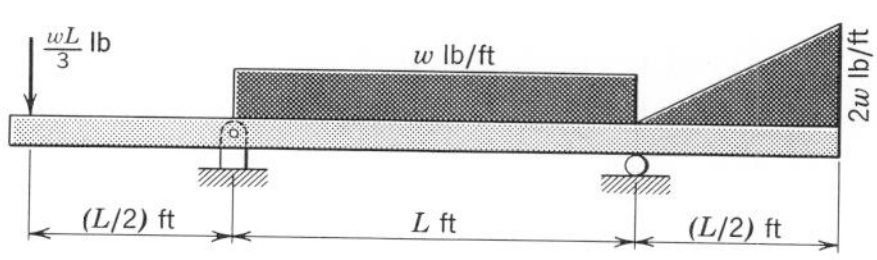


Fig. P6-62

**6–63.*** For the beam of Fig. P6–63, determine the deflection at the left end in terms of $w$, $L$, $E$, and $I$.

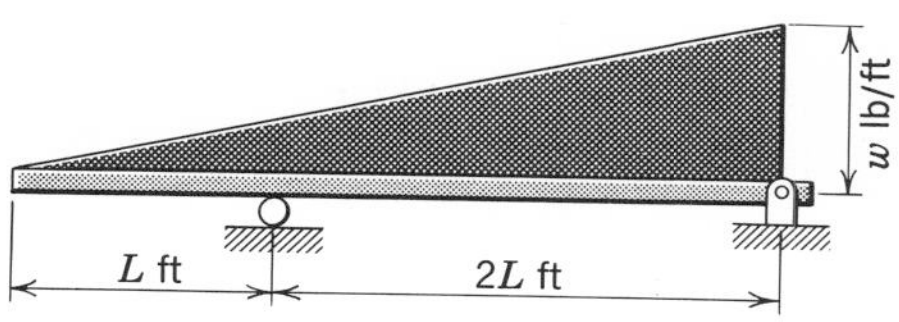


Fig. P6-63

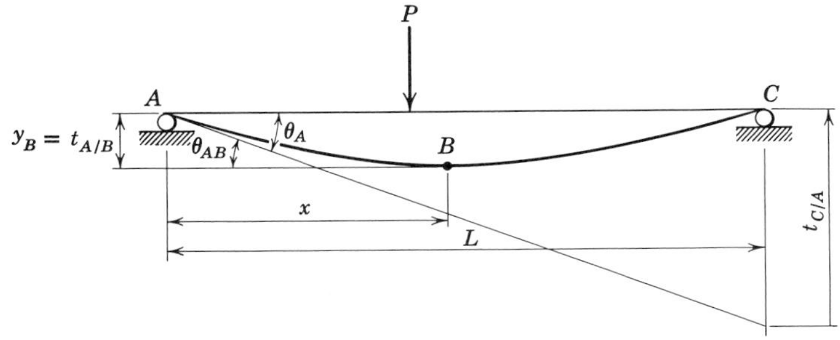


Fig. 6-19

**6–9. Maximum Deflections by Area Moments.** In most cantilever beams and in symmetrically loaded simply supported beams, the point of maximum deflection can be determined by inspection. In such instances the procedure outlined in Art. 6–8 is usually adequate for finding the maximum deflection. When the location of the point of maximum deflection is unknown, it is necessary to use both the first and second area-moment theorems for determining the maximum deflection.

Figure 6–19 shows a typical situation where the maximum deflection, at point $B$ on the elastic curve, is required. The deflection, $y_B$, or the tangential deviation, $t_{A/B}$, is wanted, but the distance $x$ along the beam is unknown and must be determined before the deflection can be obtained. Once $t_{C/A}$ is determined from the second area-moment theorem, $t_{C/A}/L$ gives $\tan \theta_A$ which is approximately equal to $\theta_A$. The angle $\theta_A$ can also be determined, in terms of $x$, from the first area-moment theorem because the tangent to the elastic curve at $B$ is horizontal and $\theta_A = \theta_{AB}$. When these two expressions for $\theta_A$ are equated, the distance $x$ is found. Finally $t_{A/B}$ is determined from the second area-moment theorem. The following example illustrates the procedure for a specific case.

**Example 6–9:** For the beam in Fig. 6–20*a*, determine the maximum deflection in terms of $w$, $L$, $E$, and $I$.

*Solution:* The equations of equilibrium give the left reaction as $wL/2$ upward and the right reaction as $wL$ upward. The moment diagram by parts is shown in Fig. 6–20*b*, *c*, and *d*. The elastic curve, very much exaggerated, of the loaded beam is shown in Fig. 6–20*e* with tangents drawn at $A$ and $B$. The second area-moment theorem, with the moment axis at $C$, gives the vertical distance to $C$ from the tangent at $A$ as

$$EIt_{C/A} = L\left(\frac{9wL^3}{4}\right) - \frac{2L}{3}(wL^3) - \frac{L}{4}\left(\frac{wL^3}{6}\right) = \frac{37}{24}wL^4.$$

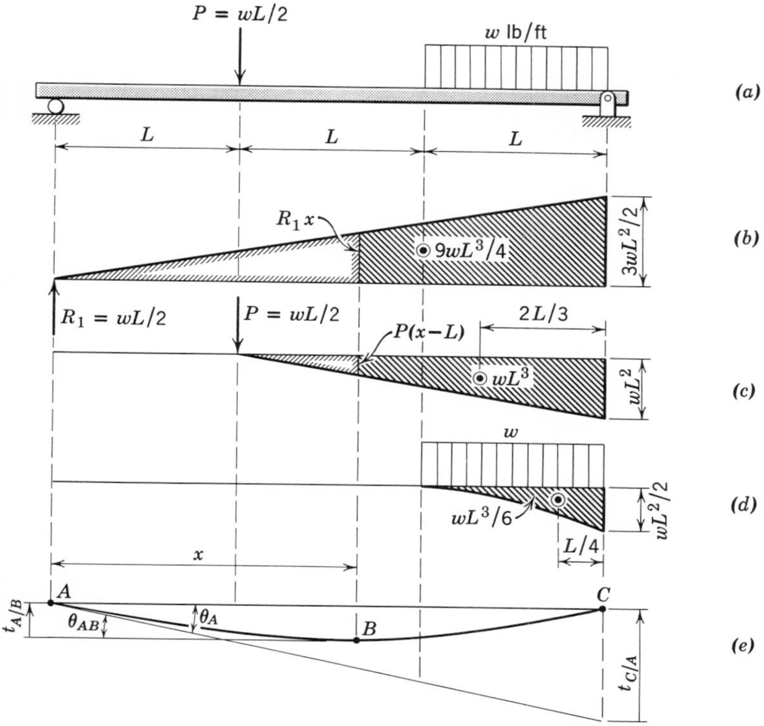


Fig. 6-20

Since the deflections are small,

$$\theta_A = \tan \theta_A = \frac{t_{C/A}}{3L} = \frac{37wL^3}{72EI}.$$

Because the tangent at $B$ is horizontal, the first area-moment theorem applied to the segment $AB$ gives

$$\theta_A = \theta_{AB} = \frac{1}{EI}\left[\frac{wL}{2}\,x\left(\frac{x}{2}\right) - \frac{wL}{2}\,(x - L)\,\frac{x - L}{2}\right] = \frac{wL}{4EI}\,(2Lx - L^2).$$

Equating the two expressions for $\theta_A$ gives

$$\frac{37wL^3}{72EI} = \frac{wL}{4EI}\,(2Lx - L^2),$$

from which

$$x = 55L/36.$$

Application of the second theorem to the segment $AB$, with the moment axis at $A$, gives the vertical distance to $A$ from the horizontal tangent at $B$ as

$$|y_B| = t_{A/B} = \frac{1}{EI}\left\{\left(\frac{2x}{3}\right)\left(\frac{wLx}{2}\right)\left(\frac{x}{2}\right) - \left[L + \frac{2}{3}(x - L)\right]\frac{wL}{2}\frac{(x-L)^2}{2}\right\}$$

$$= \frac{wL^4}{2EI}\left[\frac{1}{3}\left(\frac{55}{36}\right)^3 - \frac{1}{2}\left(\frac{19}{36}\right)^2 - \frac{1}{3}\left(\frac{19}{36}\right)^3\right],$$

where the positive result means that $A$ is above the tangent line. Therefore

$$y_B = 0.500\,\frac{wL^4}{EI}\ \text{downward.} \qquad \text{Ans.}$$

## PROBLEMS

**6–64.** Solve Prob. 6–30*a* by area moments.

**6–65.*** Solve Prob. 6–29*b* by area moments.

**6–66.** For the beam of Prob. 6–51, determine the maximum deflection between the supports.

**6–67.*** An 8-in. wide by 12-in. deep timber beam is simply supported on a span of 21 ft. A 3.50-kip load is applied 6 ft from the right support. Determine:

(*a*) The maximum deflection for a modulus of elasticity of $10^6$ psi.
(*b*) The maximum fiber stress.
(*c*) The maximum horizontal shearing stress.

**6–68.** For the beam of Fig. P6–28, locate the point of maximum deflection.

**6–69.*** A 12 I 50 steel beam is loaded and supported as shown in Fig. P6–69. Determine:

(*a*) The deflection at the right end.
(*b*) The maximum deflection between the supports.

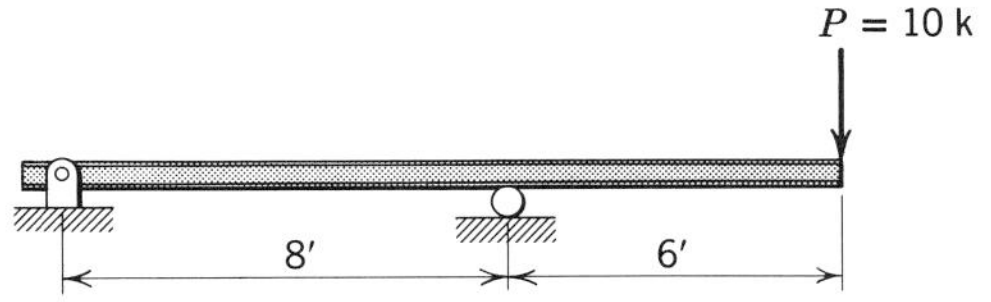


Fig. P6-69

**6–70.** For the beam of Prob. 6–25, determine:

(*a*) The deflection of the beam at the load $P$.
(*b*) The location of the point of maximum deflection.
(*c*) The maximum deflection.

**6–71.*** For the beam of Fig. P6–54, locate the point of maximum deflection between the supports.

**6–10. Deflections by Superposition.** As explained in Art. 2–5, the concept or method of superposition consists of finding the resultant effect of several loads acting on a member simultaneously as the sum of the contributions from each of the loads applied individually. The results for the separate loads are frequently available from previous work or readily determined by a previous method. In such instances the superposition method becomes a powerful concept or tool for finding stresses, deflections, etc. The method is applicable in all cases where a linear relation exists between the stresses or deflections and the applied loads.

To show that beam deflections can be accurately determined by superposition, consider the cantilever beam of Fig. 6–21 with loads $Q$, $w$, and $P$. In order to determine the deflection at any point of this beam by the double integration method, it is necessary to express the bending moment in terms of the applied loads. For each interval along the beam the value of $M$ is the algebraic sum of the moments due to the separate loads. After two successive integrations, the solution for the deflection at any point will still be the algebraic sum of the contributions from each applied load. Furthermore for any given value of $x$, the relation between applied load and resulting deflection will be linear. It is evident, therefore, that the deflection of a beam is the sum of the deflections produced by the individual loads. Once the deflections produced by a few typical individual loads have been determined by one of the methods already presented, the superposition method provides a means of rapidly solving a wide variety of more complicated problems by various combinations of known results. As more data become available, a wider range of problems can be solved by superposition.

The data in App. D are provided for use in mastering the superposition method. No attempt is made to give a large number of results as such data are readily available in various handbooks. The data given

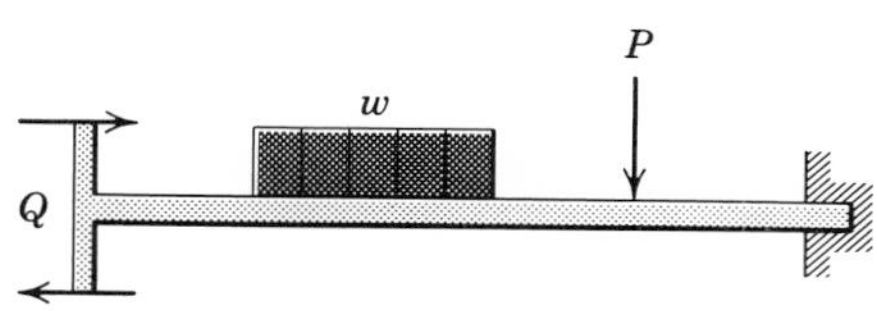


Fig. 6-21

and the illustrative examples are for the purpose of making the concept and methods clear.

**Example 6–10:** For the beam in Fig. 6–22*a*, determine the maximum deflection when $E$ is $12(10^6)$ psi and $I$ is 81 in.$^4$.

*Solution:* From symmetry and $\Sigma F_y = 0$, the reactions are each 4500 lb upward. Because of the symmetrical loading the slope of the beam is zero at the center of the span and the right (or left) half of the beam can be considered as a cantilever beam with two loads. As shown in Fig. 6–22*b*, the cantilever with two loads can be replaced by two beams (designated 1 and 2), each carrying one of the two loads.

The elastic curve (exaggerated) for part 1, as shown in Fig. 6–22*c*, gives the deflection at the right end as $y_{4.5} + y_3$, where $y_{4.5}$ is the deflection at the end of the uniformly distributed load and $y_3$ is the additional deflection of the unloaded 3 ft. From case 2 of App. D

$$y_{4.5} = -\frac{1000(4.5^4)(12^3)}{8(12)(10^6)(81)} = -\frac{9^3}{8(10^3)} = -0.0911 \text{ in.},$$

and

$$\theta_{4.5} = -\frac{1000(4.5^3)(12^2)}{6(12)(10^6)(81)} = -\frac{9}{4(10^3)} = -0.00225 \text{ rad.},$$

from which

$$y_3 = 3(12)(0.00225) = -0.0810 \text{ in.};$$

consequently the total deflection of the right end is 0.1721 in. downward.

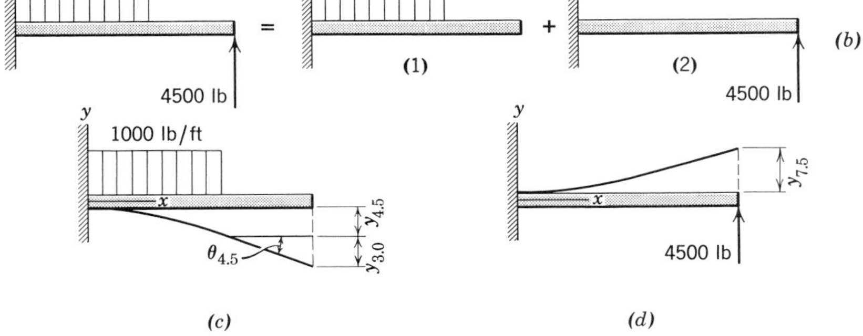


Fig. 6-22

The elastic curve (exaggerated) for part 2 is shown in Fig. 6–22*d*. From case 1 of App. D

$$y_{7.5} = +\frac{4500(7.5^3)(12^3)}{3(12)(10^6)(81)} = 1.125 \text{ in. upward.}$$

The algebraic sum of the deflections for parts 1 and 2 is +0.953 in., which means that the right end of the beam is 0.953 in. above the center. Obviously the right end does not move, and the maximum deflection is at the center and is

$$y_{max} = \underline{0.953 \text{ in. downward.}} \qquad \text{Ans.}$$

**Example 6–11:** Determine the deflection at mid-span for the steel beam ($I = 180$ in.[4]) of Fig. 6–23*a*.

*Solution:* The given loading is equivalent to the two loads, parts 1 and 2, shown in Fig. 6–23*b*. For part 1, the deflection at the center of the span is given as case 7 of App. D. It is

$$(y_C)_1 = -\frac{5(600)(16^4)(12^3)}{384(30)(10^6)(180)} = -0.1638 \text{ in.}$$

For part 2, the deflection at the center is given as a part of case 5 of App. D. It is

$$(y_C)_2 = -\frac{2000(6)[3(16^2) - 4(6^2)]12^3}{48(30)(10^6)(180)} = -0.0499 \text{ in.}$$

The algebraic sum of the deflections for mid-span is

$$y_C = -0.1638 + (-0.0499) = -0.2137 \text{ in.} = \underline{0.214 \text{ in. downward.}} \qquad \text{Ans.}$$

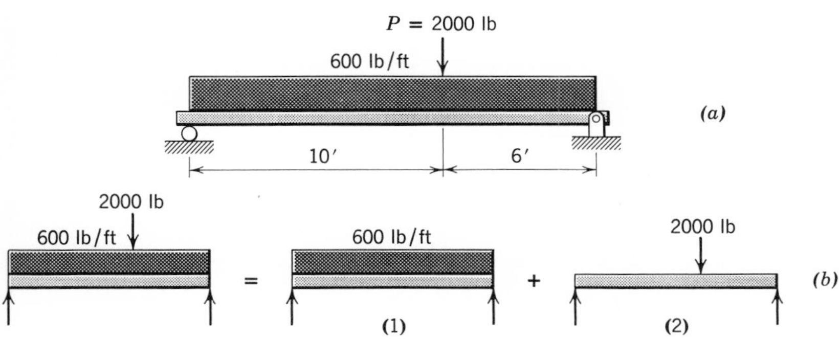


Fig. 6-23

## PROBLEMS

**6–72.*** Determine the deflection of the free end of the cantilever beam in Fig. P6–72 when $w$ is 600 lb per ft, $L$ is 10 ft, $E$ is $12(10^6)$ psi, and $I$ is 14.4 in.$^4$

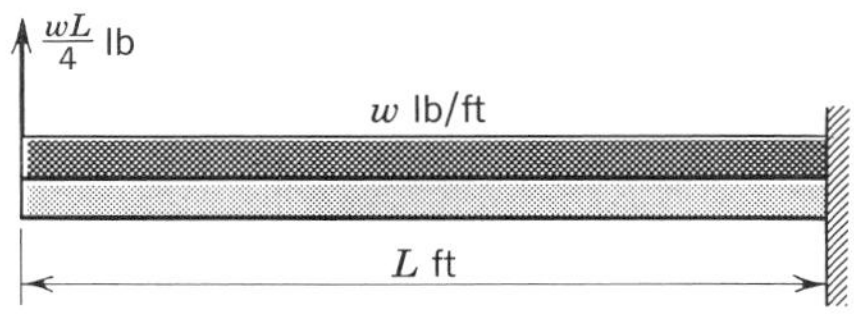


Fig. P6-72

**6–73.*** A 20-ft simply supported 10 WF 21 steel beam carries a uniformly distributed load of 400 lb per ft over the entire span and a concentrated load of 1600 lb at mid-span. Determine the maximum deflection of the beam.

**6–74.*** A 10 by 10-in. timber beam 14 ft long is simply supported at its ends. The beam carries a uniformly distributed load of 900 lb per ft over its entire length and a concentrated load of 2000 lb at the center. The modulus of elasticity of this timber is 1000 ksi. Determine the maximum deflection of the beam.

**6–75.** A timber cantilever beam 4 in. wide by 6 in. deep is 12 ft long. If an upward pull of 600 lb is applied at the free end, determine the uniformly distributed load $W$, equal to $wL$, required to bring the free end to a point ½ in. above its unloaded position. Young's modulus is equal to 1500 ksi.

**6–76.*** For the beam in Fig. P6–76 determine the maximum deflection in the right half in terms of $P$, $L$, $E$, and $I$.

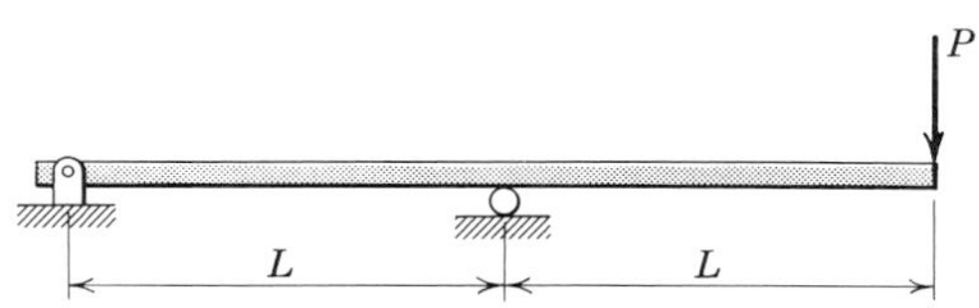


Fig. P6-76

**6–77.** A 4-in. wide by 12-in. deep timber cantilever beam supports a concentrated load of 1200 lb at the free end and a uniformly distributed load of $w$ lb per ft over the entire 10-ft span. For a modulus of elasticity of 1200 ksi, an allowable fiber stress of 1600 psi, an allowable horizontal shearing stress of 100 psi, and a maximum allowable deflection of 1.00 in., determine the maximum permissible value of $w$.

**6–78.*** Determine the deflection of the free end of the beam in Fig. P6–78 in terms of $w$, $L$, $E$, and $I$.

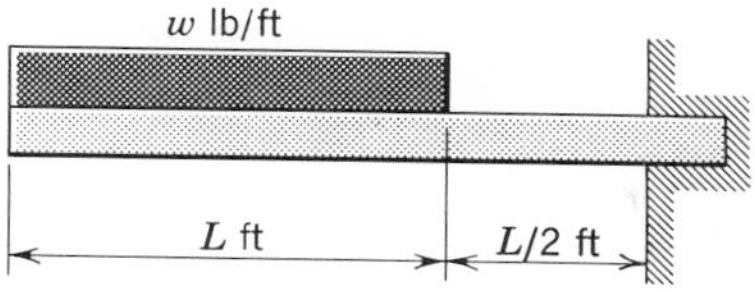


Fig. P6-78

**6–79.** Solve Prob. 6–78 with an added couple of $wL^2$ counterclockwise applied at the right end of the distributed load.

**6–80.** For the beam of Fig. P6–80, determine the mid-span deflection if $w$ is 100 lb per ft, $L$ is 12 ft, $E$ is $1.44(10^6)$ psi, and $I$ is 81 in.$^4$

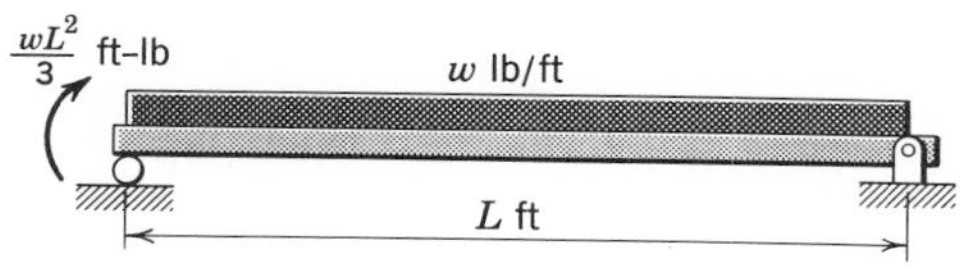


Fig. P6-80

**6–81.*** Determine the deflection of the free end of the beam in Fig. P6–81.

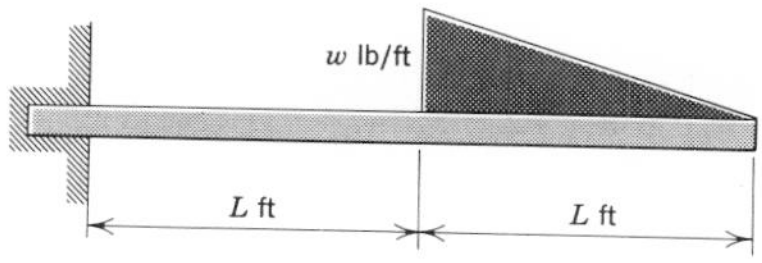


Fig. P6-81

**6–11. Deflections by Energy Methods—Castigliano's Theorem.** Strain energy techniques are frequently used to analyze the deflections of beams and structures. Of the many available methods, the application of Castigliano's theorem, to be developed here, is one of the most

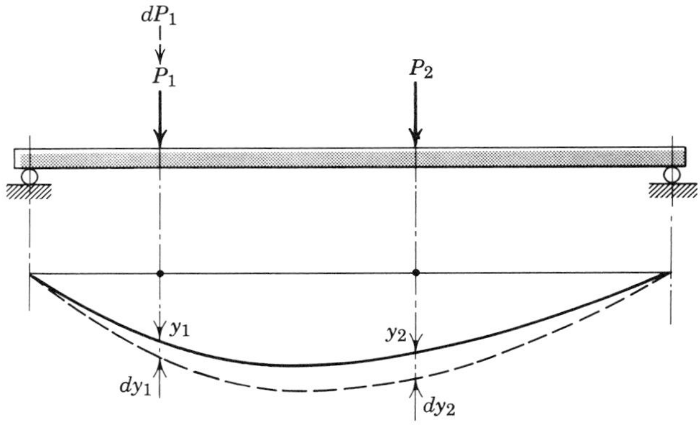


Fig. 6-24

widely used. It was presented in 1873 by the Italian engineer Alberto Castigliano (1847–1884). Although the theorem will be derived by considering the strain energy stored in beams, it is applicable to any structure for which the force-deformation relations are linear.

If the beam shown in Fig. 6–24 is slowly and simultaneously loaded by the two forces $P_1$ and $P_2$ with resulting deflections $y_1$ and $y_2$, the strain energy, $U$, of the beam is, by Clapeyron's theorem (see Art. 2–2), equal to the work done by the forces. Therefore,

$$U = \frac{1}{2} P_1 y_1 + \frac{1}{2} P_2 y_2. \tag{a}$$

Let the force $P_1$ be increased by a small amount $\Delta P_1$, and $\Delta y_1$ and $\Delta y_2$ be the changes in deflection due to this incremental load. Since the forces $P_1$ and $P_2$ are already present, the increase in the strain energy is

$$\Delta U = \frac{1}{2} \Delta P_1 \, \Delta y_1 + P_1 \, \Delta y_1 + P_2 \, \Delta y_2. \tag{b}$$

If the order of loading were reversed so that the incremental force, $\Delta P_1$, were applied first, followed by $P_1$ and $P_2$, the resulting strain energy would be

$$U + \Delta U = \frac{1}{2} \Delta P_1 \, \Delta y_1 + \Delta P_1 \, y_1 + \frac{1}{2} P_1 y_1 + \frac{1}{2} P_2 y_2. \tag{c}$$

The resulting strain energy must be independent of the order of loading; hence, by combining Eq. (a), (b), and (c)

$$\Delta P_1 \, y_1 = P_1 \, \Delta y_1 + P_2 \, \Delta y_2. \tag{d}$$

Equations (b) and (d) can be combined to give

$$\frac{\Delta U}{\Delta P_1} = y_1 + \frac{1}{2}\,\Delta y_1,$$

or upon taking the limit as $\Delta P_1$ approaches zero[3]

$$\frac{\partial U}{\partial P_1} = y_1. \tag{e}$$

For the general case in which there are many loads involved, Eq. (e) is written as

$$\frac{\partial U}{\partial P_i} = y_i. \tag{6–5a}$$

The following is a statement of Castigliano's theorem. *If the strain energy of a linearly elastic structure is expressed in terms of the system of external loads, the partial derivative of the strain energy with respect to a concentrated external load is the deflection of the structure at the point of application and in the direction of that load.* By a similar development, Castigliano's theorem can also be shown to be valid for applied moments and the resulting rotations (or changes in slope) of the structure. Thus

$$\frac{\partial U}{\partial M_i} = \theta_i. \tag{6–5b}$$

If the deflection is required either at a point where there is no unique point load or in a direction not aligned with the applied load, a dummy load is introduced at the desired point acting in the proper direction. The deflection is obtained by first differentiating the strain energy with respect to the dummy load and then taking the limit as the dummy load approaches zero. Note that if the loading consists of a number of point loads, all expressed in terms of a single parameter (e.g., $P$, $2P$, $3P$, $wL$, $2wL$), and if the deflection is wanted at one of the applied loads, one must either write the moment equation with this load as a separate identifiable term or add a dummy load at the point; so that the partial derivative can be taken with respect to this load only.

It was shown in Art. 2–2 that the strain energy per unit volume for

---

[3] The partial derivative is used because the strain energy is a function of both $P_1$ and $P_2$.

uniaxial stress is $\sigma^2/(2E)$, hence, the total strain energy under uniaxial stress is

$$U = \int_{\text{Vol.}} \frac{\sigma^2}{2E}\, dV.$$

For a beam of constant (or slowly varying) cross section subjected to pure bending, the principal stresses are parallel to the axis of the beam; therefore, using Eq. (5–1),

$$U = \frac{1}{2E} \int_{\text{Vol.}} \left(\frac{My}{I}\right)^2 dV. \tag{f}$$

By writing $dV$ as $dA\, dx$ (where $x$ is measured along the axis of the beam), Eq. (f) becomes

$$U = \frac{1}{2E} \int_0^L \left[\frac{M^2}{I^2} \int_{\text{Area}} y^2\, dA\right] dx = \frac{1}{2E} \int_0^L \frac{M^2}{I}\, dx. \tag{6–6}$$

Except for short deep beams, the strain energy in beams due to transverse shear is sufficiently small to be neglected when compared to the strain energy due to flexure and Eq. (6–6) applies even though the beam is not subjected to pure bending. In applying Eq. (6–5) to Eq. (6–6), it is usually much simpler to apply Leibnitz's rule[4] to differentiate under the integral sign so that

$$y_i = \frac{\partial U}{\partial P_i} = \frac{1}{E} \int_0^L \frac{M}{I} \frac{\partial M}{\partial P_i}\, dx. \tag{6–7}$$

**Example 6–12:** Determine the deflection at the free end of a cantilever beam of constant cross section and length $L$, loaded with a force $P$ at the free end

[4] *Advanced Calculus*, W. Kaplan, Addison-Wesley Publishing Company, Reading, Mass., 1953, p. 219.

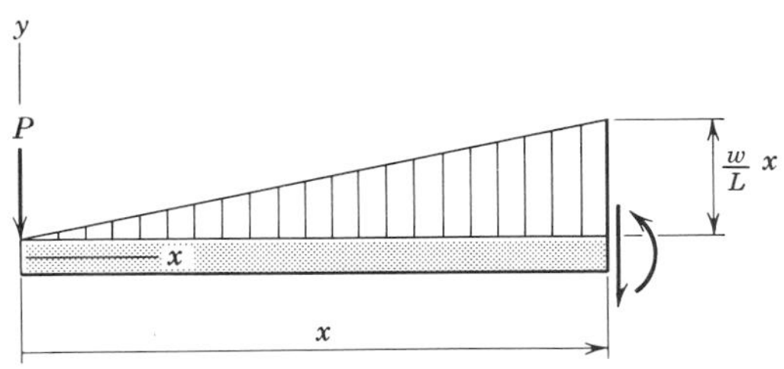


Fig. 6-25

and a distributed load which varies linearly from zero at the free end to $w$ at the support.

*Solution:* From the free-body diagram of Fig. 6–25, the moment equation is

$$M = -Px - \frac{wx^3}{6L}.$$

The partial derivative of $M$ with respect to $P$ is $-x$, and Eq. (6–7) becomes

$$EIy = \int_0^L \left(Px^2 + \frac{wx^4}{6L}\right) dx = +\frac{PL^3}{3} + \frac{wL^4}{30}.$$

The positive signs indicate deflection in the direction of the force $P$; hence

$$y = \underline{\frac{PL^3}{3EI} + \frac{wL^4}{30EI} \text{ downward.}} \qquad \text{Ans.}$$

The result may be verified by reference to App. D.

**Example 6–13:** Determine the deflection at the center of a simply supported beam of constant cross section and span $L$ carrying a uniformly distributed load $w$ lb per ft over its entire length.

*Solution:* In the free-body diagram of Fig. 6–26, the dashed force $P$ represents a fictitious load applied at the location and in the direction of the desired deflection. The moment equation is

$$M = M_w + M_P = \frac{wLx}{2} - \frac{wx^2}{2} + \frac{Px}{2} - P\left\langle x - \frac{L}{2}\right\rangle^1,$$

where the quantity $\langle x - L/2\rangle^1$ is zero for all $x \le L/2$ (see Art. 6–5). The partial derivative of $M$ with respect to $P$ is

$$\frac{\partial M}{\partial P} = \frac{x}{2} - \left\langle x - \frac{L}{2}\right\rangle^1.$$

The dummy force $P$ is equated to zero and the deflection is given by

$$EIy = \int_0^L \left(\frac{wLx}{2} - \frac{wx^2}{2}\right)\left(\frac{x}{2} - \left\langle x - \frac{L}{2}\right\rangle^1\right) dx$$

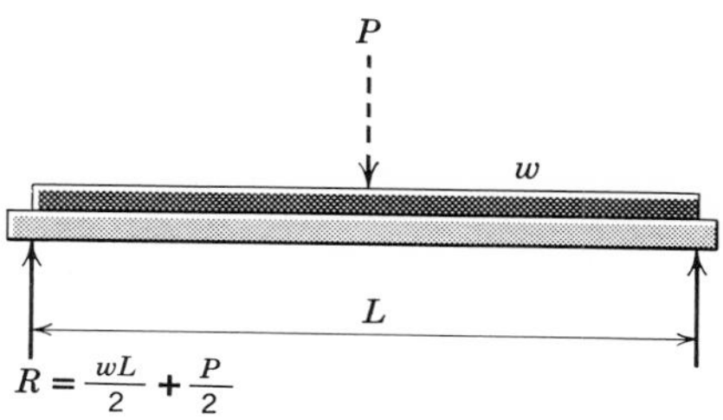


Fig. 6-26

which, for ease of integration, can be written as

$$EIy = \frac{w}{4}\int_0^L (Lx^2 - x^3)\,dx + \frac{w}{4}\int_{L/2}^L (L^2x - 3Lx^2 + 2x^3)\,dx = \frac{5wL^4}{384EI}.$$

Since the deflection is positive in the direction of the force, the deflection is downward and

$$y = \frac{5wL^4}{384EI} \text{ downward.} \qquad \text{Ans.}$$

## PROBLEMS

**Note: Solve the following problems by means of Castigliano's theorem.**

**6–82.*** Problem 6–72.

**6–83.*** A 10 I 25.4 steel beam is loaded as shown in Fig. P6–83. When the deflection at the free end is 0.48 in. downward, determine the force $P$.

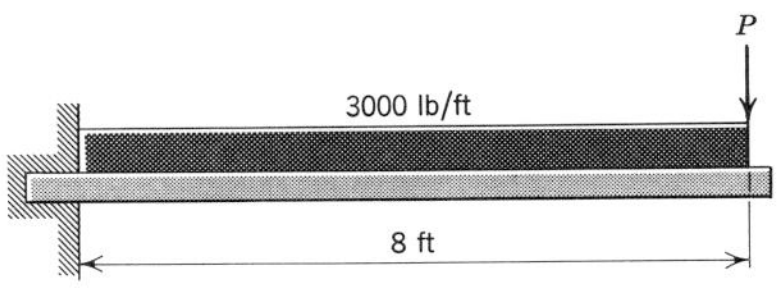


**Fig. P6-83**

**6–84.*** For the beam of Fig. P6–57, determine the deflection at the free end in terms of $w$, $L$, $E$, and $I$.

**6–85.** For the beam of Fig. P6–25, determine the deflection at $P$ in terms of $P$, $L$, $E$, and $I$.

**6–86.*** Problem 6–30*b*.

**6–87.** Problem 6–55*a*.

**6–88.*** Problem 6–74.

**6–89.*** For the beam of Fig. P6–38, determine the deflection at the concentrated load in terms of $w$, $L$, $E$, and $I$.

**6–90.*** For the beam of Fig. P6–38, determine the deflection at mid-point in terms of $w$, $L$, $E$, and $I$.

**6–91.*** Problem 6–78.

**6–92.*** Problem 6–81.

**6–93.** For the beam of Fig. P6–25, determine the slope at the right end in terms of $P$, $L$, $E$, and $I$.

**6–94.*** For the beam of Fig. P6–51, determine the slope at the left end.

**6–12. Deflections due to Shearing Stress.** As mentioned in Art. 6–3, the beam deflections calculated so far neglect the deflection produced by the shearing stresses in the beam. For short heavily loaded beams this deflection can be significant, and an approximate method for evaluating such deflections will now be developed. The deflection of the neutral surface, $dy$, due to shearing stresses in the interval $dx$ along the beam of Fig. 6–27 is

$$dy = \gamma\, dx = \frac{\tau}{G}\, dx = \frac{VQ}{GIt}\, dx,$$

from which, since the shear in Fig. 6–27 is negative,

$$\frac{GIt}{Q}\frac{dy}{dx} = -V. \tag{6–8}$$

Since the vertical shearing stress varies from top to bottom of a beam, the deflection due to shear is not uniform. This non-uniform deflection due to shear is reflected in the slight warping of the sections of a beam (see Art. 5–11). Equation (6–8) gives values too high because the maximum shearing stress (at the neutral surface) is used and also because the rotation of the differential shear element (see Art. 5–11) is ignored.

In order to obtain an idea of the relative amount of beam deflection due to shearing stress, consider a rectangular cross section for the beam of Fig. 6–27, and use the maximum stress for which

$$\frac{Q}{It} = \frac{3}{2A}.$$

The expression for $dy$ becomes

$$dy = \frac{3wx\, dx}{2AG},$$

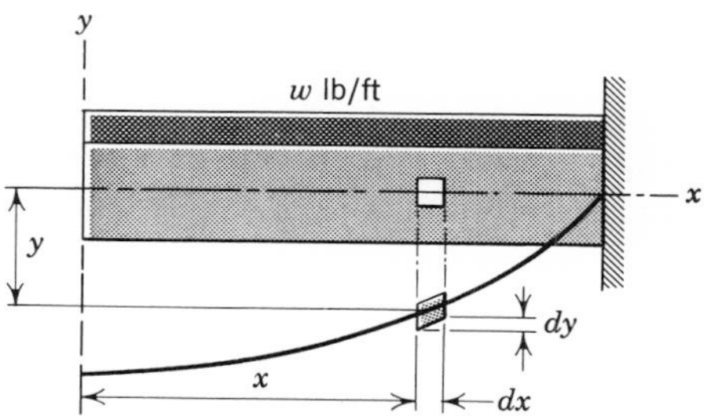


Fig. 6-27

where $V$ is replaced by its value, $-wx$. Integration along the entire beam gives the change in $y$ due to shear as

$$\Delta y = \frac{3w}{2AG}\int_0^L x\,dx = \frac{3wL^2}{4AG},$$

which equals the deflection at the left end. For this same beam the magnitude of the deflection at the left end due to flexural stresses is

$$\frac{wL^4}{8EI} = \frac{3wL^4}{2EAd^2},$$

and the magnitude of the total deflection at the left end becomes

$$y = \frac{3wL^4}{2EAd^2} + \frac{3wL^2}{4AG} = \frac{3wL^2}{4AG}\left(\frac{2L^2G}{d^2E} + 1\right). \tag{a}$$

For a material like steel the ratio $G/E$ is approximately 0.4, and, with the ratio $L/d$ selected as 6 (a very short deep beam), the deflection due to flexural stress is 28.8 times the deflection due to shear. On the other hand, for a 36 WF 160 steel beam and the same span-depth ratio, the flexural stress deflection is 4.85 times the approximate shear deflection. For a working fiber stress of 18 ksi, the two deflections are 0.389 and 0.080 in. respectively. Equation (a) indicates that the ratio of the two deflections increases as the square of $L/d$, which means that the deflection due to shear is of importance only in the case of very short deep beams.

**6–13. Closure.** Four methods for determining the deflection of a beam were presented in this chapter. The integration method consists of solving the elastic curve differential equation and evaluating the constants of integration from the boundary and matching conditions. Singularity functions are quite helpful if more than one interval is involved. The area-moment method consists of applying one or both of the area-moment theorems to the $M/(EI)$ diagram and using the geometry of the beam to obtain the required deflections. Note that the integration method can be adapted to the area-moment method by replacing the indefinite integral with a definite integral, whereby, upon integration, one obtains the change in slope and the change in tangential deviation (measured from the tangent drawn to the curve at the origin) between the limits selected. The superposition method consists of subdividing the applied loads and supports into an equivalent set of less involved components, for which solutions are readily available, and obtaining the required deflections as the algebraic sum of the deflections for the component parts. An energy method—Castigliano's theorem—consists of writing an expression for the energy, $U$, stored in the beam due to all applied loads and then evaluating the partial derivative of $U$ with respect to a real or dummy load, $P$, applied at the point where the deflection is to be obtained.

The selection of the best method for a given problem requires judgment and experience. The following miscellaneous problems are presented so that the student may obtain practice in selecting the method best adapted to a particular problem.

## PROBLEMS

**6–95.** A 10-ft horizontal cantilever beam of dry red oak is 6 in. wide by 10 in. deep. The beam is subjected to a negative couple of 1000 ft-lb and a downward concentrated load of 600 lb, both at the free end. Determine the maximum deflection of the beam.

**6–96.*** A 10-ft cantilever steel beam carries a distributed load $w$ over its entire length and a concentrated load of $5w$ at the free end. The beam is a 14 WF 43. Determine the maximum permissible value for $w$ if the fiber stress is not to exceed 20 ksi and the maximum deflection is limited to 0.5 in.

**6–97.** A timber ($E$ = 1600 ksi) cantilever beam 10 ft long has the cross section shown in Fig. P6–97. The allowable fiber stress is 1600 psi; the allowable horizontal shearing stress is 120 psi; and the maximum allowable deflection is 0.5 in. Determine the maximum permissible uniformly distributed load that the beam may carry over the entire length.

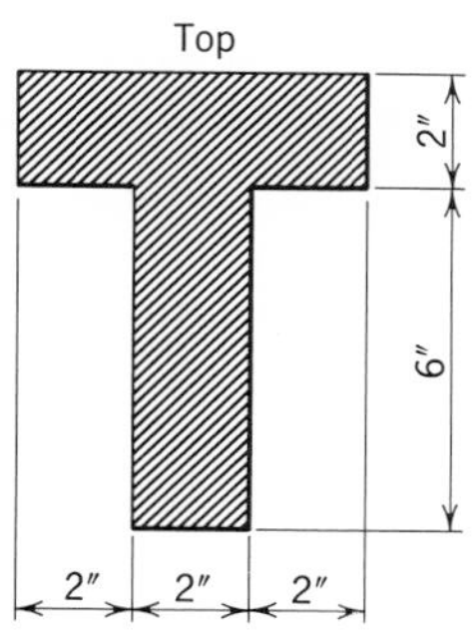


**Fig. P6-97**

**6–98.** A simple beam with a span of 20 ft is to carry a uniformly distributed load of 400 lb per ft over the entire span. The beam is to be a rectangular timber ($E = 1200$ ksi) with the depth twice the width. Determine the minimum permissible dimensions of the cross section if the allowable fiber stress is 1200 psi and the deflection is not to exceed 0.6 in.

**6–99.** A simple beam with a span of 40 ft is to carry a uniformly distributed load of 1000 lb per ft over the entire span and a concentrated load of 10 kips at the center of the span. The fiber stress is not to exceed 18 ksi and the maximum deflection is limited to 2 in. Select the lightest steel WF section for the span.

**6–100.*** For the beam of Fig. P6–100, determine the deflection at the right end in terms of $w$, $L$, $E$, and $I$.

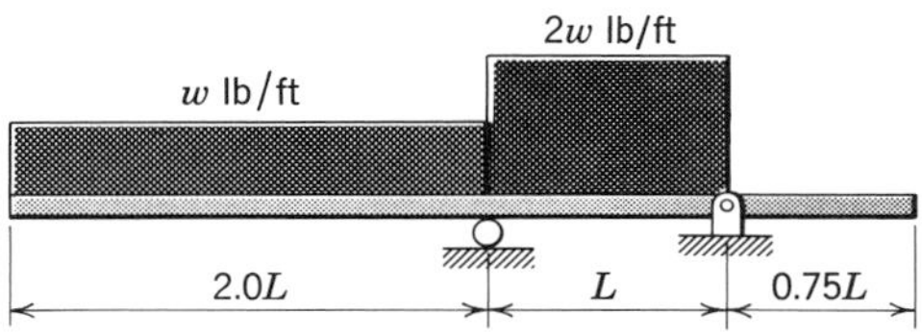


**Fig. P6-100**

**6–101.** Determine, in terms of $P$, $L$, $E$, and $I$, the deflection at the right end of the beam of Fig. P6–101.

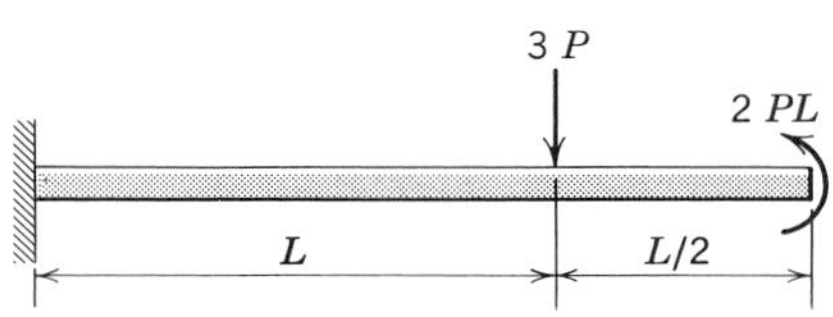


**Fig. P6-101**

**6–102.*** Determine, in terms of $w$ $L$, $E$, and $I$, the deflection at the right end of the beam of Fig. P6–102.

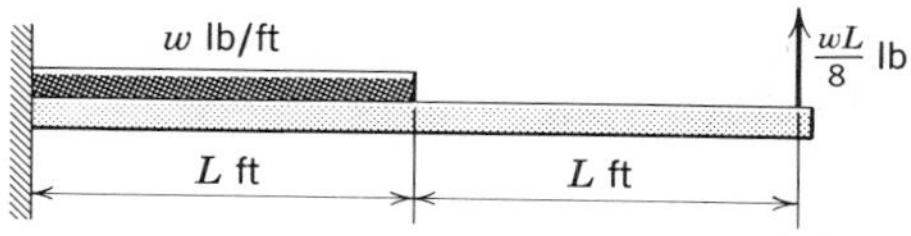


Fig. P6-102

**6–103.** Determine, in terms of $w$, $L$, $E$, and $I$, the deflection at the right end of the beam in Fig. P6–103.

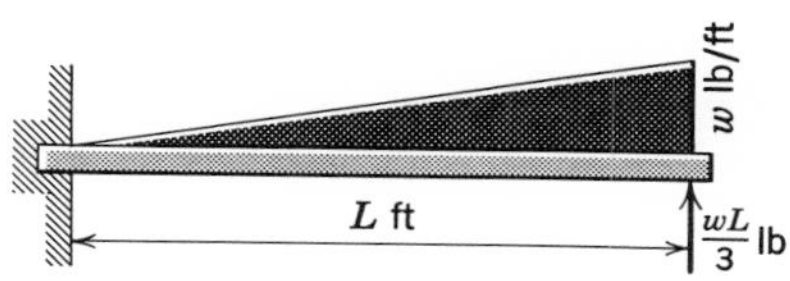


Fig. P6-103

**6–104.*** Determine, in terms of $w$, $L$, $E$, and $I$, the maximum deflection of the beam in Fig. P6–104.

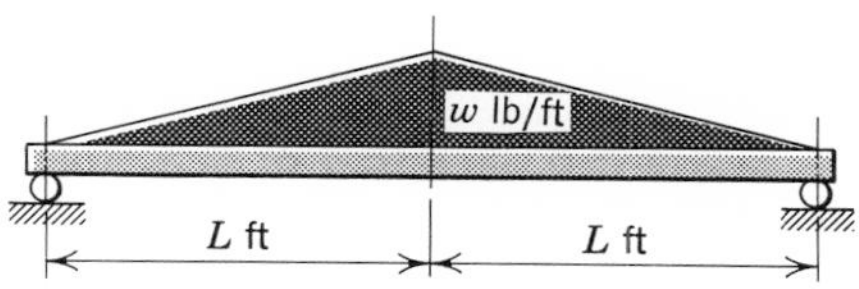


Fig. P6-104

**6–105.** A cantilever beam is loaded as shown in Fig. P6–105. The moment $Q$ is equal to $3PL/4$. Determine the maximum deflection of the beam in terms of $P$, $L$, $E$, and $I$.

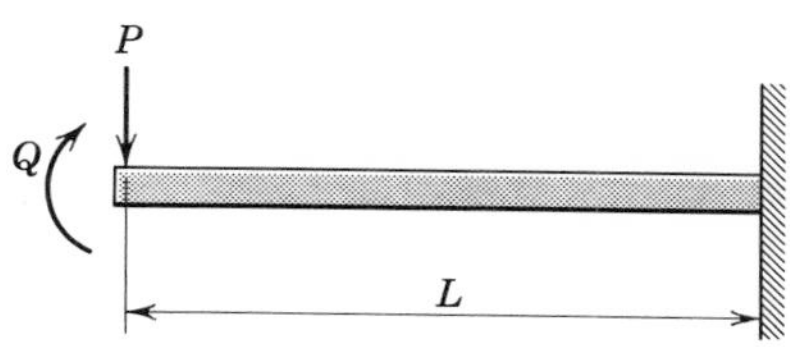


Fig. P6-105

**6–106.** In Fig. P6–106, the force $P$ is applied to a loading yoke which is connected to beam $ABC$ at $B$. The yoke and connection at $B$ are assumed to be rigid. Determine the maximum deflection of the beam in terms of $P$, $L$, $E$, and $I$.

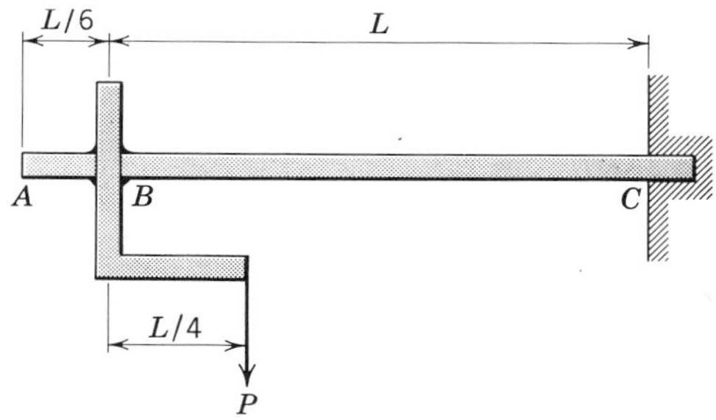


Fig. P6-106

**6–107.** For the beam of Fig. P6–107, determine the maximum deflection in terms of $P$, $L$, $E$, and $I$.

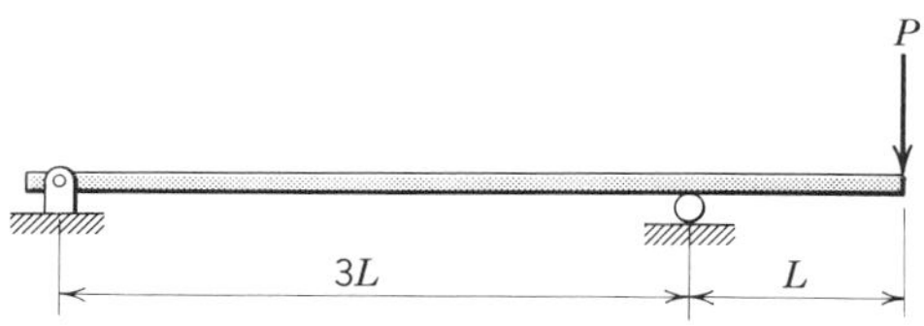


Fig. P6-107

**6–108.*** A couple $M$ is applied to the beam of Fig. P6–108 in the location shown. Determine the maximum deflection in terms of $M$, $L$, $E$, and $I$.

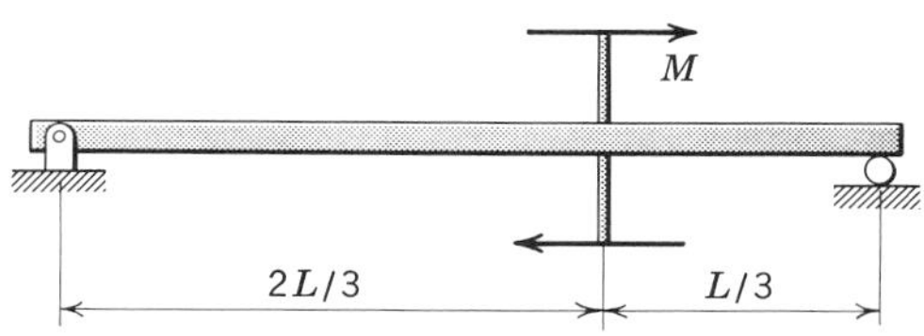


Fig. P6-108

**6–109.** Determine the maximum deflection of the beam of Fig. P6–109 in terms of $w$, $L$, $E$, and $I$.

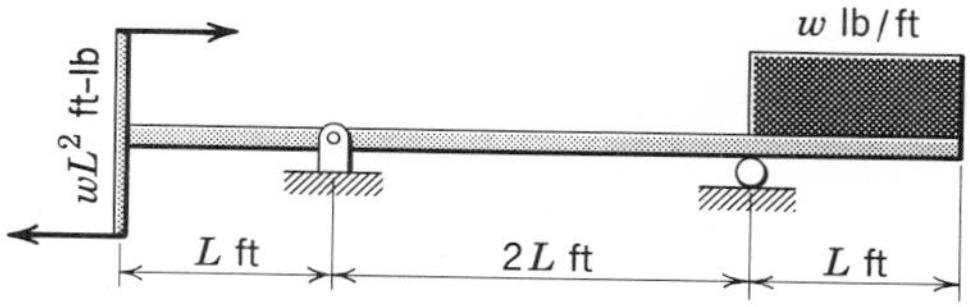


Fig. P6-109

**6–110.*** Determine the maximum deflection of the beam of Fig. P6–110 in terms of $P$, $L$, $E$, and $I$.

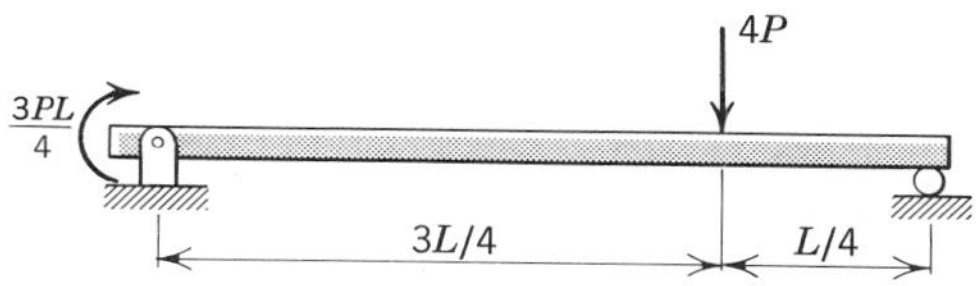


Fig. P6-110

**6–111.*** For the beam of Fig. P6–111, determine the deflection at $B$ in terms of $w$, $L$, $E$, and $I$.

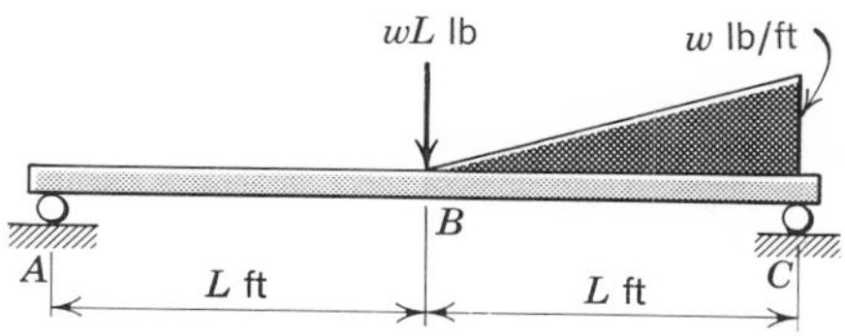


Fig. P6-111

CHAPTER 7

# *Statically Indeterminate Beams*

**7–1. Introduction.** As explained in Art. 5–1, a beam, subjected only to transverse loads, with more than two reaction components is statically indeterminate because the equations of equilibrium are not sufficient to determine all the reactions. In such cases the geometry of the deformation of the loaded beam is utilized to obtain the additional relations needed for evaluation of the reactions (or other unknown forces). This is possible because each additional constraint on a beam provides additional information concerning slopes or deflections. Such information, when utilized with the appropriate slope or deflection equations, yields expressions which supplement the independent equations of equilibrium.

Any of the methods for determining beam deflections outlined in Chap. 6 can be utilized to provide the additional equations needed to solve for the unknown reactions or loads in a statically indeterminate beam. Once the reactions and loads are known, the methods of Chap. 5 and 6 will give the required stresses and deflections.

**7–2. The Integration Method.** For statically determinate beams, known slopes and deflections were used to obtain boundary and matching conditions, from which the constants of integration in the elastic curve equation could be evaluated. For statically indeterminate beams, the procedure is identical. However, the moment equations will contain reactions or loads which cannot be evaluated from the available equations of equilibrium, and one additional boundary condition is needed for the evaluation of each such unknown. For example, if a beam is subjected to a force system for which there are two independent equilibrium equations and if there are four unknown reactions or loads on the beam, two boundary or matching conditions are needed in addition to those necessary for the determination of the constants of integration. These extra boundary conditions, when substituted in the appropriate elastic curve equations (slope or deflection), will yield the necessary additional equations. The following example illustrates the method.

**Example 7–1:** For the beam loaded and supported as shown in Fig. 7–1*a*, determine the reactions in terms of $w$ and $L$.

*Solution:* From the free-body diagram of Fig. 7–1*b* it is seen that there are three unknown reaction components ($M$, $V$, and $R$) and that only two independent equations of equilibrium are available. The additional unknown requires the use of the elastic curve equation, for which one extra boundary condition is required in addition to the two required for the constants of integration. Because three boundary conditions are available in the interval between the supports, only one elastic curve equation need be written. The origin of coordinates is arbitrarily placed at the wall, and, for the interval $0 \leq x \leq L$, the boundary conditions are: when $x = 0$, $dy/dx = 0$; when $x = 0$, $y = 0$; and when $x = L$, $y = 0$. From Fig. 7–1*c* the bending moment equation is

$$EI\,\frac{d^2y}{dx^2} = M_x = Vx + M - \frac{wx^2}{2}.$$

Integration gives

$$EI\,\frac{dy}{dx} = \frac{Vx^2}{2} + Mx - \frac{wx^3}{6} + C_1.$$

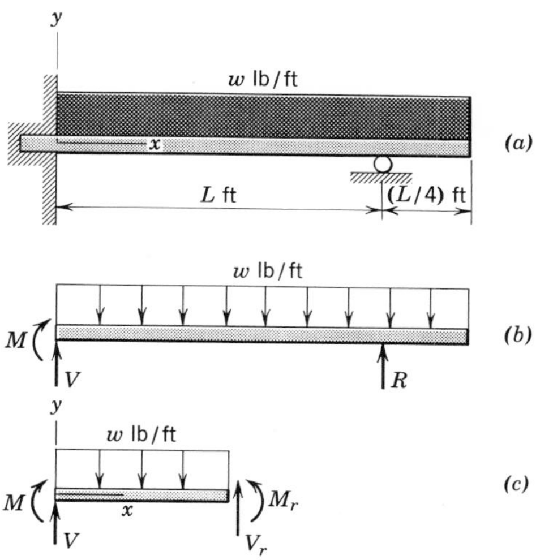


Fig. 7–1

The first boundary condition gives $C_1 = 0$. A second integration yields

$$EIy = \frac{Vx^3}{6} + \frac{Mx^2}{2} - \frac{wx^4}{24} + C_2.$$

The second boundary condition gives $C_2 = 0$, and the last boundary condition gives

$$0 = \frac{VL^3}{6} + \frac{ML^2}{2} - \frac{wL^4}{24},$$

which reduces to

$$4VL + 12M = wL^2.$$

The equation of equilibrium $\Sigma M_R = 0$ for the free-body diagram of Fig. 7–1*b* yields

$$VL + M = w\left(\frac{5L}{4}\right)\left(\frac{5L}{8} - \frac{L}{4}\right) = \frac{15wL^2}{32}.$$

The simultaneous solution of these equations gives

$$M = -\frac{7wL^2}{64} = \underline{\frac{7wL^2}{64} \text{ ft-lb counterclockwise}} \qquad \text{Ans.}$$

and

$$V = \underline{\frac{37wL}{64} \text{ lb upward.}} \qquad \text{Ans.}$$

Finally the equation $\Sigma F_y = 0$ for Fig. 7–1*b* gives

$$R + V = 5wL/4,$$

from which

$$R = \underline{\frac{43wL}{64} \text{ lb as shown.}} \qquad \text{Ans.}$$

## PROBLEMS

**7–1.*** For the beam loaded and supported as shown in Fig. P7–1, determine the left reaction.

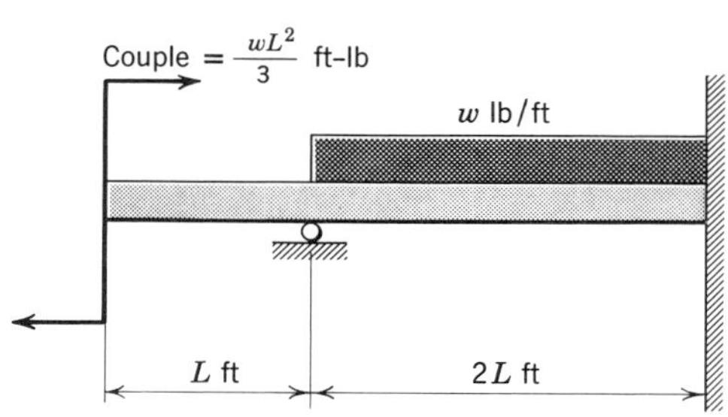


Fig. P7–1

**7–2.*** The beam in Fig. P7–2 has a couple $Q$, applied at the left end, of such magnitude that the left end remains at the same level as the right end. Determine the magnitude of the couple $Q$ in terms of $w$ and $L$.

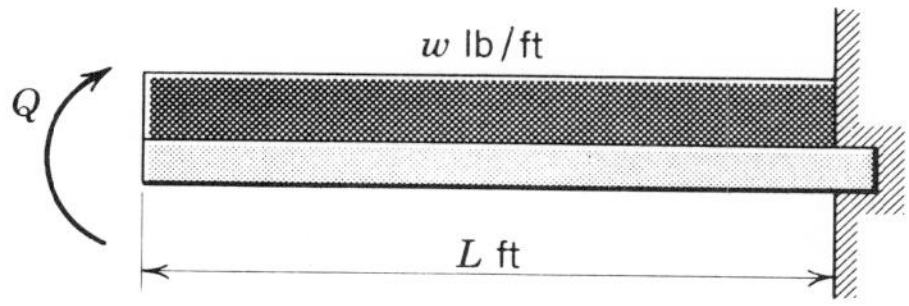


Fig. P7–2

**7–3.*** A beam is supported and loaded as shown in Fig. P7–3.

(*a*) Determine the equation of the elastic curve for the left interval in terms of $Q$, $w$, $L$, $E$, $I$, $x$, and $y$.

(*b*) Determine the value of $Q$ in terms of $w$ and $L$ which would make the slope zero over the right support.

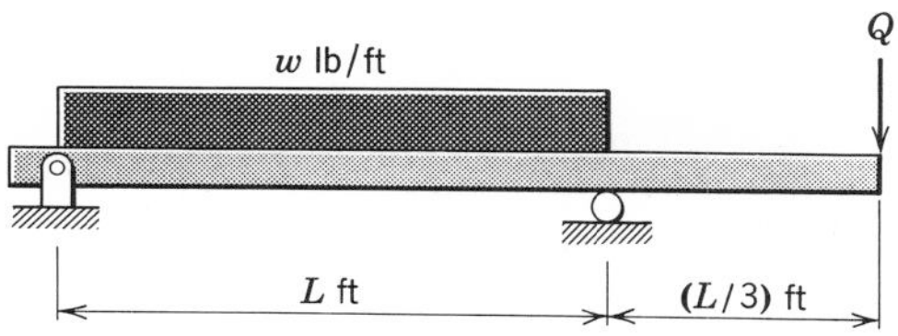


Fig. P7–3

**7–4.*** For the beam in Fig. P7–4, determine the load $P$ which will make the slope at the left support be zero.

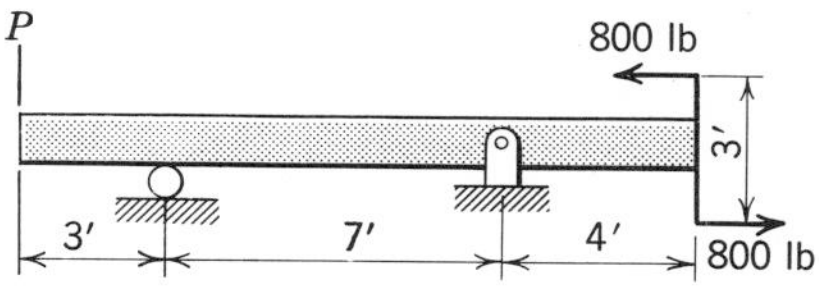


Fig. P7–4

**7–5.** A cantilever beam of length $L$ has a concentrated load $P$ at the free right end. Determine the moment, in terms of $P$ and $L$, that must be applied at the free end in order to make the deflection zero at that end.

**7–6.*** The deflection of the left end of the cantilever beam in Fig. P7–6 is $wL^4/(12EI)$ downward. Determine $Q$ in terms of $w$ and $L$.

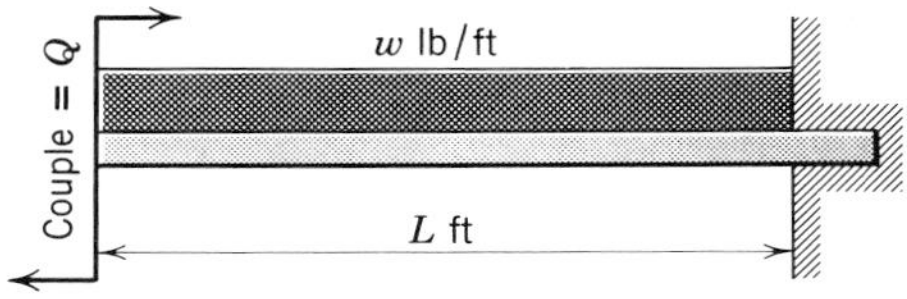


Fig. P7–6

**7–7.** The beam in Fig. P7–7 is partially fixed at the right end. The slope at this point is $wL^3/(36EI)$ upward to the right. Determine the left end reaction in terms of $L$ and $w$.

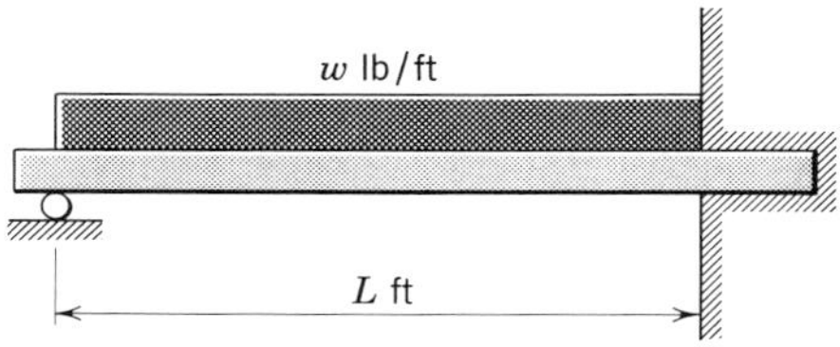


Fig. P7–7

**7–8.*** In Fig. P7–8, the slope of the beam over the support $B$ is $wL^3/(36EI)$ downward to the right. Determine all of the reactions in terms of $w$ and $L$.

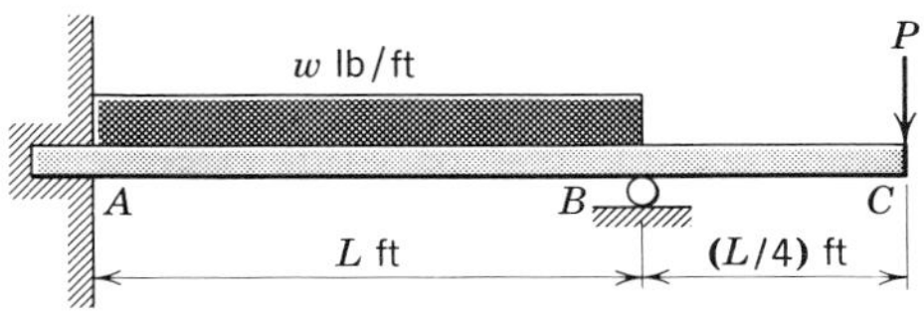


Fig. P7–8

**7–9.** Determine the maximum deflection between the supports for the beam in Fig. P7–9. Express the result in terms of $P$, $L$, $E$, and $I$.

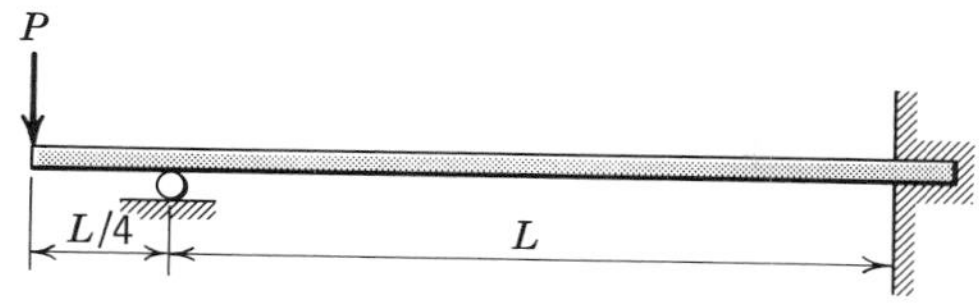


Fig. P7–9

**7–10.** In Fig. P7–9 the load $P$ is 1200 lb; the length $L$ is 8 ft; the modulus of elasticity is 1200 ksi; and the beam section is a 4-in. wide by 6-in. deep rectangle. Determine the maximum deflection between the supports and the maximum fiber and transverse shearing stresses in the beam.

**7–11.*** For the beam in Fig. P7–11, determine:

(*a*) The magnitude of the force $P$, in terms of $w$ and $L$, that will make the slope zero at point $A$.

(*b*) The deflection at point $A$ in terms of $w$, $L$, $E$, and $I$ when $P$ has the value obtained in part *a*.

(*c*) The deflection at the left end in terms of $w$, $L$, $E$, and $I$ when $P$ has the value obtained in part *a*.

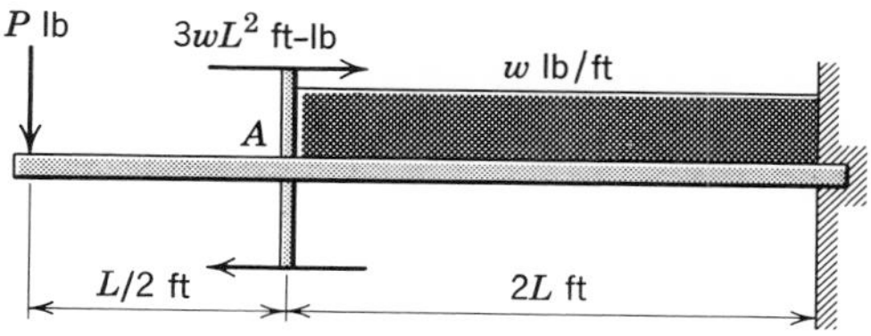


Fig. P7–11

**7–12.** The beam $AB$ of Fig. P7–12 is fixed at the end $B$. Determine the reaction at $A$ in terms of $P$.

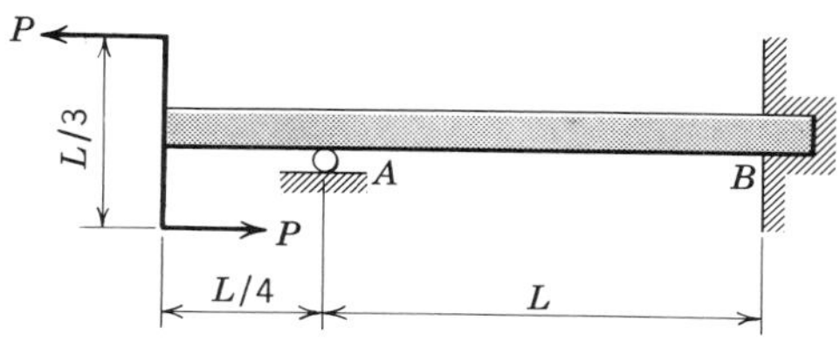


Fig. P7–12

**7–13.*** Solve Prob. 7–12 if the left support settles an amount equal to $PL^3/(12EI)$.

**7–14.** For the beam in Fig. P7–14 determine the moment $M$ in terms of $w$ and $L$ which will cause the slope of the deflected beam over the support at $A$ to be $wL^3/(6EI)$ downward to the left.

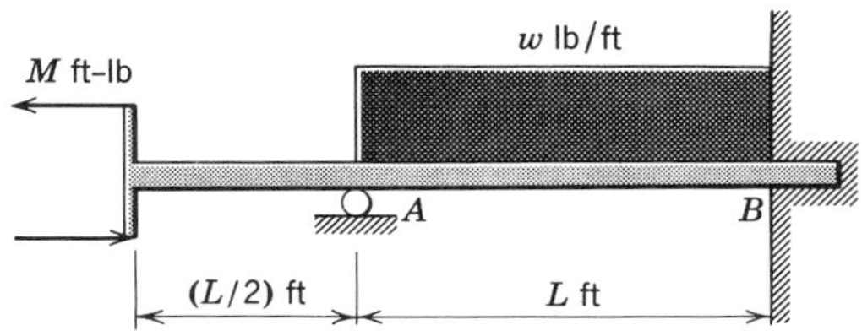


Fig. P7–14

**7–15.*** Determine, in terms of $w$ and $L$, the left reaction on the beam in Fig. P7–15.

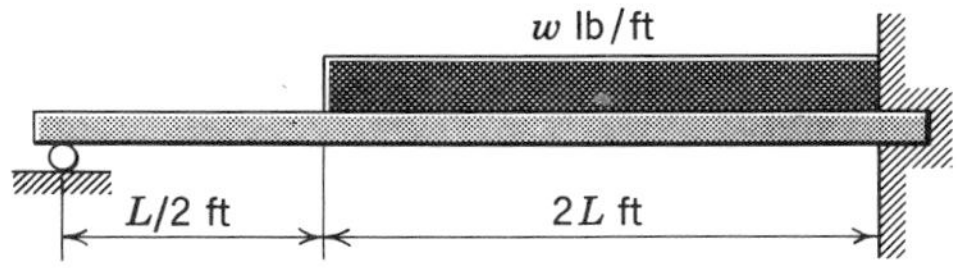


Fig. P7–15

**7–16.** The reaction at $A$ on the beam of Fig. P7–16 is $3P/2$ upward. Determine, in terms of $P$ and $L$, the reaction at $B$ and the moment, $M$, at $C$.

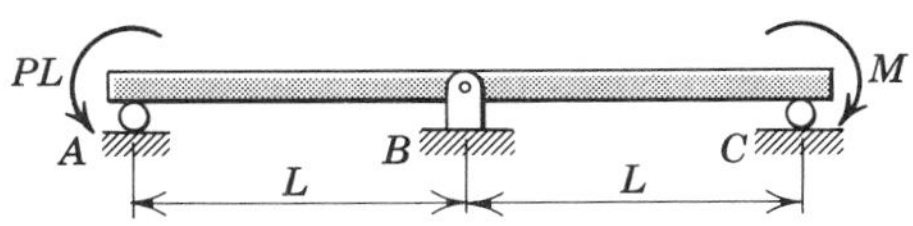


Fig. P7–16

**7–17.*** Determine $P$, in terms of $w$ and $L$, such that the slope of the elastic curve will be zero at $A$ in the beam of Fig. P7–17.

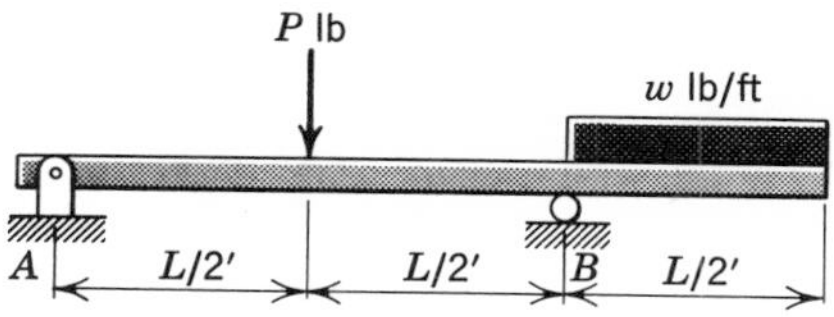


Fig. P7–17

**7–18.** Determine, in terms of $w$ and $L$, the end reactions on the continuous beam of Fig. P7–18.

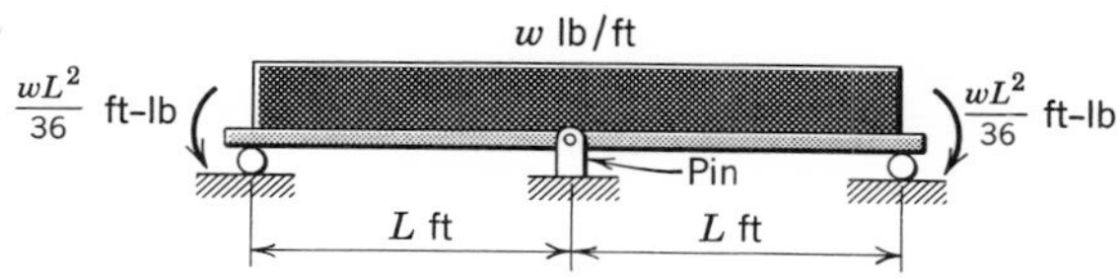


Fig. P7–18

**7–19.*** Determine, in terms of $w$ and $L$, the components of the reaction (shear and moment) at $A$ on the beam $ABC$ of Fig. P7–19.

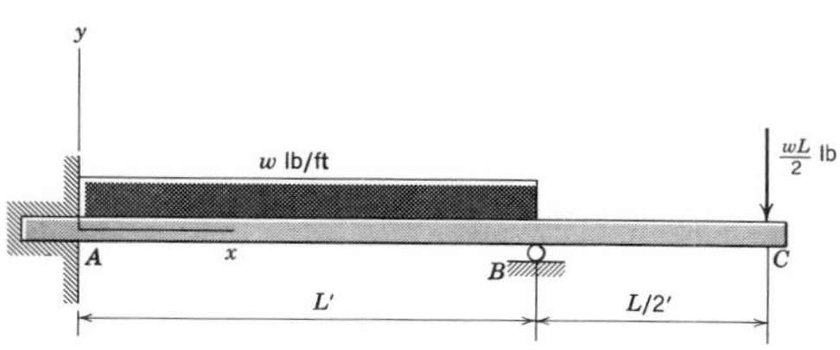


Fig. P7–19

**7–20.** A beam of span $L$ ft is fixed at the left support and restrained at the right support. Under a uniformly distributed load of $w$ lb per ft, the right support settles an amount equal to $wL^4/(48EI)$ and rotates (due to the load on an adjacent span) until the slope of the elastic curve is $wL^3/(60EI)$ downward to the right. Determine the reaction components at the left support.

**7–21.*** Under the loading shown in Fig. P7–21, the right support rotates counterclockwise an amount equal to $wL^3/(48EI)$. Both supports remain on the same level. Determine the shear at the left support in terms of $w$ and $L$.

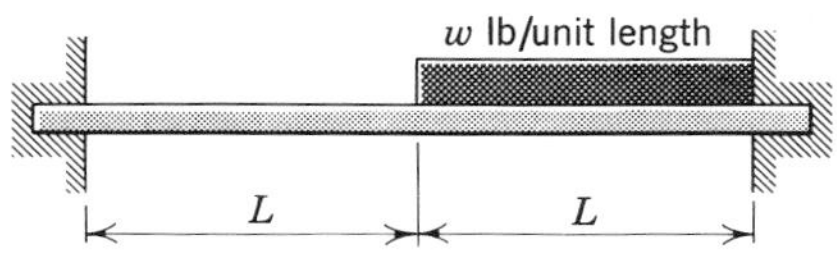


Fig. P7–21

**7–22.** The beam of Fig. P7–22 is simply supported at the left end and is framed into a column at the right end. The beam carries a distributed load which varies linearly from zero to $w$ lb per ft at the column. Under the load, the ends of the beam remain at the same level, but the right end rotates (due to loading the adjacent span) clockwise until the slope of the elastic curve is $wL^3/(24EI)$ downward to the right. Determine the left reaction on the beam in terms of $w$, $L$, $E$, and $I$.

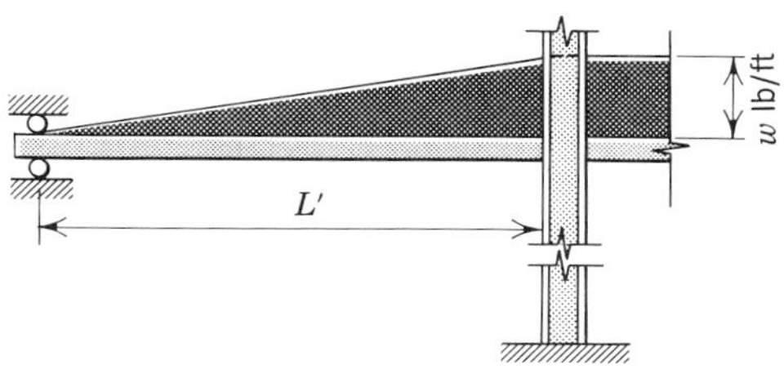


Fig. P7–22

**7–3. The Area-Moment Method.** For statically determinate beams, points with known slopes or deflections were used as tangent points and moment centers for the application of the area-moment theorems.

For statically indeterminate beams, each additional constraint provides information regarding a slope or deflection which makes it possible to write an area-moment equation to supplement the equilibrium equations. The following example illustrates the application of the concepts.

**Example 7–2:** For the beam of Fig. 7–2*a* determine the reactions in terms of $w$ and $L$.

*Solution:* There are four unknown reactions ($V_A$, $M_A$, $V_B$, and $M_B$) on the free-body diagram of Fig. 7–2*b*, and there are only two independent equations of equilibrium. Therefore two area-moment relations are necessary to supplement the equations of equilibrium.

Inspection of the elastic curve (very much exaggerated) of Fig. 7–2*c* reveals that the angle between tangents drawn at $A$ and $B$ is zero. Also the distance from a tangent drawn at $A$ to point $B$ is zero.

These geometrical statements can be expressed mathematically by the area-moment theorems. Thus, since the angle between the tangents at $A$ and $B$ is zero, the area under the moment diagram between $A$ and $B$ must be zero. Also, since the distance from the tangent drawn at $A$ to point $B$ is zero, the

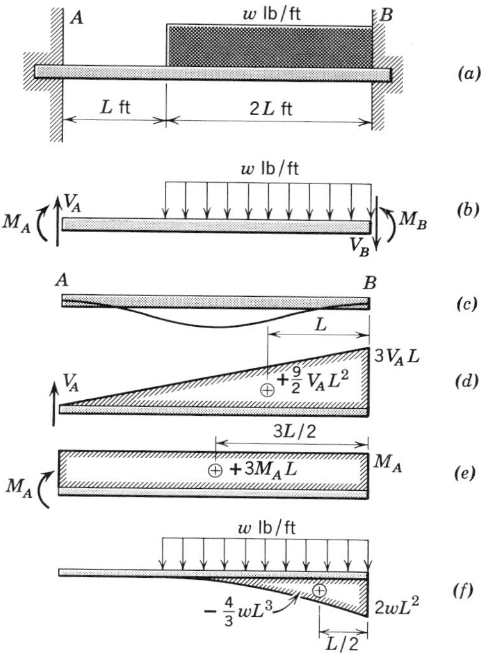


Fig. 7–2

moment of the area under the $M/(EI)$ diagram between $A$ and $B$, with respect to an axis at $B$, is zero. (In this particular problem the moment axis could also have been selected at $A$ since the deflection of $A$ from a tangent drawn at $B$ is also zero.) These two area-moment relations give

$$EI\theta_{AB} = 0 = \frac{9}{2} V_A L^2 + 3M_A L - \frac{4}{3} wL^3,$$

from which $$27V_A L + 18M_A = 8wL^2,$$

and $$EIt_{B/A} = 0 = \frac{9}{2} V_A L^2(L) + 3M_A L\left(\frac{3L}{2}\right) - \frac{4wL^3}{3}\left(\frac{L}{2}\right),$$

from which $$27V_A L + 27M_A = 4wL^2.$$

The simultaneous solution of these equations yields

$$M_A = -\frac{4wL^2}{9} = \underline{\frac{4wL^2}{9} \text{ ft-lb counterclockwise}} \qquad \text{Ans.}$$

and $$V_A = \underline{\frac{16wL}{27} \text{ lb upward.}} \qquad \text{Ans.}$$

With two reactions known, the equations of equilibrium give the other two as

$$M_B = -\frac{2wL^2}{3} = \underline{\frac{2wL^2}{3} \text{ ft-lb clockwise}} \qquad \text{Ans.}$$

and $$V_B = -\frac{38wL}{27} = \underline{\frac{38wL}{27} \text{ lb upward.}} \qquad \text{Ans.}$$

## PROBLEMS

**7–23.*** Given a beam loaded and supported as shown in Fig. P7–23, determine the force $Q$ in terms of $w$ and $L$ which will cause the beam to be horizontal at the right support.

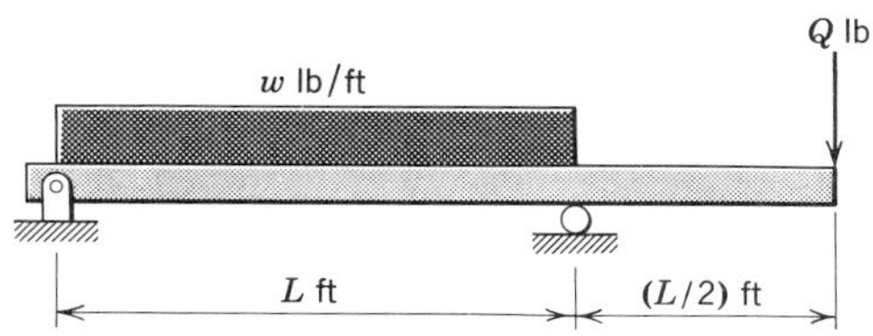


Fig. P7–23

**7–24.*** A cantilever beam of length $L$ ft carries a uniformly distributed load of $w$ lb per ft over the entire span. Determine:

(*a*) The moment in terms of $w$ and $L$ that must be applied at the free (left) end in order to make the slope zero at that end.

(*b*) The maximum deflection of the beam when the moment of part *a* is applied.

**7–25.*** Solve Prob. 7–12 by the area-moment method.

**7–26.** Solve Prob. 7–15 by the area-moment method.

**7–27.** A beam is loaded and supported as shown in Fig. P7–27. Determine:

(*a*) The value of $M$, in terms of $P$ and $L$, for which the slope of the elastic curve will be $PL^2/(36EI)$ downward to the right at the left support.

(*b*) The location, in terms of $L$, of the point of maximum deflection between the supports for the value of $M$ determined from part *a*.

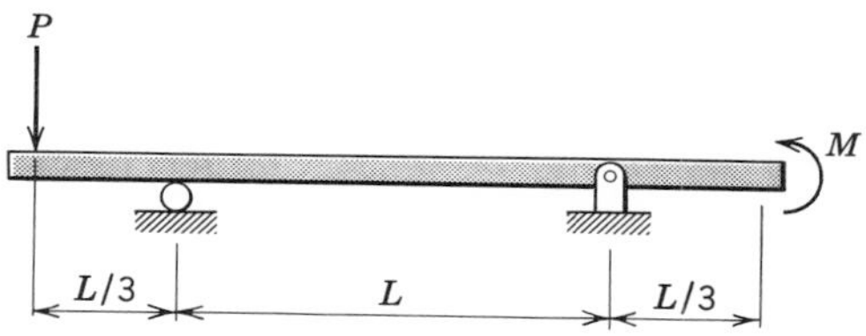


Fig. P7–27

**7–28.*** For the beam in Fig. P7–28, determine:

(*a*) The force $P$ which will give a zero slope of the elastic curve at the right support.

(*b*) The location of the point of maximum deflection between the supports for the value of $P$ as determined in part *a*.

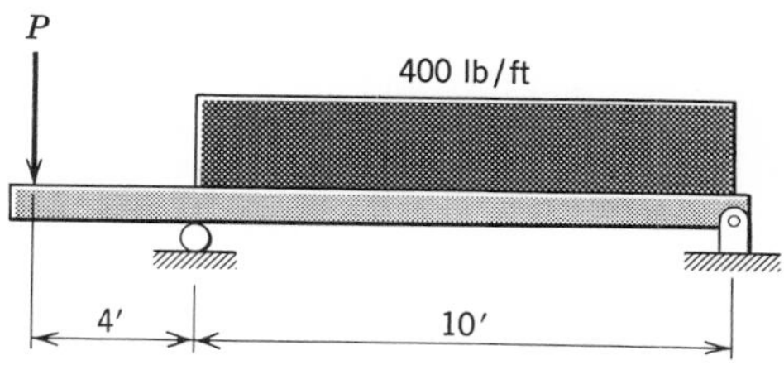


Fig. P7–28

**7–29.** Solve Prob. 7–18 by the area-moment method.

**7–30.*** Determine the reactions on the beam loaded and supported as shown in Fig. P7–30.

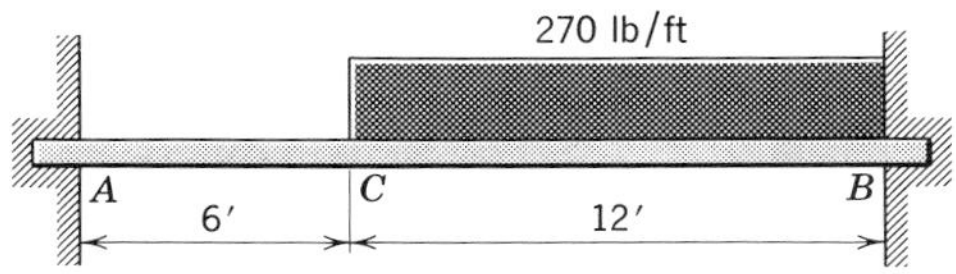


Fig. P7–30

**7–31.** Excessive deflection of a floorbeam may cause the plaster on the ceiling below to crack. A rule of thumb sometimes used to prevent the cracking of plaster is to limit the maximum deflection to 1/360th of the span. If an 8 I 23 steel beam 30 ft long and fixed at each end is used as a floorbeam to support a central concentrated load of 7200 lb, is the plaster likely to crack? Show your calculations.

**7–32.** Compute the dimensions of a timber beam that will support the loads shown in Fig. P7–32, if the width is to be one-half the depth, the flexure stress is not to exceed 1200 psi, and the deflection is not to exceed 0.60 in. The modulus of elasticity of the timber is 1200 ksi.

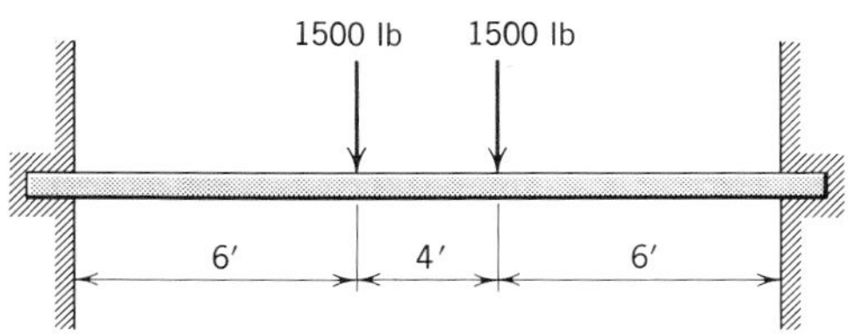


Fig. P7–32

**7–33.*** For the beam in Fig. P7–33 determine:

(*a*) The left reaction in terms of $w$ and $L$.

(*b*) The location, in terms of $L$, of the point of maximum deflection.

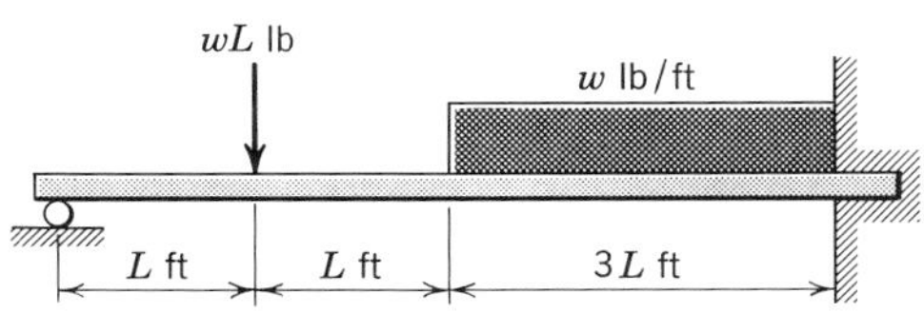


Fig. P7–33

**7–34.** The reaction at $A$ on the beam of Fig. P7–16 is $3P/2$ upward. Determine in terms of $P$, $L$, $E$, and $I$ the maximum deflection in the interval $AB$.

**7–4. The Superposition Method.** The concept (discussed in Art. 6–10) that a slope or deflection due to several loads is the algebraic sum of the slopes or deflections due to each of the loads acting individually is frequently used to provide the deformation equations needed to supplement the equilibrium equations in the solution of statically indeterminate beam problems. The general method of attack is as outlined in Art. 2–5. To provide the necessary deformation equations, selected restraints are removed and replaced by unknown loads (forces and couples); the deformation diagrams corresponding to individual loads (both known and unknown) are sketched, and the component deflections or slopes are summed to produce the known configuration. The following examples illustrate the use of superposition for this purpose.

**Example 7–3:** A steel beam 20 ft long is simply supported at the ends and at the mid-point. Under a load of 400 lb per ft uniformly distributed over the entire beam, the center support is 0.12 in. above the end supports. Determine the reactions. The moment of inertia of the cross section with respect to the neutral axis is 100 in.[4]

*Solution:* With three unknown reactions and only two equations of equilibrium available, the beam is statically indeterminate. Replace the center support with an unknown applied load. The resulting simply supported beam is equivalent to two beams with individual loads as shown in Fig. 7–3. The resulting deflection at the mid-point of the beam is the upward deflection, $y_R$, due to $R_C$ minus the downward deflection, $y_w$, due to the uniform load; that is,

$$y = y_R - y_w = 0.12 \text{ in.} \tag{a}$$

The deflection $y_w$ can be obtained from App. D, case 7, and is

$$y_w = \frac{5wL^4}{384EI} = \frac{5[400(20)][20(12)]^3}{384(30)(10^6)(100)} = 0.48 \text{ in. downward.}$$

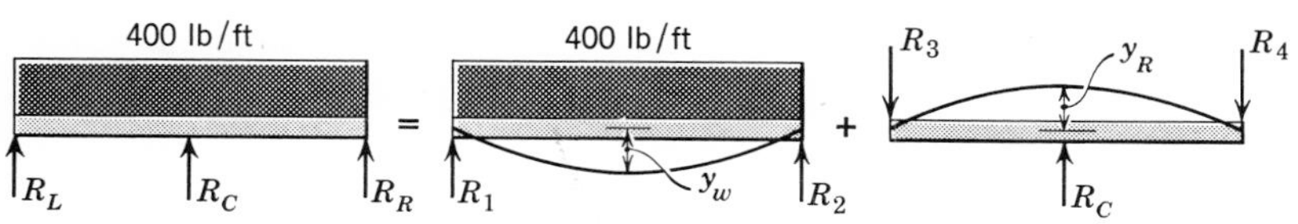


Fig. 7–3

The deflection $y_R$ can be obtained in terms of $R_C$ from App. D, case 6, and is

$$y_R = \frac{R_C L^3}{48EI} = \frac{R_C[20(12)]^3}{48(30)(10^6)(100)} = \frac{96R_C}{10^6} \text{ in. upward.}$$

When these values are substituted in Eq. (a), the result is

$$\frac{96R_C}{10^6} - 0.48 = 0.12,$$

from which

$$R_C = \underline{6250 \text{ lb upward.}} \qquad \text{Ans.}$$

The equilibrium equation, $\Sigma F_y = 0$, and symmetry give

$$R_L = R_R = \underline{875 \text{ lb upward.}} \qquad \text{Ans.}$$

Note that in determining $y_w$, $w$ lb per ft times $L$ ft gives the applied load in pounds. The remaining $L^3$ must be expressed in inches if $E$ and $I$ are in inches and the result is to be in inches. The arithmetic will frequently be simplified if the expressions for the deflections are substituted in the deflection equation in symbol form. In this example Eq. (a) becomes

$$\frac{R_C L^3}{48EI} - \frac{5wL^4}{384EI} = 0.12$$

which reduces (when multiplied by $48EI/L^3$) to

$$R_C - \frac{5wL}{8} = \frac{0.12(48EI)}{L^3}$$

or

$$R_C = \frac{5(400)(20)}{8} + \frac{0.12(48)(30)(10^6)(100)}{[20(12)]^3} = 6250 \text{ lb upward.}$$

**Example 7–4:** Determine all reactions for the beam in Fig. 7–4*a* in terms of $P$, $L$, and $a$.

*Solution:* There are four unknown reactions (a shear and moment at each end), and only two equations of equilibrium are available; therefore the beam is statically indeterminate, and two deformation equations are necessary. Replace the constraint at the right end with an unknown force and couple. The resulting cantilever beam is equivalent to three beams with individual loads as shown in Fig. 7–4*b*. Note that the unknown shear and moment at the right end are both shown as positive values so that the algebraic sign of the result will be correct. From the geometry of the constrained beam, the resultant slope and the resultant deflection at the right end are both zero. The slope and deflection at the end of each of the three replacement beams can be obtained from the expressions in App. D. Thus the first beam with load $P$ (see case 1, App. D) has a constant slope from $P$ to the end of the beam which is

$$\theta_P = -\frac{Pa^2}{2EI}.$$

The deflection, $y_P$, at the end is made up of two parts: $y_1$ for a beam of length $a$, and $y_2$ the added deflection of the tangent segment (straight line) from $P$

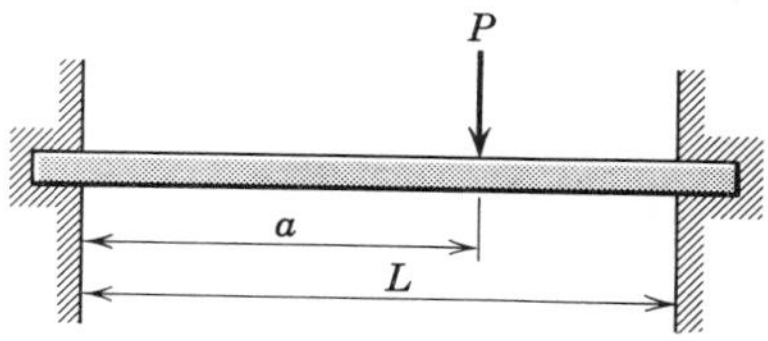


(a)

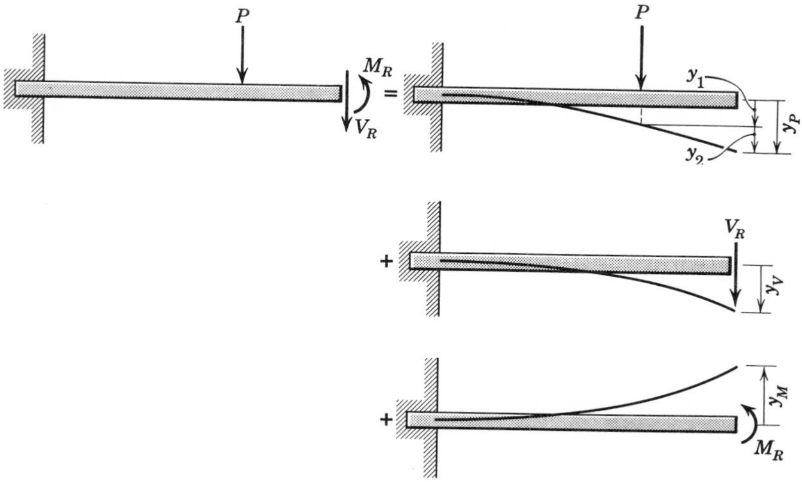


(b)

Fig. 7–4

to the end of the beam. This deflection is

$$y_P = -y_1 - y_2 = -\frac{Pa^3}{3EI} - (L - a)\theta_P$$

$$= -\frac{Pa^3}{3EI} - (L - a)\frac{Pa^2}{2EI} = \frac{Pa^3}{6EI} - \frac{Pa^2L}{2EI}.$$

The slope and deflection at the end of the beam due to the shear $V_R$ are (also from case 1, App. D)

$$\theta_V = -\frac{V_RL^2}{2EI} \quad \text{and} \quad y_V = -\frac{V_RL^3}{3EI}.$$

Finally the slope and deflection at the right end of the beam due to $M_R$ are (see case 4, App. D)

$$\theta_M = \frac{M_RL}{EI} \quad \text{and} \quad y_M = \frac{M_RL^2}{2EI}.$$

Since the resultant slope is zero,

$$\theta_P + \theta_V + \theta_M = -\frac{Pa^2}{2EI} - \frac{V_R L^2}{2EI} + \frac{M_R L}{EI} = 0.$$

Similarly

$$y_P + y_V + y_M = \frac{Pa^3}{6EI} - \frac{Pa^2 L}{2EI} - \frac{V_R L^3}{3EI} + \frac{M_R L^2}{2EI} = 0.$$

The simultaneous solution of these two equations yields

$$M_R = \underline{\frac{-Pa^2(L-a)}{L^2}} \quad \text{and} \quad V_R = \underline{\frac{-Pa^2(3L-2a)}{L^3}}. \qquad \text{Ans.}$$

When the equations of equilibrium are applied to a free-body diagram of the beam, the shear and moment at the left end are found to be

$$M_L = \underline{\frac{-Pa(L-a)^2}{L^2}} \quad \text{and} \quad V_L = \underline{\frac{+P(L^3 - 3a^2L + 2a^3)}{L^3}}. \qquad \text{Ans.}$$

## PROBLEMS

**7–35.*** A 6-in. wide by 10-in. deep timber beam 20 ft long is simply supported at the ends and at the mid-point. Under a uniformly distributed load of 400 lb per ft the middle support settles $\frac{3}{4}$ in. relative to the end supports. The flexural modulus of elasticity of the wood is 1,200,000 psi. Determine all reactions.

**7–36.*** An aluminum beam 24 ft long is simply supported at the ends and at the mid-point. The cross section of the beam is shown in Fig. P7–36. Under a uniformly distributed load of 600 lb per ft the middle support settles one inch more than the end supports. The modulus of elasticity for the aluminum is 10 million psi.

(*a*) Determine the reactions.

(*b*) Determine the maximum fiber stress.

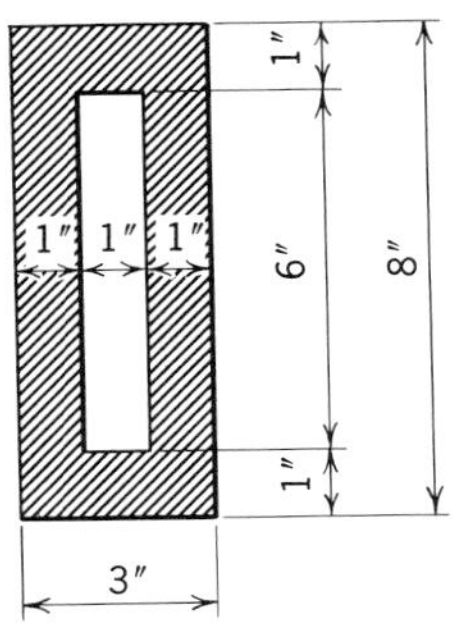


Fig. P7–36

**7–37.*** A timber beam 4 in. wide by 10 in. deep is 12 ft long. It is fixed at the left end and simply supported at the right end. It is subjected to a uniformly distributed load. If the maximum allowable fiber stress is 800 psi and the left support settles without rotation an amount equal to $wL^4/(120EI)$ where $w$ is the load per foot, determine the maximum permissible total load. The flexural modulus of elasticity for this timber is 1200 ksi.

**7–38.*** Determine, in terms of $w$ and $L$, the reaction at $B$ on the beam of Fig. P7–38.

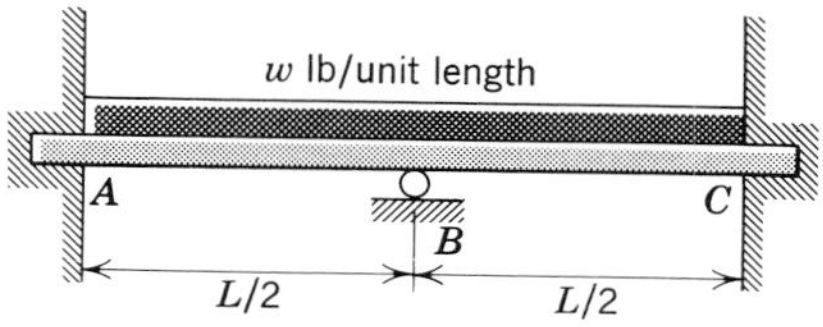


Fig. P7–38

**7–39.** The 4-in. wide by 6-in. deep wooden beam of Fig. P7–39 is fixed at the left end and supported by an aluminum alloy tie rod having a cross-sectional area of 0.15 sq in. at the right end. Use 1200 ksi and 10,000 ksi respectively for the moduli of elasticity of the wood and the aluminum, and determine the tension in the tie rod if it is unstressed before the load is applied.

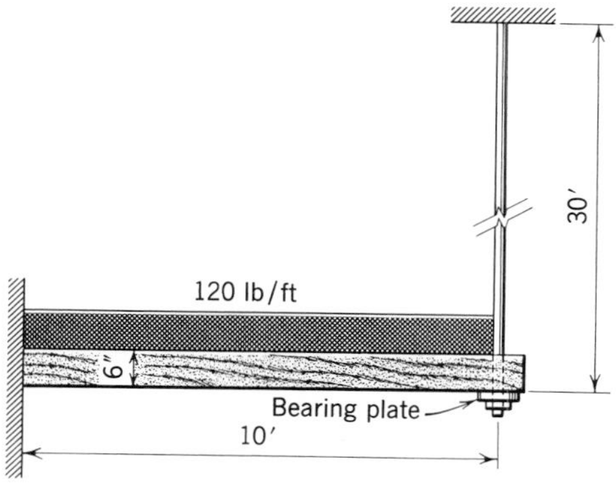


Fig. P7–39

**7–40.*** Solve Prob. 7–39 if, before loading the beam, there is a 0.30-in. gap between the bottom of the beam and the bearing plate.

**7–41.** The tie rod of Prob. 7–39 is replaced by a coil spring placed on a rigid support beneath the right end of the beam with the top of the spring $\frac{1}{4}$ in. below the bottom of the beam. The spring modulus is 600 lb per in. (i.e., the spring deflects 1 in. under a load of 600 lb). Determine the reaction of the spring on the beam.

**7–42.*** The uniformly loaded timber beam of Fig. P7–42 is simply supported at the ends and has a vertical rod support at mid-span as shown. Consider the supports at *A*, *D*, and *C* to be rigid. Construct the shear diagram for the loaded beam. Use 1500 ksi for the modulus of elasticity of the timber. The tension in rod *BD* is zero before the load is applied to the beam.

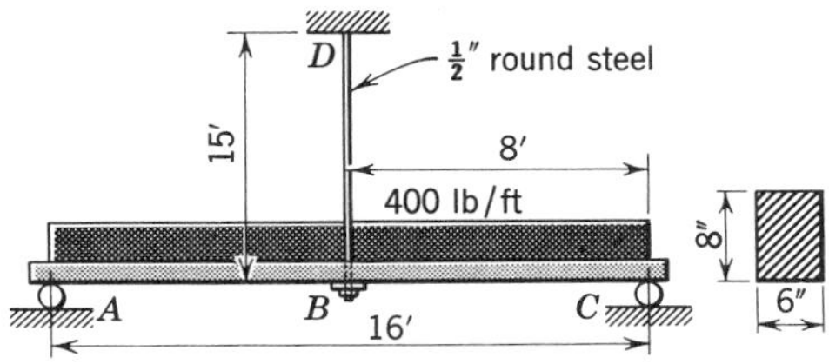


Fig. P7–42

**7–43.** Figure P7–43 shows an 8 WF 31 steel beam supported at the ends and at the center. Consider the end supports and the support at *D* (but not the post) to be rigid. The post *BD* is a 4 by 6-in. timber ($E$ = 1200 ksi) which is braced to prevent buckling. Determine the reaction at *B* due to the load of 480 lb per ft.

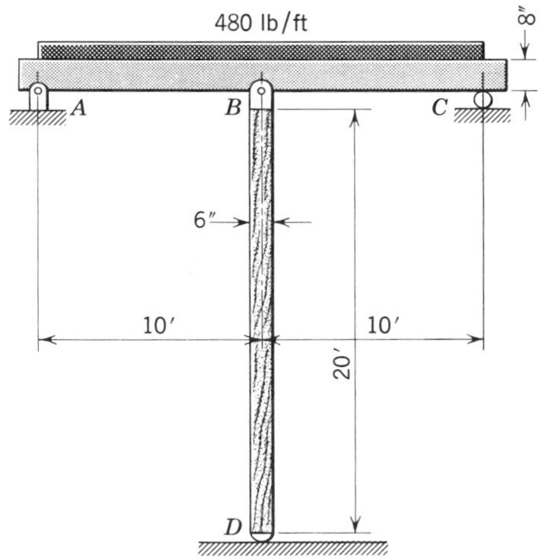


Fig. P7–43

**7–44.*** Solve Prob. 7–43 if the post $BD$ is replaced by two coil springs in series at $B$ (one spring on top of the other and a rigid support under the springs). The spring moduli are 6000 lb per in. and 10,000 lb per in. (i.e., one spring will deflect 1 in. under a load of 6000 lb and the other will deflect 1 in. under a load of 10,000 lb).

**7–45.** The 4-in. aluminum alloy beam of Fig. P7–45 is fixed at the left end, and, when unloaded, there is a 0.72-in. clearance between the right end and the support. The moment of inertia of the beam cross section is 6 in.$^4$ and $E$ may be taken as $10^7$ psi. Determine the maximum positive moment in the beam when the load of 150 lb per ft is applied.

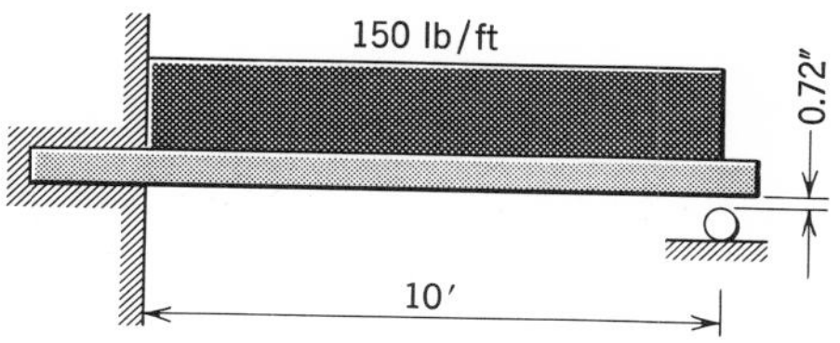


Fig. P7–45

**7–46.*** In Fig. P7–46 beam $A$ is of brass and beam $B$ is of steel. The moment of inertia of $B$ is twice that of $A$, and the modulus of elasticity of the steel is twice that of the brass. The reaction at $C$ is zero before the load $w$ is applied. Determine:

($a$) The reaction at $C$ on beam $B$ in terms of $w$ and $L$.

($b$) The deflection of beam $A$ at $C$ when $L = 12$ ft, $I$ for beam $A = 72$ in.$^4$, and $w = 2000$ lb per ft.

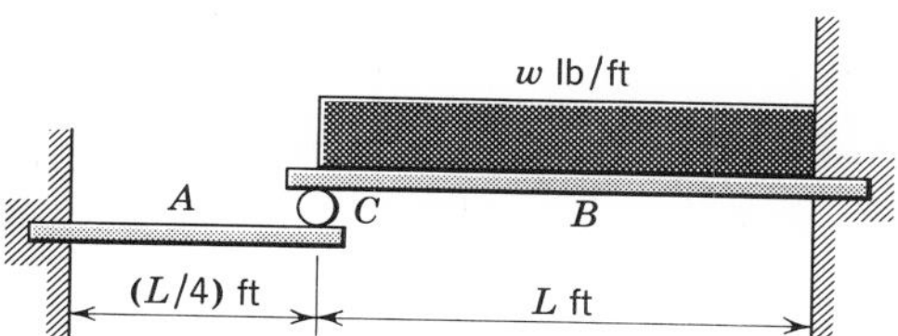


Fig. P7–46

**7–47.** With the beam loaded and supported as shown in Fig. P7–47, the slope at the wall was found to be $wL^3/(72EI)$ upward to the right, and the ends were both on the same horizontal line. Determine the reactions in terms of $w$ and $L$.

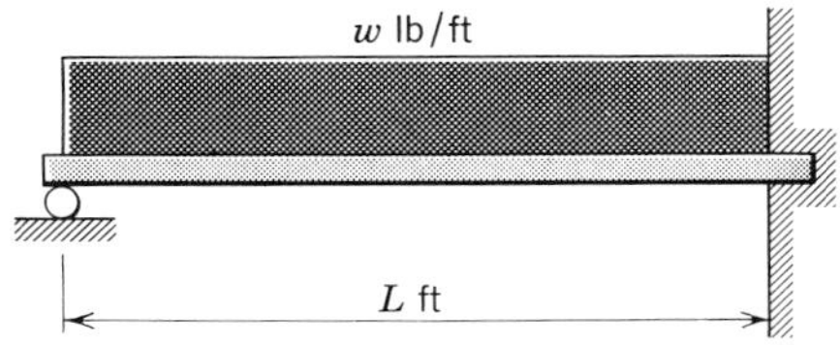


Fig. P7–47

**7–48.*** In Fig. P7–48, the aluminum alloy tie rod passes through a hole in the aluminum alloy cantilever beam and through the coil spring positioned on the end of the tie rod. Before the moment $M$ is applied to the beam, there is 0.05-in. clearance between the beam and the top of the spring. The cross-sectional area of the tie rod is 0.10 sq in., the moment of inertia of the beam cross section is 100 in.$^4$, and the spring modulus is 10,000 lb per in. Determine the axial stress in the tie rod when $M$ has a value of 240,000 in-lb as shown.

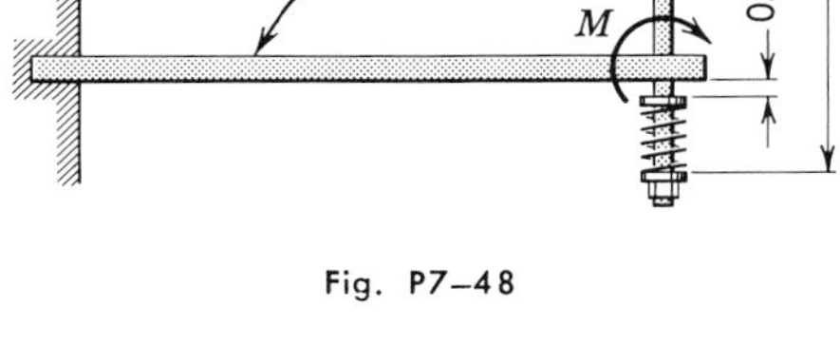


Fig. P7–48

**7–49.** The steel beam of Fig. P7–49 is fixed at the left end and, when loaded, is supported by the helical spring $B$ at the right end. Determine the reaction of the spring $B$ on the beam for the following numerical data: The cross-sectional $I = 48$ in.$^4$, $L = 20$ ft, $w = 200$ lb per ft, $y_O = 0.2$ in. (when the beam is unloaded), spring modulus = 7500 lb per in.

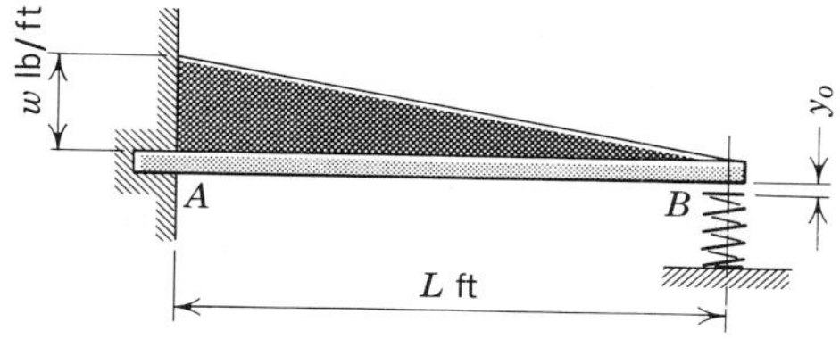


Fig. P7–49

**7–50.*** The steel beam $AB$ of Fig. P7–50 is fixed at the ends $A$ and $B$ and supported at the center by the pin-connected wooden struts $CD$ and $CE$. The cross-sectional area of each strut is 5 sq in. and the modulus of elasticity of the wood is $10^6$ psi. Determine the force in each strut due to the distributed load of 7 lb per in. The constants to be used in the expression for maximum deflection of a beam with fixed ends are $\frac{1}{384}$ for a uniformly distributed load and $\frac{1}{192}$ for a concentrated load at the center.

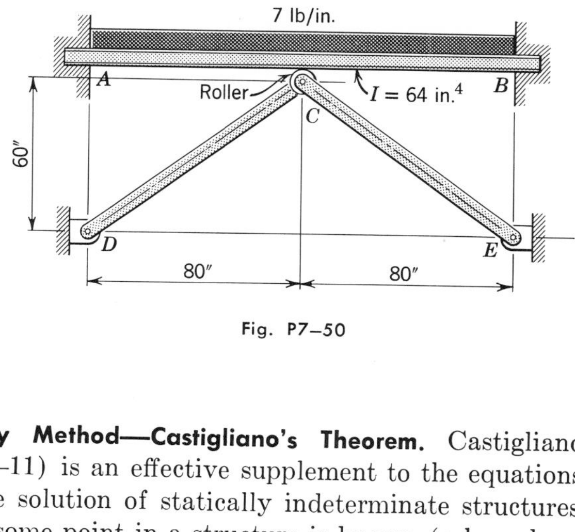


Fig. P7–50

## 7–5. Energy Method—Castigliano's Theorem.

Castigliano's theorem (see Art. 6–11) is an effective supplement to the equations of equilibrium in the solution of statically indeterminate structures. If the deflection at some point in a structure is known (a boundary condition), Eq. (6–5a) may be applied at this point and set equal to the known deflection. The resulting equation will contain one or more unknown forces. If the equation contains more than one unknown, additional equations must be written, such as equilibrium equations or the application of Eq. (6–5b). The method is illustrated in the following examples.

**Example 7–5:** Solve Ex. 7–1 by the application of Castigliano's theorem. The beam is loaded and supported as shown in Fig. 7–5*a*.

*Solution:* The free-body diagram of Fig. 7–5*b* indicates that the problem is statically indeterminate, since there are three unknown reaction components and only two independent equations of equilibrium available.

If the portion of the beam to the right of the reaction $R$ were removed and replaced by an equivalent shearing force and bending moment at the transverse section above $R$, neither $R$ nor the elastic curve in the interval $0 \leq x \leq L$ would be changed. Hence, it is necessary to deal with only the strain energy of the beam in the interval $0 \leq x \leq L$. With the coordinate system placed as shown, the resulting moment equation obtained from the free-body diagram in Fig. 7–5*c* is

$$M = Rx - \frac{w}{2}\left(x + \frac{L}{4}\right)^2$$

$$= Rx - \frac{wx^2}{2} - \frac{wLx}{4} - \frac{wL^2}{32}.$$

From Eq. (6–5*a*), the deflection at the right support is given by

$$y_O = \frac{\partial U}{\partial R} = \frac{1}{EI}\int_0^L M \frac{\partial M}{\partial R}\, dx.$$

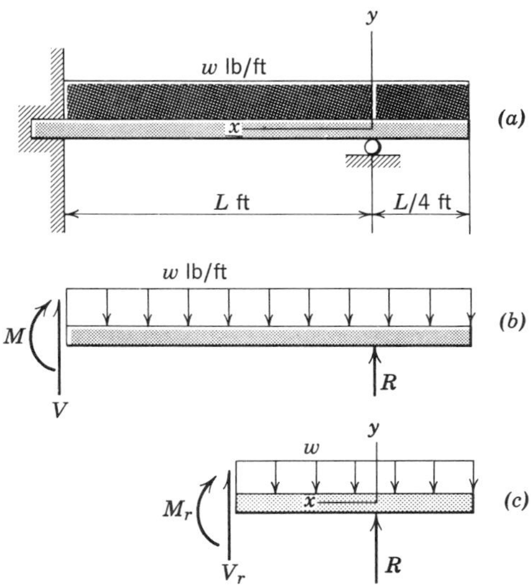


Fig. 7–5

Since the partial derivative of $M$ with respect to $R$ is $x$, the expression for the deflection is

$$y_O = \frac{1}{EI}\int_0^L \left(Rx^2 - \frac{wx^3}{2} - \frac{wLx^2}{4} - \frac{wL^2x}{32}\right) dx,$$

which upon integration and substitution of limits becomes

$$y_O = \frac{1}{EI}\left(\frac{RL^3}{3} - \frac{wL^4}{8} - \frac{wL^4}{12} - \frac{wL^4}{64}\right).$$

Since the support is unyielding ($y = 0$ when $x = 0$), the above expression can be equated to zero and solved for $R$, yielding

$$R = \frac{43wL}{64} \text{ lb upward.} \qquad \text{Ans.}$$

The other reaction components ($M$ and $V$) are obtained from the equilibrium equations, $\Sigma M_A = 0$ and $\Sigma F_y = 0$, and are

$$M = \frac{7wL^2}{64} \text{ ft-lb counterclockwise,} \qquad \text{Ans.}$$

and

$$V = \frac{37wL}{64} \text{ lb upward.} \qquad \text{Ans.}$$

**Example 7–6:** Beam $BC$ of Fig. 7–6*a* is a rolled steel 8-in., 15-lb light beam having a cross-sectional moment of inertia of 48 in.[4] The beam is fixed at $B$ and supported at $C$ by an aluminum alloy tie rod having a cross-sectional area of 0.15 sq in. Determine the force in the tie rod due to the distributed load of 800 lb per ft. Use 30,000 ksi and 10,000 ksi for the moduli of elasticity of the steel and aluminum, respectively.

*Solution:* The free-body diagram of the beam and tie rod is shown in Fig. 7–6*b* with the origin of coordinates at the top of the tie rod where the boundary condition is $x = 0$, $y = 0$. The total strain energy in the system is the sum of the energies in the tie rod and the beam, that is

$$U = \frac{\sigma^2}{2E_a} Al + \int_0^L \frac{M^2}{2E_sI} dx = \frac{P^2l}{2AE_a} + \int_0^L \frac{M^2}{2E_sI} dx.$$

The deflection at $D$ is

$$y_O = \frac{\partial U}{\partial P} = \frac{Pl}{AE_a} + \frac{1}{E_sI}\int_0^L M\frac{\partial M}{\partial P} dx = 0.$$

The moment in the beam is

$$M = Px - \frac{wx^2}{2},$$

and

$$\frac{\partial M}{\partial P} = x.$$

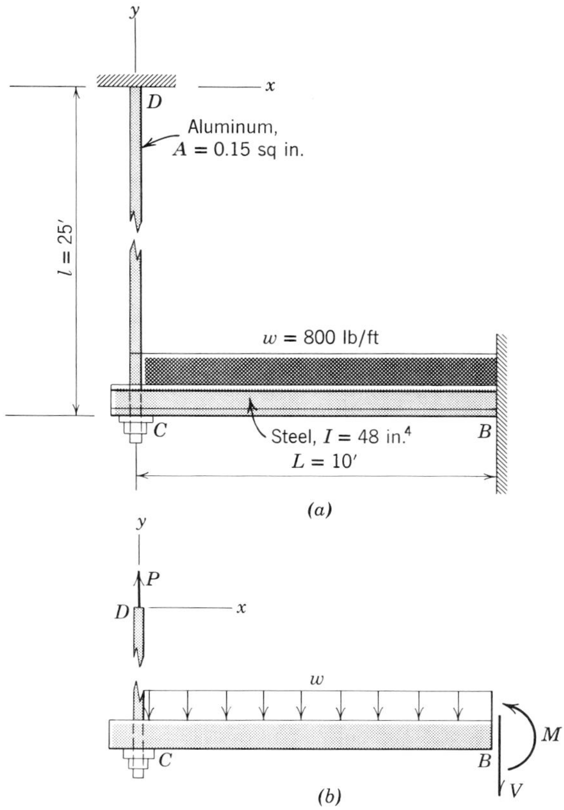


Fig. 7–6

Therefore

$$y_O = \frac{Pl}{AE_a} + \frac{1}{E_sI}\int_0^L \left(Px^2 - \frac{wx^3}{2}\right) dx = \frac{Pl}{AE_a} + \frac{PL^3}{3E_sI} - \frac{wL^4}{8E_sI}.$$

Since the structure is fixed at $D$, $y_O$ is zero and substitution of the given data yields

$$\frac{P(25)12}{15(10^{-2})(10^7)} + \frac{P(10^3)(12^3)}{3(3)(10^7)48} = \frac{800(10)(10^3)(12^3)}{8(3)(10^7)48},$$

from which

$$P = \underline{2000 \text{ lb } T}. \qquad \text{Ans.}$$

**Example 7–7:** Determine the reactions at the left end of the beam of Fig. 7–7$a$ in terms of $P$, $L$, and $a$.

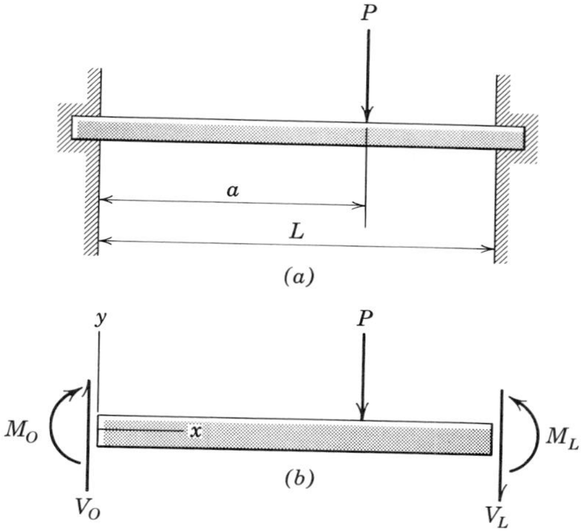


Fig. 7–7

*Solution:* The free-body diagram of Fig. 7–7*b* indicates four unknown reaction components and since there are only two independent equations of equilibrium, two supplementary equations are necessary. These are obtained by applying Eq. (6–5*a*) and (6–5*b*) and solving the resulting simultaneous equations for $M_O$ and $V_O$. With the origin at the left support, the boundary conditions are $y = 0$ and $\theta = 0$ at $x = 0$. Since the moment equation is

$$M = M_O + V_O x - P\langle x - a\rangle^1,$$

the deflection at the left end is given by Eq. (6–5a) as

$$y_O = \frac{\partial U}{\partial V_O} = \frac{1}{EI}\int_0^L M\frac{\partial M}{\partial V_O}\,dx,$$

and

$$EIy_O = \left[\frac{M_O x^2}{2} + \frac{V_O x^3}{3} - P\left\{\frac{x}{2}\langle x - a\rangle^2 - \frac{1}{6}\langle x - a\rangle^3\right\}\right]_0^L$$

where the last term was integrated by parts. Substitution of the limits gives

$$EIy_O = \frac{M_O L^2}{2} + \frac{V_O L^3}{3} - \frac{PL}{2}(L - a)^2 + \frac{P}{6}(L - a)^3.$$

The application of the boundary condition $y = 0$ at $x = 0$ gives

$$\frac{M_O L^2}{2} + \frac{V_O L^3}{3} - \frac{PL^3}{3} + \frac{PL^2 a}{2} - \frac{Pa^3}{6} = 0. \tag{a}$$

When Eq. (6-5a) is applied, the rotation at the left end is given by the expression

$$\theta_O = \frac{\partial U}{\partial M_O} = \frac{1}{EI}\int_0^L M\frac{\partial M}{\partial M_O}\,dx,$$

and

$$EI\theta_O = \int_0^L (M_O + V_O x - P\langle x - a\rangle^1)\,dx$$

$$= \left[M_O x + \frac{V_O x^2}{2} - \frac{P}{2}\langle x - a\rangle^2\right]_0^L,$$

which reduces to

$$M_O L + \frac{V_O L^2}{2} - \frac{PL^2}{2} + PaL - \frac{Pa^2}{2} = 0. \tag{b}$$

Equations (a) and (b) are solved to obtain the following results,

$$\underline{V_O = P - \frac{3Pa^2}{L^2} + \frac{2Pa^3}{L^3},} \qquad \text{Ans.}$$

and

$$M_O = \underline{-Pa + \frac{2Pa^2}{L} - \frac{Pa^3}{L^2}.} \qquad \text{Ans.}$$

## PROBLEMS

**The following problems are to be solved using Castigliano's theorem.**

**7–51.*** Solve Prob. 7–2.

**7–52.*** Solve Prob. 7–6.

**7–53.*** Solve Prob. 7–1.

**7–54.*** Solve Prob. 7–3*b*.

**7–55.*** Solve Prob. 7–8.

**7–56.*** Solve Prob. 7–14.

**7–57.** Solve Prob. 7–13.

**7–58.*** Solve Prob. 7–19.

**7–59.** Solve Prob. 7–22.

**7–60.** Solve Prob. 7–15.

**7–61.** Solve Prob. 7–33*a*.

**7–62.** Solve Prob. 7–39.

**7–63.** Solve Prob. 7–44.

**7–64.** Solve Prob. 7–48.

**7–65.*** Solve Prob. 7–49.

**7–6. Plastic Analysis.** Because of the increasing emphasis on structural analysis based upon plastic action, a brief introduction to plastic analysis of statically indeterminate beams is presented below. The analysis will be limited to ideal elastoplastic materials as defined in Art. 2–1. For practical purposes ordinary structural steel will be satisfactory, and for this material the following assumptions are reasonable and may be verified (within practical limits) by experiment:

1. As the load on the beam increases, the fiber stress will increase proportionately until the yield point for the material is attained at the section of maximum moment after which yielding at this section will progress over the entire section (see Art. 5–12 and Fig. 5–29*d*). The resisting moment will increase with the load until the stress in each fiber has reached the yield point after which the moment will remain constant. In other words, a hinge exerting a constant moment forms at the section producing a discontinuity in the slope of the elastic curve. The moment at the hinge is termed the *plastic moment,* $M_p$.

2. As the loading is increased, a second hinge will form; and additional hinges will form until the beam is reduced to what has been termed a *mechanism,* that is, a system of members, connected at their ends by pins (each pin exerting a constant moment), which will continue to deflect with no additional load.

The load which produces a mechanism is considered the *ultimate load* although strain hardening of the material might necessitate the application of a larger load to produce complete collapse. The hinge, which actually extends over a finite length of beam, is assumed to form at a discrete point in the beam.

One procedure for analyzing statically indeterminate beams consists of arbitrarily assigning the plastic moment value to the redundant moments; which procedure makes the beam statically determinate. Additional hinges are introduced at points of high moment until a mechanism develops. Finally, a check of the moments should show that nowhere does the moment exceed the magnitude of the plastic moment, $M_p$. This is not the only method used in plastic analysis[1]; however, for the simpler cases it is probably the easiest to apply. The method is illustrated by the following example.

**Example 7–8:** The continuous beam of Fig. 7–8*a* is a steel 36 WF 160 section. Consider the steel to be elastoplastic with a yield point of 36 ksi, and determine the ultimate load, $w$, (which for span $BC$ includes the weight of the beam) for the beam.

*Solution:* The free-body diagram of Fig. 7–8*b* indicates three unknown reactions, and for this force system there are only two independent equations of

[1] For additional material see *Plastic Design of Steel Frames,* L. S. Beedle, John Wiley and Sons, New York, 1958.

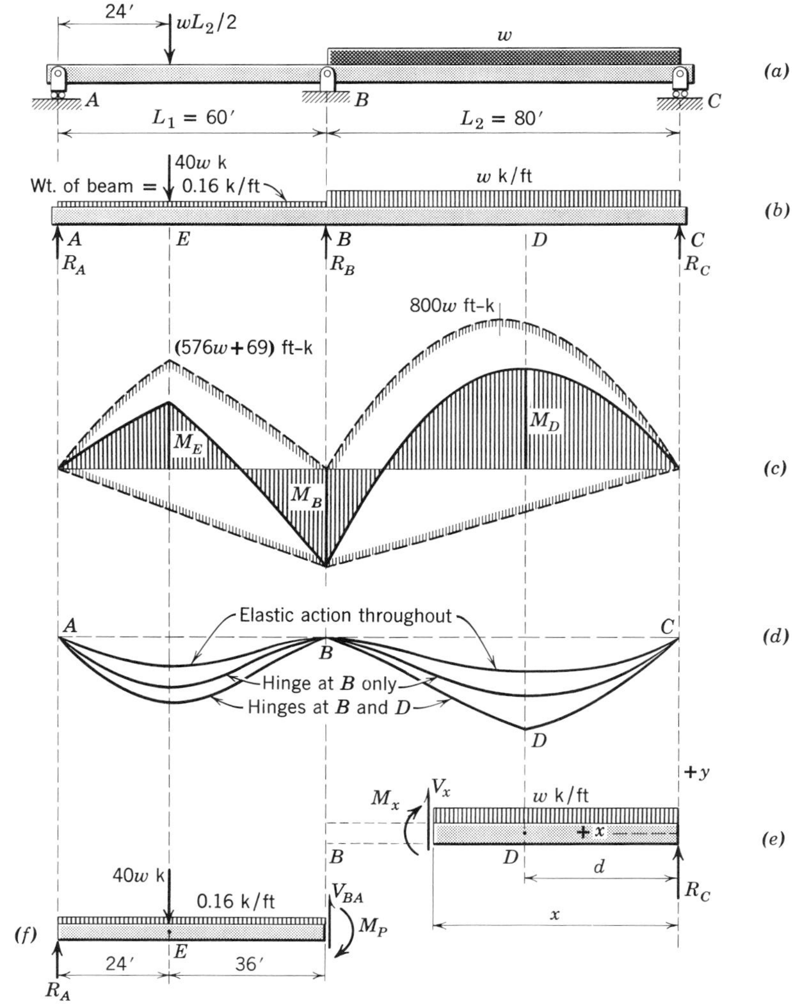


Fig. 7–8

equilibrium; hence, there is one redundant quantity. If one moment, shear, or reaction is determined, the beam becomes statically determinate. However, the ultimate load is not attained until a mechanism forms; therefore it becomes necessary to assume more than one possible hinge location. To assist in estimating the location of possible hinges, the shape of the moment diagram by parts and the composite moment diagram are shown in Fig. 7–8c, where the solid line indicates the composite diagram and the dashed lines represent the diagrams for two simple beams with a moment at $B$. This diagram indicates that hinges may possibly form at $B$, $D$, and $E$. Possible deformations are indicated (greatly exaggerated) in Fig. 7–8$d$, the top curve indicating the shape of the elastic curve when the action is entirely elastic. As the magnitude of the loading is increased beyond that necessary to produce the yield point stress in the outer fibers, a hinge will form at the section where the flexural stress is the highest. The first hinge is assumed to form at $B$, and under this condition the beam would assume the shape indicated by the middle curve of Fig. 7–8$d$. Since a mechanism has not yet formed, the loading is increased until a second hinge forms. This hinge is assumed to form at $D$, and from the lower curve of Fig. 7–8$d$ it is apparent that a mechanism has developed since, with hinges at $B$ and $D$ and with end $C$ supported on rollers, deflection can continue in the span $BC$ without an increase in the load. The second hinge may have formed at $E$ instead of at $D$; hence, the moment at $E$ must be evaluated as a check on the hinge assumptions.

The magnitude of the load necessary to produce hinges at $B$ and $D$ will first be determined, after which the moment at $E$ will be checked. It should be noted that, when the plastic moment, $M_p$, is used in either an equilibrium or a moment equation, it must have the correct sense or sign. With reference to the free-body diagram of Fig. 7–8$e$, which represents the right portion of span $BC$, the equation for the moment in span $BC$ is

$$M_x = R_C x - wx^2/2. \qquad \text{(a)}$$

When $x$ is 80 ft, $\quad M_B = -M_p = R_C(80) - w(80)^2/2,$

from which $\quad R_C = 40w - M_p/80. \qquad$ (b)

The negative sign on $M_B$ is obtained from Fig. 7–8$c$. The section of maximum positive moment in span $BC$ is at $D$ where the shear is zero and is located from the shear equation

$$V_x = -R_C + wx.$$

When $x$ is equal to $d$, $\quad V_x = 0 = -R_C + wd,$

from which $\quad d = R_C/w. \qquad$ (c)

From Eq. (a) the moment at $D$ (the plastic moment) is

$$M_D = +M_p = R_C d - wd^2/2.$$

Since $R_C$ is equal to $wd$ from Eq. (c), the plastic moment becomes

$$M_p = wd^2 - \frac{wd^2}{2} = \frac{w}{2}\left(40 - \frac{M_p}{80w}\right)^2, \qquad \text{(d)}$$

where the plus sign on $M_D$ is obtained from Fig. 7–8$c$ and $R_C$ is obtained from Eq. (b). Solving Eq. (d) for $w$ gives

$$w = \frac{5.8284M_p}{3200} \quad \text{or} \quad \frac{0.1716M_p}{3200}.$$

When these expressions are substituted into Eq. (b), only the larger value yields a positive result for $d$; therefore this expression will give the correct magnitude for $w$ unless the bending moment at $E$ exceeds $M_p$. In Art. 5–12 the plastic moment for the specified section was computed to be $616\sigma$, which amounts to 616(36)/12, or 1848 ft-kips, for this problem. Substituting this value for $M_p$ in the expression for $w$ yields

$$w = 5.828(616)(3)/3200 = 3.366 \text{ kips per ft.}$$

To check the bending moment at $E$, the reaction at $A$ will first be obtained by summing moments with respect to $B$ on the free-body diagram of Fig. 7–8$f$. The equation

$$60R_A - 40w(36) - 0.16(60)(30) + M_p = 0$$

gives $$R_A = 24(3.366) + 4.80 - 616(3)/60 = 54.8 \text{ kips.}$$

The bending moment at $E$ is

$$M_E = 54.8(24) - 0.16(24)(12) = 1269 \text{ ft-kips,}$$

which is less than 1848; therefore a hinge does not form at $E$, the moment in the beam is nowhere greater than $M_p$, and the ultimate load is

$$w = \underline{3.37 \text{ kips per ft.}} \qquad \text{Ans.}$$

Some economy in the design could be effected by reducing the size of the beam in the left span to such a section that a hinge might form at $E$; that is, a beam section could be selected for which the plastic moment is at least 1269 ft-kips. If a table of plastic section moduli were available,[2] the section could be determined from the expression, $M_p = Z_p\sigma_{yp}$, where $Z_p$ is the plastic section modulus.[3] If such a table is not available, the following method may be used. It has been shown[4] that the ratio $Z_p/Z$ (where $Z$ is the elastic section modulus) for WF and I-sections ranges from 1.10 to 1.23 with 1.14 as the average for all WF sections and 1.18 as the average for I-sections. Using the average value of 1.14 gives

$$Z = \frac{Z_p}{1.14} = \frac{M_E}{1.14\sigma_{yp}} = \frac{1269(12)}{1.14(36)} = 371 \text{ in.}^3.$$

From App. C, a 33 WF 130 with a section modulus of 404.8 appears to be a likely choice, and it will be checked as outlined in Art. 5–12. When the cross-sectional dimensions given in App. C are used, the plastic moment is

$$[\overbrace{11.51(0.855)(16.12)}^{\text{one flange}} + \overbrace{15.69(0.580)(15.69/2)}^{\text{one-half web}}][2][36/12] = 1379 \text{ ft-kips,}$$

which is more than adequate. Therefore one could splice the 33 WF 130 to the 36 WF 160 at some section between $B$ and $E$. Theoretically, the splice could be made where the bending moment in span $AB$ has a value of $-1379$ ft-kips (which is approximately 5.3 ft left of $B$). Practically, to avoid splicing at a

[2] See App. 2 in *Plastic Design of Steel Frames*, L. S. Beedle, John Wiley and Sons, New York, 1958.

[3] The plastic modulus is the arithmetic sum of the absolute values of the first moments, with respect to the neutral axis, of the cross-sectional areas on each side of the neutral axis.

[4] See Fig. 7.1 of reference 2.

section of large moment, the joint might be located some distance to the left, the distance depending on the judgement of the designer or a pertinent specification. The resulting beam will develop hinges at $B$ and $D$ and will just begin to yield at $E$ under a load of 3.37 kips per ft.

The maximum value for $w$, based on elastic analysis and with the assumption that the $I_{AB}$ for the full length of 60 ft is $(7/10)I_{BC}$, is found to be 2.76 kips per ft; thus, the plastic analysis for this problem gives a loading approximately 22 per cent higher than that obtained from elastic analysis. Part of this increase is due to the plastic moment being greater (about 14 per cent) than the elastic moment, and part is due to the increase in load necessary to cause formation of the second plastic hinge. These loads would be considered failure loads which, when divided by an appropriate factor of safety,[5] will yield working loads.

The discussion presented in this article is very limited in scope and, hence, does no more than present some of the fundamental concepts of plastic design involving an ideal elastoplastic material. A complete analysis would involve the rotation capacity of the beam sections used, to ascertain if the first hinge to form will sustain sufficient plastic deformation to permit the formation of successive hinges. Furthermore, shearing stresses or local buckling may be quite significant for some designs. These and other design problems are considered to be beyond the scope of this book.

## PROBLEMS

**Note: The following problems are to be solved by plastic analysis, and for simplification the weight of the beam is to be neglected. All beams are of structural steel having a yield point and proportional limit of 36 ksi. For WF beam design, use $Z_p = 1.14\,Z$.**

**7–66.*** A two-span continuous beam with simple supports is subjected to a uniformly distributed load of 1 kip per ft on both spans. The span lengths are 32 ft and 40 ft respectively, and the beam is to have a constant section throughout its length. Determine the lightest WF beam required if a factor of safety of 2 is specified.

**7–67.*** A 24 WF 76 ($Z_p = 200$ in.$^3$) is used as a two-span simply supported continuous beam with spans of 30 ft and 50 ft respectively. The beam carries uniformly distributed loads of $2w$ kips per ft on the 30-ft span and $w$ kips per ft on the 50-ft span. Determine the ultimate load $w$ for the beam.

**7–68.** The beam of Prob. 7–67 will develop a plastic hinge at the intermediate support when the value of $w$ is 2.80 kips per ft. For this loading, determine the lightest WF beam required for the 30-ft span, and locate the theoretical splice point.

[5] Here, the factor of safety is defined as the ratio of the ultimate (failure) load to the computed (or specified) applied load. This definition, which differs from that of Art. 2–4, will be used in all plastic analysis problems in this book.

**7–69.*** The beam of Fig. P7–69 is a 12 WF 40 ($Z_p$ = 57.6 in.$^3$), and the span, $L$, is 24 ft. Determine the magnitude of the ultimate load $w$.

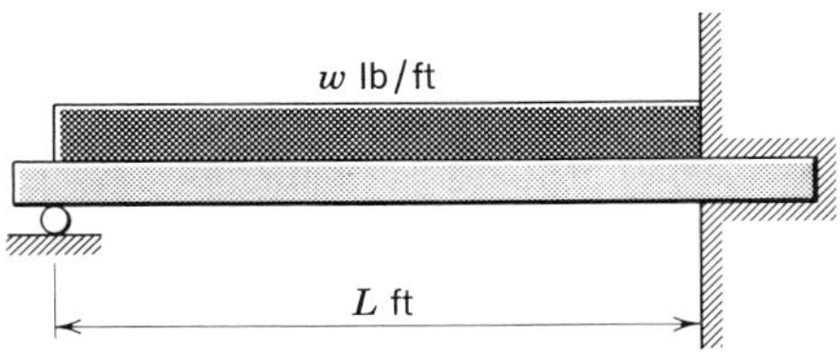


Fig. P7–69

**7–70.** The beam of Fig. P7–69 has a span, $L$, of 20 ft; it is to carry a load of 2.5 kips per ft with a factor of safety of 2. Determine the lightest WF section required.

**7–71.*** The beam of Fig. P7–71 is a 24 WF 76 ($Z_p$ = 200 in.$^3$). Determine the ultimate load $w$.

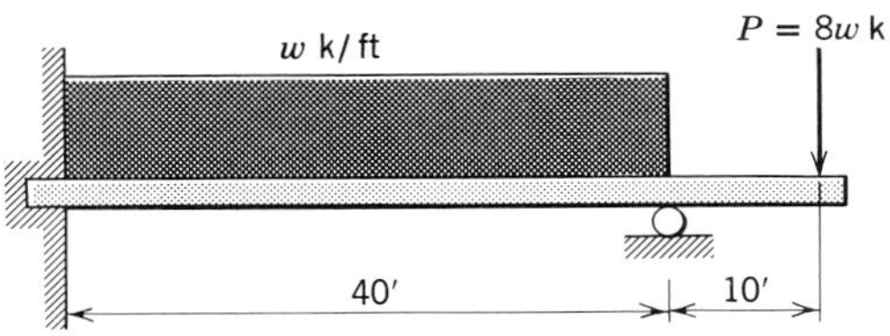


Fig. P7–71

**7–72.*** The continuous beam of Fig. P7–72 is a 12 WF 27 ($Z_p$ = 38 in.$^3$). Determine the ultimate load $w$.

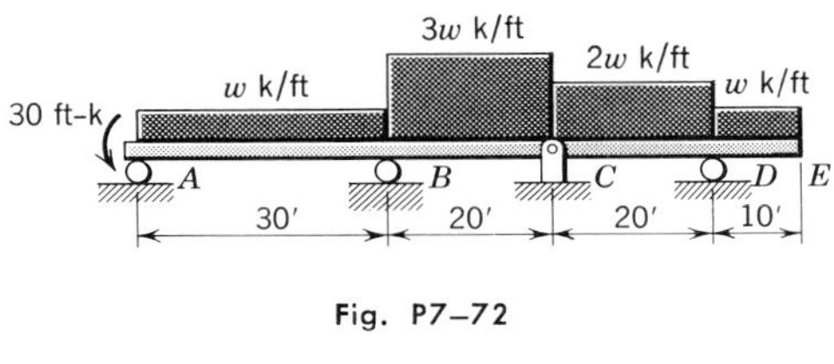


Fig. P7–72

**7–73.** Under a load $w$ of 1.52 kips per ft applied as shown in Fig. P7–72, a plastic hinge forms at $C$. Determine the lightest WF section required for the portion of the beam from near $C$ to $E$. Locate the theoretical splice point.

**7–7. Closure.** In this chapter were presented four methods of elastic analysis and one method of plastic analysis of statically indeterminate beams. When the analysis is to be based on elastic action, the choice of method will be governed by the particular problem and the analyst's familiarity with certain of the four methods presented. As a possible aid to selection of the method, the following outline is submitted:

1. The integration method is the fundamental method and may be used to solve any problem although its application may sometimes be laborious. This method requires that, in addition to the boundary conditions necessary for the determination of constants of integration, there be additional boundary (or matching) conditions equal in number to the difference between the number of unknowns and the number of independent equations of equilibrium.

2. The area-moment method provides a neat and somewhat less laborious technique to supplement or replace the integration approach. The method requires the application of an area-moment theorem for each unknown in excess of the number of independent equations of equilibrium. The area-moment equations to be written will be dictated by the geometry of the deflected beam. If the expressions for the areas and centroids of the moment diagrams are not known, the areas and their first moments may be obtained by integration. The method is applicable to any problem and is readily applied to a beam with variable moment of inertia by drawing an $M/I$ diagram and subdividing it into finite areas.

3. For many problems the superposition method provides the most rapid solution. The application of this method depends on the availability of a summary (or a memorization) of slope and deflection expressions for various spans and loadings and, except for the simpler cases, depends on the ingenuity of the analyst.

4. The energy method is preferred by many, and is frequently much shorter than other methods. The method presented in this book requires one or more applications of Castigliano's theorem, Eq. (6–5a) or (6–5b), at points where the deflections or slopes are known. The use of singularity functions to provide a single moment equation for the entire span is frequently advantageous.

The following miscellaneous problems are presented to provide practice in the selection of the method best adapted to a particular problem.

## PROBLEMS

**7–74.*** A cantilever beam of length $L$ carries a uniformly distributed load of $w$ lb per ft and an upward concentrated force $P$ lb at the free end such that the deflection of the free end is $wL^4/(9EI)$ downward. Determine $P$.

**7–75.** A beam 12 ft long is simply supported at the ends and carries a uniformly distributed load of 150 lb per ft over its entire length. A third support is added at mid-span. This support is $4L^3/(EI)$ lower than the end supports, where $L$ is the total length (12 ft). Determine the amount the end reactions are changed by adding this third support.

**7–76.*** A cantilever beam is loaded as shown in Fig. P7–76. Determine $Q$ in terms of $w$ and $L$ such that the beam will have zero slope at $x = L/3$.

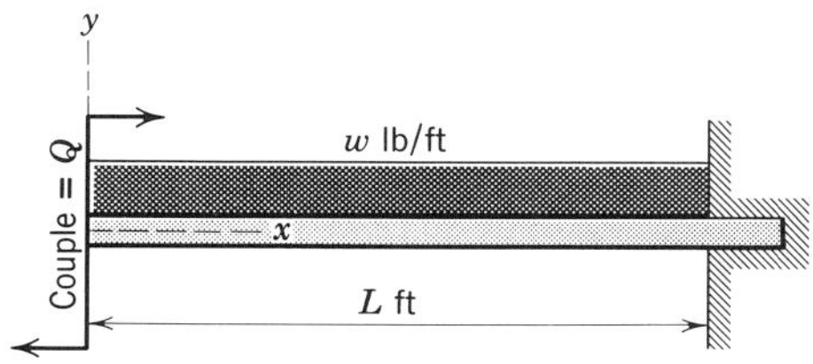


Fig. P7–76

**7–77.** The bending moment at support $B$ in the continuous beam of Fig. P7–77 is $-wL^2/8$. The load $P$ is large enough to insure an upward reaction at $C$. Determine:

(*a*) The shear to the left of support $C$ in terms of $w$ and $L$.
(*b*) The maximum deflection in span $BC$ in terms of $w$, $L$, $E$, and $I$.

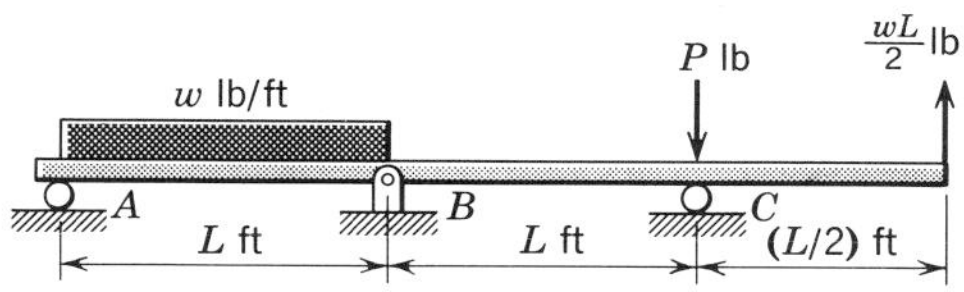


Fig. P7–77

**7–78.*** The bending moment at support $B$ in the continuous beam of Fig. P7–78 is $-wL^2/60$. Determine:

(a) The shear to the right of support $B$ in terms of $w$ and $L$.

(b) The maximum deflection in the span $BC$ in terms of $w$, $L$, $E$, and $I$.

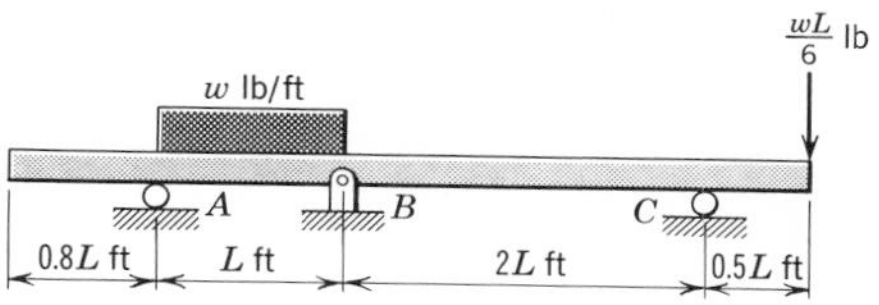


Fig. P7–78

**7–79.** For the beam of Prob. 7–78, determine:

(a) The reaction at $A$ in terms of $w$ and $L$.

(b) The deflection at the extreme left end in terms of $w$, $L$, $E$, and $I$.

**7–80.*** The slope at $A$ of the beam of Fig. P7–80 is $wL^3/(36EI)$ downward to the right. Determine the reaction on the beam at $A$ in terms of $w$ and $L$.

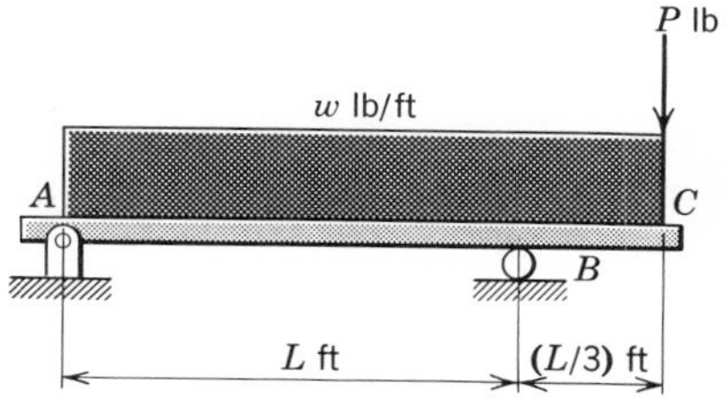


Fig. P7–80

**7–81.** In Fig. P7–81 the steel beam $DE$ and the aluminum alloy ($E = 10^7$ psi) beam $AC$ are pinned together by the rigid strut $BD$ and loaded as shown. Determine the reaction at $B$ on beam $AC$.

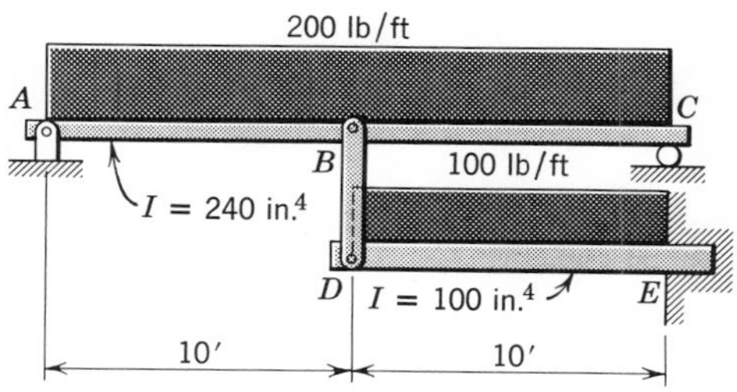


Fig. P7–81

**7–82.*** A 100-in. long steel beam is supported at the right end by a helical spring having a modulus of 3000 lb per in. and fixed at the left end. The beam supports a uniformly distributed load of 20 lb per in. over the entire span and $I = 24$ in.$^4$ Determine the deflection of the spring.

**7–83.** Solve Prob. 2–55 with $EF$ and $CD$ treated as steel beams with $I = 40$ in.$^4$ and $ED$ an aluminum alloy ($E = 10^7$ psi) member with a cross-sectional area of 2 sq in. and a length of 3 in.

CHAPTER 8

# *Combined Static Loading —Dynamic Loading*

**8–1. Introduction.** The stresses and strains produced by three fundamental types of loads (centric, torsional, and flexural) have been analyzed in the preceding chapters. Also combinations of centric and torsional loads were presented in Art. 4–4. Many machine and structural elements are subjected to a combination of any two or all three of these types of loads, and a procedure for calculating the stresses on a given plane at a point is required. One method is to replace the given force system by an equivalent system of forces and couples, each of which is selected in such a manner that the stresses produced by it can be calculated by methods already developed. The composite effect can be obtained by the principle of superposition as explained in Art. 2–5 provided that the combined stresses do not exceed the proportional limit. Various combinations of loads that can be analyzed in this manner are discussed in the following articles.

**8–2. Centric and Flexural Loads Combined.** A common example of a centric load combined with a bending load is a structural or machine part with a force acting parallel to but not along the axis of the member, for example, the vertical portion $AB$ of the loading hook of Fig. 8–1. Such an eccentric load can be replaced by a centric load (a force acting along the centroidal axis of the part $AB$) and one or more couples. If the couples are parallel or perpendicular to a plane of symmetry, the procedures developed in Chap. 5 can be applied directly.[1]

[1] For sections having no plane of symmetry see *Advanced Mechanics of Materials*, 2nd ed., F. B. Seely and J. O. Smith, John Wiley and Sons, New York, 1952, pp. 113–119.

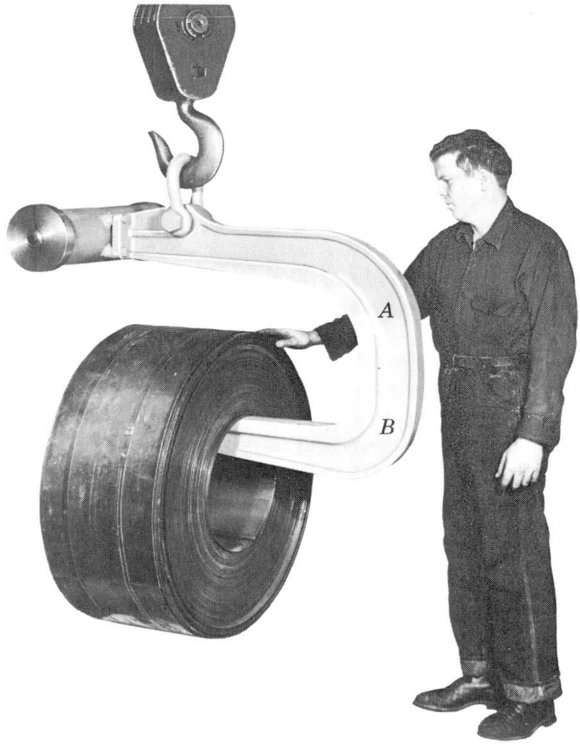


Fig. 8–1. 5-Ton Littell Coil Loading Hook. Courtesy: Steel Founders Society of America

The normal stress at any point on a transverse plane through $AB$ is the algebraic sum of the normal stresses due to the centric load and that due to the bending load (the couples). Since the transverse shear is zero, this stress will be a principal stress. The procedure is illustrated in the following example.

**Example 8–1:** The cast iron frame of a small press is shaped as shown in Fig. 8–2*a*. Figure 8–2*b* represents the cross section *a–a*, and the moment of inertia of the area is 1000/3 in.$^4$ with respect to the centroidal axis *c–c*. For a load $P$ of 16 kips, determine:

(*a*) The normal stress distribution on section *a–a*.

(*b*) The principal stresses for the critical points on section *a–a*.

*Solution:* (*a*) When the free-body diagram of the top portion of the frame was drawn, two equal opposite collinear forces $P$ through the centroid of the section *a–a* were added as shown in Fig. 8–2*c*. The force $P_1$ is a centric load

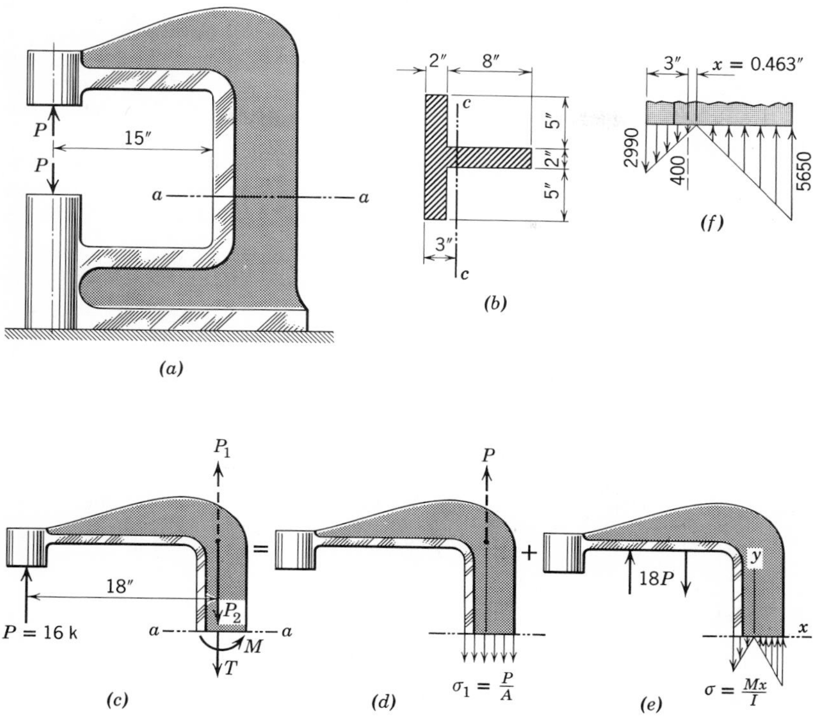


Fig. 8–2

equal to the resisting force $T$ on section $a$–$a$. The other two forces $P$ and $P_2$ constitute an $18P$-in-kip couple equal to the resisting moment, $M$, on section $a$–$a$.

The axial stress uniformly distributed over section $a$–$a$, as shown in Fig. 8–2$d$, is

$$\sigma_1 = \frac{P}{A} = \frac{16{,}000}{40} = 400 \text{ psi } T.$$

The maximum tensile fiber stress occurs at the left edge of the section $a$–$a$ and is

$$\sigma_2 = \frac{Mc}{I} = \frac{16{,}000(18)(3)}{1000/3} = 2592 \text{ psi } T.$$

The maximum compressive fiber stress occurs at the right edge of the section $a$–$a$ and is

$$\sigma_3 = \frac{Mc}{I} = \frac{16{,}000(18)(7)}{1000/3} = 6048 \text{ psi } C.$$

The distribution of fiber stresses is shown in Fig. 8–2*e*.

When the principle of superposition is applied, the total normal stress at any point on section *a–a* is the algebraic sum of the component stresses at the point. When the stress distributions of Fig. 8–2*d* and *e* are added, the resultant stress distribution for section *a–a* is as shown in Fig. 8–2*f*, which is the answer for part *a*.

(*b*) With no shearing stresses on section *a–a*, the normal stresses are principal stresses, and the critical points, as observed from the stress distribution of Fig. 8–2*f*, are the left and right edges of the section. The principal stresses for the right edge are

$$\sigma_p = \underline{5650 \text{ psi } C} \quad \text{and} \quad \underline{0}. \qquad \text{Ans.}$$

The principal stresses for the left edge are

$$\sigma_p = \underline{0} \quad \text{and} \quad \underline{2990 \text{ psi } T}. \qquad \text{Ans.}$$

Although not requested in this example, the location of the neutral axis (line of zero stress) can be determined as the place where the compressive fiber stress is 400 psi because this stress will just balance the axial tensile stress of 400 psi. The distance on Fig. 8–2*f* can be obtained by writing

$$400 = \frac{16{,}000(18)x}{1000/3},$$

from which

$$x = 0.463 \text{ in.}$$

to the right of the centroidal axis of the cross section.

**Example 8–2:** A gray cast iron compression member is subjected to a vertical load $P$ as shown in Fig. 8–3*a*. The allowable normal stresses on plane $ABCD$ are 20,000 psi $C$ and 5000 psi $T$. Determine the maximum allowable load $P$.

*Solution:* When two vertical forces $P$ (labeled $P_1$ and $P_2$ for clarity) of opposite sense are introduced at the center of the top section, $EFGH$, the eccentric force $P$ is replaced by a centric force $P_2$ and a couple $PP_1$ as shown in Fig. 8–3*b*.

The stresses produced by the centric load are easily calculated; however, the couple does not act in a plane of symmetry of the member, and the flexural stresses produced are not readily computed for this orientation. This difficulty can be eliminated by replacing the resultant couple $PP_1$ by its components in or parallel to planes of symmetry of the compression member. When another pair of vertical forces equal to $P$ (labeled $P_3$ and $P_4$ for clarity) of opposite sense are introduced at the center of the edge $EH$, the couple $PP_1$ is replaced by two component couples $PP_3$ and $P_4P_1$ as shown in Fig. 8–3*c*. The component couple $P_4P_1$ acts in a plane of symmetry, and the flexural stresses produced are given by the flexure formula. Since no transverse shearing stresses occur to cause twisting (see Art. 5–13), the flexural stress due to couple $PP_3$ can also be determined by means of the flexure formula even though $PP_3$ is not in a plane of symmetry.

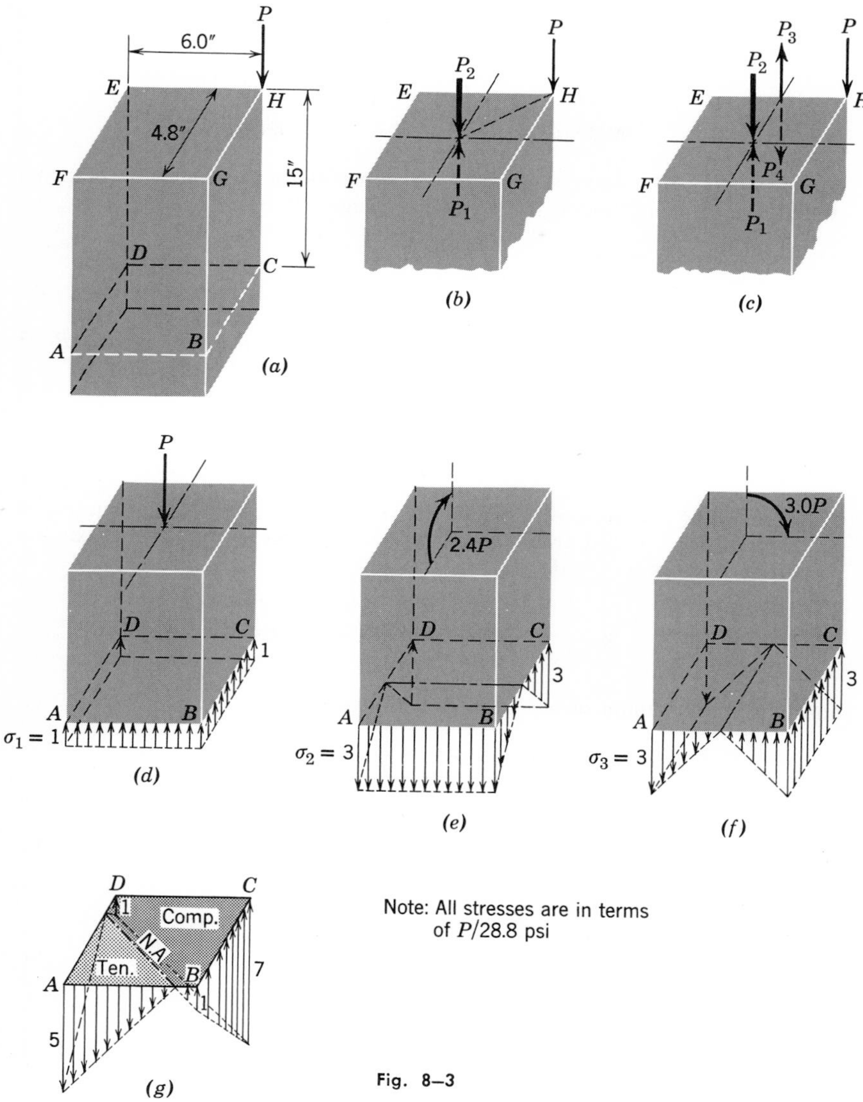
P
6.0″
E
H
4.8″
15″
F
G
D
C
A
B
(a)
P
$P_2$
E
H
F
G
$P_1$
(b)
$P_3$
P
$P_2$
E
H
$P_4$
F
G
$P_1$
(c)
P
D
C
1
A
B
$\sigma_1 = 1$
(d)
2.4P
D
C
3
A
B
$\sigma_2 = 3$
(e)
3.0P
D
C
3
A
B
$\sigma_3 = 3$
(f)
D
C
1
Comp.
N.A.
Ten.
7
A
B
1
5
(g)
Note: All stresses are in terms of $P/28.8$ psi

Fig. 8–3

The centric load $P_2$ gives a uniformly distributed compressive stress on plane $ABCD$, and the value is

$$\sigma_1 = \frac{P}{A} = \frac{P}{28.8} \text{ psi.}$$

The couple $P_4P_1$ gives a compressive stress along the edge $CD$ of plane $ABCD$ and an equal tensile stress along the edge $AB$. These stresses are

$$\sigma_2 = \frac{Mc}{I} = \frac{2.4P(2.4)}{(6)(4.8)^3/12} = \frac{P}{9.6} \text{ psi.}$$

The couple $PP_3$ gives a compressive stress along edge $BC$ of plane $ABCD$ and an equal tensile stress along the edge $AD$ equal to

$$\sigma_3 = \frac{Mc}{I} = \frac{3P(3)}{(4.8)6^3/12} = \frac{P}{9.6} \text{ psi.}$$

The component stresses are shown in the three stress distribution diagrams of Fig. 8–3*d*, *e*, and *f*, and from the diagrams one observes that the maximum compressive stress occurs at $C$ and is

$$\sigma_C = \frac{P}{28.8} + \frac{P}{9.6} + \frac{P}{9.6} = \frac{7P}{28.8} \text{ psi,}$$

and that the maximum tensile stress occurs at $A$ and is

$$\sigma_T = \frac{P}{9.6} + \frac{P}{9.6} - \frac{P}{28.8} = \frac{5P}{28.8} \text{ psi.}$$

When the allowable stresses are substituted in the above expressions, the values of $P$ become

$$P_C = \frac{28.8}{7}(20{,}000) = 82{,}300 \text{ lb}$$

and

$$P_T = \frac{28.8}{5}(5000) = 28{,}800 \text{ lb.}$$

Since $P_T$ is less than $P_C$, the maximum allowable load is

$$P = \underline{28{,}800 \text{ lb.}} \qquad \text{Ans.}$$

The distribution of the combined stresses is shown in Fig. 8–3*g* as an aid to the visualization of the stress pattern due to the given loading.

Another example involving combined centric and flexural loads is a beam with a force or a component of a load acting parallel to the axis of the beam. In this case the beam shearing stress is not zero everywhere and may be an important factor in some instances.

This type of combined load is complicated by having the deflection of the beam and the amount of bending moment dependent upon each other. As a beam deflects, the centroid of a specific cross section of the beam moves relative to the line of action of any longitudinal force. This movement alters the moment or couple applied to the beam due to the eccentricity of the longitudinal force, as further explained in Ex. 8–3. For this reason a complete solution suggests the use of successive approximations.[2] Fortunately the change in stress due to the deflection of the beam can normally be neglected as shown by the following example.

**Example 8–3:** A 5-ft long timber cantilever beam with the cross section of Fig. 8–4*a* is loaded as shown in Fig. 8–4*b*. The modulus of elasticity of the timber is 1200 ksi. Determine:

(*a*) The normal stress distribution on a vertical plane through the beam at the wall.

(*b*) The principal stresses and maximum shearing stress at a point 2.50 in. above the centroidal axis on a cross section at the wall.

---

[2] It is recognized that the differential equation expressing the moment as a function of both $x$ and $y$ has a solution (see Art. 9–6). However, for many practical problems, the method of successive approximations is more efficient.

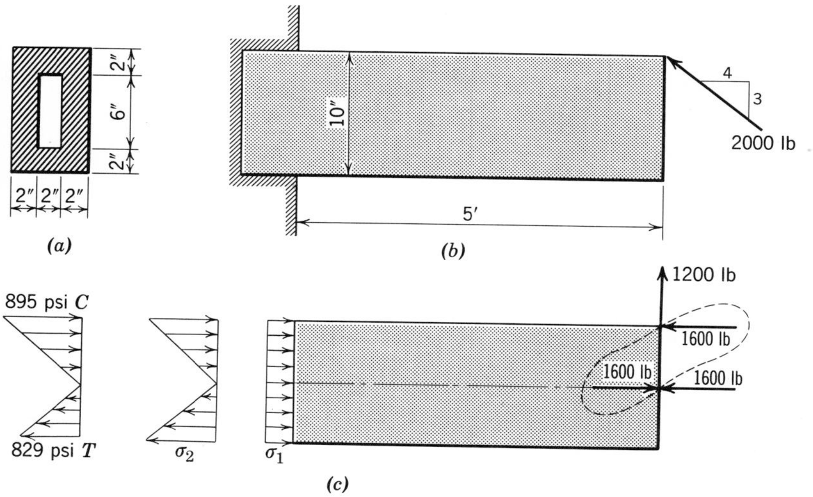


Fig. 8–4

*Solution:* The 2000-lb load can be replaced by a 1200-lb vertical force, a 1600-lb horizontal force through the centroid of the cross section of the beam at the wall, and a couple with 1600-lb horizontal forces as shown in Fig. 8–4*c*.

(*a*) The axial compressive stress produced by the 1600-lb horizontal force on the section at the wall is

$$\sigma_1 = \frac{P}{A} = \frac{1600}{60 - 12} = 33.3 \text{ psi } C$$

as shown in Fig. 8–4*c*.

If the deflection at the right end of the beam is neglected, the bending moment at the support is +1200(60) + 1600(5) which is equal to +80,000 in-lb. If the deflection of the beam is appreciable, the couple at the free end will be increased since, for the load to be centric, the 1600-lb force must pass through the centroid of the resisting section (which is at the support for this problem). The deflection of the right end due to the 1200-lb load as given by case 1 of App. D is

$$y_1 = \frac{PL^3}{3EI} = \frac{1200(60^3)}{3(12)(10^5)[6(10^3) - 2(6^3)]/12} = \frac{9}{58} \text{ in. upward.}$$

The deflection at the right end due to the 8000-in-lb couple as given by case 4 of App. D is

$$y_2 = \frac{ML^2}{2EI} = \frac{1600(5)(60^2)}{2(12)(10^5)(464)} = \frac{1.5}{58} \text{ in. upward.}$$

The sum of these deflections is 0.1810 in. upward; and, with this deflection included, the moment of the couple becomes 1600(5.18) or 8288, thus making the total moment 80,290 in-lb or 0.36 per cent higher, which is so slight an increase that it will be neglected. Hence, the flexural stresses at the wall are

$$\sigma_2 = \frac{Mc}{I} = \frac{80{,}000(5)}{464} = 862 \text{ psi } T \text{ and } C.$$

The distributions of the component stresses are shown in Fig. 8–4*c*. Also shown is the distribution of the combined stresses which is the answer to part *a*.

(*b*) The point specified is half way from the centroidal axis to the outermost fiber; therefore the normal stress on the transverse plane is

$$\sigma = \frac{862}{2} + 33 = 464 \text{ psi } C.$$

The horizontal and vertical shearing stresses at the specified point are

$$\tau = \frac{VQ}{It} = \frac{1200[12(4) + 2(2.75)]}{464(4)} = 34.6 \text{ psi.}$$

From Eq. (1–4) the principal stresses and maximum shearing stress at the specified point are

$$\sigma_p = -232 \mp \sqrt{232^2 + 34.6^2}$$
$$= -232 \mp 235$$
$$= \underline{467 \text{ psi } C} \text{ and } \underline{3 \text{ psi } T} \qquad \text{Ans.}$$

and

$$\tau_{\max} = \underline{235 \text{ psi.}} \qquad \text{Ans.}$$

## PROBLEMS

**Note: In the following problems, the effect of deflection on the eccentricity is to be neglected, unless otherwise noted.**

**8–1.*** The cross section of the aluminum beam ($E = 10^7$ psi) of Fig. P8–1 is a rectangle 4 in. wide by 6 in. deep. Determine the maximum normal stress on a vertical section at the wall:

(*a*) Neglecting the deflection of the member.

(*b*) Including the approximate increase in stress due to the deflection of the 50-in. horizontal portion of the member.

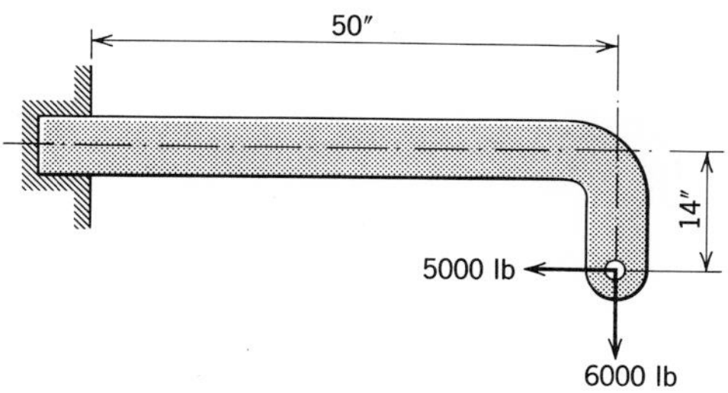


**Fig. P8–1**

**8–2.*** Determine the maximum normal stress on a transverse plane of the beam in Fig. P8–2. Neglect stress concentration, and assume lateral buckling is prevented by restraint at support *B*.

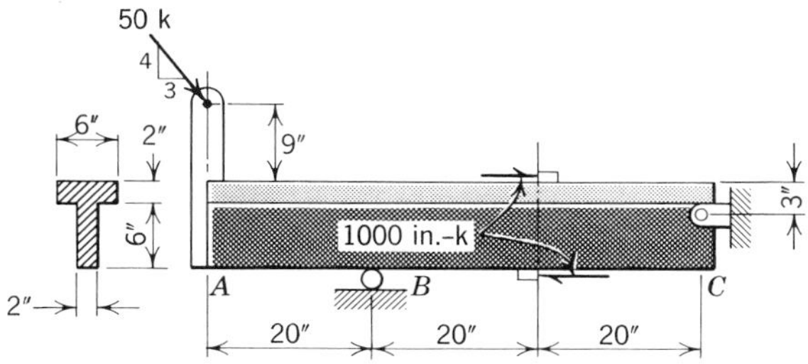


Fig. P8–2

**8–3.** Determine the maximum normal stress on a transverse plane in the straight portion of the structure in Fig. P8–3, and indicate clearly where it occurs. The member is braced in the plane perpendicular to the plane of symmetry.

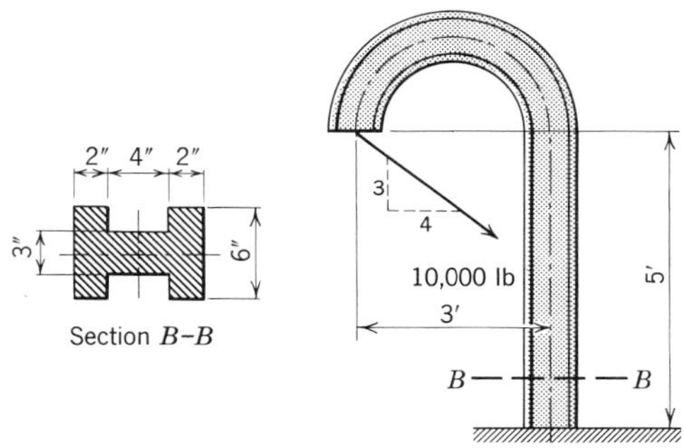


Fig. P8–3

**8–4.** The straight portion, *AB*, of the cast steel machine part of Fig. P8–4 has a hollow rectangular cross section with outside dimensions 12 by 10 in. and walls 2 in. thick. For the indicated loading determine the maximum normal stress on a transverse plane within the portion *AB*.

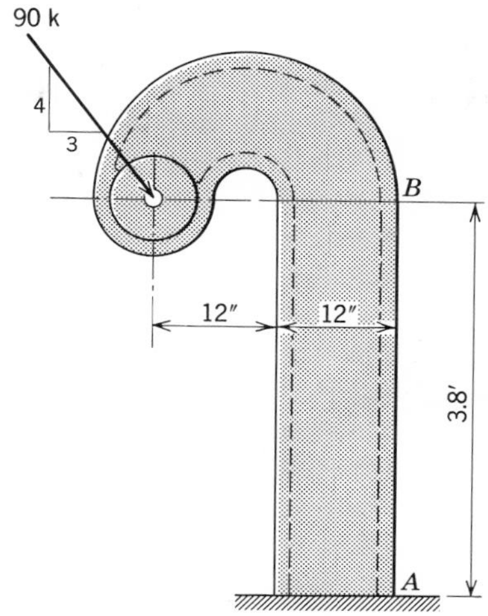


Fig. P8–4

**8–5*.** A short post is loaded by a vertical force $P$ of 4800 lb and a horizontal force $H$ of 1000 lb as shown in Fig. P8–5. Determine the vertical normal stresses at the corners $A$, $B$, $C$, and $D$ and locate the line of zero normal stress in plane $ABCD$.

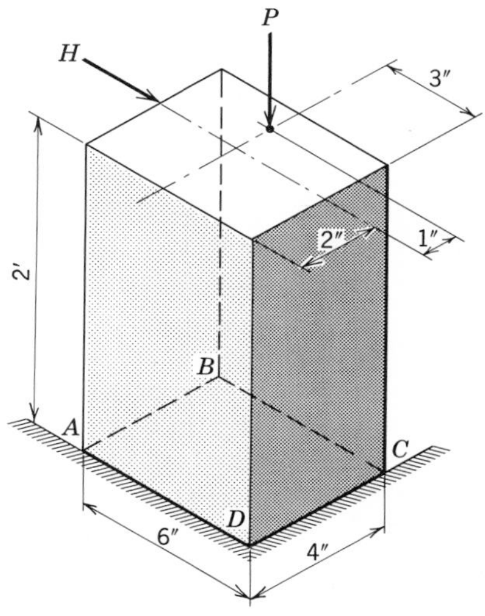


Fig. P8–5

**8–6.** The structure in Fig. P8–6 supports the load $W$ as indicated. Member $AB$ is made up of two $\frac{1}{2}$ by 3-in. bars which are connected at 4-in. intervals to prevent buckling. The weight of $W$ is 18 kips. Determine the maximum tensile and compressive stresses on any transverse plane through $AB$ and state where each occurs.

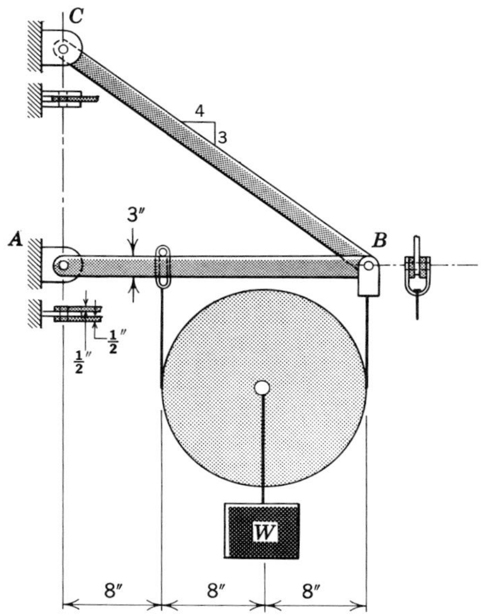


Fig. P8–6

**8–7.*** For the beam of Fig. P8–7 determine the maximum normal stress on plane $a$–$a$.

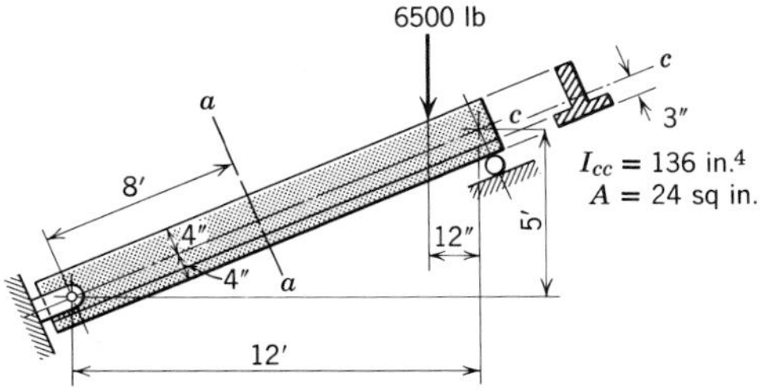


Fig. P8–7

**8–8.*** The member in Fig. P8–8 has a solid circular cross section of radius $r$. Determine the minimum permissible radius of the member if the maximum tensile stress on section $a$–$a$ is not to exceed 20,000 psi.

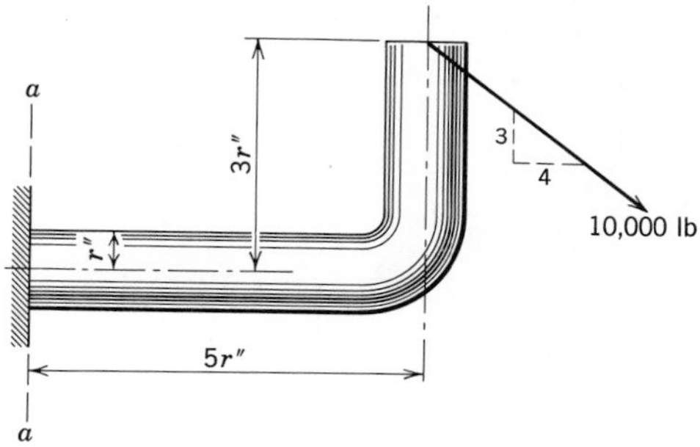


Fig. P8–8

**8–9.** For the member in Fig. P8–9 determine the maximum allowable value of $P$ for a limiting tensile stress of 5000 psi and a limiting compressive stress of 16,000 psi on plane $AB$.

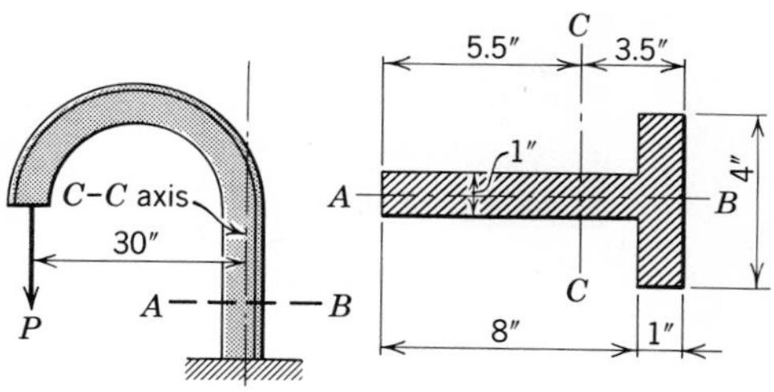


Fig. P8–9

**8–10.*** The cast iron frame of a small press is shaped as shown in Fig. P8–10. The allowable stresses on plane *A–A* are 4000 psi *T* and 20,000 psi *C*. Determine the capacity, *P*, of the press.

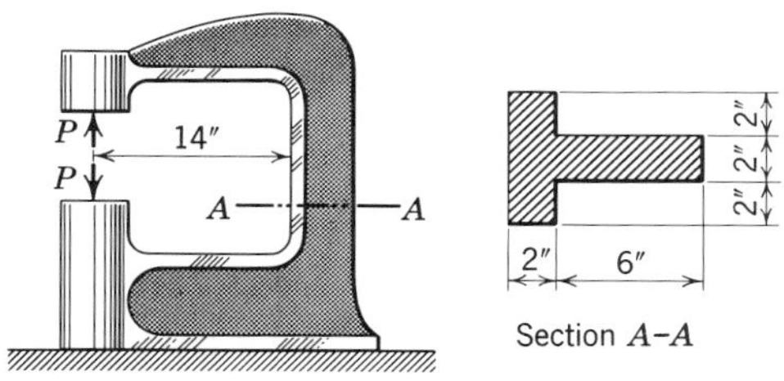


Fig. P8–10

**8–11.** The cross section of the straight vertical portion of the Littell coil loading hook of Fig. 8–1 is approximately as shown in Fig. P8–11. The horizontal distance from the line of action of the applied load to the inside face *AB* of the hook cross section is 22.0 in. Determine the maximum principal stresses on section *ABCD* for a 5-ton coil.

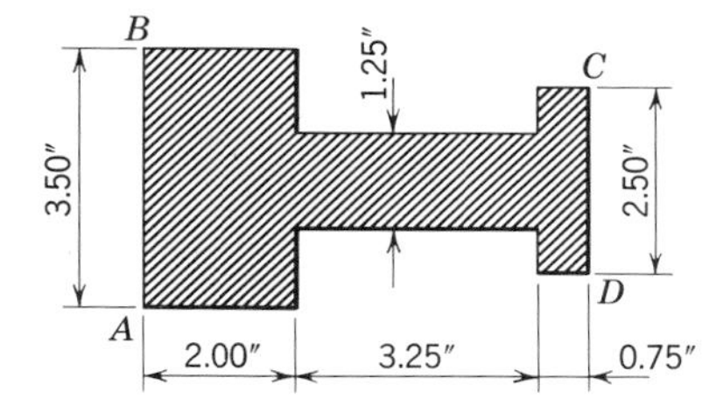


Fig. P8–11

**8–12*** Determine the force *P* in Fig. P8–12 which will result in a maximum compressive stress on plane *A–A* of 15 ksi and the corresponding average longitudinal shearing stress at the centroidal axis.

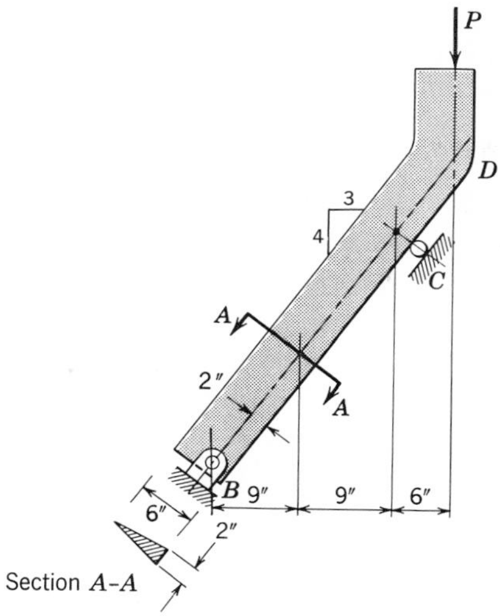


Fig. P8–12

**8–13.** The beam in Fig. P8–13 has a 3 by 8-in. rectangular cross section and is loaded in the plane of symmetry. Determine, and show on a sketch, the principal and maximum shearing stresses at point $A$.

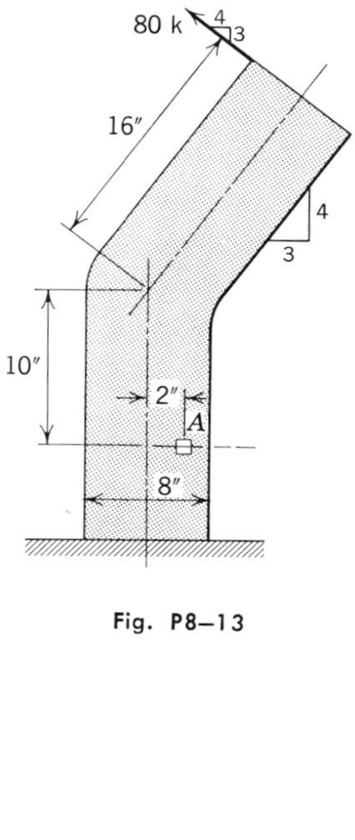


Fig. P8–13

**8–14.*** The frame in Fig. P8–14 is constructed of full-sized 4 by 4-in. timbers. Determine and show on sketches, the principal and maximum shearing stresses at points $G$ and $H$.

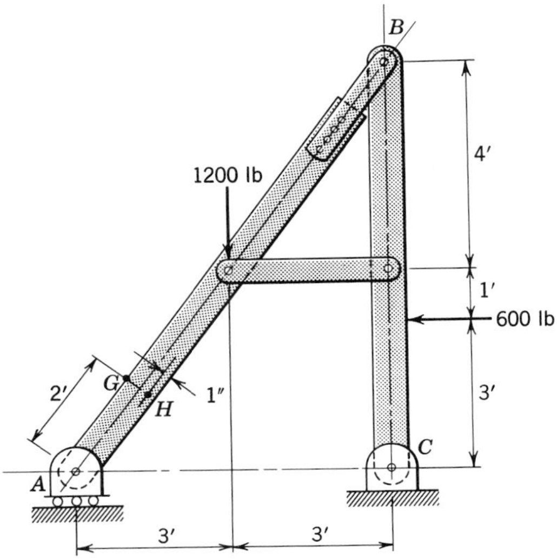


Fig. P8–14

**8–15.** The following data apply to the cantilever beam of Fig. P8–15: $P = 5600$ lb, $T = 2400$ lb, $e = 7$ in., and $a = 28$ in. Determine, and show on a sketch, the principal and maximum shearing stresses at point $A$ (in the web).

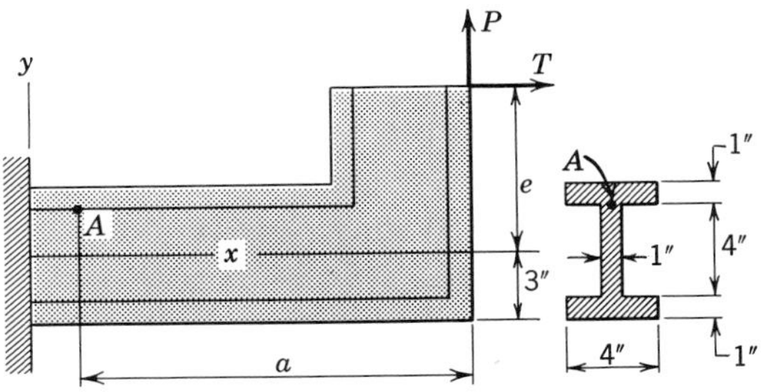


Fig. P8–15

**8–16.*** In Fig. P8–15, $e = 7$ in., $a = 24$ in., and the stresses at point $A$ (in the web) are $\sigma_x = 4800$ psi $C$ and $\tau_{xy} = 1000$ psi.

($a$) Determine the magnitudes of $P$ and $T$.

($b$) Determine, and show on a sketch, the principal and maximum shearing stresses at point $A$.

**8–17.** The magnitude of $P$ in Fig. P8–15 is two thirds of $T$, and $a$ is equal to four $e$. The stresses at $A$ (in the web) are $\sigma_x = 4000$ psi $C$ and $\tau_{xy} = 1200$ psi.

($a$) Determine the values of $P$ and $e$.

($b$) Determine $\sigma_{p1}$, $\sigma_{p2}$, and $\tau_{\max}$ at point $A$.

**8–18.*** The principal stresses at point $A$ (in the web) in Fig. P8–15 are 2 ksi $T$ and 8 ksi $C$. In addition, $T$ is three-fourths of $P$, and $a$ is equal to three $e$. Determine the values of $P$ and $a$.

**8–19.** Determine, and show on a sketch, the principal and maximum shearing stresses at point $B$ on section $A$–$A$ of Fig. P8–19. Assume the transverse shearing stress to be uniformly distributed along any horizontal line in the section.

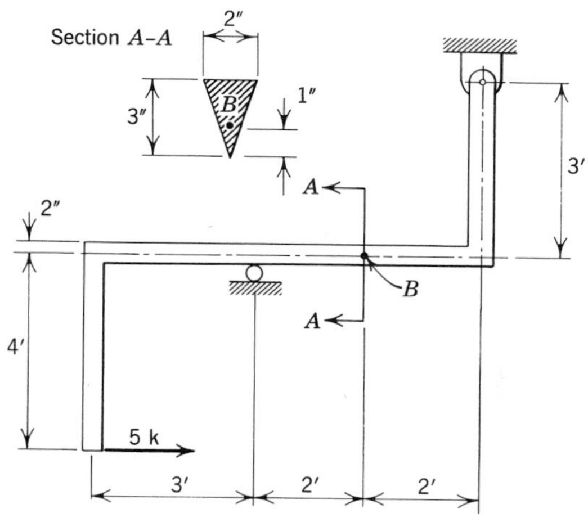


Fig. P8–19

**8–20.*** Determine, and show on a sketch, the principal stresses and the maximum shearing stress at point $A$ in the roof beam in Fig. P8–20.

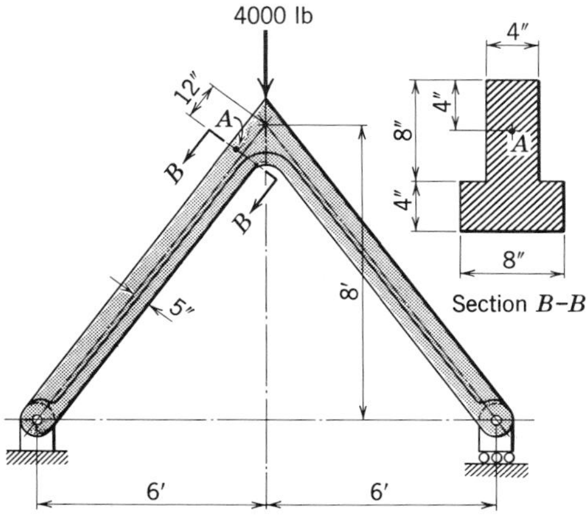


Fig. P8–20

**8–21.** The machine element of Fig. P8–21 is loaded in a plane of symmetry. Determine, and show on a sketch, the principal and maximum shearing stresses at point $A$, which is in the web at the junction of flange and web. Neglect stress concentration.

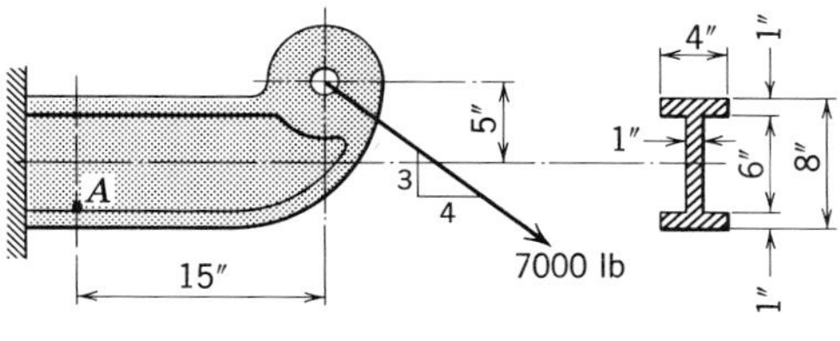


Fig. P8–21

**8–3. Centric, Torsional, and Flexural Loads Combined.** In numerous industrial situations, machine members are subjected to all three general types of loads studied so far. When a flexural load is combined with torsional and centric loads, it is frequently difficult to locate the point or points where the most severe stresses occur. The following review or summary statements may be helpful in this regard.

The longitudinal and transverse shearing stresses are maximum at the centroidal axis of the section where $V$ is maximum. The flexural stress is maximum at the greatest distance from the centroidal axis on the section where $M$ is maximum. The torsional shearing stress is maximum at the surface of a shaft at the section where $T$ is maximum. With these facts in mind, it is normally possible to locate one or more possible points of high stress. Even so, it may be necessary to calculate the various stresses at more than one point of a member before locating the most severely stressed point. For stresses below the proportional limit of the material, the superposition method can be used to combine stresses on any given plane at any specific point of a loaded member.

After the stresses on a pair of mutually perpendicular planes at a specific point are determined, the methods of Art. 1–6 and 1–7 can be used to find the principal stresses and the maximum shearing stress at the point.

The following example illustrates the procedure for the solution of combined stress problems.

**Example 8–4:** The 4-in. diameter shaft of Fig. 8–5*a* is subjected to the loads shown. Determine the maximum permissible value of $P$ for allowable stresses of 16,000 psi $T$ and 9000 psi shear.

*Solution:* When two equal opposite forces $P$ are introduced along line $EF$ as in Fig. 8–5*b*, it is seen that the shaft is subjected to a couple of $20P$ in-lb clockwise looking toward the wall and a concentrated load $P$ acting at $E$. In other words, the shaft acts as a cantilever beam with a concentrated load $P$ at the right end, a torsion member subjected to a $20P$-in-lb torque, and an axially loaded member with a load of $48P$ lb.

The couple produces a torsional shearing stress at all surface points of the shaft of

$$\tau_{xt} = \frac{Tc}{J} = \frac{20P(2)}{\pi(2^4)/2} = \frac{5P}{\pi}$$

as shown in Fig. 8–5*c*.

The axial tensile stress at all points of the shaft is

$$\sigma_{x_1} = \frac{P}{A} = \frac{48P}{\pi(2^2)} = \frac{12P}{\pi}$$

as shown in Fig. 8–5*d*.

The cantilever beam load $P$ produces a maximum tensile fiber stress at point $A$ and an equal compressive stress on the opposite side of the shaft. The axial stress is tensile, and therefore the compressive flexural stress is not critical. The tensile fiber stress at $A$ is (see Fig. 8–5*e*)

$$\sigma_{x_2} = \frac{Mc}{I} = \frac{24P(2)}{\pi(2^4)/4} = \frac{12P}{\pi}.$$

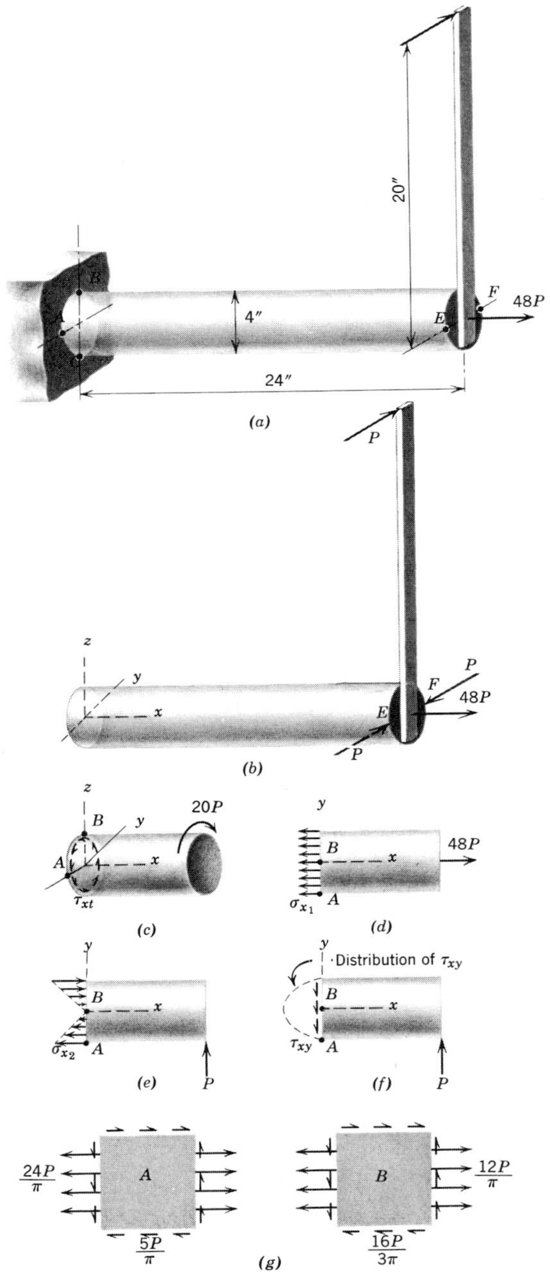

20″
4″
24″
48P
B
A
C
E
F
(a)
P
z
y
x
48P
P
E
F
P
(b)
z
B
y
20P
A
x
τxt
(c)
y
B
x
48P
σx1
A
(d)
y
B
x
σx2
A
P
(e)
y
Distribution of τxy
B
x
τxy
A
P
(f)
24P/π
A
5P/π
B
12P/π
16P/3π
(g)


Fig. 8–5

The transverse shearing stress is equal to the longitudinal shearing stress, which is maximum on the $xz$ plane including such surface points as $B$ and $C$ where the torsional shearing stress is also maximum. This maximum shearing stress is (see Fig. 8–5$f$)

$$\tau_{xy} = \frac{VQ}{It} = \frac{P\left[\dfrac{\pi(2^2)}{2}\right]\left[\dfrac{4(2)}{3\pi}\right]}{\dfrac{\pi(2^4)}{4}(4)} = \frac{P}{3\pi}.$$

The torsional and transverse shearing stresses are in the same direction along the top of the shaft and in opposite directions along the bottom of the shaft. Therefore any critical situation due to these two stresses will develop along the top of the shaft at some point such as $B$.

All four stresses are maximum at some point or points in the section of the shaft at the wall since the fiber stress is not maximum at any other section of the shaft. Therefore a complete analysis for the member can be made on the section at the wall. A complete analysis including interior points of this section is beyond the scope of this book. However, a fairly good solution can be obtained by investigating the combined stresses at two surface points $A$ and $B$. At $A$ the fiber stress, axial stress, and torsional shearing stress are all maximum, whereas the transverse shearing stress is zero. At $B$ the axial stress, torsional shearing stress, and transverse shearing stress are all maximum, whereas the fiber stress is zero. The fiber stress and transverse shearing stress are never maximum at the same point. Furthermore the torsional shearing stress is never maximum at an interior point. Therefore the chance of finding a more severe stress at an interior point is unlikely.

The stresses on orthogonal planes through $A$ and $B$ are as shown in Fig. 8–5$g$. When these values are substituted in Eq. (1–4) of Art. 1–6 and the resulting expressions compared, it is evident that the combination of stresses at $A$ is more severe than that at $B$. The maximum stresses at $A$ are

$$\sigma_A = \frac{12P}{\pi} + \sqrt{\left(\frac{12P}{\pi}\right)^2 + \left(\frac{5P}{\pi}\right)^2} = \frac{25P}{\pi}$$

and

$$\tau_A = \sqrt{\left(\frac{12P}{\pi}\right)^2 + \left(\frac{5P}{\pi}\right)^2} = \frac{13P}{\pi}.$$

The allowable values of $P$ from the allowable stresses and these relations are

$$16{,}000 = 25P_\sigma/\pi$$

or

$$P_\sigma = 640\pi = 2010 \text{ lb}$$

and

$$9000 = 13P_\tau/\pi$$

or

$$P_\tau = 9000\pi/13 = 2170 \text{ lb}.$$

The smaller value controls and the maximum permissible value is

$$P = \underline{2010 \text{ lb.}}$$ Ans.

## PROBLEMS

**In the following problems, stress concentrations and the effect of beam deflection on the eccentricity are to be neglected unless otherwise noted.**

**8–22.*** Determine, and show on a sketch, the principal and maximum shearing stresses at point $P$ on the solid shaft in Fig. P8–22.

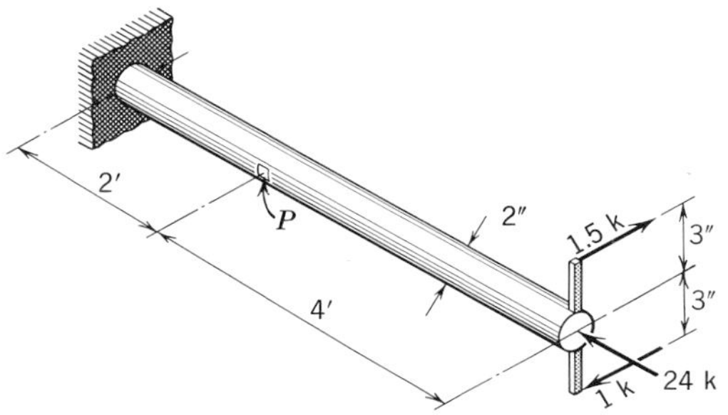


Fig. P8–22

**8–23.*** A solid shaft 4 in. in diameter is acted on by forces $P$ and $Q$ as shown in Fig. P8–23. Determine, and show on a sketch, the principal stresses and the maximum shearing stress at point $A$ on the surface of the shaft.

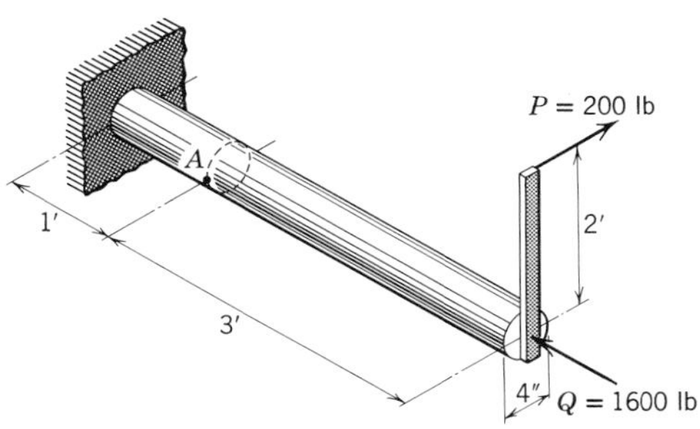


Fig. P8–23

**8–24.*** (*a*) At what point on the surface of the shaft in Fig. P8–24 is the horizontal tensile stress maximum?

(*b*) Determine and show on a sketch the principal stresses and the maximum shearing stress at this point.

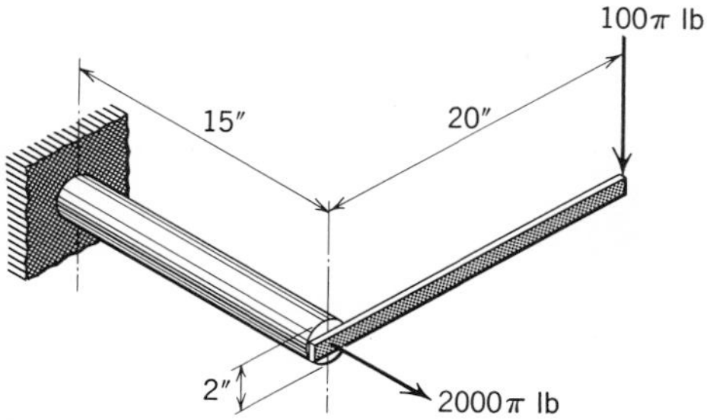


Fig. P8–24

**8–25.** A steel shaft 4 in. in diameter is supported in flexible bearings at its ends. Two pulleys each 2 ft in diameter are keyed to the shaft. The pulleys carry belts as shown in Fig. P8–25.

Determine, and show on a sketch, the principal stresses and the maximum shearing stress at point *A*.

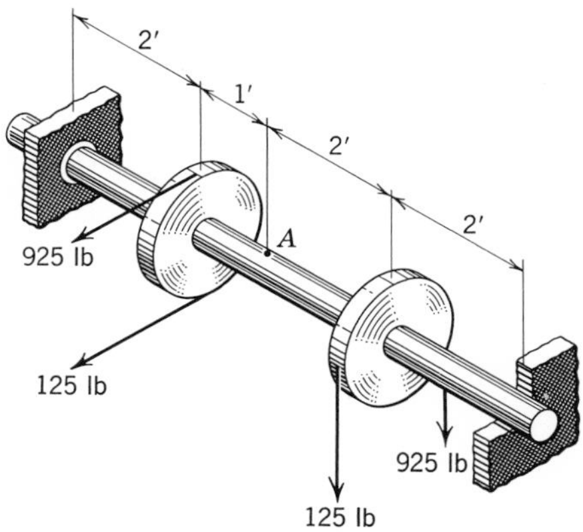


Fig. P8–25

**8–26.*** The shaft in Fig. P8–26 is subjected to the loads shown. If the tensile and shearing stresses at point $A$ (on the surface of the shaft) may not exceed 18,000 psi and 12,000 psi, respectively, determine the value of $R$.

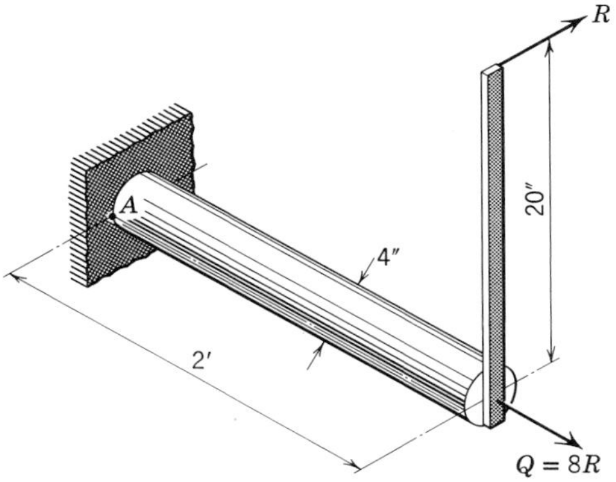


Fig. P8–26

**8–27.** Determine the minimum permissible diameter for the shaft of Fig. P8–27 if the shearing and tensile stresses (at point $A$) are not to exceed 11,000 psi and 18,000 psi, respectively.

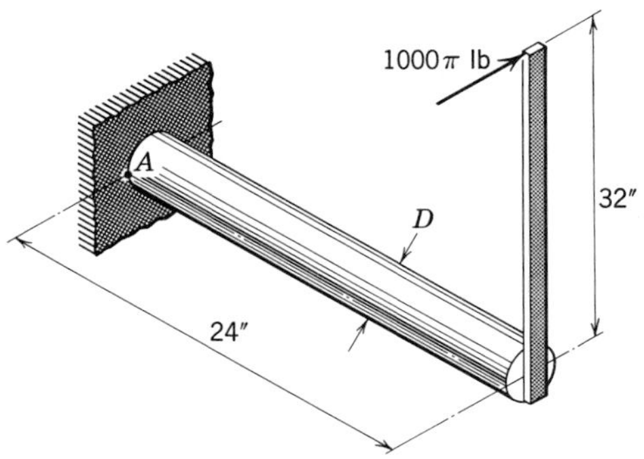


Fig. P8–27

**8–28.** A 2-in. diameter shaft is subjected to an axial tensile force of $8P$ kips, a bending moment of $22P$ in-kips, and a torque of $10P$ in-kips. The maximum normal stress in the shaft is 20 ksi $T$. Determine the maximum shearing stress and the magnitude of $P$.

**8–29.** Figure P8–29 is a free-body diagram of a closed cylindrical pressure vessel with an inside diameter of 60 in. and a wall thickness of 0.20 in. The tank is subjected to an internal pressure of 100 psi, and the external loads shown are applied to rigid plates attached to the ends of the cylinder. Determine the principal and maximum shearing stresses acting on an element at the inside of the tank at $A$ and show them on a sketch. Neglect the weight of the tank and assume that the walls do not buckle.

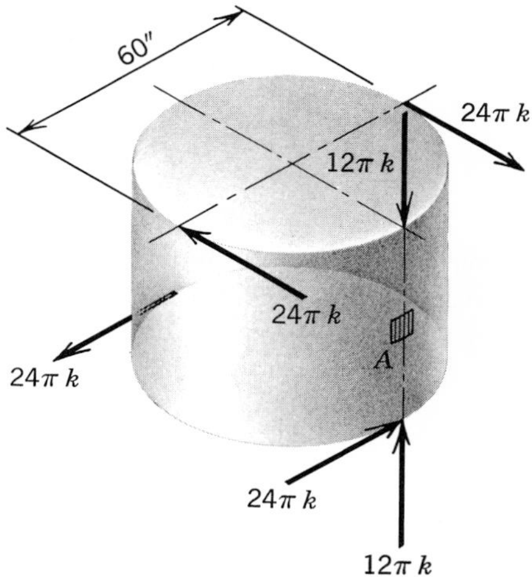


Fig. P8–29

**8–30.*** Because of discontinuities at points $A$ and $B$ on the shaft of Fig. P8–30, the normal and shearing stresses at these points are limited to 18 ksi $T$ and 10 ksi, respectively. Determine the maximum permissible value of $P$.

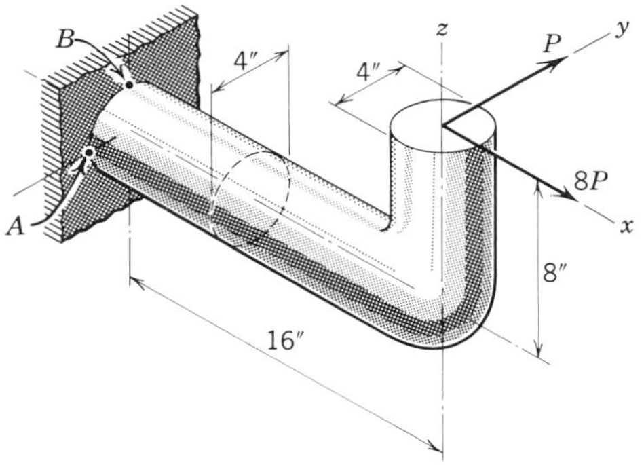


Fig. P8–30

**8–4. Dynamic Loading—Equivalent Static Load.** When the motion of a body is changed (accelerated), the force necessary to produce this acceleration is called a *dynamic force* or *load*. For example, the force which an elbow in a pipeline exerts on the fluid in the pipe to change its direction of flow, the pressure on the wings of an airplane pulling out of a dive, the collision of two automobiles, and a man jumping on a diving board are all examples of dynamic loading. A suddenly applied load is called an *impact load*. The last two of the preceding examples are considered impact loads. Under impact loading, if there is elastic action, the loaded system will vibrate until equilibrium is established. A dynamic load may be expressed in terms of mass times the acceleration of the mass center, in terms of the rate of change of the momentum, or in terms of the change of the kinetic energy of the body.

For many problems in mechanics of materials, the energy approach, introduced in Art. 2–2, is most effective. The magnitude of the dynamic load is often expressed in terms of the kinetic energy delivered to the loaded system; hence, it is often referred to as an *energy load*. For example, a particle of mass $m$ moving with a speed $v$ possesses a kinetic energy $mv^2/2$; and, if this particle is stopped by a body, the energy absorbed by the body (the loaded system) is some fractional part of $mv^2/2$, the balance of the energy being converted into sound, heat, and permanent deformation of the striking particle.

Dynamic loading produces, in the loaded system, stresses and strains, the magnitude and distribution of which will depend not only on the usual parameters encountered previously but also upon the velocity of propagation of the strain waves through the material of which the system is composed. This latter consideration, although very important when loads are applied with high velocities, may often be neglected when the velocity of application of the load is low. The velocities are considered to be low when the loading time permits the material to act in the same manner as it does under static loading; that is, the relations between stress and strain and between load and deflection are essentially the same as those already developed for static loading.

For many engineering applications, the velocities of the applied dynamic loads are small enough to justify the assumption that the material behaves in the same manner as it does under slowly applied loads. Using this assumption, an *equivalent static load* can be determined which will produce the same maximum deflection as that due to the dynamic load. This assumption implies that the stress-strain diagram for any point in the loaded system, as well as the strain distribution, will be identical under either the *dynamic* or the *equivalent static* load.

As an illustration, consider the steel cantilever beam of Fig. 8–6*a* on which the weight, $W$, is dropped as indicated. The variation of the end deflection of the beam with respect to time is shown in Fig. 8–6*b* where $\Delta$, equal to $\delta_{max}$, is the maximum deflection resulting from the falling weight transferring part of its kinetic energy ($mv^2/2$) to the beam and where $\delta_{stat}$ is the static deflection resulting when the beam and weight finally reach equilibrium. The static load, $P$, applied as in Fig. 8–6*c* produces the same maximum deflection, $\Delta$; and the deflection curve and stress-strain diagrams are assumed to be the same in both cases. Hence, the strain energy in the beam is the same in both cases, and, since the strain energy is equal to the effective applied energy (work done on the beam), the following relation may be stated,

$$\begin{bmatrix}\text{Effective}\\ \text{energy}\\ \text{applied}\end{bmatrix} = \begin{bmatrix}\text{Work done by}\\ \text{the equivalent}\\ \text{static load}\end{bmatrix}. \tag{a}$$

The effective energy is that portion of the applied energy which produces the maximum deflection. In addition to energy lost in the form of sound, heat, and local distortion, some is absorbed by the supports.

From Art. 2–2, the strain energy in terms of the static load equals $\int P\,d\delta$ or $P_{avg.}(\delta_{max})$. Equation (a) may then be written as

$$U = P_{avg.}(\Delta),$$

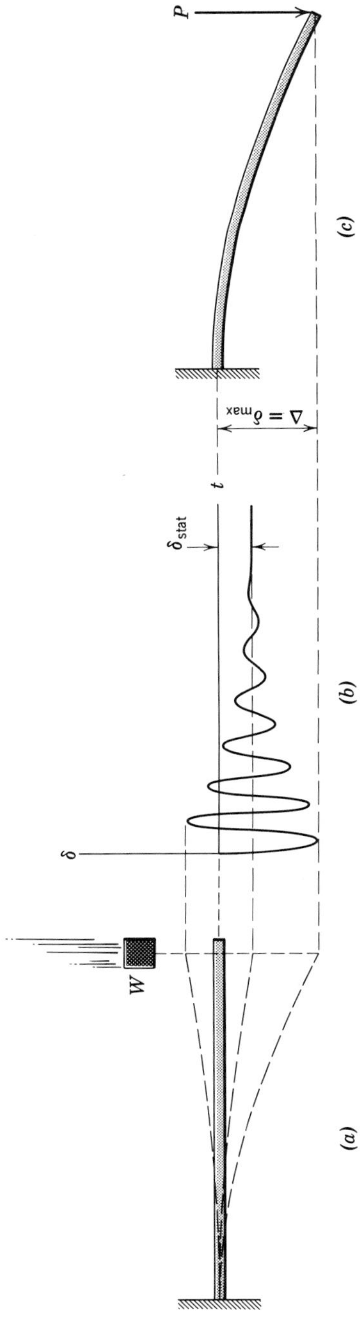


Fig. 8–6

and, if stress and strain are proportional (load-deflection diagram is a straight line), $P_{avg.}$ is equal to one-half $P_{max}$. Hence, for a material possessing a linear stress-strain relationship,

$$U = P\Delta/2. \tag{8–1}$$

The right-hand side of Eq. (8–1) involves the relations between load, stress, and strain previously encountered under static loading and applies to any type of loading. If the effective energy applied, $U$, is due to a change of velocity of a translating body, the left side of Eq. (8–1) becomes

$$\eta m v^2/2,$$

where $\eta$ is that portion of the applied energy which is effective, $m$ is the mass of the body, and $v$ is the velocity of the moving body just before impact. If the effective energy applied comes from a body falling from rest and being stopped by the resisting structure, the left-hand side of Eq. (8–1) comes from the change in potential energy of the body and is

$$\eta W h,$$

where $W$ is the weight of the body and $h$ is the total vertical distance through which it falls.

For torsional energy loading, Eq. (a) can be written as

$$U = T\theta/2, \tag{8–2}$$

where $T$ is the slowly applied torque which will produce the same maximum angle of twist, $\theta$, as the suddenly applied torque if the maximum shearing stress is less than the proportional limit.

A term sometimes encountered in working with dynamic loading is load factor (or impact factor), defined as the *ratio of the equivalent static load to the magnitude of the force which the loading mass would exert on the loaded system if slowly applied*. For example, the equivalent static loading on the wings of an airplane pulling out of a dive divided by the weight of the plane (loading on horizontal tail surfaces neglected) is the load factor.

Examples 8–5 and 8–6 illustrate the application of Eq. (8–1).

**Example 8–5:** The upper 3 ft of the steel bar $AB$ in Fig. 8–7$a$ is 2 in. wide by ½ in. thick, and the lower 2 ft is 1 in. wide by ½ in. thick. When the weight, $W$, is released from rest from a height of 20 in. onto the end $A$ of the bar, the axial stress in the bar (neglecting stress concentration) is not to ex-

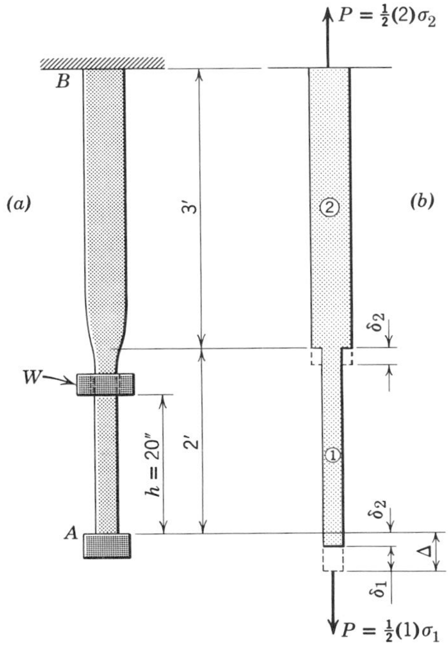


Fig. 8–7

ceed 16 ksi. Assume that 80 per cent of the energy of the falling block is absorbed elastically by member $AB$. Determine:

(*a*) The maximum permissible weight for block $W$.

(*b*) The maximum permissible weight for $W$ if the bar $AB$ is 1 in. wide through its entire length.

*Solution:* (*a*) Figure 8–7*b* indicates the static load applied to the bar and the resulting maximum deflections. When Eq. (8–1) is used, the following expression is obtained.

$$\eta Wh = (0.8)W(20 + \Delta) = P\Delta/2. \qquad \text{(b)}$$

The quantities $P$ and $\Delta$ can be calculated directly from the given data. From Fig. 8–7*b*

$$1(\sigma_2) = 0.5(\sigma_1) \quad \text{or} \quad \sigma_1 = 2\sigma_2.$$

This expression indicates that $\sigma_1$ (equal to 16 ksi) is the limiting stress and that $\sigma_2$ must be 8 ksi.

The load $P$ is

$$P = \sigma_1 A_1 = 16(1)(0.5) = 8 \text{ kips},$$

and the deflection $\Delta$ is

$$\Delta = \delta_2 + \delta_1 = 36\epsilon_2 + 24\epsilon_1 = (36\sigma_2 + 24\sigma_1)/30{,}000$$

$$= \frac{36(8) + 24(16)}{30{,}000} = 0.0224 \text{ in.}$$

Substituting these values of $P$ and $\Delta$ into Eq. (b) gives the expression

$$0.8(W)(20.0224) = (8)(0.0224)/2,$$

from which $W = \underline{5.59 \text{ lb.}}$ Ans.

(*b*) If the bar is 1 in. wide for its entire length,

$$\Delta = \frac{(36 + 24)(16)}{3(10^4)} = 0.032 \text{ in.},$$

and $0.8(W)(20.032) = (1/2)(8)(0.032),$

from which $W = \underline{7.99 \text{ lb.}}$ Ans.

In view of the development of Art. 2–2 which indicates that the energy-absorbing capacity of a member is a function of the amount of material present, the results in parts *a* and *b* seem paradoxical since the strength of the member is increased by removing some material. The apparent discrepancy is probably best explained by referring to the stress-strain diagram of Fig. 8–8. For part *a* of the problem, the energy absorbed per unit volume by the lower 2 ft of the bar is represented by the area *ocd* which gives

$$u = \frac{1}{2}\left(\frac{16}{E}\right)16 = \frac{8(16)}{3(10^4)} = 0.00427 \text{ in-kip per cu in.}$$

The energy absorbed per unit volume by the upper 3 ft of the bar is represented by area *oab* of Fig. 8–8 and is seen to be only one-fourth as much

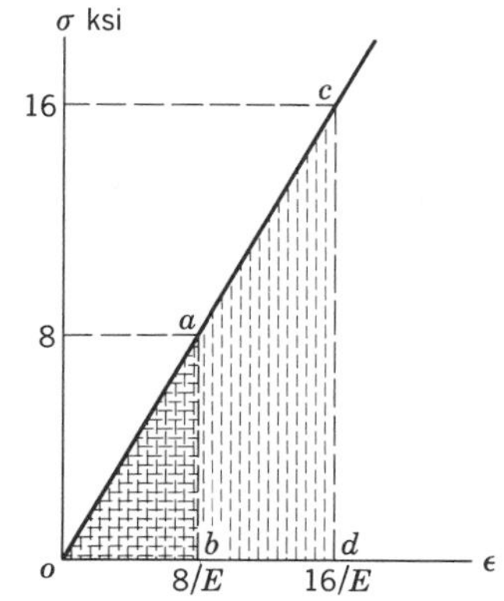


Fig. 8–8

because both the stress and the strain are one-half as much as in the lower 2 ft. When one-half of the volume of the upper 3 ft is removed, the energy absorbed by each unit volume of the remaining material is quadrupled (area *ocd* as compared to area *oab*), resulting in a net gain in energy-absorbing capacity. For this reason, bolts designed to carry dynamic loads are sometimes turned down so that the diameter of the shank of the bolt is the same as the diameter at the root of the threads. In this way the stress distribution in the bolt is approximately uniform, and the capacity of the bolt to absorb energy is a maximum.

**Example 8–6:** The simply supported 6061-T6 aluminum alloy beam $A$ of Fig. 8–9$a$ is 3 in. wide by 1 in. deep. The center support is a helical spring with a modulus of 100 lb per in. The spring is initially unstressed and in contact with the beam. When the 50-lb block drops 2 in. onto the top of the beam, 90 per

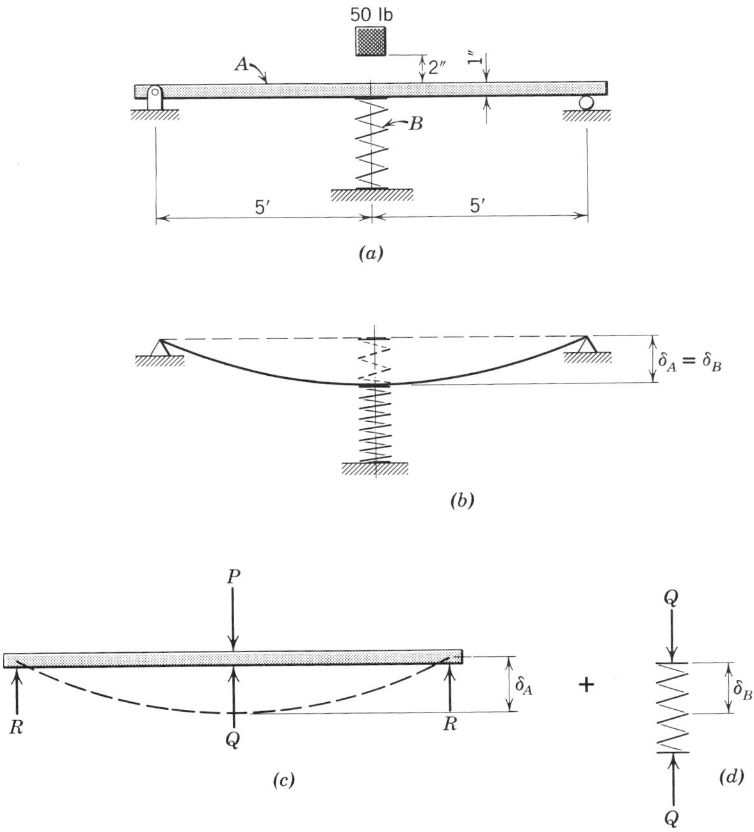


Fig. 8–9

cent of the energy is absorbed elastically by the system consisting of beam $A$ and spring $B$. Determine:

(*a*) The maximum flexural stress in the beam for this loading.

(*b*) The load factor.

(*c*) The proportion of the effective energy absorbed by each of the components $A$ and $B$.

*Solution:* In Fig. 8–9*b*, *c*, and *d* are shown a deflection diagram for the system and free-body diagrams for the component parts. Since the action is assumed to be elastic and the materials obey Hooke's law, the load-deflection relation will be linear, and Eq. (8–1) is valid; hence,

$$50(2 + \Delta)(0.9) = P\Delta/2. \qquad \text{(c)}$$

The fiber stress is wanted; therefore the forces $P$ and $Q$ will be evaluated, and from these forces and Fig. 8–9*c* the fiber stress can be determined. From Fig. 8–9*c* and *d* and App. D

$$\delta_A = \frac{(P - Q)L^3}{48EI} = \frac{(P - Q)(10^3)(12^3)}{48(10^7)(3)(1^3)/12} = \frac{(P - Q)(144)}{(10^4)},$$

and $$\delta_B = \frac{Q}{100}.$$

From Fig. 8–9*b*,

$$\delta_A = \delta_B = \Delta,$$

or $$\frac{(P - Q)(144)}{10{,}000} = \frac{Q}{100},$$

from which $$Q = \frac{36P}{61};$$

then $$\Delta = \frac{Q}{100} = \frac{36P}{6100},$$

and Eq. (c) becomes $$50\left(2 + \frac{36P}{6100}\right)0.9 = \left(\frac{1}{2}\right)\frac{36P^2}{6100}$$

or $$P^2 - 90P = 30{,}500$$

from which $$P = 225.3 \quad \text{and} \quad -135.3.$$

The positive root is the magnitude of $P$ necessary to produce the maximum downward deflection of the beam, whereas the negative root is the force necessary to produce the maximum upward deflection (see Fig. 8–6*b*) of the vibrating system.

When the value of 225.3 lb for $P$ is used,

$$Q = \frac{36(225.3)}{61} = 133.0 \text{ lb},$$

and $$P - Q = 92.3 \text{ lb}.$$

(*a*) From Fig. 8–9*c*, $R$ is 46.15 lb, and the maximum bending moment occurs at the center of the beam and is 46.15(5)(12) in-lb; therefore, the maximum flexural stress is

$$\sigma = \frac{Mc}{I} = \frac{46.15(60)/2}{3(1^3)/12}$$

$$= \underline{5540 \text{ psi } T \text{ and } C,} \qquad \text{Ans.}$$

which is well below the proportional limit, and, therefore, the action is elastic.

(*b*) The load factor is

$$\frac{225.3}{50} = \underline{4.51.} \qquad \text{Ans.}$$

(*c*) The resultant equivalent static load at the center of beam $A$ is $(P - Q)$ and is 92.3 lb, and the energy absorbed elastically by the beam is

$$U_A = \frac{(92.3)(\Delta)}{2}.$$

The elastic energy absorbed by the helical spring is

$$U_B = \frac{(133.0)(\Delta)}{2}.$$

The total energy absorbed elastically by the system is

$$U = \frac{(225.3)(\Delta)}{2};$$

therefore the proportions absorbed by each component are

$$\frac{92.3}{225.3} = \underline{41 \text{ per cent for } A} \qquad \text{Ans.}$$

and

$$\frac{133.0}{225.3} = \underline{59 \text{ per cent for } B.} \qquad \text{Ans.}$$

To recapitulate, Eq. (8–1) is applicable only to systems which have a linear relationship between load and deformation, and only when the velocity of application of the dynamic load is not too high.

## PROBLEMS

**In the following problems, the mass of the loaded members is to be neglected, and, unless stated otherwise, all of the applied energy is to be assumed to be absorbed elastically by the loaded members. Neglect stress concentration.**

**8–31.*** A load of 30 lb is dropped from a height of 2 in. onto a helical spring. The modulus of the spring is 120 lb per in. Determine:

(*a*) The maximum deflection of the spring.

(*b*) The static load that would produce the same deflection.

**8–32.*** A load of 1000 lb compresses a given car spring 0.1 in. What static load would be required to produce the same deflection as would be produced by a load of 350 lb dropped from a height of 1 ft?

**8–33.*** A weight of 30 lb is dropped from a height of 4 ft onto the center of a small rigid platform as shown in Fig. P8–33. The two steel rods supporting the platform are 1 by 2 in. by 8 ft long. If one-half of the energy is effective in producing stress in the rods, determine:

(*a*) The equivalent static load.

(*b*) The deflection of the platform.

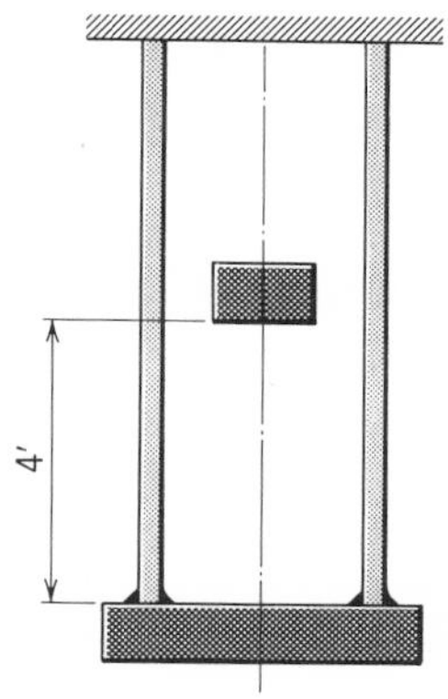


Fig. P8–33

**8–34.*** A hot-rolled Monel bar with a cross-sectional area of 3.5 sq in. is to be used as a tension member subject to axial energy loads. The allowable axial tensile stress is 10 ksi. Determine the minimum length of bar required for an axial energy load of 50 ft-lb when eight-tenths of the energy is effective in producing stress in the bar. See Appendix A.

**8–35.** A flat bar of steel 2 in. thick and 5 ft long has a width of 6 in. for 2 ft of its length and a width of 4 in. for the remaining 3 ft. Determine the maximum axial stress developed in the bar by an axial energy load of 100 ft-lb applied at one end of the bar. Assume that four-fifths of the energy produces stress in the bar.

**8–36.*** A weight, $W$, is dropped 10 in. onto the end of a steel bar as shown in Fig. P8–36. If 80 per cent of the energy is assumed to be effective in producing stress in the bar and the maximum stress is not to exceed 18,000 psi tension, determine:

($a$) The weight of $W$ if the bar is as shown in Fig. P8–36.

($b$) The weight of $W$ if the bar has a uniform cross section of 2 sq in. throughout its entire length.

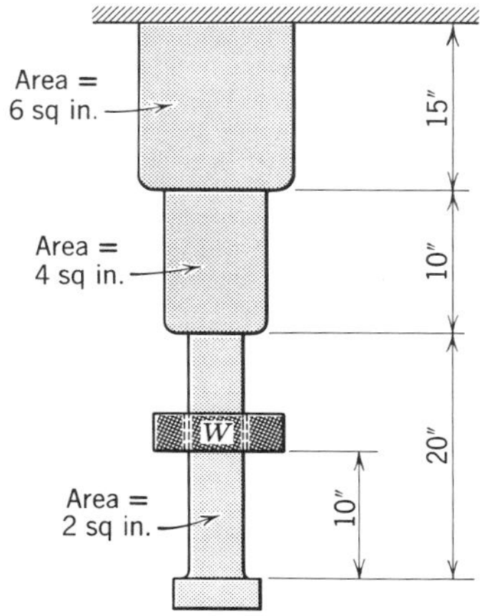


Fig. P8–36

**8–37.** Derive an expression for the total energy that may be absorbed by a round cantilever beam of length $L$ when it is loaded by a concentrated load $P$ gradually applied at the free end. Express the result in terms of $P$, $L$, $E$, and $I$.

**8–38.*** A 12 I 40.8 steel section is used as a simply supported beam 10 ft long. What weight falling on the center of the span from a height of 1 in. will cause a maximum fiber stress of 18,000 psi in this beam? Consider three-fourths of the energy to be effective in producing the given stress. The web of the I-beam is vertical.

**8–39.** A block weighing 1.00 lb falls 4.0 in. and strikes the middle of a simply supported wooden beam 1.0 in. wide, 0.5 in. deep, on a 30-in. span. Assume that the beam absorbs eight-tenths of the energy load elastically. Using $E$ for this timber as $1.5(10^6)$ psi, determine:

($a$) The equivalent static load.

($b$) The maximum deflection due to the falling weight.

**8–40.*** A weight of 80 lb is dropped from a height of 2 in. onto the free end of a steel cantilever beam. The beam is 2 in. wide, 3 in. deep, and 30 in. long. Assume that only one-half of the energy of the falling weight is effective in producing deflection in the beam. Determine:

(*a*) The maximum deflection of the beam due to the falling weight.

(*b*) The static load necessary to produce the same deflection.

(*c*) The maximum fiber stress caused by the falling weight.

**8–41.** A timber cantilever beam is 6 in. wide, 3 in. deep, and 5 ft long. Young's modulus is $1.2(10^6)$ psi. Determine the height from which a 27-lb weight should be dropped upon the free end of the beam in order to produce a maximum deflection three times as great as that produced by the same weight gradually applied.

**8–42.*** A 10-ft steel shaft 3 in. in diameter is subjected to a 100-ft-lb energy load.

(*a*) Determine the magnitude of the maximum normal stress in the shaft when the load is axial.

(*b*) Determine the maximum shearing stress when the load is a torsional energy load.

**8–43.** A hollow steel shaft is 15 ft long and has an outside diameter of 10 in. and an inside diameter of 8 in. A 5-ft length of the 15-ft section is solid.

(*a*) Determine the torsional energy load required to produce a maximum shearing stress of 10,000 psi in the shaft.

(*b*) Determine the magnitude of the angle of twist caused by the above energy load.

**8–44.*** An effective torsional energy load of $6\pi$ ft-lb is applied to the steel shaft shown in Fig. P8–44. Determine the maximum shearing stress in the shaft. Neglect stress concentration.

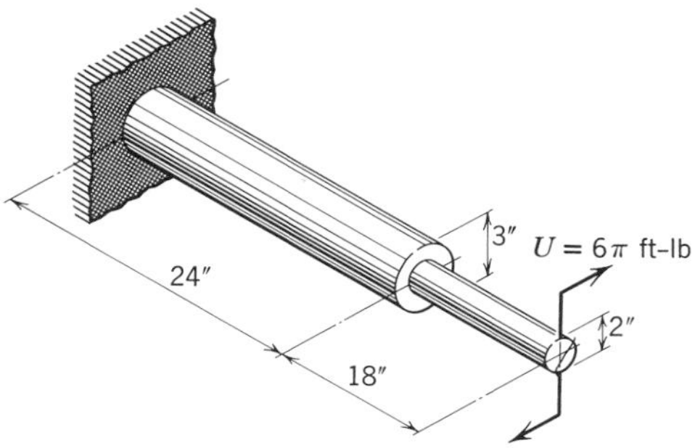


Fig. P8–44

**8–45.** The beam of Fig. P8–45 is 3 in. wide and 1 in. deep and the supporting coil springs each have a modulus of 50 lb per in. When the block $W$ is slowly applied to the center of the beam the ends of the beam were observed to move downward 0.10 in. and the center of the beam moved an additional 0.12 in. downward. Determine:

(*a*) The weight of the block $W$.

(*b*) The height $h$ from which $W$ must be dropped to produce a fiber stress in the beam of 2400 psi.

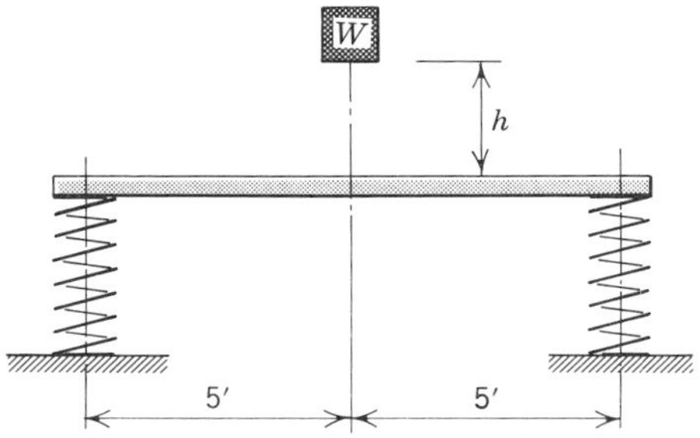


Fig. P8–45

**8–46.*** When the 50-lb block $B$ of Fig. P8–46 was dropped from a point 2 in. above the top of the coil spring $C$ (modulus is 150 lb per in.), the point $D$ on the steel cantilever beam $A$ was observed to deflect 1.44 in. downward. Determine:

(*a*) The percentage of the energy in block $B$ that was absorbed by the coil spring and beam combination ($A$ and $C$).

(*b*) The load factor.

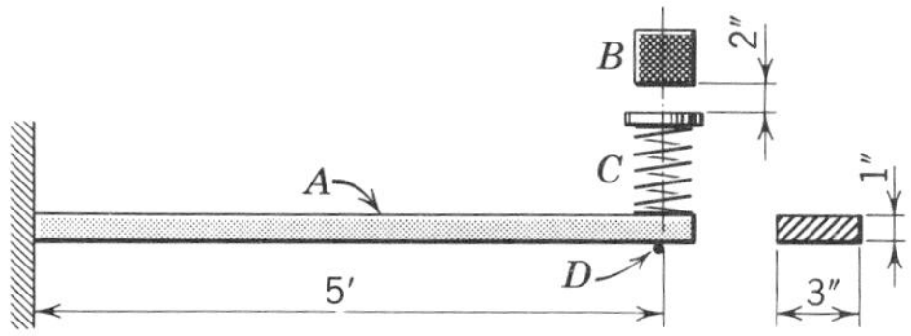


Fig. P8–46

**8–47.** The timber beam $A$ in Fig. P8–47 is 6 in. wide and 2 in. deep and is connected to the rigid arm $B$ which rotates about the horizontal pin at $O$. The modulus of elasticity of the timber is 1200 ksi and moduli of the helical springs are given on the figure. Body $W$ weighs 10 lb.

($a$) Determine the maximum height, $h$, from which $W$ can be dropped on the beam without producing a flexural stress in the beam of more than 1.2 ksi. Assume that 80 per cent of the energy of $W$ is effective in producing elastic deflections in the beam-spring system.

($b$) Determine the load factor corresponding to the $h$ of part ($a$).

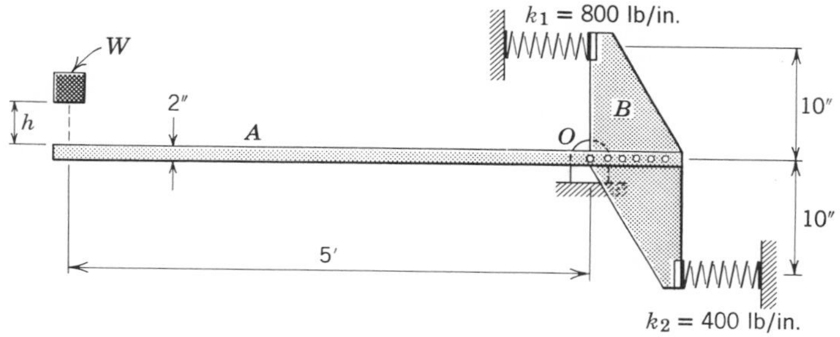


Fig. P8–47

**8–48.*** Figure P8–48 shows a steel beam with a box cross section 6 in. deep and 8 in. wide with a coil spring on the end of it. The spring modulus is 2500 lb per in. and the cross-sectional moment of inertia of the beam is 96 in.[4] Determine the maximum height from which the 250-lb weight, $W$, may be dropped upon the spring without exceeding a fiber stress of 15,000 psi in the beam. Assume that 80 per cent of the energy of the falling weight is absorbed elastically by the beam-spring combination.

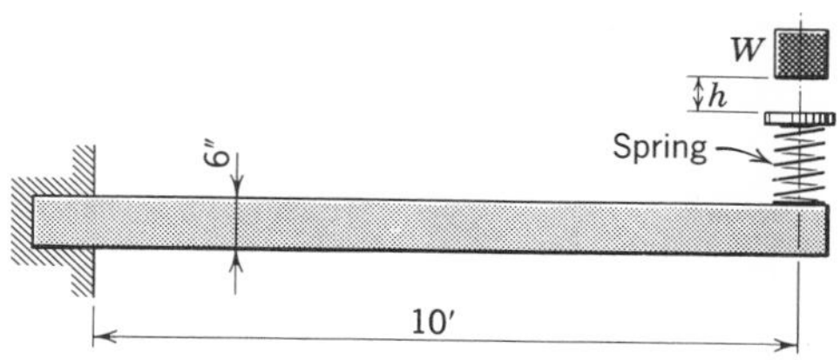


Fig. P8–48

**8–49.** Solve Prob. 8–48 if the coil spring is removed from the top of the beam and placed on a rigid support beneath the right end of the beam (the beam just touches the spring before the weight is dropped). All other data remain unchanged.

**8–50.** If the coil spring of Prob. 8–48 is removed from the top of the beam and placed on a rigid support beneath the right end of the beam (with the beam just touching the top of the spring) and if the 250-lb weight is dropped 5 in. onto the beam, what is the maximum fiber stress in the beam? All other data given in Prob. 8–48 remain unchanged.

**8–51.*** Beams $A$ and $B$ of Fig. P8–51 are of wood ($E$ is 1200 ksi) and are 2 in. deep by 6 in. wide. When the 20-lb weight, $W$, is dropped on the beam $A$, 90 per cent of the energy is absorbed elastically by beams $A$ and $B$, and the maximum deflection is 1.2 in. Determine the distance $h$, the load factor, and the amount of energy absorbed by each beam.

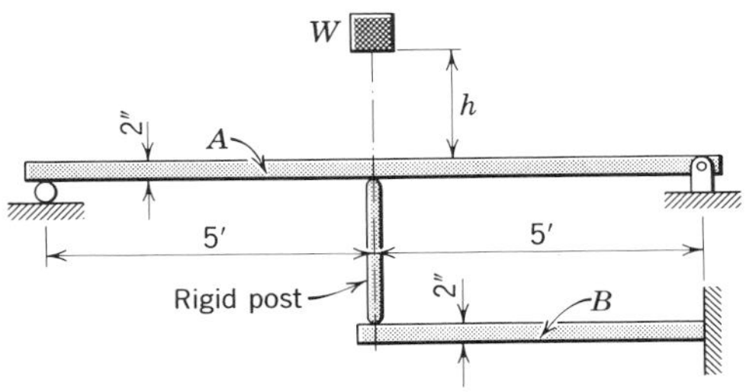


Fig. P8–51

**8–52.*** The aluminum alloy 6061-T6 beams $A$ and $B$ of Fig. P8–52 are 1 in. deep by 4 in. wide, and the helical spring $C$ has a modulus of 80 lb per in. The 10-lb block is dropped onto beam $A$ with a load factor of 4. If 80 per cent of the energy is absorbed elastically by $A$, $B$, and $C$:

($a$) What is the distance, $h$?

($b$) How much energy is absorbed elastically by each of the components?

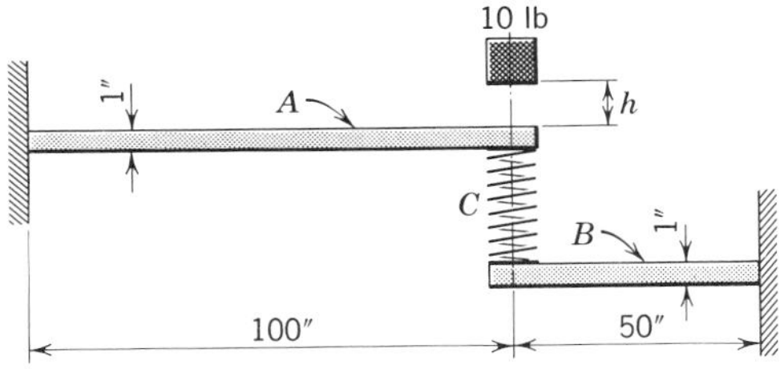


Fig. P8–52

**8–53.** Beam $AB$ of Fig. P8–53 has a flexural rigidity of $EI$ and is securely fastened to the rigid arm $C$. The modulus of the helical spring is $6EI/L$.[3] The structure rotates about a shaft at $B$. Determine the height $h$, in terms of $W$, $L$, $E$, and $I$, from which $W$ must be dropped to have a load factor of 4. Assume 75 per cent of the energy of $W$ is effective in producing elastic deformation in the beam-spring system.

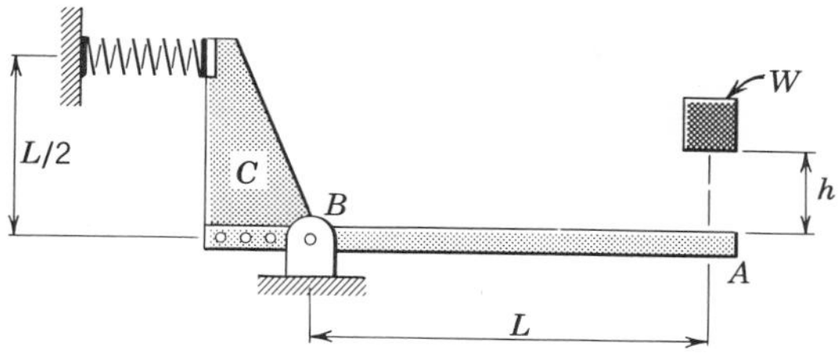


Fig. P8–53

**8–54.*** The structure in Fig. P8–54 consists of a solid steel rod $AB$ rigidly connected to the steel beam $BC$. The rod is fixed at $A$ and supported by a smooth bearing $D$ at $B$. The 100-lb weight $W$ is dropped one inch onto the bar and 80 per cent of the energy of the falling weight is absorbed elastically in $AB$ and $BC$. Determine the maximum fiber stress in $BC$ and the maximum normal stress on the top surface of $AB$ at the wall.

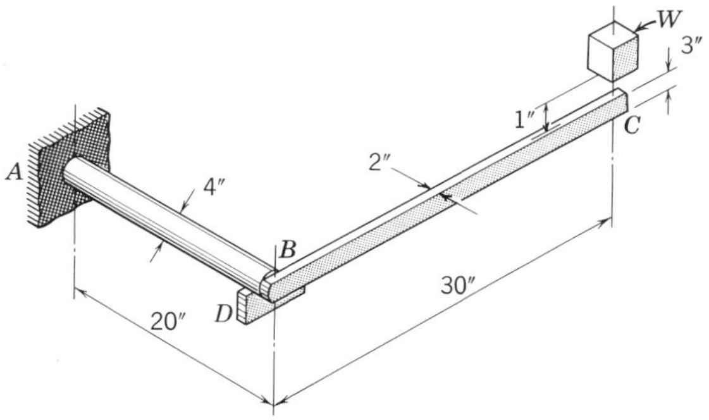


Fig. P8–54

**8–55.** Solve Prob. 8–54 if the bearing $D$ is removed.

**8–5. Theories of Failure.** A tension test of an axially loaded member is easy to conduct, and the results, for many types of materials, are well known. When such a member fails, the failure occurs at a specific principal (axial) stress, a definite axial strain, a maximum shearing stress of one-half the axial stress, and a specific amount of strain energy per unit volume of stressed material. Since all of these limits are reached simultaneously for an axial load, it makes no difference which criterion (stress, strain, or energy) is used for predicting failure in another axially loaded member of the same material.

For an element subjected to biaxial or triaxial loading, however, the situation is more complicated because the limits of normal stress, normal strain, shearing stress, and strain energy which existed at failure for an axial load are not all reached simultaneously. In other words, the cause of failure, in general, is unknown. In such cases it becomes important to determine the best criterion for predicting failure because test results are difficult to obtain and the combinations of loads are endless. Several theories have been proposed for predicting failure of various types of material subjected to many combinations of loads. Unfortunately none of the theories agree with test data for all types of materials and combinations of loads. Several of the more common theories of failure are presented and briefly explained in the following paragraphs.

**Maximum-normal-stress theory.**[3] The maximum-normal-stress theory predicts failure of a specimen subjected to any combination of loads when the maximum normal stress at any point reaches the axial failure stress as determined by an axial tensile or compressive test of the same material.

The maximum-normal-stress theory is presented graphically in Fig. 8–10*b* for an element subjected to biaxial principal stresses in the $x$ and $y$ directions as shown in Fig. 8–10*a*. The limiting stress $\sigma_f$ is the failure stress for this material when loaded axially. Any combination of biaxial principal stresses, $\sigma_x$ and $\sigma_y$, which is represented by a point inside the square of Fig. 8–10*b*, is safe according to this theory; whereas any combination of stresses which is represented by a point outside of the square will cause failure of the element on the basis of this theory.

**Maximum-shearing-stress theory.**[4] The maximum-shearing-stress theory predicts failure of a specimen subjected to any combination of

[3] Often called Rankine's theory after W. J. M. Rankine (1820–1872), an eminent engineering educator at the university level in England.

[4] Sometimes called Coulomb's theory because it was originally stated by him in 1773. More frequently called Guest's theory or law because of the work of J. J. Guest in England in 1900.

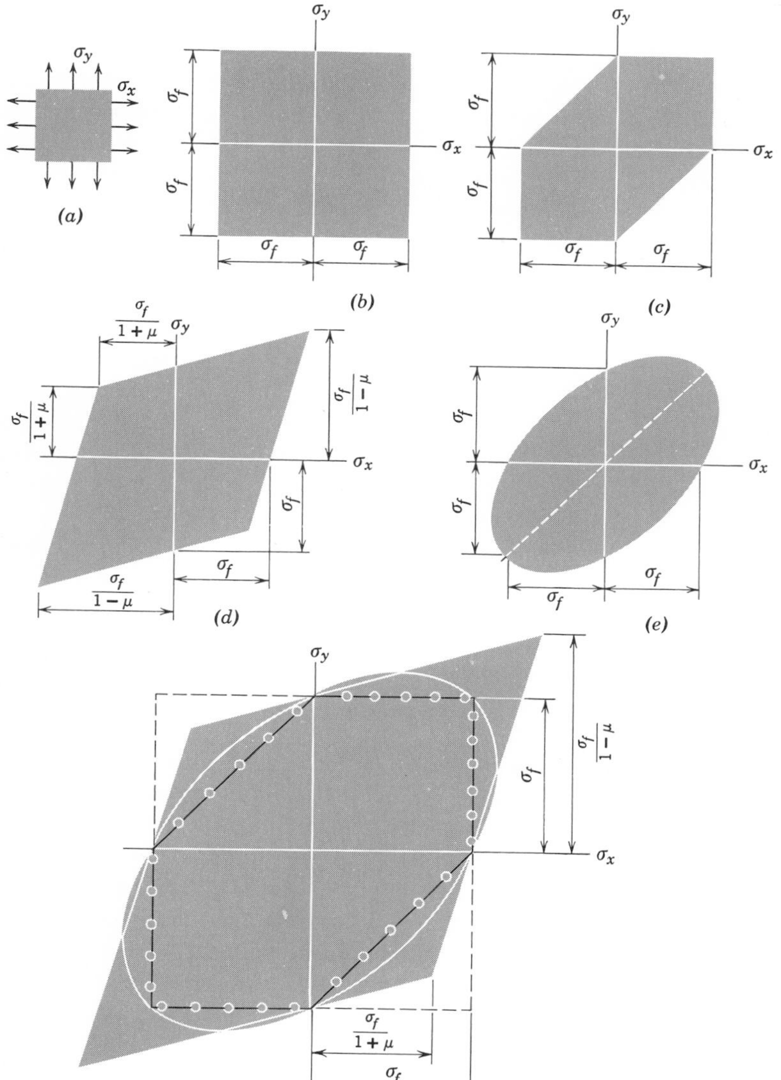

$\sigma_y$
$\sigma_x$
(a)
$\sigma_f$
(b)
(c)
$\frac{\sigma_f}{1+\mu}$
$\frac{\sigma_f}{1-\mu}$
(d)
(e)
(f)


Fig. 8–10

loads when the maximum shearing stress at any point reaches the failure stress, $\tau_f$ equal to $\sigma_f/2$, as determined by an axial tensile or compressive test of the same material. For ductile materials the shearing elastic limit, as determined from a torsion test (pure shear), is greater than one-half the tensile elastic limit (with an average value of $\tau_f$ about $0.57\sigma_f$). This means that the maximum-shearing-stress theory errs on the conservative side by being based on the limit obtained from an axial test.

The maximum-shearing-stress theory is presented graphically in Fig. 8–10*c* for an element subjected to biaxial ($\sigma_z$ is equal to zero) principal stresses as shown in Fig. 8–10*a*. In the first and third quadrants $\sigma_x$ and $\sigma_y$ have the same sign, and the maximum shearing stress is half of the numerically larger principal stress $\sigma_x$ or $\sigma_y$ as explained in Art. 1–6. In the second and fourth quadrants, where $\sigma_x$ and $\sigma_y$ are of opposite sign, the maximum shearing stress is half of the arithmetical sum of the two principal stresses. In the fourth quadrant the equation of the boundary, or limit stress, line is

$$\sigma_x - \sigma_y = \sigma_f,$$

and in the second quadrant the relation is

$$\sigma_x - \sigma_y = -\sigma_f.$$

**Maximum-normal-strain theory.**[5] The maximum-normal-strain theory predicts failure of a specimen subjected to any combination of loads when the maximum normal strain at any point reaches the failure strain, $\sigma_f/E$, at the proportional limit, as determined by an axial tensile or compressive test of the same material.

The maximum-normal-strain theory is presented graphically in Fig. 8–10*d* for an element subjected to biaxial principal stresses (not greater than the proportional limit) as shown in Fig. 8–10*a*. The limiting strain in the positive $x$ direction is

$$\epsilon_x = \frac{\sigma_f}{E} = \frac{\sigma_x - \mu\sigma_y}{E},$$

from which

$$\sigma_y = \frac{1}{\mu}(\sigma_x - \sigma_f),$$

[5] Often called Saint Venant's theory because of the work of Barre de Saint Venant (1797–1886), a great French mathematician and elastician.

which is observed to be a straight line through the point $(\sigma_f, 0)$ with a slope of $1/\mu$. The limiting strain in the positive $y$ direction is

$$\epsilon_y = \frac{\sigma_f}{E} = \frac{\sigma_y - \mu\sigma_x}{E},$$

from which

$$\sigma_y = \mu\sigma_x + \sigma_f,$$

which is observed to be a straight line through the point $(0, \sigma_f)$ with a slope equal to $\mu$.

**Maximum-strain-energy theory.** The maximum-strain-energy theory predicts failure of a specimen subjected to any combination of loads when the strain energy per unit volume (see Chap. 2 for the derivation of the relation between $\sigma_f$ and the elastic strain energy per unit volume) of any portion of the stressed member reaches the failure value of the strain energy per unit volume as determined by an axial tensile or compressive test of the same material. The maximum-strain-energy theory has been largely replaced by the maximum-distortion-energy theory.

**Maximum-distortion-energy theory.**[6] This theory differs from the maximum-strain-energy theory in that the portion of the strain energy producing volume change is considered ineffective in causing failure by yielding. Supporting evidence comes from experiments which show that homogeneous materials can withstand very high hydrostatic stresses without yielding. The portion of the strain energy producing change of shape of the element is assumed to be completely responsible for the failure of the material by inelastic action.

The strain energy of distortion is most readily computed by determining the total strain energy of the stressed material and subtracting the strain energy corresponding to the volume change. Article 2–2 defined the quantity $\sigma^2/(2E)$ as the strain energy per unit volume for a member subjected to a slowly applied axial load. This expression can also be written as

$$u = \frac{\sigma^2}{2E} = \frac{\sigma\epsilon}{2}$$

---

[6] Frequently called the Huber-Hencky-von Mises theory as it was proposed by M. T. Huber of Poland in 1904 and independently by R. von Mises of Germany in 1913. The theory was further developed by H. Hencky and von Mises in Germany and the United States.

where $u$ is the strain energy per unit volume and $\sigma$ and $\epsilon$ are the slowly applied axial stress and strain. This equation assumes that the stress does not exceed the proportional limit.

When an elastic element is subjected to triaxial loading, the stresses can be resolved into three principal stresses such as $\sigma_x$, $\sigma_y$, and $\sigma_z$, where $x$, $y$, and $z$ are the principal axes. These stresses will be accompanied by three principal strains which are related to the stresses by Eq. (2–2), Art. 2–3. If it is assumed that the loads are applied simultaneously and gradually, the stresses and strains will increase in the same manner. The total strain energy per unit volume is the sum of the energies produced by each of the stresses (energy is a scalar quantity and can be added algebraically regardless of the directions of the individual stresses) thus,

$$u = \left(\frac{1}{2}\right)(\sigma_x\epsilon_x + \sigma_y\epsilon_y + \sigma_z\epsilon_z).$$

When the strains are expressed in terms of the stresses this equation becomes

$$u = [1/(2E)][(\sigma_x^2 + \sigma_y^2 + \sigma_z^2) - 2\mu(\sigma_x\sigma_y + \sigma_y\sigma_z + \sigma_z\sigma_x)].$$

The strain energy can be resolved into two components $u_v$ and $u_d$, resulting from a volume change and a distortion, respectively, by considering the principal stresses to be made up of two sets of stresses as indicated in Fig. 8–11*a*, *b*, and *c*. The state of stress in Fig. 8–11*c* will result in distortion only (no volume change) if the sum of the three normal strains is zero. That is,

$$\begin{aligned} E(\epsilon_x + \epsilon_y + \epsilon_z)_d = {} & [(\sigma_x - p) - \mu(\sigma_y + \sigma_z - 2p)] \\ & + [(\sigma_y - p) - \mu(\sigma_z + \sigma_x - 2p)] \\ & + [(\sigma_z - p) - \mu(\sigma_x + \sigma_y - 2p)] = 0 \end{aligned}$$

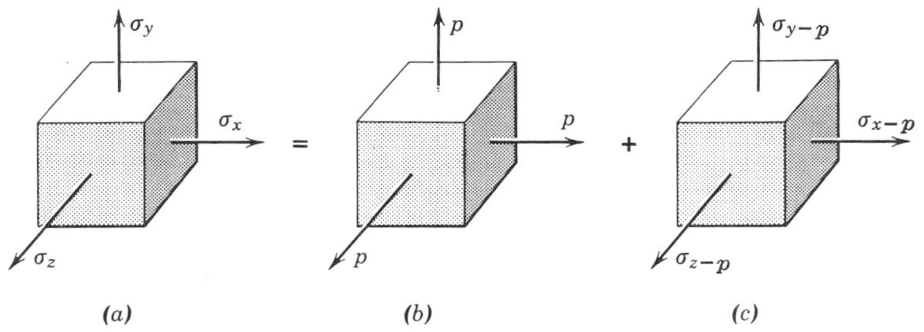


Fig. 8–11

which reduces to

$$(1 - 2\mu)(\sigma_x + \sigma_y + \sigma_z - 3p) = 0.$$

Therefore

$$p = (\sigma_x + \sigma_y + \sigma_z)/3.$$

The three normal strains due to $p$ are, from Eq. (2–2),

$$\epsilon_v = (1 - 2\mu)p/E$$

and the energy resulting from the hydrostatic stress (the volume change) is

$$u_v = 3\left(\frac{p\epsilon_v}{2}\right) = \frac{3}{2}\,\frac{1 - 2\mu}{E}\,p^2$$

$$= \frac{1 - 2\mu}{6E}(\sigma_x + \sigma_y + \sigma_z)^2.$$

The energy resulting from the distortion (change of shape) is

$$u_d = u - u_v$$

$$= \frac{1}{6E}[3(\sigma_x^2 + \sigma_y^2 + \sigma_z^2) - 6\mu(\sigma_x\sigma_y + \sigma_y\sigma_z + \sigma_z\sigma_x) - (1 - 2\mu)(\sigma_x + \sigma_y + \sigma_z)^2].$$

When the third term in the brackets is expanded, the expression can be rearranged to give

$$u_d = \frac{1 + \mu}{6E}[(\sigma_x^2 - 2\sigma_x\sigma_y + \sigma_y^2) + (\sigma_y^2 - 2\sigma_y\sigma_z + \sigma_z^2) + (\sigma_z^2 - 2\sigma_z\sigma_x + \sigma_x^2)]$$

$$= \frac{1 + \mu}{6E}[(\sigma_x - \sigma_y)^2 + (\sigma_y - \sigma_z)^2 + (\sigma_z - \sigma_x)^2]. \qquad \text{(a)}$$

The maximum-distortion-energy theory of failure assumes that inelastic action will occur whenever the energy given by Eq. (a) exceeds the limiting value obtained from a tensile test. For this test, only one of the principal stresses will be non-zero. If this stress is called $\sigma_f$, the value of $u_d$ becomes

$$(u_d)_f = \frac{1 + \mu}{3E}\sigma_f^2$$

and when this value is substituted in Eq. (a) it becomes

$$2\sigma_f^2 = (\sigma_x - \sigma_y)^2 + (\sigma_y - \sigma_z)^2 + (\sigma_z - \sigma_x)^2 \qquad \text{(b)}$$

for failure by slip.

When a state of plane stress exists, assume $\sigma_z$ equals zero, Eq. (b) becomes

$$\sigma_x^2 - \sigma_x\sigma_y + \sigma_y^2 = \sigma_f^2.$$

This last expression is the equation of an ellipse with its major axis along the line $\sigma_x$ equals $\sigma_y$ as shown in Fig. 8–10*e*.

An alternate approach to the distortion-energy theory of failure consists of determining the shearing stress on the octahedral planes, defined as any plane whose normal makes equal angles with the three principal axes. Figure 8–12 is a free-body diagram of an element bounded by the three principal planes and the octahedral plane. Since

$$\theta_x = \theta_y = \theta_z$$

$$\cos^2\theta_x + \cos^2\theta_y + \cos^2\theta_z = 3\cos^2\theta_x = 1$$

or

$$\cos\theta_x = \cos\theta_y = \cos\theta_z = 1/\sqrt{3}.$$

If the area of the octahedral plane is $\Delta A$, the other three areas are $\Delta A/\sqrt{3}$. The equilibrium equation along the $n$ axis gives

$$\Sigma F_n = \sigma_{\text{oct}}\,\Delta A - \frac{1}{\sqrt{3}}\left(\sigma_x \frac{\Delta A}{\sqrt{3}}\right) - \frac{1}{\sqrt{3}}\left(\sigma_y \frac{\Delta A}{\sqrt{3}}\right) - \frac{1}{\sqrt{3}}\left(\sigma_z \frac{\Delta A}{\sqrt{3}}\right) = 0$$

or

$$\sigma_{\text{oct}} = (\sigma_x + \sigma_y + \sigma_z)/3.$$

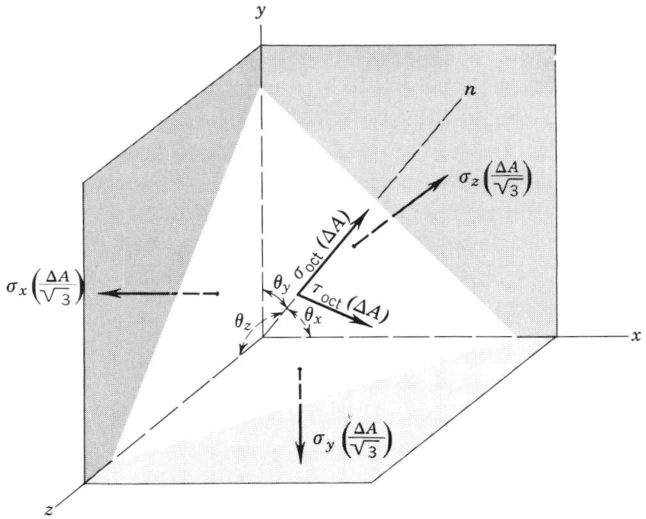


Fig. 8–12

Note that $\sigma_{oct}$ is the same as $p$ (the hydrostatic stress) in Fig. 8–11*b*. The resultant of the normal and shearing forces on the octahedral plane must equal the resultant of the forces on the other three planes, thus

$$(\sigma_{oct}\,\Delta A)^2 + (\tau_{oct}\,\Delta A)^2 = \left(\sigma_x \frac{\Delta A}{\sqrt{3}}\right)^2 + \left(\sigma_y \frac{\Delta A}{\sqrt{3}}\right)^2 + \left(\sigma_z \frac{\Delta A}{\sqrt{3}}\right)^2$$

from which

$$\begin{aligned}(\tau_{oct})^2 &= \frac{\sigma_x^2 + \sigma_y^2 + \sigma_z^2}{3} - \left[\frac{\sigma_x + \sigma_y + \sigma_z}{3}\right]^2 \\ &= \frac{1}{9}\,[3(\sigma_x^2 + \sigma_y^2 + \sigma_z^2) - (\sigma_x^2 + \sigma_y^2 + \sigma_z^2) \\ &\quad - 2(\sigma_x\sigma_y + \sigma_y\sigma_z + \sigma_z\sigma_x)] \\ &= \frac{1}{9}\,[(\sigma_x - \sigma_y)^2 + (\sigma_y - \sigma_z)^2 + (\sigma_z - \sigma_x)^2]. \qquad \text{(c)}\end{aligned}$$

When Eq. (b) and (c) are compared it is seen that

$$2\sigma_f^2 = (3\tau_{oct})^2$$

or

$$\tau_{oct} = \frac{\sqrt{2}}{3}\,\sigma_f = 0.471\sigma_f.$$

The octahedral shearing stress is thus related directly to the tensile failure stress and can be used as a criterion of the distortion energy at any point.

The graphical representations of Fig. 8–10*b*, *c*, *d*, and *e* are superimposed in Fig. 8–10*f* for convenient comparison of the different theories. The maximum-distortion-energy theory has been found to agree best with available test data for ductile materials, and its use in design practice is increasing. The maximum-normal-strain theory is unsafe for ductile materials. The maximum-normal-stress theory is also unsafe for ductile materials with principal stresses of opposite sign. The maximum-shearing-stress theory is conservative for ductile materials and is used in some design codes. The maximum-normal-stress theory agrees with test data for brittle materials and is generally accepted in design practice for such materials.

The following example illustrates the application of the theories of failure in predicting the load-carrying capacity of a member.

**Example 8–7:** The solid circular shaft of Fig. 8–13 has a proportional limit of 64 ksi and a Poisson's ratio of 0.30. Determine the value of the load $R$ for failure by slip as predicted by each of the four given theories of failure. Assume that point $A$ is the most severely stressed point.

*Solution:* When two equal opposite forces $R$ are introduced along line $EF$ of Fig. 8–13, it is evident that the bending moment at point $A$ is $36R$ in-kips, the

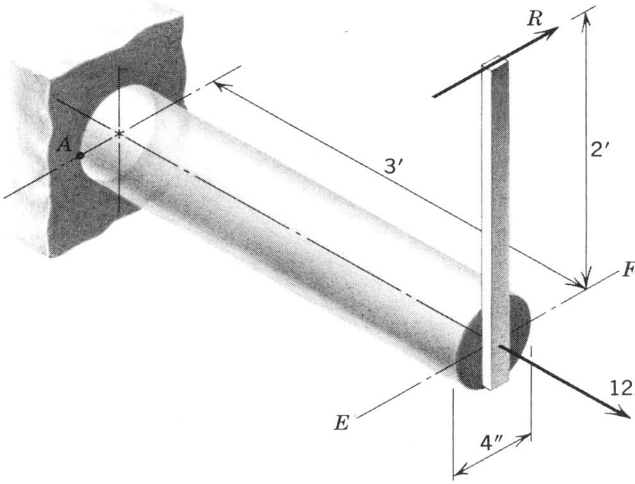


Fig. 8–13

torsional moment is $24R$ in-kips, and the axial load is $12R$ kips. The flexural stress at $A$ is

$$\sigma_1 = \frac{Mc}{I} = \frac{36R(2)}{\pi(2^4)/4} = \frac{18R}{\pi}\ T.$$

The direct stress at $A$ is

$$\sigma_2 = \frac{P}{A} = \frac{12R}{\pi(2^2)} = \frac{3R}{\pi}\ T,$$

and the sum of the two tensile stresses at $A$ becomes

$$\sigma = \sigma_1 + \sigma_2 = \frac{21R}{\pi}.$$

The torsional shearing stress at $A$ is

$$\tau = \frac{Tc}{J} = \frac{24R(2)}{\pi(2^4)/2} = \frac{6R}{\pi}.$$

The maximum stresses at $A$ are

$$\tau_{\max} = \sqrt{\left(\frac{10.5R}{\pi}\right)^2 + \left(\frac{6R}{\pi}\right)^2} = \frac{12.1R}{\pi},$$

$$\sigma_{p1} = \frac{10.5R}{\pi} + \frac{12.1R}{\pi} = \frac{22.6R}{\pi}\ T,$$

and

$$\sigma_{p2} = \frac{10.5R}{\pi} - \frac{12.1R}{\pi} = -\frac{1.6R}{\pi} = \frac{1.6R}{\pi}\ C.$$

**8–66.*** The bar of Fig. P8–30 is to be made of 8630 steel alloy (see stress strain diagram in App. A) and the loading is to be $8P$ in the $x$ direction as shown, and $3P$ in the $y$ direction (replacing the force $P$). The critical point is at the top of a vertical diameter at the wall. Using the maximum-normal-stress, the maximum-shearing-stress, and the maximum-distortion-energy theories of failure, determine the maximum permissible magnitude for $P$ to give a minimum factor of safety of 2 with respect to failure by slip.

CHAPTER 9

# *Columns*

**9–1. Introduction.** The principles developed in Chap. 1 apply to any member subjected to an axial tensile load, regardless of its dimensions. Compressive loading of long members, called *columns* or *struts,* may result in *buckling* in which the members experience lateral deflections that are excessive. This behavior can be demonstrated by placing one end of a yard stick on the floor and, while holding it vertical, pushing down on the top end. Buckling is a *stability* failure and the resulting lateral deflection is not related to the applied load by a direct proportion as in the case of laterally loaded beams. Consequently, the relations between load, dimensions, material properties, and conditions of loading must be re-examined and new expressions developed for column analysis. The following discussion of the nature of stability will point out the circumstances under which systems become unstable.

The classical example for demonstrating the classes of equilibrium is that of a ball at rest on a surface as indicated in Fig. 9–1. If the ball in Fig. 9–1*a* is perturbed from the equilibrium position shown, the action of gravity will cause it to return to its equilibrium position. This condition is known as *stable equilibrium.* The ball in Fig. 9–1*b* can be moved to any position on the level surface and it will remain at rest in its new position, tending neither to seek nor to move further away from the original equilibrium position. This is called *neutral equilibrium.* However, if the ball of Fig. 9–1*c* is perturbed, the action of gravity will cause it to move still further away from the position shown until it eventually finds some new stable equilibrium position. The condition, illustrated in Fig. 9–1*c,* is known as *unstable equilibrium.*

Now consider the prismatic bar pinned at the bottom and loaded axially at the top, as shown in Fig. 9–2. In addition to the smooth pin at the bottom, there is attached a torsion spring which offers a resistance, $K\theta$, to rotation about the pin. If the bar is displaced (by a

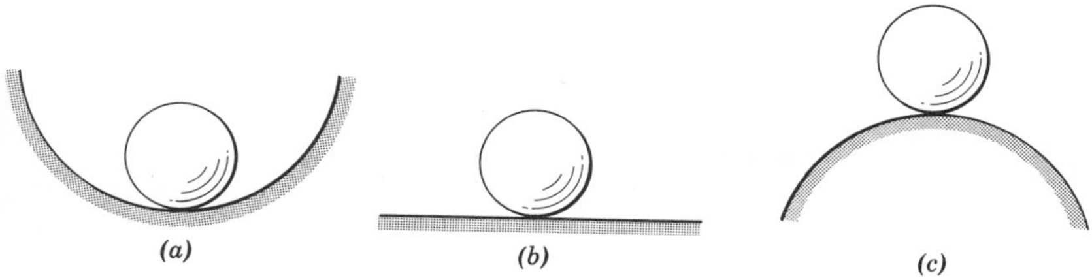

Fig. 9–1

momentary-lateral force) to the position shown in the free-body diagram, Fig. 9–2*b*, the torsion spring at the bottom will exert a restoring couple which will return the bar to the original (vertical) position provided the restoring couple exceeds the moment of the force *P* with respect to the pin. The mathematical statement for this condition is

$$K\theta > PL \sin \theta,$$

and the system is comparable to the system of Fig. 9–1*a*. If the system is to be in equilibrium for any displacement (neutral equilibrium),

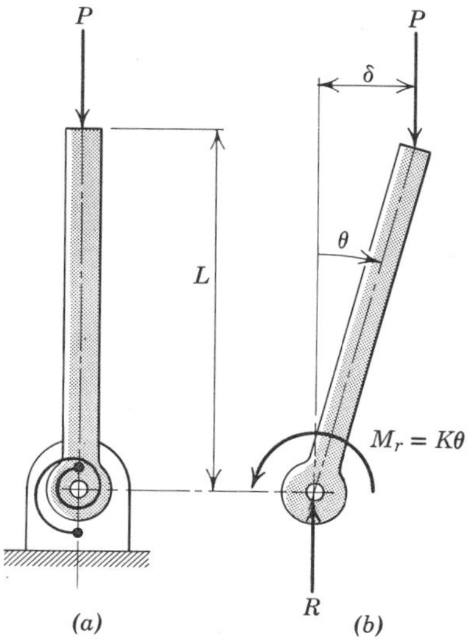


Fig. 9–2

the moment equation with respect to the pin is

$$K\theta = PL \sin \theta$$

from which

$$P = \frac{K}{L} \frac{\theta}{\sin \theta}. \tag{a}$$

The plot of Eq. (a) is given in Fig. 9–3. A study of this curve leads to the observation that if $P$ is less than or equal to $K/L$, a condition of stable equilibrium exists, since it takes a force greater than $K/L$ to hold the bar in a displaced position. If the force $P$ is increased beyond the value $K/L$, the equilibrium becomes unstable, since the bar, given a small accidental displacement, will continue to rotate until Eq. (a) is satisfied. Observe that an increase of only eleven per cent in the magnitude of $P$ above $K/L$ will result in a displacement of 45 degrees before equilibrium is re-established. Thus, there is a *critical load* (at $\theta = 0$, $P = K/L$ where the curve of Fig. 9–3 branches, or bifurcates with one branch the vertical line) below which the system is stable and above which the system is unstable.

If the bar and torsion spring system of Fig. 9–2 is replaced by an elastic bar fixed at the bottom as in Fig. 9–4, the analogy between the action of a column and the mechanical system is apparent. The entire column is now the spring which resists lateral deflection. The lateral

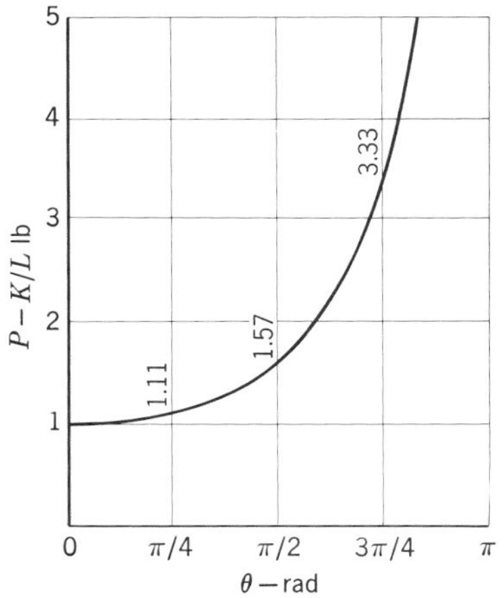


Fig. 9–3

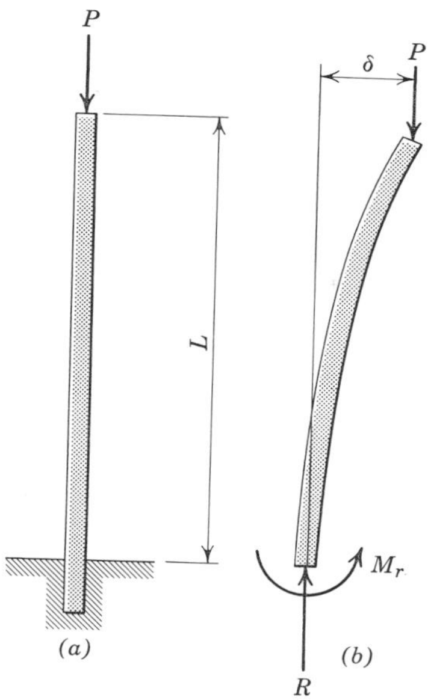


Fig. 9–4

deflection can be initiated by accidental eccentricities due to (1) deviations from absolute straightness of the column; (2) non-uniformity of stress distribution within the member due to non-homogeneity of the material or residual stresses; or (3) error in alignment when the load is applied. Only a small increase in load above the critical load will significantly increase the initial deflection—possibly enough to induce inelastic action in the material on the concave side of the deflected column, with the further possibility that such inelastic action could be sufficient to cause collapse (stability failure). The action would be analogous to failure of the torsion spring of Fig. 9–2.

As suggested previously, the problem is to develop an expression for the limiting compressive load which may be applied to an elastic prismatic bar; in other words, an expression comparable to Eq. (a) which will apply to members similar to Fig. 9–4. The first recorded attempt at such a development was that of Leonhard Euler (1707–1783), a Swiss mathematician, who in 1757 published an analysis of long slender columns which was derived from the elastic curve equation for a beam. This relationship, known as *Euler's formula,* was developed a number of years before a complete and correct analysis of the flexure problem was available.

Euler's solution is limited to slender columns, whereas short compression members can be treated as explained in Chap. 1. Many real columns, however, lie between these two extremes, and neither solution is applicable. Considerable effort has been made to obtain a reasonably simple solution which will apply to real columns of any length. However, because of the uncertain nature of many of the factors involved, which defy all but statistical or empirical analyses, most columns are designed by either various empirical formulas or a modified form of the Euler formula. However, in the development of even an empirical formula, it is helpful to start with some rational mathematical expression, and since, for any real column, the load will be applied with an eccentricity, either accidental or deliberate, the development in the next article will take such eccentricity into account. Later it will be observed that the resulting expression for an ideal column may, with the introduction of certain empirical modifications, be used to predict the critical load for a real column.

**9–2. The Secant Formula.** Figure 9–5*a* represents a member of length $L$ subjected to longitudinal loads which act a distance $e$ from the center line of the unloaded member. When the loads are applied, the member deflects to the position shown in Fig. 9–5*b*. Transverse loading is neglected in this analysis, which requires either that the weight of the member be negligible, that the member be placed in a vertical position, or that Fig. 9–5 represent top views of a member resting on a smooth horizontal plane. Axes are selected as shown for convenience.

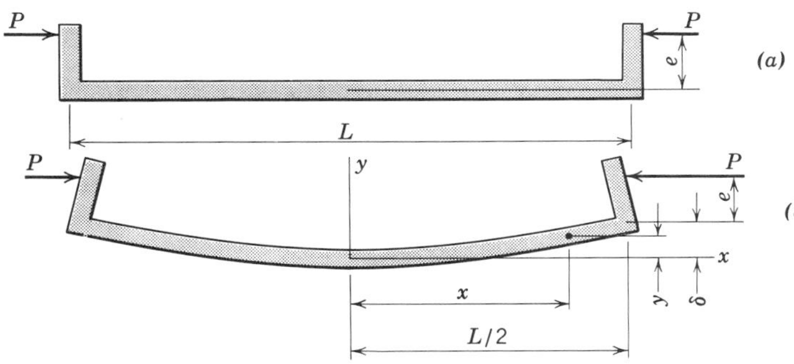


Fig. 9–5

The bending moment at any section is

$$M = P(e + \delta - y),$$

and, if the stress does not exceed the proportional limit and deflections are small, the differential equation of the elastic curve can be written (see Chap. 6) as

$$EI\frac{d^2y}{dx^2} = P(e + \delta - y)$$

or

$$\frac{d^2y}{dx^2} + \frac{P}{EI}y = \frac{P}{EI}(e + \delta). \tag{9–1}$$

This is an ordinary second-order differential equation with constant coefficients which can be solved by several methods. Probably the most common method is to assume a solution, and verify it by substitution in the differential equation. Since $y$ and its second derivative with respect to $x$ are involved, a function of $x$ whose second derivative is the original function multiplied by a negative constant is desired. Certain exponential and trigonometric functions have this property, and a trial solution is

$$y = A \sin px + B \cos px + C, \tag{a}$$

where $A$ and $B$ are constants of integration (determined by the boundary conditions) and $p$ and $C$ are constants determined by the differential equation. Substitution of Eq. (a) in Eq. (9–1) gives

$$-Ap^2 \sin px - Bp^2 \cos px + \frac{AP}{EI}\sin px + \frac{BP}{EI}\cos px + \frac{CP}{EI} = \frac{P}{EI}(e + \delta),$$

which shows that Eq. (9–1) is satisfied for all values of the constants $A$ and $B$ if

$$C = e + \delta \quad \text{and} \quad p^2 = \frac{P}{EI}.$$

Therefore a general solution of the differential equation is

$$y = A \sin \sqrt{P/(EI)}\, x + B \cos \sqrt{P/(EI)}\, x + e + \delta. \tag{b}$$

The constants $A$ and $B$ can be obtained from the two boundary conditions that the slope and deflection of the elastic curve are zero at the origin; that is, at $x = 0$, $y = 0$ and $dy/dx = 0$. When these expressions are substituted in Eq. (b), they give

$$0 = B + (e + \delta)$$

and

$$\left.\frac{dy}{dx}\right]_{x=0} = A\sqrt{P/(EI)}\cos\sqrt{P/(EI)}\,x - B\sqrt{P/(EI)}\sin\sqrt{P/(EI)}\,x\Big]_{x=0} = 0$$

or

$$A\sqrt{P/(EI)}\,(1) = 0,$$

from which $A = 0$ and $B = -(e + \delta)$.

The solution to Eq. (9–1) thus becomes

$$y = (e + \delta)[1 - \cos\sqrt{P/(EI)}\,x]. \tag{9–2}$$

The maximum deflection can be determined from Eq. (9–2) since $y = \delta$ at $x = L/2$. Thus

$$\delta = e + \delta - e\cos\sqrt{P/(EI)}\,(L/2) - \delta\cos\sqrt{P/(EI)}\,(L/2)$$

or

$$\delta = e\left[\frac{1 - \cos\sqrt{P/(EI)}\,(L/2)}{\cos\sqrt{P/(EI)}\,(L/2)}\right] = e[\sec\sqrt{P/(EI)}\,(L/2) - 1]. \tag{c}$$

The maximum compressive stress can be obtained by superposition since it has already been assumed that the stress will not exceed the proportional limit (when the elastic curve equation was written). The maximum stress is

$$\sigma_{\max} = \frac{P}{A} + \frac{M_{\max}c}{I} = \frac{P}{A} + \frac{P(e + \delta)c}{Ar^2}$$

in which $r$ is the radius of gyration of the column cross section.

From Eq. (c) $$e + \delta = e\sec\sqrt{\frac{P}{EI}}\frac{L}{2};$$

therefore $$\sigma_{\max} = \frac{P}{A}\left[1 + \frac{ec}{r^2}\sec\sqrt{\frac{P}{EI}}\frac{L}{2}\right]$$

or

$$\frac{P}{A} = \frac{\sigma_{\max}}{1 + \dfrac{ec}{r^2}\sec\left(\dfrac{L}{2r}\sqrt{\dfrac{P}{EA}}\right)}. \tag{9–3}$$

Equation (9–3) is known as the secant formula and relates the average unit load, $P/A$, to the dimensions of the column, the properties of the column material, and the eccentricity, $e$. The quantity $L/r$ is called the *slenderness ratio,* and it is extremely important in predicting column behavior. The quantity $ec/r^2$ is called the *eccentricity ratio* and is seen to depend on the eccentricity of the load and the dimensions of the column. If the column were loaded axially, $e$ would presumably be zero, and the maximum stress would be equal to $P/A$. It is virtually impossible, however, to eliminate all eccentricity which might result from various factors such as the initial crookedness of the column, minute flaws in the material, and a lack of uniformity of the cross section as well as accidental eccentricity of the load. A committee of the American Society of Civil Engineers[1] made an extensive study of the results of a great many column tests and came to the conclusion a value of 0.25 for $ec/r^2$ would give results with the secant formula which would be in good agreement with experimental tests on axially loaded columns of structural steel in ordinary structural sizes. A more complete discussion of eccentrically loaded columns is included in Art. 9–5. When Eq. (9–3) is used, $\sigma_{max}$ is the proportional limit although the yield point of structural steel may be used for convenience, and $P$ is the load at which inelastic action would start to occur. This load should be divided by a factor of safety for design purposes. Note that $P/A$ occurs on both sides of the equation, and thus it is not directly proportional to $\sigma$. Consequently, the load resulting from reducing $\sigma$ to some proportion of the proportional limit will not be reduced by the same amount. In order to obtain a given "factor of safety," the factor must be applied to the load each time it occurs in the formula and not to the value of $\sigma$; that is, in Eq. (9–3) $kP_w$ should be substituted for $P$, where $P_w$ is the "working load" (maximum safe load) and $k$ is the factor of safety.

In order to make effective use of Eq. (9–3), curves showing $P/A$ and $L/r$ can be drawn for various values of $ec/r^2$ for any given material. Figure 9–6 is such a set of curves for a material with $\sigma = 40{,}000$ psi and $E = 30(10^6)$ psi. The curve labeled "Euler curve" will be discussed later.

Equation (c) indicates that, for a given column in which $E$, $I$, and $L$ are fixed, the maximum deflection $\delta$ is a function of the load $P$ and the eccentricity $e$. For any value of $e$, the quantity

$$\sec \sqrt{P/(EI)}\,(L/2)$$

approaches positive or negative infinity as the argument $\sqrt{P/(EI)}\,(L/2)$ approaches $\pi/2$, $3\pi/2$, $5\pi/2$, . . . , and the deflection $\delta$ increases in-

[1] Final Report of Special Committee on Steel Column Research, *Trans. Am. Soc. of Civil Engrs.,* Vol. 98, 1933.

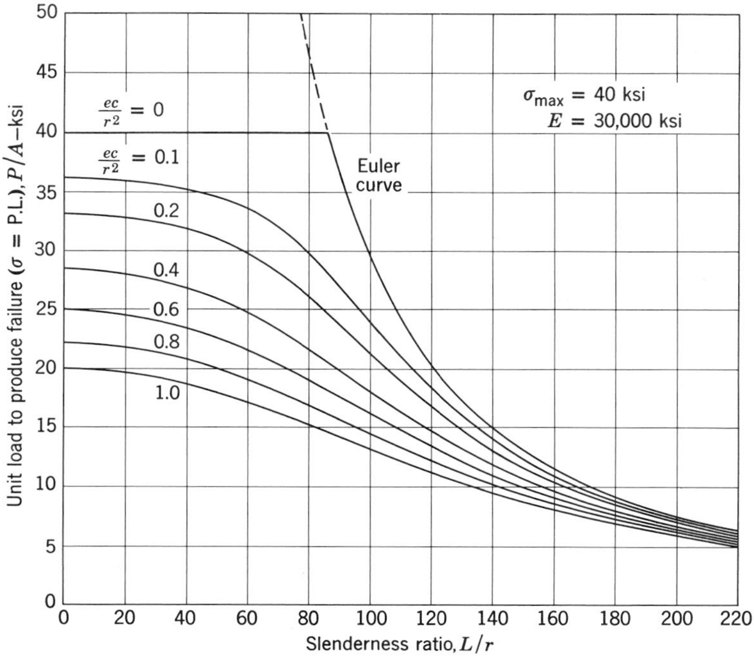


Fig. 9–6

definitely, indicating that the critical load corresponds to one of these angles. If $\pi/2$ is chosen, then

$$\sqrt{P/(EI)}\ (L/2) = \pi/2,$$

or

$$\sqrt{P/(EI)} = \pi/L \tag{d}$$

from which

$$P_{cr} = \frac{\pi^2 EI}{L^2}. \tag{9–4}$$

If Eq. (d) is substituted into Eq. (9–2), the deflection curve of Fig. 9–5*b* is obtained. If the argument $3\pi/2$ were chosen, the curve would be a cosine curve with three segments and nodes at the third-points. Similarly, the argument $5\pi/2$ would result in a curve with five segments, and so on. All these are valid mathematical solutions to Eq. (9–2); however, the smallest argument will give the least magnitude for the critical load and experience indicates that a column loaded statically as in Fig. 9–5 with no intermediate restraints will bend as shown in the figure.

The quantities $E$, $I$, and $L$ can be eliminated from Eq. (c) by using Eq. (9–4), in which case the maximum deflection becomes

$$\delta = e[\sec (\pi/2) \sqrt{P/P_{cr}} - 1]. \tag{e}$$

The graphs of Eq. (e) for various values of $e$ are shown in Fig. 9–7. Note that the deflection approaches infinity as $P$ approaches $P_{cr}$ regardless of the eccentricity, $e$. As $e$ approaches zero, however, the deflection is extremely small until the load reaches the critical value, at which time the deflection increases very rapidly. This limiting case represents an axially loaded column, and the curves illustrate how the deflection under a supposedly axial load can be initiated by a small accidental eccentricity. Equation (e) suggests that for $e$ equal to zero

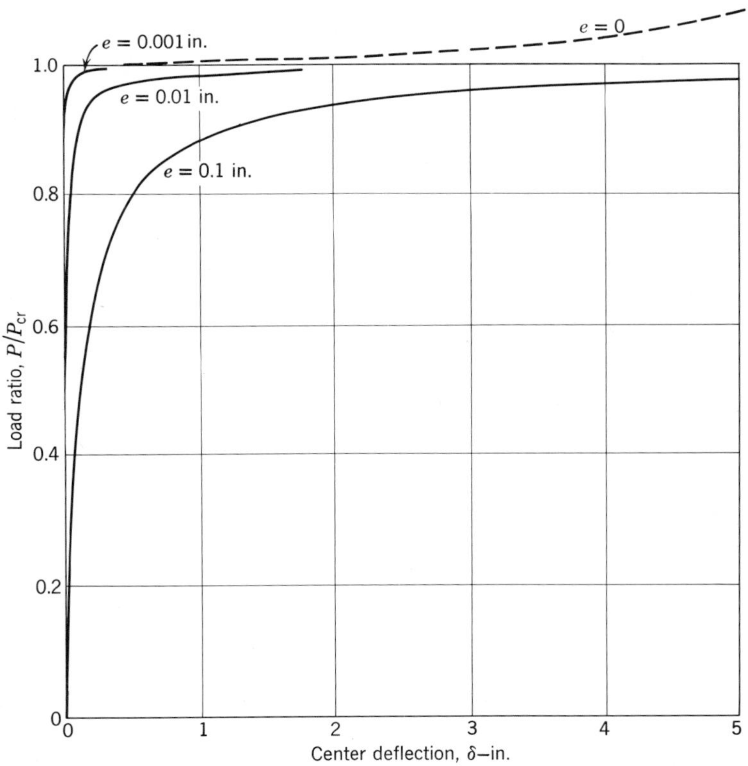


Fig. 9–7

and $P$ equal to $P_{cr}$, the deflection $\delta$ is indeterminate. However Eq. (e), having been derived from the differential equation for *small* deflections, is not valid for large deflections. In Fig. 9–7, the curve shown in dashed lines for $e$ equal to zero is a plot of the exact solution[2] which indicates that a load only slightly greater than $P_{cr}$ would probably cause collapse of the column.[3] Note the similarity between this curve and that of Fig. 9–3.

When $I$ is replaced by $Ar^2$ in Eq. (9–4), the expression for the critical load becomes

$$\frac{P_{cr}}{A} = \frac{\pi^2 E}{(L/r)^2}; \qquad (9\text{–}4a)$$

in this form it is known as the *Euler formula*. The radius of gyration in the quantity $L/r$, the slenderness ratio, is determined with respect to the *axis about which bending tends to occur;* and, for a centric load, pivot ends, and no intermediate bracing, it is always the least radius of gyration. A *pivot-ended* column is supported in such a manner that both the bending moment and lateral movement are zero at the ends.

Equation (9–4$a$) was used to plot the Euler curve of Fig. 9–6. Since this equation was derived on the assumption of elastic action, it is not valid for values of $P/A$ greater than 40,000 psi at which value $L/r$ is equal to 86. As mentioned previously, it is nearly impossible to obtain zero eccentricity in an actual column, but Fig. 9–6 indicates that the effect of the eccentricity decreases as the slenderness ratio increases. Consequently, Euler's equation can be used to predict the buckling loads for axially loaded columns provided that the slenderness ratio is large (more than about 150 for the steel represented in Fig. 9–6). Note that the slenderness ratio must be computed to decide whether or not the Euler equation is valid for the problem under consideration.

## PROBLEMS

**In the following problems assume that all columns are pivot ended and that Euler's formula is applicable for steel with $L/r$ greater than 140 and for aluminum or timber with $L/r$ greater than 80.**

**9–1.** Derive the Euler formula, Eq. (9–4) or (9–4$a$), by solving the differential equation of the elastic curve for an axially loaded pivot-ended column. Locate the origin of coordinates at one end of the column and integrate the differential equation (instead of assuming a solution and checking).

---

[2] See *Theory of Elastic Stability,* 2nd Ed., S. Timoshenko and J. Gere, McGraw-Hill, New York, 1961, pp. 77–82.

[3] The curve is valid for any deflection only until the proportional limit of the material is exceeded at which time the curve drops.

**9–2.*** A hollow circular steel column 10 ft long has an inner diameter of 3 in. and an outer diameter of 4 in. Determine the slenderness ratio.

**9–3.*** The cross section of a 15-ft column is shown in Fig. P9–3. Determine the slenderness ratio.

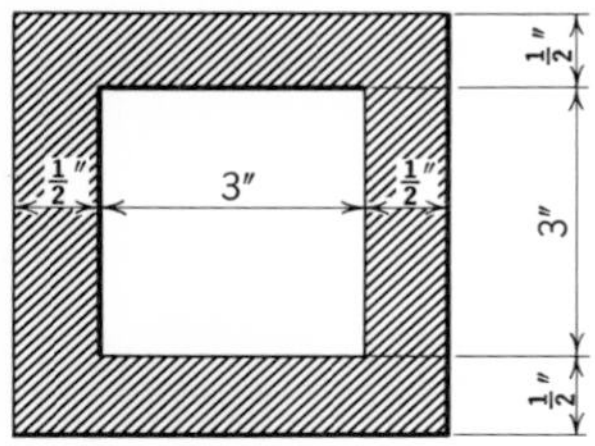


Fig. P9–3

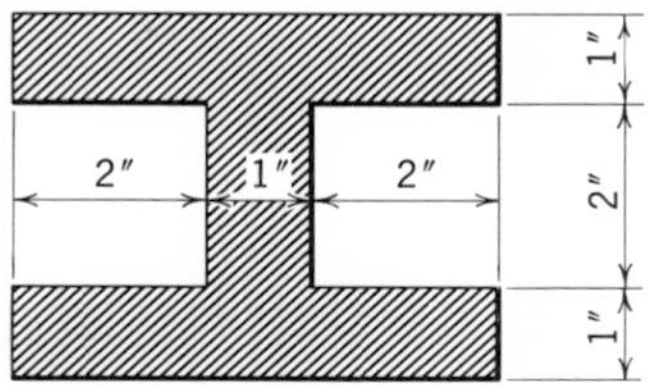


Fig. P9–4

**9–4.*** The cross section of an 8-ft column is shown in Fig. P9–4. Determine the slenderness ratio.

**9–5.** The cross section of a 20-ft column is shown in Fig. P9–5. Determine the slenderness ratio.

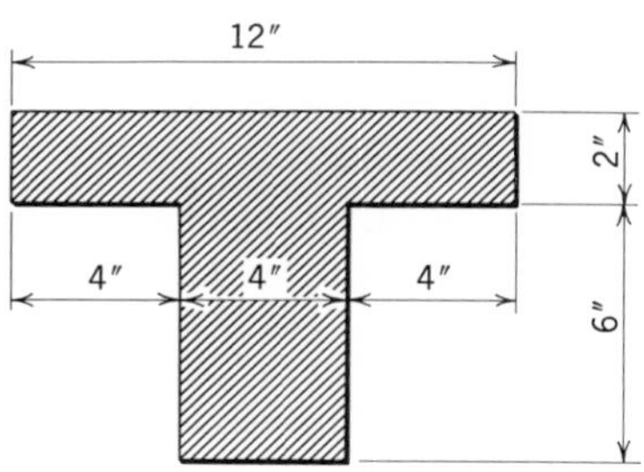


Fig. P9–5

**9–6.*** Two 4 by 3.5 by 0.5-in. angles 8 ft long are to support an axial compression load. Determine the slenderness ratio if:

(*a*) The angles are fastened together to form a T-section with the 4-in. legs connected.

(*b*) The angles are fastened together as in (*a*), except that the 3.5-in. legs are connected.

(*c*) The angles are not fastened together (each can act as an independent axially loaded member).

**9–7.** A column 20 ft long is made of two 10 ⊔ 15.3 sections latticed 5 in. back to back (see Fig. 9–11*a*).

(*a*) Determine the slenderness ratio of this column.

(*b*) What is the slenderness ratio of one of the channels?

(*c*) Determine the correct distance between the backs of the channels for a column with equal resistance to bending about either axis.

**9–8.*** Determine the slenderness ratio of a 5-ft column having the cross section of Fig. P9–8.

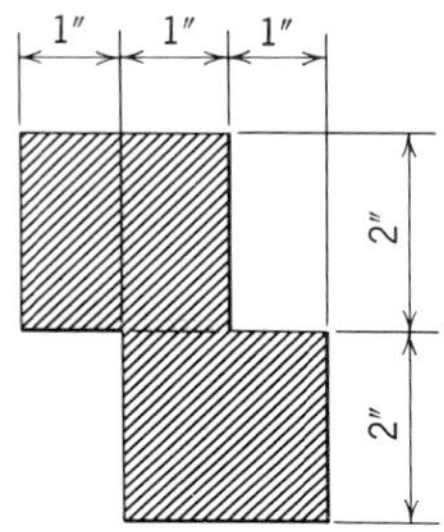


Fig. P9–8

**9–9.** Use Fig. 9–6 to estimate the maximum safe axial compression load for a steel strut having a rectangular cross section 2 in. by 1 in. and a length of 25 in. Use a factor of safety of 2.5 and assume $ec/r^2$ to be 0.20. What is the corresponding value of the eccentricity?

**9–10.*** A strut in a machine is made of steel having the same properties used in preparing Fig. 9–6. The rectangular cross section of the strut is 0.500 by 1.000 in., and it is 15 in. long. The strut buckles about the long axis of the section at a load of 9000 lb. Use Fig. 9–6 to estimate the value of the eccentricity ratio, $ec/r^2$, and the resulting eccentricity of the applied load.

**9–11.** The column of Prob. 9–2 is made of the steel for which Fig. 9–6 was prepared. When loaded in compression, the column buckles at a load of 121 kips. Use Fig. 9–6 to determine the value of the eccentricity ratio and the corresponding eccentricity of the applied load.

**9–12.*** Compute the maximum permissible length for a 4-in. diameter solid steel (the steel of Fig. 9–6) shaft which is to support an axial compressive load of 80 kips with a factor of safety of 2.5. Assume $ec/r^2 = 0.20$. Check the computed length with Fig. 9–6.

**9–13.** Compute the value of $L/r$ corresponding to $P/A = 30$ ksi and $ec/r^2 = 0.30$ for the data used to plot Fig. 9–6. Check this point to see that it is in reasonable agreement with Fig. 9–6.

**9–14.*** A 5-ft length of 1-in. diameter stainless steel tubing with a wall thickness of 0.050 in. is tested as a round-ended column. The critical load was found to be 1590 lb. Show that this column can be considered to be an *Euler* column (assume the 1.000-in. diameter to be the mean diameter), and determine Young's modulus for the material.

**9–15.** Determine the maximum load that a 4 by 6-in. dry Douglas fir (see App. A) column 10 ft long can support with a factor of safety of 3.

**9–16.*** Two 6 ⊔ 3.63 aluminum alloy 6061-T6 (see App. A) channels are fabricated into a box section by welding together the outstanding edges of the flanges. Such a section 15 ft long is to be used as a column. Determine the maximum load which may be applied with a factor of safety of 2. The cross-sectional properties of the channel are the same as those for a steel 6 ⊔ 10.5 (see App. C).

**9–17.** Determine the allowable axial compressive load, with a factor of safety of 2, for a steel 8 WF 40 which is 25 ft long (see App. C for cross-sectional properties).

**9–18.*** Determine the maximum axial compressive load that a 3 by 3 by ¼-in. steel angle 10 ft long can support with a factor of safety of 2.5 (see App. C for cross-sectional properties).

**9–3. Effect of End Conditions.** When the secant formula, Eq. (9–3), and the Euler formula, Eq. (9–4*a*), were derived, a zero bending moment at the ends of the column was assumed. The length $L$ in the column formulas is therefore the distance between two points of zero bending moment. Figure 9–8 shows four common end conditions for columns. The first is a pivot- or round-ended column, and the distance between points of zero moment is the total length of the column and should be used in the column formulas. The ends of the column in Fig. 9–8*b* are fixed or built in, and, since the deflection curve is symmetrical, the distance between points of zero moment (inflection points) is half the total length of the column. This distance, called the *effective length* of the column, is to be used in the column formulas for columns with both ends fixed. The member in Fig. 9–8*c* is fixed at one end, free at the other, and equivalent to half of the column in Fig. 9–8*a*. Thus, the effective length $L'$ to be used in the formulas is twice the length of the column. The column in Fig. 9–8*d* is fixed at one end and pinned at the other. The distance from the top to the inflection point is approximately 0.7 the length of the column and can be used in Eq. (9–3), (9–4), and (9–4*a*).

A pin-ended column is usually loaded through a pin which, as a result of friction, is not competely free to rotate; therefore there will

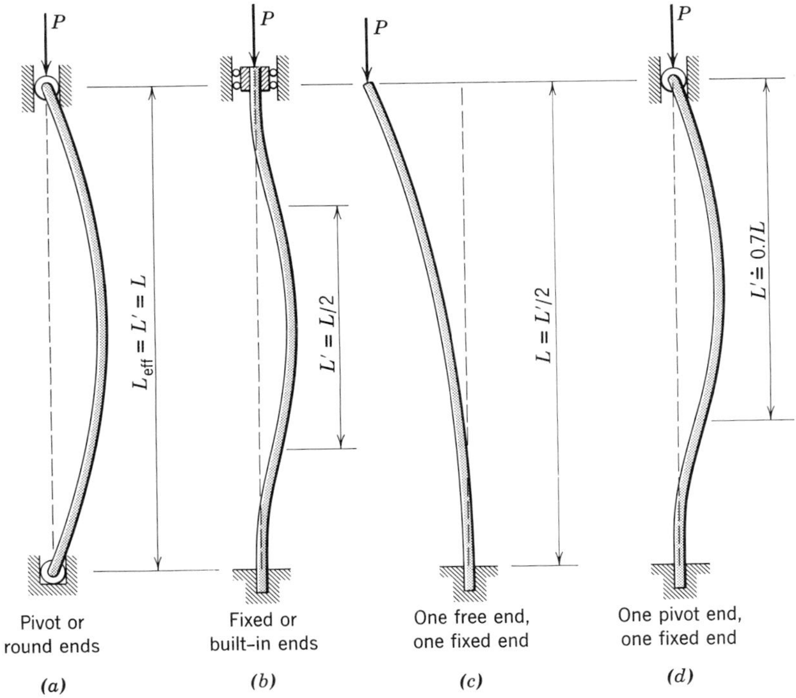


Fig. 9–8

always be an indeterminate moment at the ends of a pin-connected column which will reduce the distance between the inflection points to a value less than $L$. Also it is impossible to support a column so that all rotation is eliminated, and so the effective length of the column in Fig. 9–8*b* will be somewhat greater than $L/2$. As a result, it is usually necessary to modify the effective column lengths indicated by the ideal end conditions. The amount of the corrections will depend on the individual application. In summary, the term $L/r$ in all column formulas is interpreted to mean the effective slenderness ratio, $L'/r$. In the problems in this book, the length given for a member is assumed to be the effective length unless otherwise noted.

## PROBLEMS

**9–19.** By solving the differential equation of the elastic curve for a column with an applied axial load and equal moments applied at each end, verify the effective length given for the column shown in Fig. 9–8*b*.

**9–20.** Consider the rigid beam supported by two columns built in at both ends as shown in Fig. P9–20. Assuming that buckling would occur as shown, determine the effective lengths of the columns by solving the differential equation of the elastic curve.

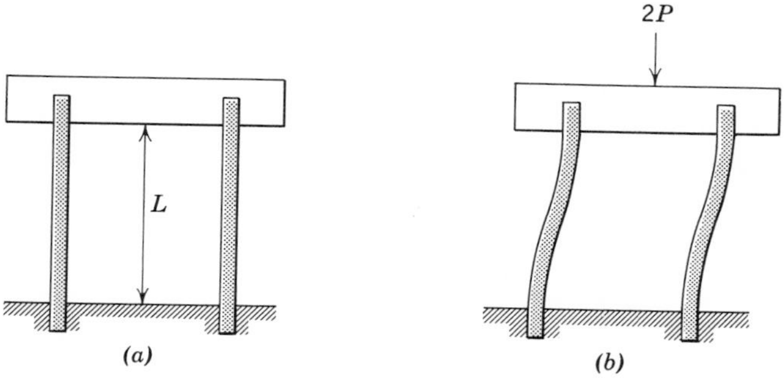


Fig. P9–20

**9–4. Empirical Column Formulas: Centric Loading.** Referring to Fig. 9–6, one observes that the envelope of the family of curves is the horizontal line $P/A = 40$ ksi up to $L/r$ of 86 and the plot of the Euler formula, Eq. (9–4*a*), for $L/r$ greater than 86. This curve represents the failure unit loads for *ideal* columns of any length under centric loading. However, for real columns (which will always have built-in eccentricities), the true curve will be an S-shaped curve similar to those shown in Fig. 9–6. Numerous column experiments indicate that the Euler formula, Eq. (9–4) or (9–4*a*), is reliable for designing axially loaded columns, provided the slenderness ratio is within the range where the eccentricity has relatively little effect. This range is called the *slender* range, which in Fig. 9–6 would begin in the region of $L/r$ equal to 120–140 (the region of transition is dependent on the material). At the low end of the slenderness scale, the column would behave essentially as a compression block and the failure unit load would be the compressive strength of the material (for metals, usually the elastic strength, $\sigma_y$). The extent of this range, the *compression block* range, is a matter of judgment, or is dictated by specifications. In Fig. 9–6, the small eccentricity curves seem to indicate that the range 0 to 40 could be considered the compression block range. The range between the compression block and the slender ranges is known as the *intermediate* range and the only equation of a rational nature which applies to real columns in this range is the secant formula, Eq. (9–3). Since the application of the secant formula to centric loading requires an estimate of the accidental eccentricity ratio, the equation acquires an empirical nature. Furthermore, the application is so involved that simpler empirical formulas have been developed which give results that are in reasonable agreement with experimental results within the intermediate range. Although a great many equations have been proposed, most of them come under one of three major classifications. These three general formulas are the *Gordon-Rankine*, the *straight-line*, and the *parabolic* formulas, described in the following paragraphs.

The *Gordon-Rankine* formula, one of the first empirical formulas, was developed by Gordon and Rankine, English engineers, during the middle of the nineteenth century. The formula is usually written as

$$\frac{P}{A} = \frac{\sigma_0}{1 + C_1(L/r)^2}, \tag{a}$$

in which $\sigma_0$ and $C_1$ are experimentally determined constants. The equation plots as a curve similar to the secant curve.

The *straight-line* formula, proposed by T. H. Johnson about 1890, is written in general form as

$$\frac{P}{A} = \sigma_0 - C_2\frac{L}{r}, \tag{b}$$

where again $\sigma_0$ and $C_2$ are experimentally determined material dependent constants.

At about the same time that T. H. Johnson proposed the straight-line formula, Professor J. B. Johnson proposed a *parabolic* formula of the general form

$$\frac{P}{A} = \sigma_0 - C_3\left(\frac{L}{r}\right)^2, \tag{c}$$

where $\sigma_0$ and $C_3$ are the experimentally determined constants. This latter formula with the proper choice of constants can be made to agree quite closely in the intermediate range with the secant formula with certain prescribed numerical constants. See, for example, the close correlation between curves ② and ⑦ in Fig. 9–9.

For design purposes, the entire range of unit loads for a given material is covered by an appropriate set of specifications known as a *column code*. The code will specify the empirical formula for the intermediate range, and the limits of the intermediate range; usually, it will also specify the Euler formula in the slender range. If the code is written for *safe* (or *allowable*) loads, the factor of safety, $k$, will be specified for the Euler formula, and will either be specified or be included in the constants for the empirical formula. A few representative codes are listed in Table 9–1, in which the Euler formula is specified for the slender range except for Code No. 5. Some codes, such as Nos. 3, 4, and 5, are for ultimate (buckling) loads. When these codes are used to determine safe loads, the values obtained from the formulas must be divided by an appropriate factor of safety. The data in Table 9–1 are taken or adapted from handbooks published by the American Institute of Steel Construction (A.I.S.C.), the American Association of State Highway Officials (A.A.S.H.O.), the Armed Forces Supply Support Center (Mil-Hdbk-5), the Aluminum Company of America (Alcoa Handbook), and the Forest Products Laboratory (U.S.D.A. Hdbk #72). A graphic comparison of selected codes is provided in Fig. 9–9.

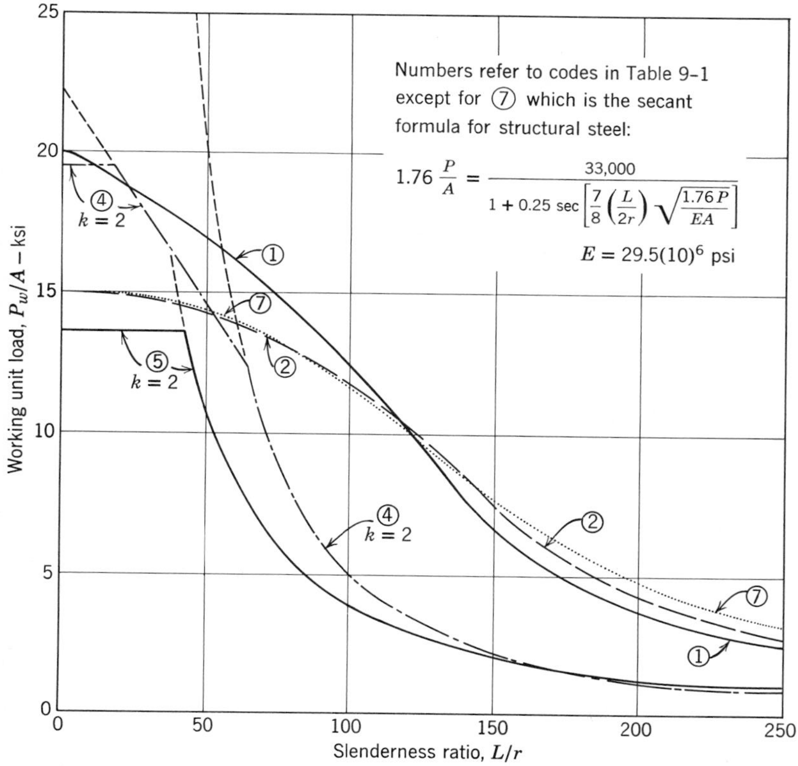


Fig. 9–9

## TABLE 9–1 Some Representative Column Codes

| Code No. | Source | Material and Modulus of Elasticity, ksi | Intermediate Range Formulas and Limitations<br>Note; $\frac{L}{r}$ refers to the effective ratio $\frac{L'}{r}$ | Slender Range Factor of Safety |
|---|---|---|---|---|
| 1 | A.I.S.C. Specification 1962 | Steel with Yield Point of $\sigma_y$ 29(10³) | $\frac{P_w}{A} = \frac{1}{k}\left[\sigma_y - \frac{\sigma_y}{2C_c^2}\left(\frac{L}{r}\right)^2\right]$<br>$C_c^2 = \frac{2\pi^2 E}{\sigma_y}$<br>$k = \frac{5}{3} + \frac{3(L/r)}{8C_c} - \frac{(L/r)^3}{8C_c^3}$<br>$0 \le \frac{L}{r} \le C_c$ | 1.92 |
| 2 | A.A.S.H.O. Specification 1961 | Structural Steel 29(10³) | $\frac{P_w}{A} = 15{,}000 - \frac{1}{3}\left(\frac{L}{r}\right)^2$, psi<br>$k$ included<br>$0 \le \frac{L}{r} \le 140$ | 1.70* |
| 3 | Adapted from Mil-Hdbk-5 1959 | Aluminum Alloy 7075-T6 10.5(10³) | $\frac{P_u}{A} = 75{,}300 - 13.7\left(\frac{L}{r}\right)^2$, psi<br>Ultimate load; use $k$ as specified<br>$19.7 \le \frac{L}{r} \le 52.4$ | As specified |
| 4 | Adapted from Alcoa Structural Handbook 1960 | Aluminum Alloy 2024-T4 10.6(10³) | $\frac{P_u}{A} = 44{,}800 - 313\left(\frac{L}{r}\right)$, psi<br>Ultimate load; use $k$ as specified<br>$18.5 \le \frac{L}{r} \le 64$ | As specified |
| 5 | Adapted from Mil-Hdbk-5 1959 | Magnesium Alloy ZK60A-T5 6.5(10³) | $\frac{P_u}{A} = \frac{76(10^5)}{(L/r)^{1.5}}$, psi<br>Ultimate load; use $k$ as specified<br>$43 < \frac{L}{r}$ | No Euler Equation |
| 6 | Adapted from U.S.D.A. Wood Handbook #72 1955 | Douglas Fir (coast type) 1.6(10³) | For square or round solid cross sections<br>$\frac{P_w}{A} = 1450 - 1.656(10^{-5})\left(\frac{L}{r}\right)^4$, psi<br>$k$ included<br>$38 \le \frac{L}{r} \le 73.5$<br>Use 1450 for $\frac{L}{r} < 38$ | 3 |

* Not in A.A.S.H.O. specification.

In order to provide a smooth transition from the intermediate to the slender range, the constants may be adjusted so that the intermediate curve becomes tangent to the Euler curve, thus defining the upper limit of the intermediate range. For safe loads, the factor of safety to be included in the Euler formula can be adjusted to make this curve tangent to any of the formulas for safe loads in the intermediate range. This is done by equating the two expressions for $P/A$, and equating the two expressions for the first derivatives of $P/A$ with respect to $L/r$. The resulting two equations can be solved for the factor of safety to be used in the Euler formula and the $L/r$ at the transition from intermediate to slender range. This technique will be demonstrated later. The practice of making the intermediate and Euler curves tangent is not universal; many column codes plot as curves similar to that for curve ④ of Fig. 9–9.

The formulas in Table 9–1 are representative of a large number of column equations which have been incorporated in various design codes. The slenderness ratio is always the effective slenderness ratio, $L'/r$. If no end conditions are specified in the problems presented later, the stated length is the effective length. Note that the use of high strength materials will increase the allowable $(P/A)$ value for short columns but will have little, if any, effect on the strength of long columns, since the critical load, the Euler load, depends on Young's modulus, not on the elastic strength of the material. Note also that the use of fixed or restrained ends, which has the effect of reducing the length of the column, materially increases the strength of slender columns, but has much less influence on short compression members.

Instead of working with sundry codes for particular materials, it has been suggested that, if only one or two general types of curves would be satisfactory for all materials, the equations of the curves could be presented in dimensionless form to apply to columns made of any material. The usual technique is to divide all values of $P/A$ by a stress parameter, $\sigma_0$, which is usually (but not always) taken as the elastic strength, $\sigma_y$, of a material, and all values of $L/r$ by a slenderness ratio parameter, usually the slenderness ratio, which when applied to the Euler equation will give unity for the ratio, $(P/A)/\sigma_0$ (a one-to-one ratio). The resulting curve is illustrated by Fig. 9–10 where, for convenience, the parameters are labeled $R$ and $S$. The general equation for the Euler part of the curve is

$$R = \frac{1}{S^2}, \qquad (S \geq S_t) \tag{9–5}$$

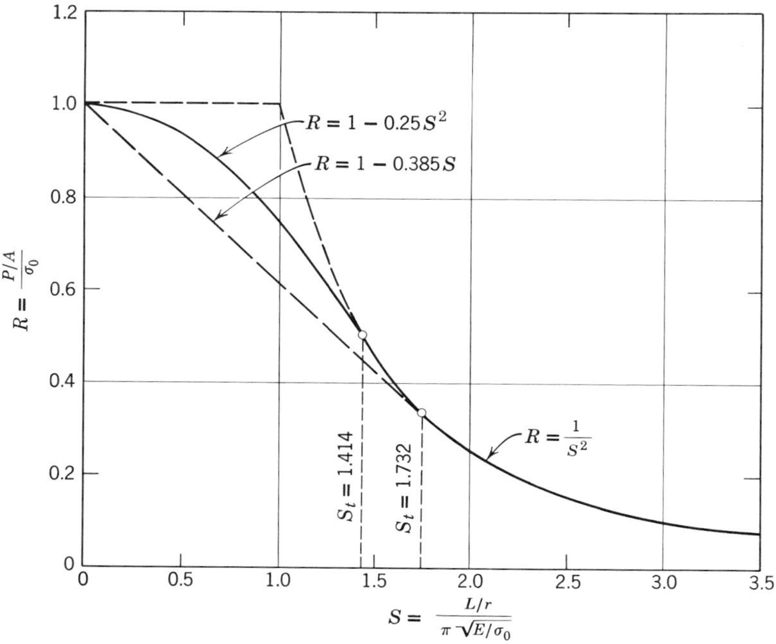


Fig. 9–10

and the general equation for the parabola, which is tangent to the Euler curve, is

$$R = 1 - C_3S^2. \tag{d}$$

In order to find the values of $C_3$ and $S_t$, the technique previously described will be employed as follows: Equating the two expressions above yields

$$1 - C_3S_t^2 = 1/S_t^2,$$

and equating the first derivatives of these expressions yields

$$-2C_3S_t = -2/S_t^3.$$

The solution of these two equations for $C_3$ and $S_t$ yields the following: $C_3 = 0.250$ and $S_t = \sqrt{2} = 1.414$. Hence, the equation of the parabola is

$$R = 1 - 0.250S^2. \qquad (0 \le S \le 1.414) \tag{9–6}$$

In similar fashion, using the general straight-line equation,

$$R = 1 - C_2 S, \tag{e}$$

and proceeding as before, the constants are evaluated as, $C_2 = 0.385$ and $S_t = \sqrt{3} = 1.732$. Hence, the equation of the straight line which is tangent to the Euler curve is

$$R = 1 - 0.385\,S. \qquad (0 \leq S \leq 1.732) \tag{9–7}$$

To recapitulate, the non-dimensional curves presented here apply to any material provided that the behavior of the column can be reliably predicted by one of the two curves illustrated. Note that for a straight-line formula, Eq. (b), the value of $\sigma_0$ is often an experimentally determined stress which is larger than the compressive strength, $\sigma_y$. In such cases, the straight line curve in Fig. 9–10 will have a horizontal segment, $R_0 = \sigma_y/\sigma_0$, at the low $L/r$ end; so that $P/A$ will never exceed $\sigma_y$ (see, for example, curve ④ in Fig. 9–9). Note also that if the best straight line for a given material is not tangent to the Euler curve, Eq. (9–7) may not be reliable.

All of the discussion so far has been concerned with *primary* instability, where the column deflects as a whole into a smooth curve. No discussion of compression loading is complete without reference to *local instability* in which the member fails locally by crippling of thin sections. Thin open sections such as angles, channels, and H-sections are particularly sensitive to crippling failure. The design of such members to avoid crippling failure is usually governed by specifications controlling the width-thickness ratios of outstanding flanges. Closed section members of thin material (thin-walled tubes, for example) must also be examined for crippling failure when the members are short (see Mil-Hdbk-5).

**Example 9–1:** Two steel 10 ⊔ 25 are latticed 5 in. back to back, as shown in Fig. 9–11*a* and *b*, to form a column. Determine the maximum allowable axial load for effective lengths of (1) 25 ft and (2) 40 ft, using:

(*a*) Code No. 1 for structural steel (see App. A).

(*b*) Equations (9–5) and (9–6) for structural steel. Use the same factors of safety as in (*a*).

*Solution:* (*a*) Both $I_x$ and $I_y$ (see Fig. 9–11) or $r_x$ and $r_y$ must be known in order to determine the minimum radius of gyration. Properties of the channel section are given in App. C. The value of $I_y$ for two channels is obtained by using the parallel axis theorem; thus

$$I_y = 2I_c + 2Ad^2 = 2Ar_c^2 + 2Ad^2,$$

and

$$r_y = \sqrt{I_y/(2A)} = \sqrt{\frac{2A(r_c^2 + d^2)}{2A}} = \sqrt{r_c^2 + d^2}.$$

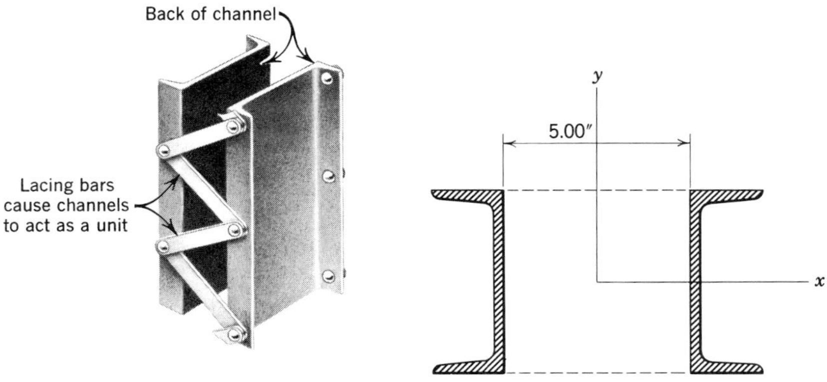


(a) (b)

Fig. 9–11

The above expression indicates that the radius of gyration for the two channels is the same as that for one channel; a fact which is obtained directly from the definition of radius of gyration. Then

$$r_y = \sqrt{(0.68)^2 + (2.50 + 0.62)^2} = 3.19 \text{ in.},$$

which is less than the tabular value of 3.52 in. for $r_x$. This means that the column tends to buckle with respect to the $y$ axis, and the slenderness ratios are

$$\frac{L}{r} = \frac{12(25)}{3.19} = 94.0, \quad \text{and} \quad \frac{12(40)}{3.19} = 150.5.$$

From Code No. 1,

$$C_c^2 = 2\pi^2(29)(10^3)/36 = 15{,}900; \qquad C_c = 126.$$

The slenderness ratios above indicate that the 25-ft column is in the intermediate range; hence

$$k\frac{P}{A} = 36 - \frac{36(94^2)}{2(15{,}900)} = 26.0.$$

The factor of safety is given by

$$k = \frac{5}{3} + \frac{3(94/126)}{8} - \frac{(94/126)^3}{8} = 1.89.$$

Hence, the safe load for the 25-ft column is

$$P = \frac{26.0(2)(7.33)}{1.89} = \underline{202 \text{ kips.}}$$ Ans.

The 40-ft column having a slenderness ratio of 150.5 is in the slender range; hence

$$P = \frac{\pi^2 EA}{k(L/r)^2} = \frac{\pi^2(29)(10^3)(2)(7.33)}{1.92(150.5^2)} = \underline{96.5 \text{ kips.}} \qquad \text{Ans.}$$

(b) Reference to Fig. 9–10 shows that the parameter $\pi\sqrt{E/\sigma_0}$ is evaluated as

$$\pi\sqrt{E/\sigma_y} = \pi\sqrt{29(10^3)/36} = 89.16.$$

For the 25-ft column, $S = 94/89 = 1.055$ which is less than 1.414; therefore, the parabolic formula applies and

$$R = 1 - 0.25(1.055^2) = 0.722 = (P/A)/\sigma_y$$

from which

$$P = 0.722(36)(2)(7.33)/1.89 = \underline{202 \text{ kips.}} \qquad \text{Ans.}$$

For the 40-ft column, $S = 150.5/89 = 1.69$ which is greater than 1.414; hence, the Euler formula applies and

$$R = \frac{1}{S^2} = \frac{1}{1.69^2} = 0.350 = \frac{P/A}{\sigma_y}$$

from which

$$P = 0.350(36)(2)(7.33)/1.92 = \underline{96.3 \text{ kips.}} \qquad \text{Ans.}$$

**Example 9–2:** Determine the dimensions necessary for a 20-in. rectangular strut to carry an axial load of 1500 lb with a factor of safety of 2. The material is aluminum alloy 2024-T4, and the width of the strut is to be twice the thickness.

(*a*) Use Code No. 4.
(*b*) Use Eq. (9–5) and (9–7).

*Solution:* (*a*) The code is represented by two different equations which depend upon the value of $L/r$, which in turn will depend on the equation used. Thus it will be necessary to assume that one of the equations applies and use it to obtain the dimensions of the column, after which the value of $L/r$ must be calculated and used to check the validity of the equation used. Assume $L/r$ is less than 64, in which case the straight-line equation is valid. The cross section is shown in Fig. 9–12, and the least moment of inertia, $I_x$, is

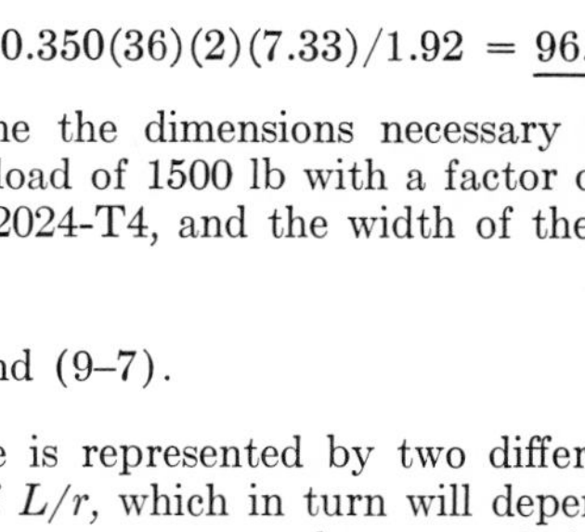


Fig. 9–12

equal to $bt^3/12$; the area, $A$, is equal to $bt$ or $2t^2$; and the least radius of gyration is

$$r = \left(\frac{bt^3/12}{bt}\right)^{1/2} = \frac{t}{\sqrt{12}}.$$

The slenderness ratio is

$$\frac{L}{r} = \frac{20\sqrt{12}}{t},$$

and, when this value and the expression for the area are substituted in the straight-line formula of Code 4, it becomes

$$k\left(\frac{P}{A}\right) = \frac{2(1500)}{2t^2} = 44{,}800 - 313\left(\frac{20\sqrt{12}}{t}\right),$$

from which

$$t = 0.546 \text{ in.}$$

The value of $L/r$ for this thickness is

$$\frac{L}{r} = \frac{20\sqrt{12}}{0.546} = 127$$

which is greater than 64 and indicates that the straight-line formula is not valid. The problem must be solved again using the Euler equation. Thus

$$\frac{P}{A} = \frac{1500}{2t^2} = \frac{\pi^2 E}{2(L/r)^2} = \frac{9.87(10.6)(10^6)}{2[(20)\sqrt{12}/t]^2},$$

from which

$$t^4 = 0.06882 \quad \text{and} \quad t = 0.512 \text{ in.}$$

The value of $L/r$ is 135.3 for this thickness which confirms the use of the Euler formula. The dimensions of the cross section are

$$t = \underline{0.512 \text{ in.}} \quad \text{and} \quad b = \underline{1.024 \text{ in.}} \qquad \text{Ans.}$$

(*b*) Assume $S$ is less than 1.732, in which case the straight-line formula applies. The parameter $\pi\sqrt{E/\sigma_0}$ becomes

$$\pi\sqrt{10.6(10^3)/44.8} = 48.3;$$

consequently,

$$S = \frac{L/r}{\pi\sqrt{E/\sigma_0}} = \frac{20\sqrt{12}}{48.3t} = \frac{1.433}{t}.$$

When this value is substituted into Eq. (9–7), $R$ is found to be

$$R = 1 - 0.385\left(\frac{1.433}{t}\right) = \frac{k(P/A)}{\sigma_0} = \frac{2(1500)}{2t^2(44{,}800)} = \frac{0.0335}{t^2},$$

from which

$$t = 0.607 \text{ in.}$$

Upon substituting this value for $t$ in the equation for $S$,

$$S = \frac{1.433}{0.607} = 2.36,$$

which is greater than the 1.732 assumed and, therefore, the slender-range formula must apply. Equation (9–5) yields

$$R = \frac{k(P/A)}{\sigma_0} = \frac{0.0335}{t^2} = \frac{1}{S^2} = \left(\frac{t}{1.433}\right)^2$$

from which

$$t^4 = 0.0688 \quad \text{and} \quad t = 0.512 \text{ in.}$$

Therefore, the dimensions of the cross section are

$$t = \underline{0.512 \text{ in.}} \quad \text{and } b = \underline{1.024 \text{ in.}} \qquad \text{Ans.}$$

Note that using Eq. (9–7) resulted in a larger value for $t$ (0.607) than that obtained from Code No. 4 (0.546). The reason for this is that the straight line in Fig. 9–10 is tangent to the Euler curve at (1.732, 0.333); whereas the straight line of code 4, if reduced to non-dimensional form, would intersect the Euler curve at (1.333, 0.544). Thus, for any value of $S$ less than 1.732 in Fig. 9–10, the value of $R$, hence $P/A$, would be more conservative than that obtained from code 4.

When a rolled section is to be selected to support a specified load, it is usually necessary to make several trial solutions since there is no direct relationship between areas and radii of gyration for different structural shapes. The best section is usually the section with the least area (lightest weight) which will support the load. A minimum area can be obtained by assuming $L/r = 0$. The load-carrying capacity of various sections with areas larger than this minimum can then be calculated, using the proper column formula, to determine the lightest one which will carry the specified load.

**Example 9–3:** Select the lightest WF steel section listed in App. C to support a centric load of 150 kips as a 15-ft column. Use Code No. 2.

*Solution:* If $L/r$ were small, $P/A$ would be 15,000 psi, and the area would be

$$A = 150{,}000/15{,}000 = 10 \text{ sq in.}$$

A column should be selected from Table 1, App. C, with an area greater than 10.00 sq in. for the first trial. In this case try an 8 WF 40 section for which $A$ is 11.76 sq in. and $r_{min}$ is 2.04 in. The value of $L/r$ for this column is

$$\frac{L}{r} = \frac{180}{2.04} = 88.2,$$

and the load that it can support is

$$P = A\left[15{,}000 - \frac{(L/r)^2}{3}\right] = 11.76\left[15{,}000 - \frac{88.2^2}{3}\right] = 146{,}000 \text{ lb.}$$

This load is less than the design load; therefore a column with either a larger area, a larger radius of gyration, or both must be investigated. As a second trial value use a 10 WF 45 section for which $A$ is 13.24 sq in. and $r$ is 2.00 in. For this section $L/r$ is equal to 90, and the load is

$$P = 13.24[15{,}000 - 90^2/3] = 163{,}000 \text{ lb.}$$

Since the 45-lb column is stronger than necessary and the 40-lb column is not strong enough, any other sections investigated should weigh between 40 and 45 lb per ft. The only WF section in App. C which might satisfy the requirements is a 14 WF 43 section for which $A$ is 12.65 sq in. and $r$ is 1.89 in. The slenderness ratio and allowable load for this column are

$$(L/r) = 95.2 \quad \text{and} \quad P = 151{,}500 \text{ lb.}$$

Thus the lightest column is a 14 WF 43 section. Ans.

The above is not necessarily the best procedure. Different designers have different approaches to the trial and error procedure, and for certain problems one approach may be better than another. The important point to make here is that the problem of design involving rolled shapes (other than simple geometric shapes) is, in general, solved by trial and error.

## PROBLEMS

**Note: Refer to App. A for necessary properties. When applying Eq. (9–6) and (9–7), use the elastic strength for $\sigma_0$.**

**9–21.*** Use Code No. 1 and structural steel for this problem.

(*a*) Determine the maximum allowable total load that three rectangular steel bars 1 by 4 in. by 5 ft long will carry as columns, each acting independently of the other two.

(*b*) Determine the maximum allowable load that these three bars can carry when they are welded together to form an H column. The factor of safety is 1.84.

**9–22.*** Four $2\frac{1}{2}$ by $2\frac{1}{2}$ by $\frac{1}{4}$-in. aluminum alloy 2024-T4 angles 5 ft long are to serve as a column. Using Code No. 4 with a factor of safety of 2, determine the maximum allowable load:

(*a*) When the angles are fastened together as in Fig. P9–22 to act as a unit.

(*b*) When the angles are not connected and each acts as an independent axially loaded member.

Fig. P9–22

**9–23.*** A 25-ft column is composed of two steel 10 ⊔ 30 sections. Using Code No. 2, determine the maximum permissible load:

(*a*) If the channels are latticed 5.5 in. back to back as in Fig. 9–11(*a*).

(*b*) If the channels are not connected and each acts as an independent axially loaded member.

**9–24.*** Two 5 by 3½ by ½-in. steel angles 12 ft long are to serve as a column. Using Code No. 2, determine the maximum allowable load:

(*a*) When the angles are fastened together to form a T-section with the 5-in. legs connected.

(*b*) When the angles are not connected and each acts as an independent axially loaded member.

**9–25.** A connecting rod made of SAE 4340 heat treated steel has the cross section of Fig. P9–25. The pins at the ends of the rod are parallel to the $x$ axis and are 5 ft apart. Assume that the pins offer no restraint to bending about the $x$ axis but provide complete fixity for bending about the $y$ axis. Using Code No. 1, determine the maximum allowable axial compression load for this rod. Use 1.92 for the factor of safety.

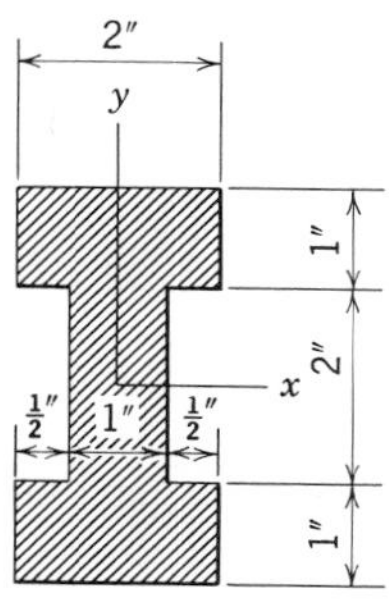


Fig. P9–25

**9–26.*** A column of aluminum alloy 2024-T4 is composed of two 4 by 4 by ½-in. angles riveted together as shown in Fig. P9–26. The length between end connections is 8 ft, and the end conditions are such that there is no restraint to bending about the $y$ axis; but restraint to bending about the $x$ axis reduces the effective length to $8/\sqrt{2}$ ft. Using Code No. 4 and a factor of safety of 2, determine the maximum permissible load.

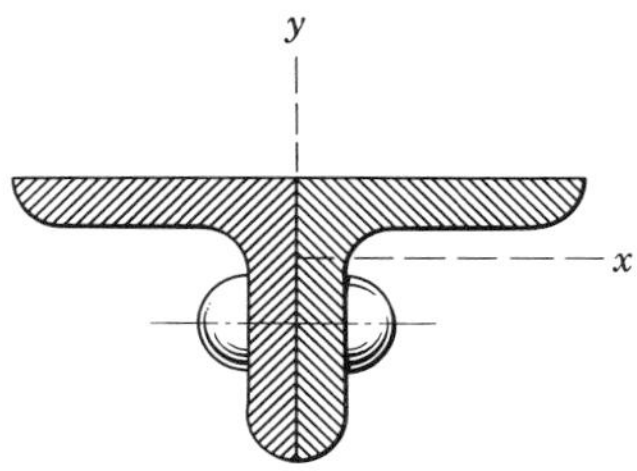


Fig. P9–26

**9–27.** Figure P9–27 shows an airplane part of aluminum alloy 7075-T6 which is to carry an axial compression load. The pins at the ends offer practically no restraint to bending about the axis of the pin, but the restraint about the perpendicular axis reduces the effective length to $L/\sqrt{3}$. Using Code No. 3 and a factor of safety of 2, determine the maximum allowable load.

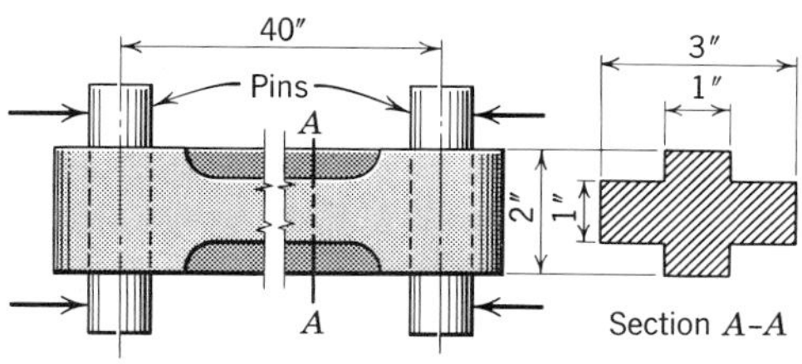


Fig. P9–27

**9–28.*** Solve Prob. 9–27 for a machine part 60 in. long (between pins) made of nodular cast iron. Use Eq. (9–5) and (9–6) and a factor of safety of 2.5.

**9–29.** A cast steel column (properties are the same as those for structural steel) has the cross section of Fig. P9–29. The column is 10 ft long and is fixed at the base and free at the top. Using Eq. (9–5) and (9–6) and a factor of safety of 3, determine the maximum allowable load.

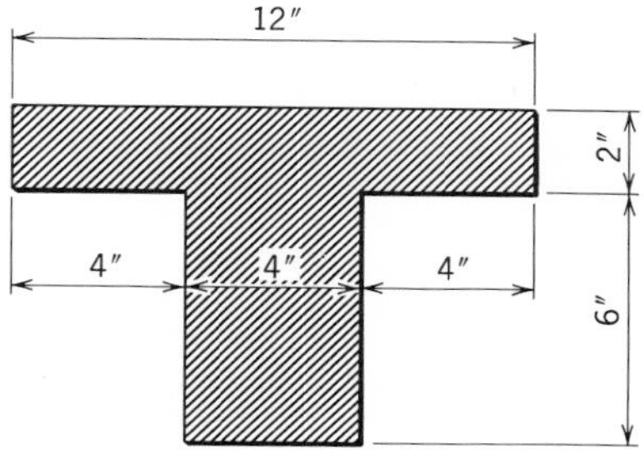


Fig. P9–29

**9–30.*** Two 6 by $3\frac{1}{2}$ by $\frac{1}{2}$-in. angles of aluminum alloy 2014-T4 are welded together as shown in Fig. P9–30 to form a 13-ft pin-ended column. The pins provide no restraint to bending about the $x$ axis but do reduce the effective length to 10 ft for bending about the $y$ axis. The sectional properties of one angle are given in the figure where point $C$ is the centroid of one angle. Using Eq. (9–5) and (9–7), determine the maximum permissible axial load for a factor of safety of 2.

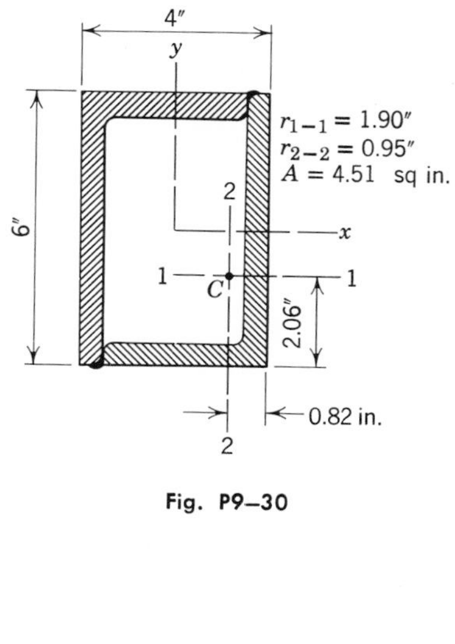


Fig. P9–30

**9–31.** Four 6061-T6 aluminum alloy 6 ⊔ 2.83 are fabricated as in Fig. P9–31 to form a pin-ended column 15 ft long. The pins offer no restraint to bending about the $y$ axis, but for bending about the $x$ axis, the pins provide restraint sufficient to reduce the effective length to 10 ft. The cross-sectional properties of one channel are the same as those for a steel 6 ⊔ 8.2 given in App. C. Using Eq. (9–5) and (9–7), determine the maximum permissible axial load for a factor of safety of 2.5.

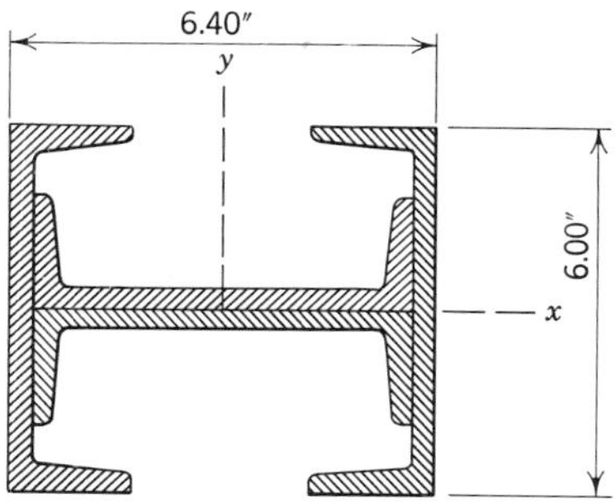


Fig. P9–31

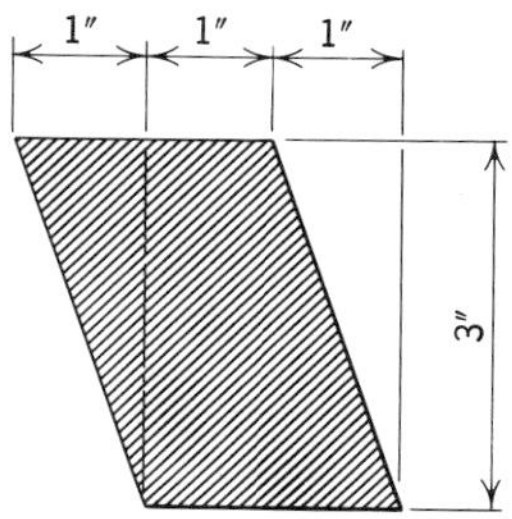


Fig. P9–32

**9–32.*** A 6-ft strut having the section of Fig. P9–32 is made of aluminum alloy 6061-T6. Assume the strut to have fixed ends. Using Eq. (9–5) and (9–7) and a factor of safety of 2, determine the maximum permissible axial compressive load.

**9–33.*** A 6-ft strut having the cross section of Fig. P9–33 is made of T-1 steel having an elastic strength of 90 ksi. Assume the strut to be fixed at both ends. Using Eq. (9–5) and (9–6), determine the maximum allowable axial compressive load that may be applied with a factor of safety of 2.

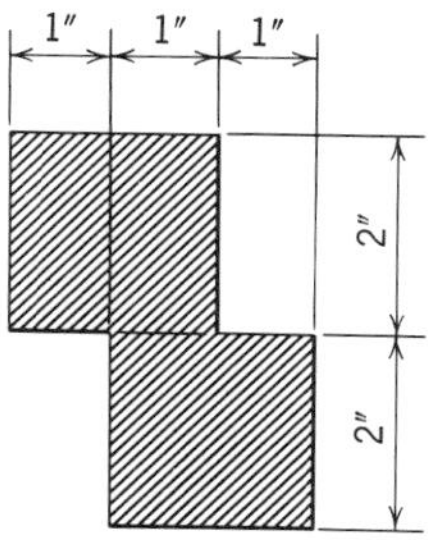


Fig. P9–33

**9–34.** Using Code No. 1 for low-alloy steel ($\sigma_y = 50$ ksi) and a factor of safety of 1.92 (both ranges), determine the maximum allowable load that a 10 WF 60 member can carry as a 35-ft column:

(*a*) If it is unbraced throughout its length.

(*b*) If it is laterally braced 20 ft from one end in the direction parallel to the $x$ axis; that is, the effective length is 20 ft for bending about the $y$ axis.

**9–35.** The compression member $AB$ of the truss in Fig. P9–35 is an 8 WF 40 steel section with the $x$–$x$ axis lying in the plane of the truss. The member is continuous from $A$ to $B$. Consider all connections to be the equivalent of pivot ends, and use Code No. 2.

(*a*) Determine the maximum permissible load in $AB$.

(*b*) Determine the maximum permissible load if the bracing member $CD$ is removed.

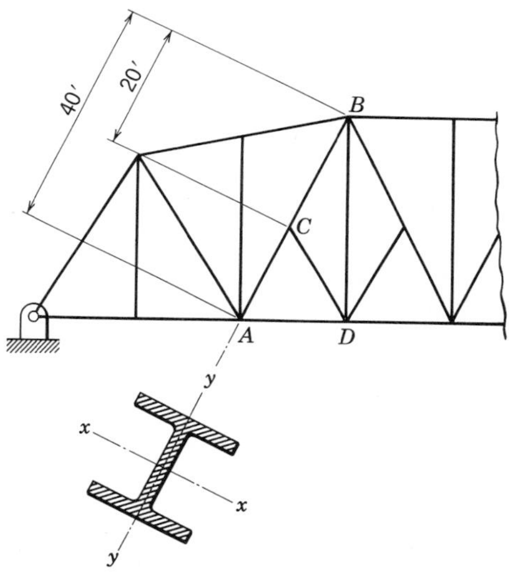


Fig. P9–35

**9–36.*** Solve Prob. 9–35 assuming that the joints at $A$ and $B$ provide enough restraint that the effective length $A$ to $B$ is $40/\sqrt{1.5}$. Assume the connection at $C$ offers no restraint to bending.

**9–37.*** A 40-ft column is composed of two structural steel 12 ⊔ 20.7 latticed together with the flanges turned out as in Fig. 9–11*a*. The column is fixed at the base and the top is pinned to a roller which slides in a vertical slot.

(*a*) What should be the spacing of the channels (back to back) in order to provide a column with equal resistance to bending about either principal axis?

(*b*) Determine the maximum allowable load according to Code No. 1.

**9–38.** A 5 by 5 by $\frac{3}{4}$-in. aluminum alloy 7075-T6 angle is used as a fixed-ended column. Determine the maximum allowable length of this column which will support a load of 120 kips. Use Code No. 3 with a factor of safety of 2.5.

**9–39.** A strut of aluminum alloy 2024-T4 having the cross section of Fig. P9–39 is to carry an axial compressive load of 100 kips with a factor of safety of 2. The strut is 45 in. long and is fixed at the bottom and pivoted at the top. Using Code No. 4, determine the value of $d$.

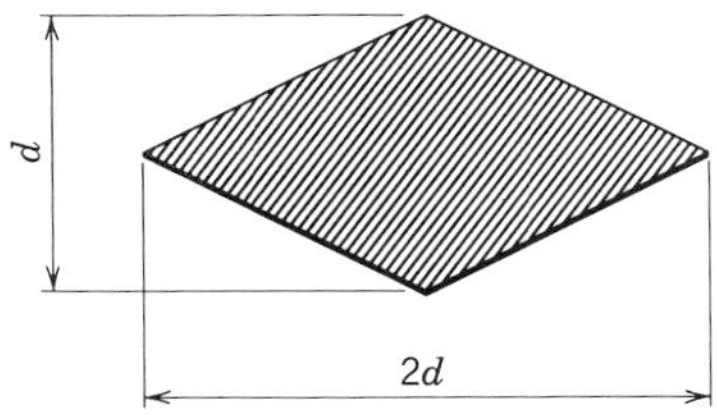


Fig. P9–39

**9–40.** A sand bin weighing 200 tons is to be supported by four square timber columns 15 ft long. What size dressed Douglas fir (dry) timbers should be bought to comply with Code No. 6? Assume that the load is to be equally divided among the columns and that the column load is axial. Dressed timber posts are $\frac{1}{2}$ in. smaller than nominal.

**9–41.** A magnesium alloy (ZK60A-T5) strut 60 in. long is to transmit an axial compression load of 20 kips. Restraint at the ends of the strut make the effective length one-third greater than that for completely fixed ends. The strut is to be rectangular in cross section, one dimension being fixed at 2.00 in. Using Code No. 5 and a factor of safety of 2, determine the other dimension.

**9–42.** Design a square column 50 in. long to carry a load of 6000 lb with a factor of safety of 3. The material is annealed 18-8 stainless steel. Use Eq. (9–5) and (9–6).

**9–43.** Design a hollow circular column 50 in. long to carry a load of 15 kips with a factor of safety of 2. The material is aluminum alloy 6061-T6 and the ratio of inside to outside diameters is to be 0.8. Use Eq. (9–5) and (9–7).

**9–44.** Using Code No. 1, select a structural steel WF section to serve as a column 12 ft long carrying an axial load of 200 kips with a factor of safety of 1.92.

**9–45.** Using Code No. 2, select two steel channels to form a column 25 ft long to carry an axial load of 120 kips. The channels are to be spaced and latticed so that the resistance to bending with respect to either axis of symmetry is the same.

**9–46.** Design an 8-ft column to carry a load of 100 kips with a factor of safety of 2. Use two aluminum alloy 2024-T4 angles with unequal legs as shown in Fig. P9–46.

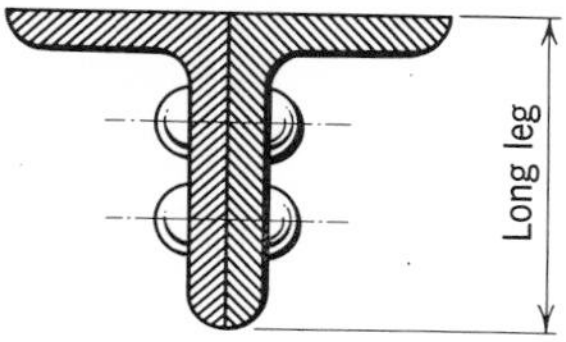


Fig. P9–46

**9–47.** Solve Prob. 9–46 if the angles are spaced so that resistance to bending is the same about either principal axis.

**9–48.** Derive a parabolic formula for aluminum alloy columns with fixed ends subject to the following conditions: $\sigma_0 = 20{,}000$ psi; $E = 10^7$ psi; factor of safety for the Euler equation = 2.00. Derive the equation so that the two curves will be tangent.

**9–49.** The straight line of Code No. 4 is not tangent to the Euler curve. Using the value of $\sigma_0$ given in the code, determine:

(*a*) The value of the constant $C_2$ which will make the two curves tangent.
(*b*) The value of $L/r$ at the point of tangency.

**9–5. Eccentric Loads.** Although a given column will support the maximum load when the load is applied centrically, it is sometimes necessary to apply an eccentric load to a column. For example, a beam supporting a floor load in a building may in turn be supported by an angle riveted or welded to the side of a column as shown in Fig. 9–13. Frequently, the floorbeam is framed into the column with a stiff connection, and the column is then subjected to a bending moment due to the continuity of floorbeam and column. The eccentricity of the load, or the bending moment, will increase the stress in the column or reduce its load-carrying capacity. Three methods will be presented for com-

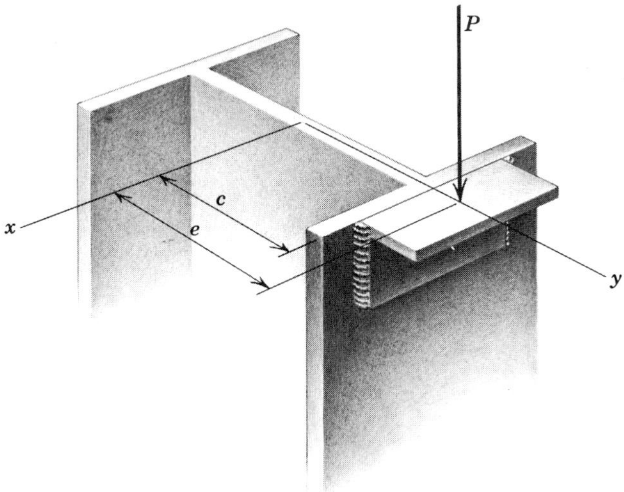


Fig. 9–13

puting the allowable load on a column subjected to an eccentric load (or an axial load combined with a bending moment).

**The secant formula.** The secant formula, Eq. (9–3), was derived on the assumption that the applied load had an initial eccentricity, $e$; this value can be substituted in the equation to determine the failure load (the load to cause incipient inelastic action). As mentioned previously, there is usually a small amount of unavoidable eccentricity which must be approximated when using this formula for centric loads. This unavoidable eccentricity should be added to the intentional eccentricity, particularly if the intentional eccentricity is small. The use of the secant formula is quite time consuming, unless a series of curves, such as those in Fig. 9–6, is available.[4] The secant formula is specified by the American Railway Engineering Association and the American Association of State Highway Officials.

**Modified column formulas.** The second solution is based on the arbitrary specification that the sum of the direct and bending stresses ($P/A + Mc/I$) shall not exceed the average unit load prescribed by

[4] See, for example, App. B of the A.A.S.H.O. Standard Specifications for Highway Bridges.

the appropriate column formula. For example, when this approach is used with the parabolic formula, the equation becomes

$$\frac{P}{A} + \frac{Mc}{I} = \sigma_0 - C_3\left(\frac{L}{r}\right)^2. \qquad \text{(a)}$$

In effect, this merely replaces the low $L/r$ end of the secant curve with a parabola (see Fig. 9–9), and if the constants in the parabola are properly chosen, the agreement between the two curves is quite good. Since the parabola will deviate from the secant curve very rapidly beyond the point of reverse curvature, Eq. (a) is limited to the intermediate range. This leaves only the secant formula valid for columns in the slender range with eccentric loading. Although it is possible to use other types of intermediate formulas for the right side of Eq. (a), the use of the parabola only is recommended here, since it is an attempt to approximate a formula known to be rational.

The application of the secant formula or Eq. (a) to an eccentrically loaded column such as that of Fig. 9–13 must be accompanied by an investigation of the stability of the column as an axially loaded member. Referring to Fig. 9–13, when the load is applied in the plane containing the axis about which the moment of inertia is least (the $y$ axis of Fig. 9–13), the column might fail by buckling about either axis, and both possibilities should be checked. (1) It can fail by buckling about the axis of least radius of gyration (the $y$ axis in Fig. 9–13) because $L/r$ is largest about this axis. In this case, all formulas developed for centric loading apply[5] and the radius of gyration is $r_y$. (2) The column can fail by buckling about the axis of maximum radius of gyration (the $x$ axis) due to the combined effect of the bending moment and the direct load. In this case, the values required are $r_x$, $I_x$, and the distance $c$ as shown in Fig. 9–13. Note that $c$ is the maximum distance to the compression side of the centroidal axis. The smaller of the two loads is, of course, the limiting load.

**The interaction formula.** One of the modern expressions for treating combined loads is known as the interaction formula, of which several types are in use. The analysis for compression members subjected to

[5] The load is assumed to be axial; that is, the eccentricity in the $y$ direction does not decrease resistance to buckling about the $y$ axis of Fig. 9–13. For columns with I-sections in the slender range, it can be shown that such an assumption may be unwarranted. See "Lateral Buckling of I-Section Columns with Eccentric End Loads in Plane of the Web," B. Johnston, *Journal of Applied Mechanics*, December 1941, p. A-176.

bending and direct stress may be derived as follows. It is assumed that the stress in the column can be written as

$$\frac{P}{A} + \frac{Mc}{I} = \sigma_w,$$

and, when this expression is divided by $\sigma_w$, it becomes

$$\frac{P/A}{\sigma_w} + \frac{Mc/I}{\sigma_w} = 1. \tag{b}$$

When considering eccentrically loaded columns, the value of $\sigma_w$ will, in general, be different for the two terms. In the first term, $P/A$ represents an axial unit load on a column; therefore, the value of $\sigma_w$ should be the average working unit load on an axially loaded column as obtained by an empirical formula such as those presented in the preceding article. In the second term, $Mc/I$ represents the bending stress induced in the member as a result of the eccentricity of the load or an applied bending moment; therefore the corresponding value of $\sigma_w$ should be the allowable flexural stress. Since the two values of $\sigma_w$ are different, one recommended form of Eq. (b) is the following interaction formula[6]

$$\frac{P/A}{\sigma_a} + \frac{Mc/I}{\sigma_b} = 1, \tag{c}$$

in which $P/A$ = the average unit load on the eccentrically loaded column,
$\sigma_a$ = the *allowable* average unit load for an axially loaded column (note that the greatest value of $L/r$ should be used to calculate $\sigma_a$),
$Mc/I$ = the bending stress in the column,
and $\sigma_b$ = the *allowable* bending stress.[7]

When Eq. (c) is used to select the most economical section, it will usually not be possible to obtain a section which will satisfy the equation exactly. Any section which makes the sum of the terms on the left side of the equation less than unity is considered to be safe, and the *safe* section which gives the largest sum (less than unity) is the most efficient section.

[6] The 1961 AISC specifications permit this formula if $(P/A)/\sigma_a \leq 0.15$; otherwise, the second term of the equation is amplified to provide for secondary moments due to lateral displacement of any cross section.

[7] The 1961 AISC specifications gives $\sigma_b = 0.66\sigma_y$ with a reduced value for compression flanges in some cases. See *The Steel Construction Manual.*

**Example 9–4:** A 25-ft 18 WF 96 steel column supports an eccentric load which is applied 5 in. from the center of the section measured along the web. Determine the maximum safe load the column can support (using Code No. 2):

(*a*) According to the modified column formula.

(*b*) According to the interaction formula with $\sigma_b$ equal to 18 ksi.

*Solution:* (*a*) The following properties are obtained from App. C:

$$A = 28.22 \text{ sq in.}, \quad r_{\min} = 2.71 \text{ in.},$$

$$r_{\max} = 7.70 \text{ in.}, \quad \text{and} \quad c = 18.16/2 = 9.08 \text{ in.}$$

Assume first that the column tends to buckle about the axis parallel to the web. Note that there is no eccentricity about this axis. The value of $L/r$ is

$$\frac{L}{r} = \frac{25(12)}{2.71} = 110.7,$$

which is in the intermediate range. The average unit load is

$$P/A = 15{,}000 - 0.333(110.7^2) = 10{,}920 \text{ psi}.$$

For bending about the axis perpendicular to the web, the radius of gyration is 7.70 in. ($L/r$ will be much less than 110.7), and the modified column formula becomes

$$\frac{P}{A} + \frac{Mc}{I} = 15{,}000 - 0.333(L/r)^2.$$

The bending moment is $Pe$, and, when numerical values are substituted in the equation, it becomes

$$\frac{P}{A} + \frac{P(5)9.08}{A(7.70^2)} = 15{,}000 - 0.333\left(\frac{300}{7.70}\right)^2,$$

from which

$$P/A = 8210 \text{ psi}.$$

Since this last unit load is less than that for buckling about the web, the maximum allowable load is

$$P = 8210(28.22) = \underline{232{,}000 \text{ lb.}}$$ Ans.

(*b*) The interaction formula for eccentrically loaded columns is

$$\frac{P/A}{\sigma_a} + \frac{Mc/I}{\sigma_b} = \frac{P/A}{\sigma_a} + \frac{Pec/(Ar^2)}{\sigma_b} = 1.$$

The value of $\sigma_a$ was found to be 10,920 psi in part *a*, and, when numerical data are substituted in the expression, it becomes

$$\frac{P}{A}\left[\frac{1}{10{,}920} + \frac{5.00(9.08)/(7.70^2)}{18{,}000}\right] = 1,$$

which gives

$$P/A = 7456 \text{ psi}.$$

The allowable load using this specification is

$$P = 7456(28.22) = \underline{210{,}000 \text{ lb.}}$$ Ans.

The two results are about 10 per cent apart in this case.

## PROBLEMS

**Use the table of codes of Art. 9–4 modified for eccentricity unless otherwise specified. Refer to App. C for properties. For $\sigma_0$ in Eq. (9–6), use the elastic strength unless specified otherwise.**

**9–50.*** A hollow square steel member 4 in. in outside dimension and with ½-in. thickness of walls functions as a column 15 ft long. Determine the maximum allowable load that the column will carry when the load is applied with a known eccentricity of ⅝ in. along an axis of symmetry parallel to a pair of sides. Use Fig. 9–6 and a factor of safety of 2. Assume an accidental eccentricity ratio of 0.20.

**9–51.*** A 2-in. diameter steel strut is subjected to an eccentric compression load, $P$, located as shown in Fig. P9–51. The effective length for bending about the $x$ axis is 60 in. but for bending about the $y$ axis, the end conditions reduce the effective length to 40 in. Determine the load $P$ which may be applied with a factor of safety of 1.5. Use Fig. 9–6 and assume an accidental eccentricity ratio of 0.10.

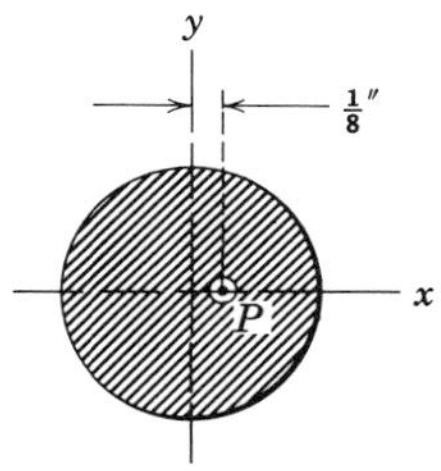


Fig. P9–51

**9–52.** A hollow square steel member with outside and inside dimensions of 4 in. and 3 in. (walls are ½ in. thick) functions as a column 15 ft long. The load is applied with a known eccentricity of ⅝ in. along one of the diagonals of the square, and the end conditions are such that for bending about the other diagonal the effective length is reduced to 10 ft. Determine the load which may be applied with a factor of safety of 2. Use Fig. 9–6 and assume an accidental eccentricity ratio of 0.15.

**9–53.*** An 8 WF 31 structural steel section acts as a 30-ft column. Determine the maximum allowable load it can carry if the resultant load acts along the web 3 in. from the column axis. Use Code No. 2.

**9–54.*** Two 8 ⊔ 18.75 structural steel sections 30 ft long are latticed 3 in. back to back as shown in Fig. P9–54 to make a column. Determine the maximum working load which the column may carry if the load acts at point $A$. Use Code No. 2.

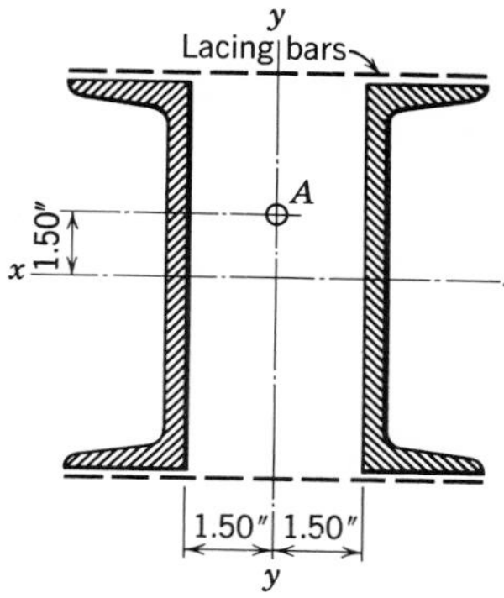


Fig. P9–54

**9–55.** A structural steel T-section having the properties given in Fig. P9–55 is to be used as a compression member with an effective length of 24 ft. Using Code No. 1 and a factor of safety of 1.92 for both ranges, determine:

($a$) The maximum permissible axial load.

($b$) The maximum permissible bending moment, applied to the ends of the member in the $yz$ plane (the $z$ axis is the axis of the member), that may be superimposed on the load of part ($a$) if the moment induces compression in the flange.

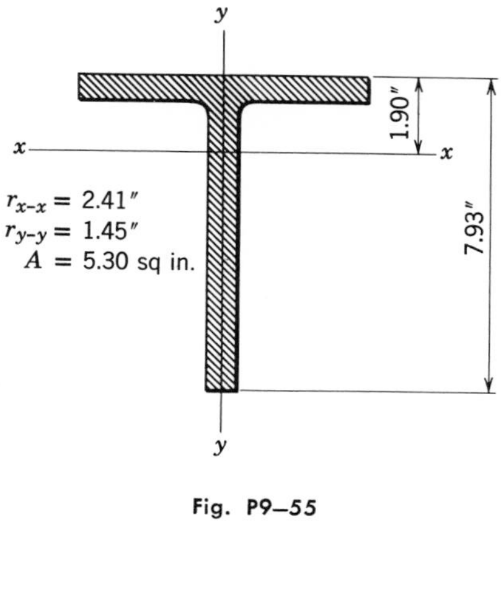


Fig. P9–55

**9–56.*** A 16 WF 64 alloy steel ($\sigma_y$ = 50 ksi) section is to function as a 25-ft column fixed at the base. The column is to be considered free at the top for bending with respect to the $x$ axis (see App. C) and pivot-ended at the top for bending with respect to the $y$ axis. The load is applied on the web 3 in. from the center of the cross section. Using Code No. 1 and a factor of safety of 1.92 for both ranges, determine the maximum permissible value for the eccentric load.

**9–57.*** A 7075-T6 aluminum alloy compression member with an effective length of 5 ft has the tee cross section of Fig. P9–57. Using Code No. 3 and a factor of safety of 2, determine:

(*a*) The maximum permissible axial load.

(*b*) The maximum permissible bending moment in the $yz$ plane which can be superimposed as shown on the axial load of part (*a*).

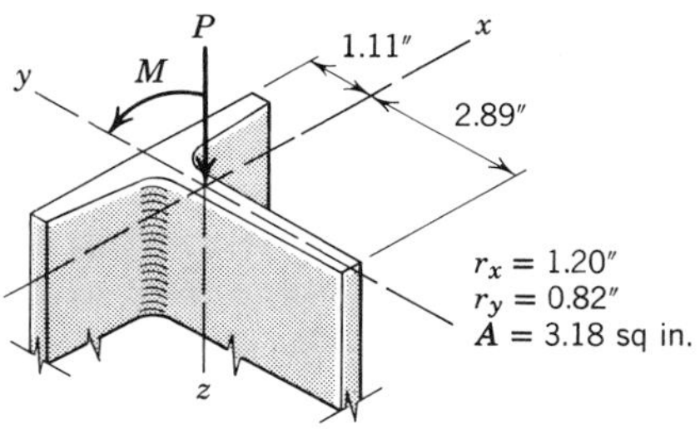


Fig. P9–57

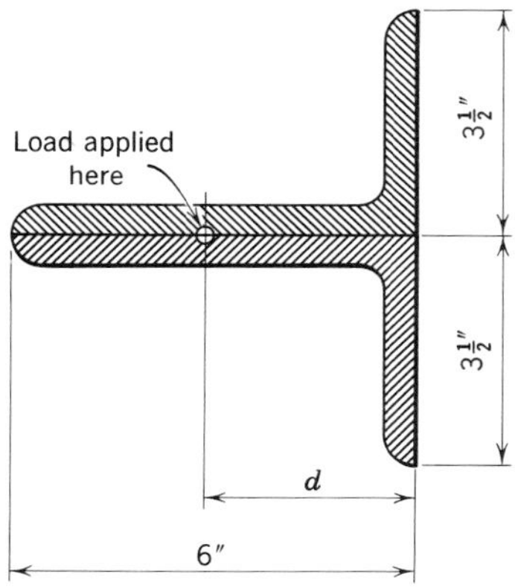


Fig. P9–58

**9–58.** Two 6 by 3½ by ⅜-in. structural steel angles are welded together, as shown in Fig. P9–58, to act as a column with an effective length of 18 ft. Using Code No. 2, determine:

(*a*) The maximum safe axial load.

(*b*) The maximum and minimum values for the distance, $d$, when the load of part (*a*) is applied as shown.

**9–59.*** An 8 WF 31 structural steel compression member has an effective length of 23 ft and is subjected to an axial load $P$ and a bending moment of 12 ft-kips with respect to the $x$ axis of the cross section (see App. C). Determine the maximum safe load $P$. Use Code No. 1 and the interaction formula with an allowable flexural stress of $0.66\sigma_y$.

**9–60.** Solve Prob. 9–54 using Code No. 1 and the interaction formula for eccentric loads with an allowable flexural stress of 0.66 $\sigma_y$.

**9–61.*** A 16 WF 64 steel column 15 ft long carries an eccentric load of 150,000 lb at a distance $e$ from the center of the section measured along the web. Determine the maximum permissible value of $e$. Use Code No. 2 and the interaction formula for eccentric loads with an allowable flexural stress of 18,000 psi.

**9–62.** Determine the maximum load, $F$, that can be applied with a factor of safety of 5 to the gray cast iron frame of Fig. P9–62. Use Eq. (9–5) and (9–6) for the compressive strength, and since gray iron is a brittle material, use the ultimate compressive strength for $\sigma_0$. Note that the flange is in tension and the load may be limited by the tensile strength of the iron.

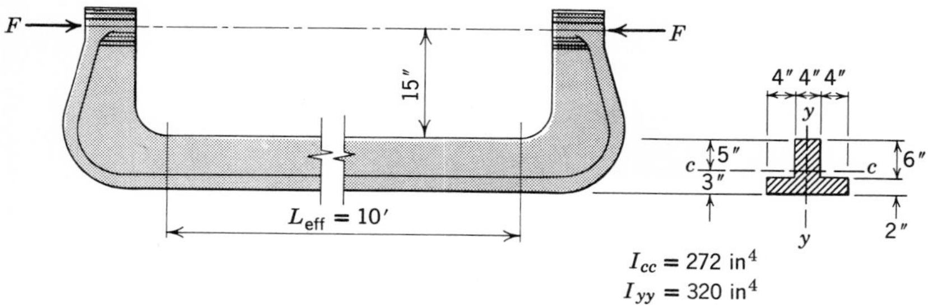


Fig. P9–62

**9–63.*** The gray cast-iron compression member of Fig. P9–63 has an effective length of 70 in. and is subjected to a 5000-ft-lb bending moment as shown. Determine the maximum safe load, $P$, that may be applied (in addition to the bending moment). Use Eq. (9–5) and (9–6) with a factor of safety of 5. For gray iron, use the ultimate compressive strength for $\sigma_0$.

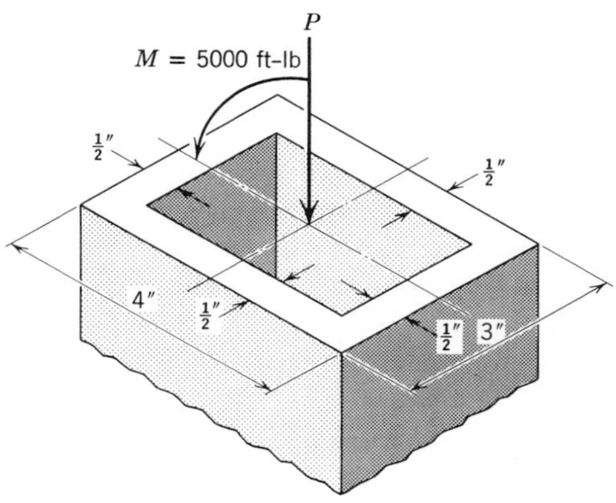


Fig. P9–63

**9–64.** Determine the maximum allowable load for a 12-ft dry Douglas fir column with the cross section and loading showing in Fig. P9–64. Use Eq. (9–5) and (9–6) with a factor of safety of 3.

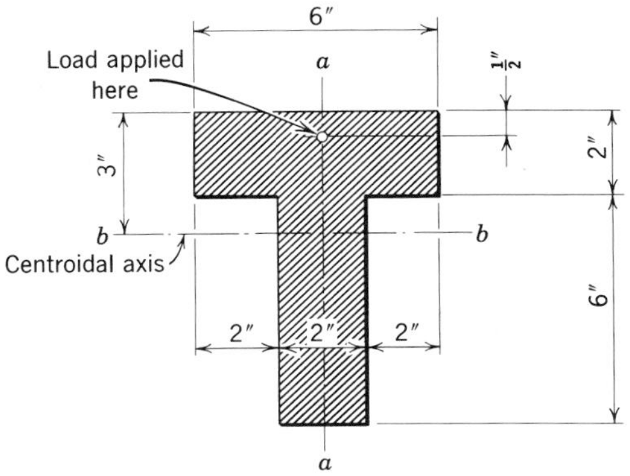


Fig. P9–64

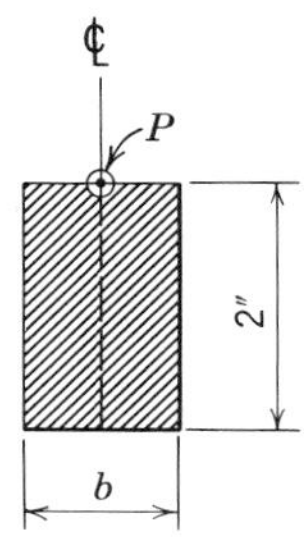


Fig. P9–65

**9–65.** A 3-ft 6061-T6 aluminum alloy strut with a rectangular cross section is to be subjected to an eccentric load, $P$, as shown in Fig. P9–65. Determine the maximum permissible value of $P$ and the value of the dimension $b$ if the strut is to have equal resistance to buckling about either principal axis. Use Eq. (9–5) and (9–6) and a factor of safety of 2.

**9–6. Special Column Problems.** Although the secant formula and its various empirical approximations are adequate for treating a wide variety of engineering problems involving columns, there are certain types of problems involving column action for which the secant formula does not apply. One example is the class of problems known as beam columns; members which are subjected to lateral loading in addition to compressive axial loading. Unless the problem has been treated previously,[8] it is necessary to revert to the differential equation of the elastic curve to find a solution for the particular problem. It may be difficult to develop an explicit solution for $P/A$ as a function of $L/r$ (as was done in Art. 9–2); in fact, such a solution may be un-

[8] There is considerable literature available on this subject. See, for example, Chap. II of *Strength of Materials Part II,* S. Timoshenko, 3rd Ed., McGraw-Hill, New York, 1956.

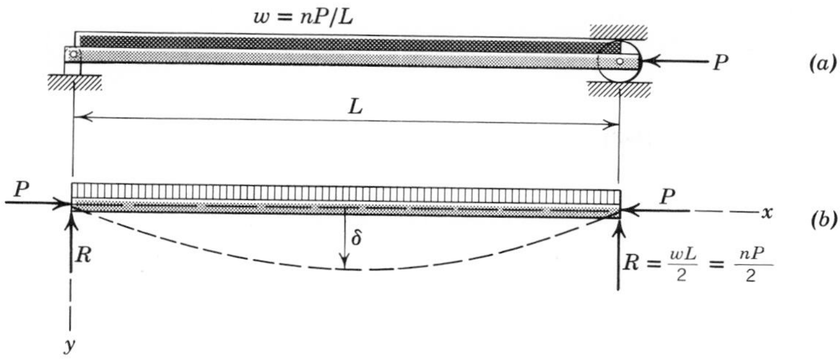


Fig. 9–14

necessary. Much of engineering design is done by iteration; hence, if an expression for the maximum deflection in a beam column can be found, the maximum moment can be determined and the maximum normal stress is $P/A + Mc/I$, where $M$ is a function of the compressive as well as the transverse loads. With an assumed trial section for the member, the maximum stress can be computed to check the adequacy of the section. The following example will illustrate one method of solving the differential equation.

**Example 9–5:** Develop the expression for the maximum deflection of the beam column of Fig. 9–14*a*. The column ends are free to rotate, but are restrained against lateral displacement.

*Solution:* The free-body diagram is shown in Fig. 9–14*b*, with the $y$ axis chosen positive downward to facilitate the solution. For this choice of axes, the curvature indicated is negative, and the differential equation is

$$EI\frac{d^2y}{dx^2} = -Py - \frac{nP}{2}x + \frac{nP}{2L}x^2.$$

The boundary conditions are: $y = 0$ at $x = 0$ and, from symmetry, $dy/dx = 0$ at $x = L/2$. By re-arranging terms and letting $P/(EI) = k^2$ the equation becomes

$$\frac{d^2y}{dx^2} + k^2y = \frac{nk^2}{2L}x^2 - \frac{nk^2}{2}x. \qquad \text{(a)}$$

The complementary solution for this equation is

$$y_c = A\sin kx + B\cos kx,$$

and the particular solution is of the form

$$y_p = Cx^2 + Dx + E.$$

When this last equation is substituted into Eq. (a) and coefficients of like powers of $x$ are matched, the constants are found to be

$$C = \frac{n}{2L}, \qquad D = -\frac{n}{2}, \qquad \text{and} \quad E = -\frac{n}{Lk^2}.$$

Combining the solutions gives

$$y = y_c + y_p = A \sin kx + B \cos kx + \frac{nx^2}{2L} - \frac{nx}{2} - \frac{n}{Lk^2}. \tag{b}$$

Substituting the boundary conditions gives

$$y \bigg|_{x=0} = B - \frac{n}{Lk^2} = 0;$$

$$\frac{dy}{dx}\bigg|_{x=L/2} = Ak \cos \frac{kL}{2} - Bk \sin \frac{kL}{2} = 0,$$

from which

$$B = \frac{n}{Lk^2}; \qquad A = \frac{n}{Lk^2} \tan \frac{kL}{2}.$$

The maximum deflection, $\delta$, is found by substituting these constants into Eq. (b) and assigning $x$ the value $L/2$; thus,

$$\delta = \frac{n}{Lk^2}\left(\sec \frac{kL}{2} - 1\right) - \frac{nL}{8};$$

or in terms of $P$ and the distributed load, $w$,

$$\underline{\delta = \frac{wEI}{P^2}\left[\sec \frac{L}{2}\sqrt{\frac{P}{EI}} - 1 - \frac{PL^2}{8EI}\right].} \qquad \text{Ans.}$$

## PROBLEMS

**9–66.*** Develop, in terms of $P$, $E$, $I$, $L$, and $n$, the expression for the maximum deflection in the beam column of Fig. P9–66.

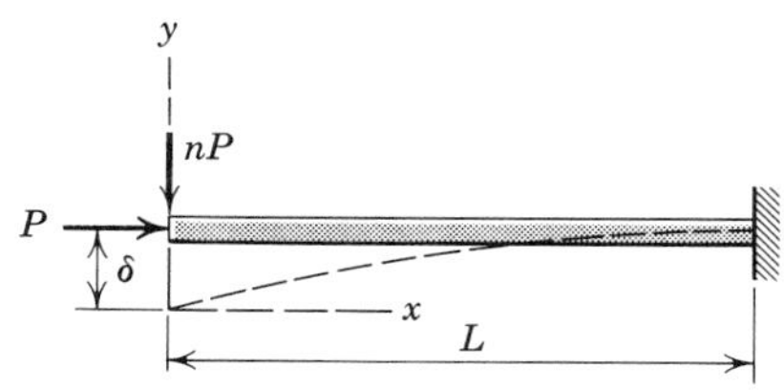


Fig. P9–66

**9–67.** Develop, in terms of $P$, $E$, $I$, $L$, and $n$, the expression for the maximum deflection in the beam column of Fig. P9–67.

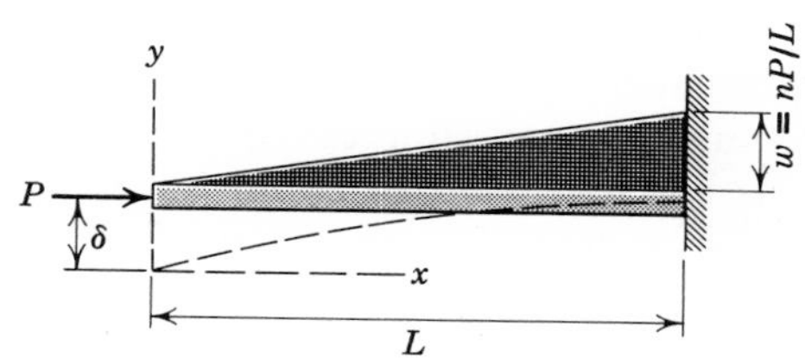


Fig. P9–67

**9–68.*** Develop the equation of the elastic curve for the beam column of Fig. P9–68 in terms of $n$, $P$, $L$, $E$, $I$, $x$, and $y$.

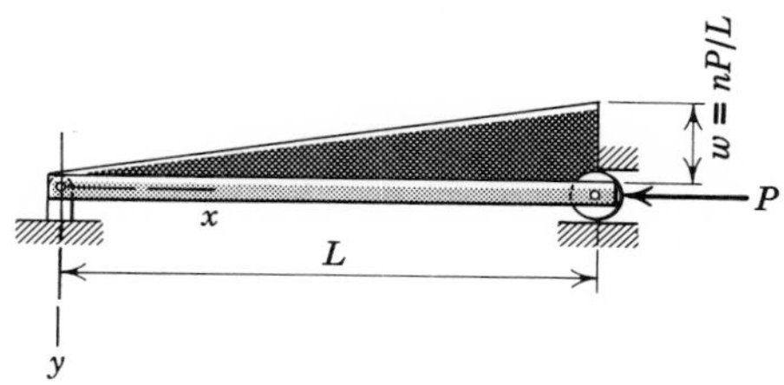


Fig. P9–68

**9–69.** By solving the differential equation of the elastic curve, verify the effective length given for the column of Fig. 9–8*d*.

**9–70.*** Determine in terms of $E$, $I$, and $L$, the critical load for the column of Fig. P9–70.

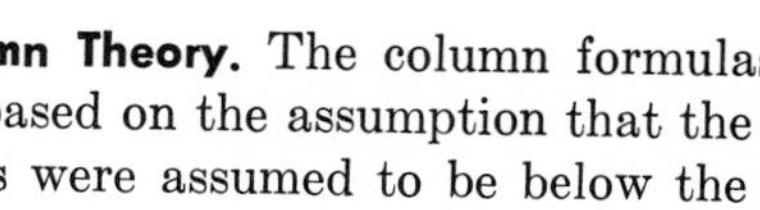


Fig. P9–70

**9–7. Inelastic Column Theory.** The column formulas discussed in Art. 9–2 and 9–4 were based on the assumption that the action was elastic; that is, the stresses were assumed to be below the proportional limit. The Euler formula can thus be used for an *ideal, axially* loaded column for any value of $P/A$ less than the proportional limit of the material.

The first theory for predicting the buckling loads of columns loaded in the inelastic range was proposed by F. R. Engesser about 1890 and is known as the tangent-modulus theory. It is based on the assumption that the Euler formula can be used in the inelastic range provided that Young's modulus is replaced by the tangent modulus of the material. The tangent modulus is defined as the slope of the stress-strain diagram for a material at *a particular* stress, and it is thus a function of the stress (and strain) for stresses more than the proportional limit. For stresses less than the proportional limit, the tangent modulus, of course, is the same as Young's modulus. For an ideal elastoplastic material the tangent modulus becomes zero at the proportional limit since the stress-strain diagram becomes horizontal. If $E_T$ represents the tangent modulus, the buckling load, according to the tangent-modulus theory, is

$$\frac{P}{A} = \frac{\pi^2 E_T}{(L/r)^2}. \tag{a}$$

In order to use Eq. (a), values of $E_T$ corresponding to various values of $P/A$ must be obtained from a stress-strain diagram. It is difficult to determine precise values of the slope at specific points on the curve; therefore, it is usually desirable to draw a supplementary curve showing $E_T$ plotted against $P/A$ and use this curve to obtain values for $E_T$. When corresponding values of $E_T$ and $P/A$ are available, values of $L/r$ can be computed from Eq. (a) and used to plot a graph indicating the load-carrying capacity of columns when loaded in the plastic region of the stress-strain diagram. The Euler formula was derived on the assumption that the modulus of elasticity was constant over the cross section of the column whereas the tangent-modulus formula assumes a variable modulus; therefore, the latter formula cannot be considered to provide a rigorous solution. If the assumption is made that, for any given value of the load, the *variation* of strain across any section of the column is small, the slope of the stress-strain diagram (the tangent modulus) may be considered a constant for the material at all points in any given section. Since the strains will vary linearly (plane section remains plane), the result will be a linear stress distribution, thus justifying the use of the differential equation for bending. On the basis of this reasoning, it appears that the modified Euler (tangent-modulus) formula should give reasonable results, particularly for materials for which the tangent modulus does not vary too abruptly.

The shape of the tangent-modulus curve is influenced to a considerable degree by the shape of the stress-strain diagram near the proportional limit. Three idealized stress-strain diagrams are shown in Fig.

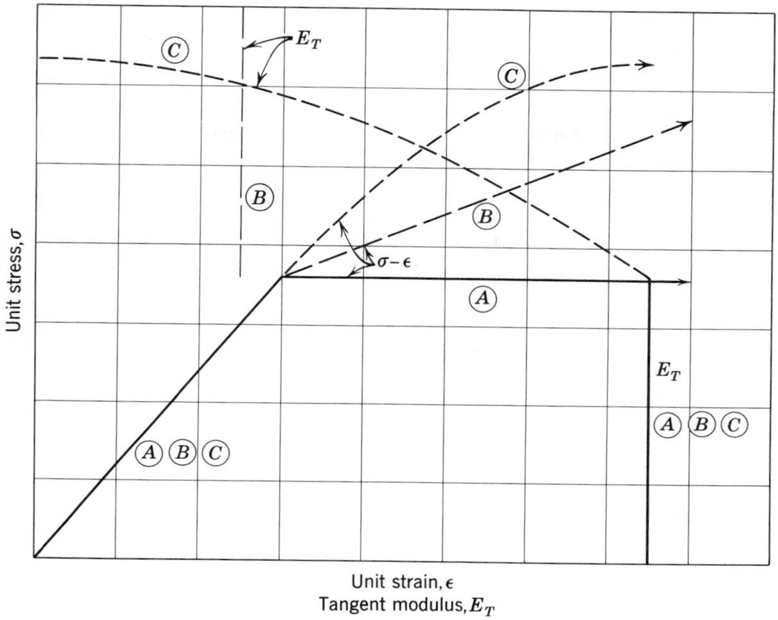


Fig. 9–15a

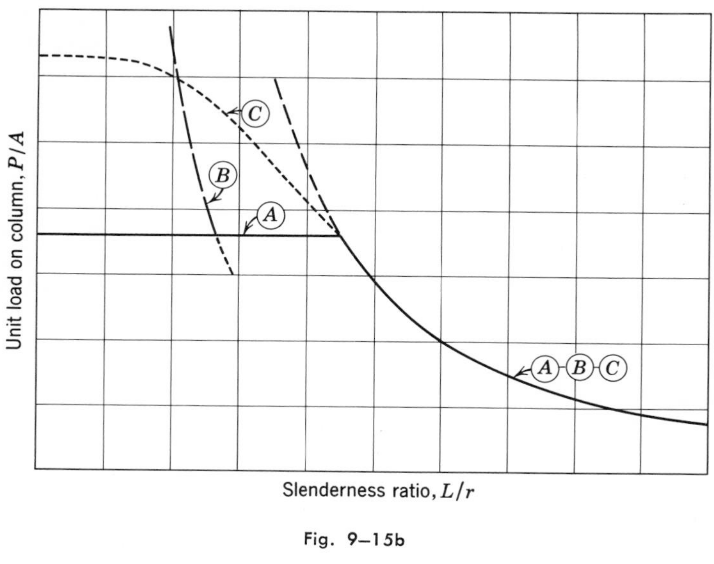


Fig. 9–15b

9–15*a* for three hypothetical materials which have identical diagrams to the proportional limit. Material *A* is elastoplastic, *B* has a diagram represented by two straight lines, and *C* has a gradual curve at the knee. The variation of the tangent modulus with the stress is shown on the same figure, with the modulus being represented by horizontal distances. The buckling loads, as predicted by the tangent modulus, are shown on Fig. 9–15*b*. Note that the buckling load predicted by the tangent modulus is the same as that predicted by the Euler formula in the elastic range and that this load is usually greater than loads obtained by actual experiment. However, when sufficient care is exercised to assure a centric load, the buckling loads indicated by the tangent-modulus formula are fairly close to experimental results[9] for columns which are loaded in the elastic and plastic range.

For designing compression members by the tangent-modulus theory, it is almost essential that a $P/A$ vs $L/r$ curve, drawn to scale, be available. If for a given problem, $L/r$ is known and the buckling (or safe) load is wanted, the value of $P/A$ corresponding to the $L/r$ can be read directly from the curve. If the problem is one of selecting a section to carry a specified load, the trial-solution approach is necessary. In other words, the techniques of Art. 9–4 are followed, with the formulas replaced by graphs.

## PROBLEMS

**9–71.** The data in the accompanying table were obtained from a stress-strain diagram for aluminum alloy 7075-T6. Young's modulus for the alloy is 10,600 ksi, and the proportional limit is 47 ksi. Plot a curve showing the relation of the buckling load, $P/A$, to $L/r$, according to the tangent-modulus formula, for columns made of this material. Determine the buckling load for a 7-ft column made of a 3-in. diameter solid round rod of 7075-T6.

| $\sigma$, ksi | $E_T$, ksi |
|---|---|
| 50 | 10,300 |
| 55 | 9,000 |
| 60 | 7,000 |
| 65 | 4,800 |
| 70 | 2,000 |
| 72 | 1,400 |

**9–72.** Plot the tangent-modulus curve for a 9.25 per cent nickel-steel alloy. Use the stress-strain diagram of App. A, and assume the diagram applies to compressive loading as well as tensile loading.

[9] Evaluation of Aeroplane Metals, J. A. Van den Broek, *Journal of the Royal Aeronautical Society,* 1946.

# CHAPTER 10

# *Repeated Loading*

**10–1. Introduction.** The discussion in previous chapters of this book, except Chap. 8, has been limited to static loading, for which the behavior of a structure under load can be predicted on the basis of the static properties of the material. However, many machine and structural elements are subjected to a type of loading known as *repeated loading,* under which the load is applied and removed, or the magnitude (and frequently the direction) is changed, many thousands of times during the life of the member. For example, the diesel engine rocker arms of Fig. 10–1, as well as the valves, push rods, connecting

Fig. 10–1. Fatigue failures of exhaust valve rocker levers. Courtesy: Cleveland Diesel Engine Division, General Motors Corp.

rods, and crankshaft, are subjected to repeated loading. In many instances the repeated loading is accidental, as a result of vibrations induced by unpredictable or uncontrollable circumstances.

If the applied load changes from any magnitude in one direction to the same magnitude in the opposite direction, the loading is termed *completely reversed;* whereas, if the load changes from one magnitude to another (the direction does not necessarily change), the load is said to be a *fluctuating load.*

The physical effect of a repeated load on a material is different from that of a static load, the failure always being a fracture (or brittle type of failure) regardless of whether the material is brittle or ductile. Under repeated loading, a small crack forms in a region of high localized stress, and very high stress concentration accompanies the crack. As the load fluctuates, the crack opens and closes and progresses across the section. Frequently this crack propagation continues until there is insufficient cross section left to carry the load and the member ruptures, the failure being given the very descriptive term *fatigue*

Fig. 10–2. Courtesy: Prof. T. J. Dolan, University of Illinois

*failure*. It is unfortunate that fatigue failures occur at stresses well below the static *elastic* strength of the material.[1]

Figures 10–2, 10–3, and 10–4 show characteristic fatigue failures. Figure 10–2 is a photograph of a torsional fatigue failure at the front

[1] For some ductile metals in the fully annealed condition (low elastic strength), the endurance limit may be slightly higher than the elastic strength.

Fig. 10–3

end of a crankshaft. This failure started in an area of fretting corrosion (point $A$) under a hub which was assembled at this point. From here the fracture progressed to points $B$ and $C$, before complete separation occurred. Figure 10–3$b$ shows the fatigue failure of one of the turbine blades of Fig. 10–3$a$. Here the failure started at the trailing edge (point $A$) and progressed to $BC$, the final fracture showing as the dark area.

The failure of one of the compressor blades, such as $B$, of Fig. 10–4$a$ is shown in Fig. 10–4$b$. The fracture appears to have started at $A$

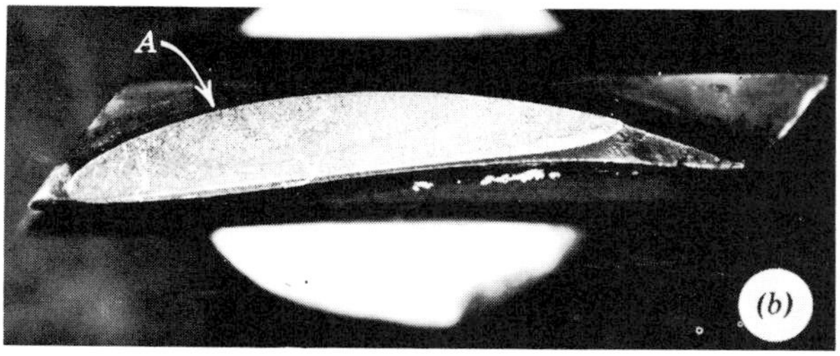


Fig. 10–4

and continued as indicated by the light area. The material to fracture last is identified by the dark area.

The importance of repeated loading was recognized in the first half of the nineteenth century. It is thought[2] that J. V. Poncelot (1788–1867), a French professor of engineering mechanics, mentioned the failure of a spring by fatigue in a book published in 1839. Also W. J. M. Rankine (1820–1872), a Scotch engineer, in a paper published in 1843, discusses the fracture of railroad locomotive axles. Through the years 1843 to 1870, the railroad wheel and axle assembly occupied much attention and evoked considerable discussion, including the debate regarding whether the metal crystallized in service. The greatest single contribution to the investigation is that of A. Wöhler (1819–1914), a German engineer, who designed fatigue-testing machines and conducted many fatigue tests, which included the effect of stress concentration due to changes in cross section. The argument that the iron changed from a fibrous to a crystalline structure under the repeated loading apparently was successfully refuted during this same period. Although the idea of change in the metal structure under repeated stressing persists to this day with some individuals, metallurgical investigation has shown that the crystalline structure of solid metals undergoes no extensive change under such conditions.

Much attention has been devoted to the question of what causes fatigue failure, and, although the question is not completely answered, the electron microscope and X-ray techniques have done much to promote an understanding of the mechanism of fatigue damage. Apparently, in a region of high localized stress (usually at or near the surface), localized slip bands form, and under repeated stressing these bands increase in number and width, rapidly at first but more slowly later. As the damage (slip) progresses, microscopic cracks are formed, and eventually these grow and join to form visible cracks.

Although the mechanism of progressive damage seems to be understood, the fundamental reasons for such action are not known. One explanation advanced is that strain hardening accompanies the local inelastic deformation (slip) and that the restraint of adjacent crystals prevents the localized strain from becoming zero as the applied load becomes zero. This results in the localized stress becoming higher with each cycle and the localized material becoming more brittle. For some combination of stress and material, it would be possible for the magnitude of the localized stress to eventually reach the static ultimate

[2] *History of Strength of Materials,* S. Timoshenko, McGraw-Hill Book Co., New York, 1953.

strength of the material, and a local crack would result. Another theory suggests that certain dislocations (defects in atomic structure) are displaced (localized slip) with relative ease and, under cyclic loading, migrate through the material until a number of them coalesce at some region where further movement is inhibited. Such a collection of dislocations in a region at or near the surface may be responsible for the action described in the preceding paragraph.

Although it is possible to have fatigue failure in a region of compressive stress, the usual failures occur under tensile stress and at stress levels considerably lower than those in a compressive stress failure.

**10–2. Resistance to Failure under Completely Reversed Loading.** The index of resistance to failure under completely reversed repeated loading is known as the *endurance limit,* defined as *the maximum completely reversed stress to which a material may be subjected millions of times without fracture.* The magnitude of the endurance limit is determined from repeated load (or fatigue) tests. One commonly used machine for determining endurance limit is the R. R. Moore rotating beam machine shown schematically in Fig. 10–5. Observe that the loading simulates that imposed on a railroad car axle between the wheels, the specimen being subjected to pure bending and the flexural stresses becoming completely reversed each one-half revolution. There are many other types of machines available including rotating cantilever beams, non-rotating simple and cantilever beams, and axially loaded members. For the last three, the load may be applied by

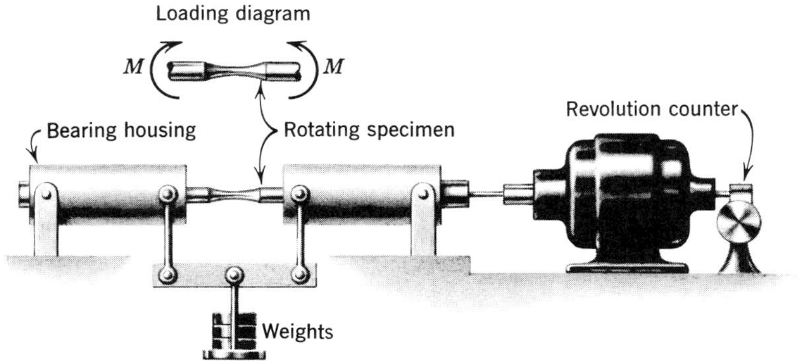


Fig. 10–5

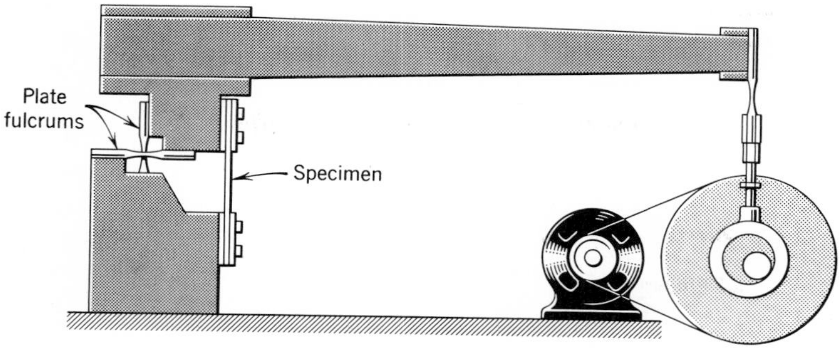


Fig. 10–6

means of a crank and connecting rod, one machine being indicated diagrammatically in Fig. 10–6. For such experimental setups, the loading may be varied so that fluctuating loads of various ranges may be imposed.

The testing technique is to subject a series of identical specimens to loads of different magnitudes and note the number of cycles of stress (or load), $n$, necessary to fracture the specimen. The data are plotted on a stress-cycle ($\sigma$–$n$) diagram, and the endurance limit can be determined from the diagram. In order to space the points on the diagram far apart for clarity, it is customary to plot the data on semilogarithmic paper, the stress being plotted to a linear scale and the cycles to a logarithmic scale as indicated in Fig. 10–7. For some materials (particularly the ferrous materials) the curve seems to break sharply as in curves $A$ and $B$ of Fig. 10–7, indicating that, if the stress has the magnitude of $\sigma_{EA}$, material $A$ could probably be subjected to an unlimited number of cycles without failure.[3] On the other hand, the alloys of aluminum and magnesium exhibit curves like $C$ and $D$ of Fig. 10–7 which are apparently asymptotic to the $n$ axis. For such materials, the stress corresponding to some arbitrary number of cycles, such as $5(10^8)$ for the aluminum alloys, is taken as the endurance limit.

It must be observed that the very nature of the fatigue failure—dependent as it is upon local imperfections or peculiar configuration of grain structure as well as upon the precision necessary for machining identical specimens—would dictate that the results obtained from a

[3] The data for Fig. 10–7 were obtained from polished specimens. Some endurance limits in App. A are for the material in the as-rolled condition and, therefore, have lower values.

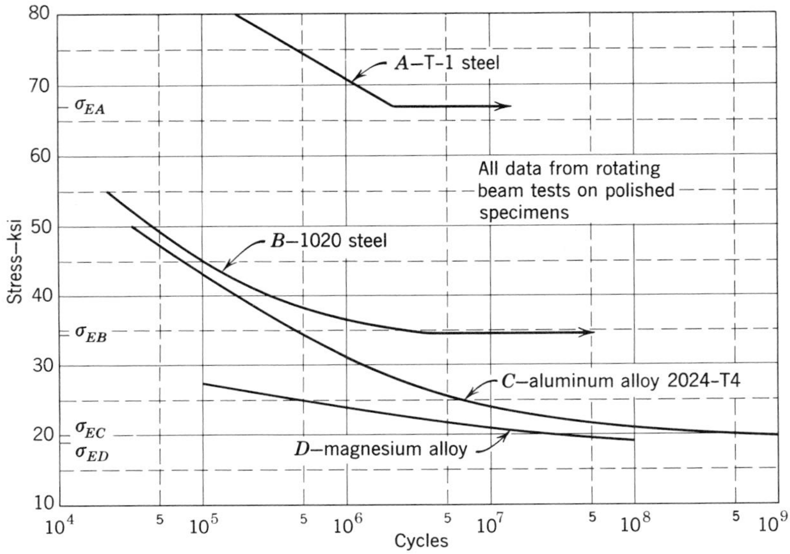


Fig. 10–7

large number of specimens might show considerable scatter. The results could be better represented by a series of curves representing different probabilities of failure or as a band such as that in Fig. 10–8.[4] The customary $\sigma$–$n$ diagram (such as shown in Fig. 10–7) is quite likely an approximate 50 per cent probability of failure curve; that is, a specimen tested at a particular maximum stress will have a fifty-fifty chance of failure when subjected to the number of cycles indicated by the intersection of the line representing the test stress with the 50 per cent curve.

If a machine part is to be designed for a limited life, a strength index higher than the endurance limit can be used. Theoretically, one can estimate the number of cycles of stress to which the member will probably be subjected during its lifetime, enter the $\sigma$–$n$ diagram with this value, and read the corresponding stress from the stress scale. If the member were stressed to this value, one could expect a fracture failure at the end of the expected lifetime, and the design of the machine part could be based on this value reduced by an appropriate factor of safety. In this connection, however, the statistical nature of

[4] "Effect of Stress Amplitude on Statistical Variability in Fatigue Life of 75S-T6 Aluminum Alloy," G. M. Sinclair and T. J. Dolan, *Trans. A.S.M.E.*, Vol. 75, 1953.

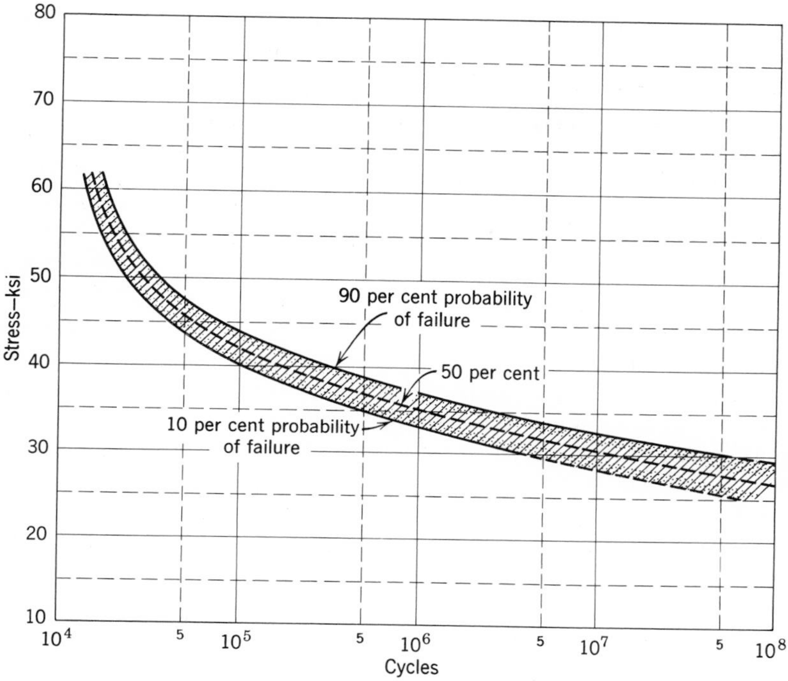


Fig. 10–8

the $\sigma$–$n$ data should be emphasized since the life of any particular manufactured part may be much less than that indicated by the ordinary $\sigma$–$n$ diagram.

**10–3. Design Factors Affecting Fatigue Strength of Structural Elements.** Since the fatigue failure starts at a point of high localized stress, any discontinuity, accidental or intended, is very apt to initiate such a failure. The accidental discontinuities must be taken care of by the factor of safety; however, the discontinuities due to deliberate design must be included in stress computations by an appropriate stress concentration factor (see Art. 1–8). This factor modifies the nominal equations for stress computations; thus

$$\sigma = \frac{KP}{A}, \qquad \tau = \frac{KT\rho}{J}, \qquad \sigma = \frac{KMy}{I},$$

where $K$ is a factor greater than unity and may be obtained from the curves of Fig. 10–9, 10–10, and 10–11.[5] In general, the factors may be based on either the gross or the net section (the minimum cross section at the discontinuity), and it is necessary for the designer to know whether the factors at his disposal apply to the gross or the net section. The factors $K_t$ of Fig. 10–9, 10–10, and 10–11 are all based on the net section except those of curve ④ of Fig. 10–9, which are to be applied to the gross section. Sometimes the solution of a problem is expedited by the use of the factor based on the gross section ($K_g$), and, for this purpose, conversion expressions are given with the various curves.

Instead of the theoretical stress concentration factor, a similar experimental factor known as the *endurance limit reduction factor* may be used. The endurance limit reduction factor is determined from repeated load tests and is the ratio of the endurance limit as determined from standard test specimens to the endurance limit as determined from specimens containing a particular discontinuity. The magnitude of this factor is dependent upon the size of the specimen, upon the kind of material, and upon the care exercised in machining of the test specimen. In general, the endurance limit reduction factor is smaller than the theoretical stress concentration factor; however, the larger the specimen and the stronger and more fine grained the material, the less will be the difference between the two factors. Naturally, if a designer has at his disposal an endurance limit reduction factor which is appropriate (that is, the factor was obtained from a specimen of comparable size and material to that of the proposed design), such a factor should be used. In case the size and material limitations are not met or are unknown, the stress concentration factor should be used since it will be larger than the other factor and the resulting design will be more conservative.

---

[5] Curves 1, 2, 5, 6, and 8: "Photoelastic Studies in Stress Concentration," M. M. Frocht, *Mechanical Engineering,* August, 1936.

Curve 3: "Stress Concentration Factors Around a Central Circular Hole in a Plate Loaded Through Pin in the Hole," M. M. Frocht and H. N. Hill, *Trans. A.S.M.E. Journal,* March, 1940.

Curve 4: "Plane Stress and Plane Strain in Bipolar Co-ordinates," G. B. Jeffery, *Philosophical Transactions of the Royal Society of London,* Vol. 221, 1921.

Curve 7: "How Holes and Notches Affect Flat Springs," A. M. Wahl, *Machine Design,* May, 1940.

Curve 9: "Stress Concentration Factors and Their Effect on Design," G. H. Neugebauer, *Product Engineering,* Feb., 1943.

Curve 10: "Torsional Stress Concentrations in Shafts of Circular Cross-Section and Variable Diameter," L. S. Jacobsen, *Trans. A.S.M.E.,* Vol. 47, 1925.

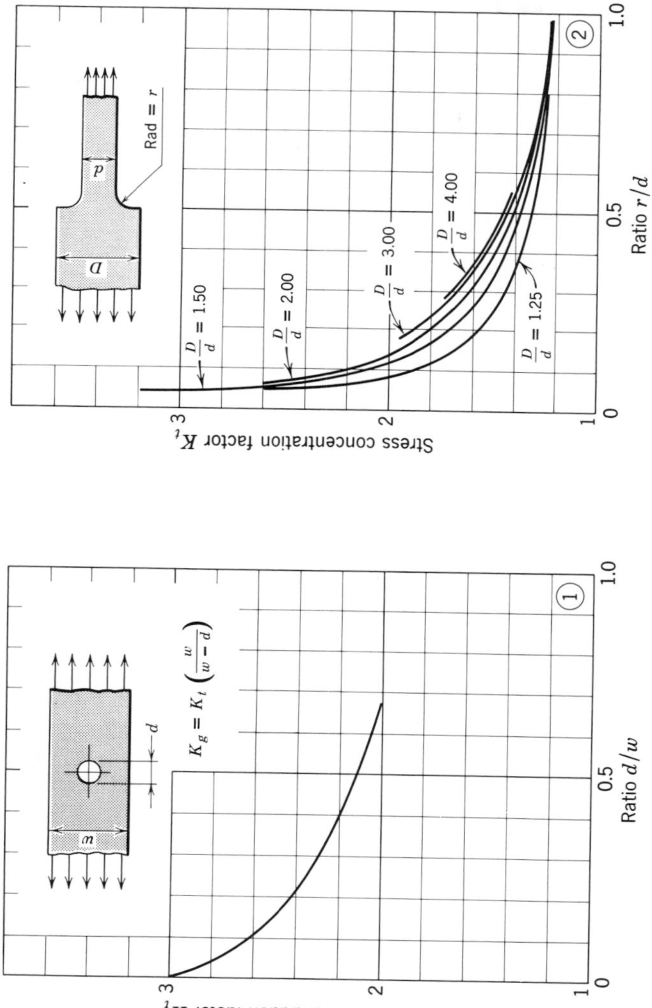


Fig. 10–9a. Stress concentration factors for axial loading on flat bars.

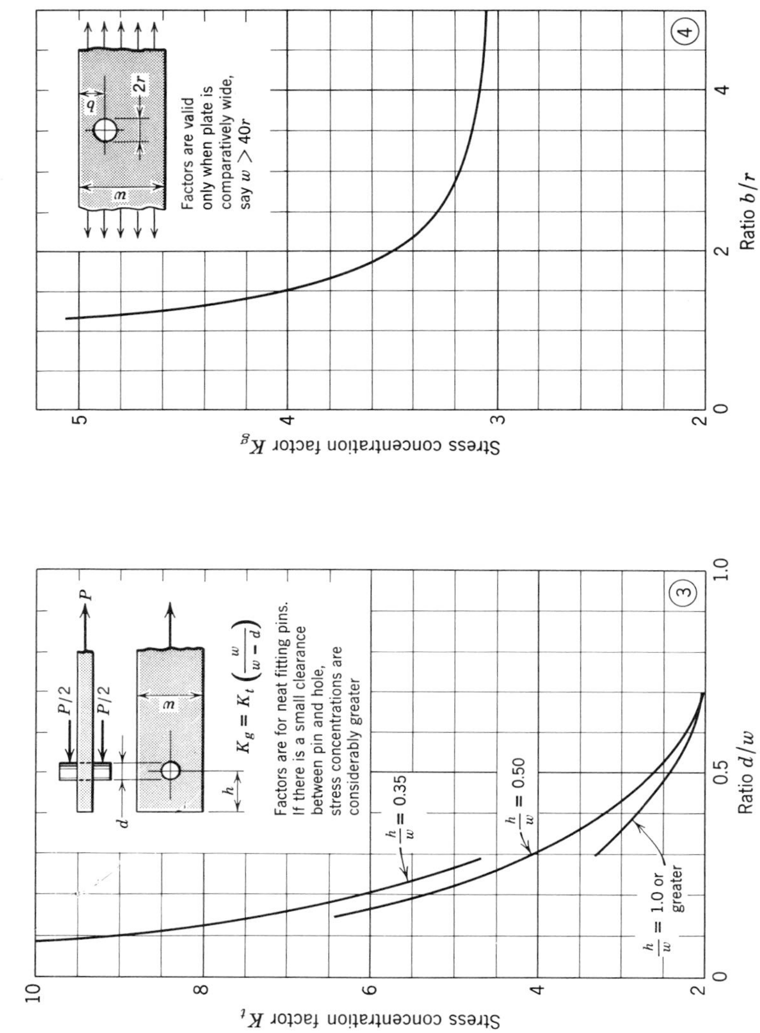


Fig. 10–9b. Stress concentration factors for axial loading on flat bars.

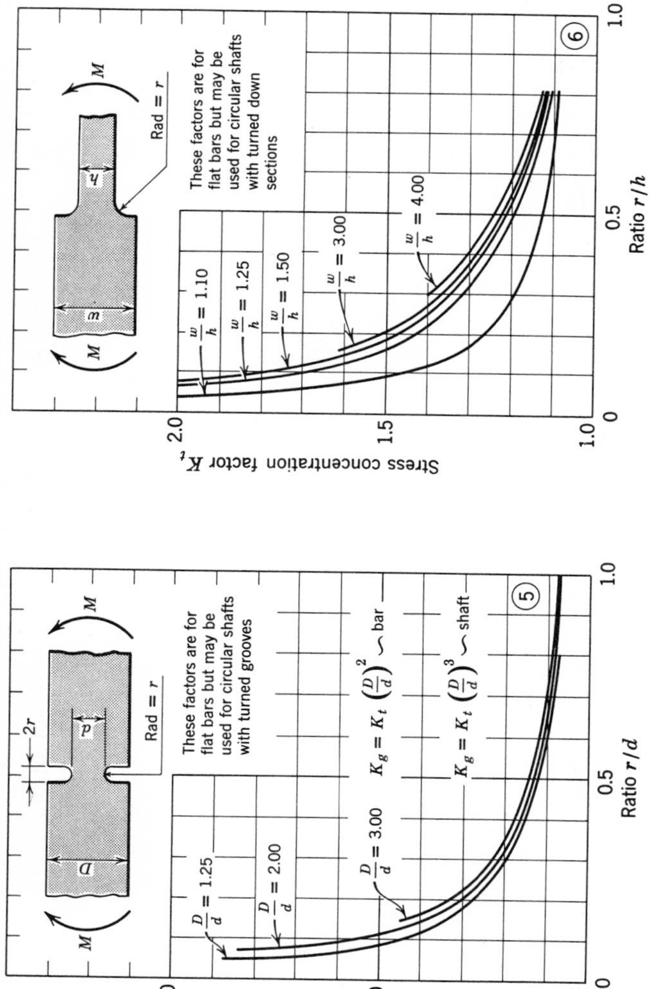


**Fig. 10–10a. Stress concentration factors for flexural loading.**

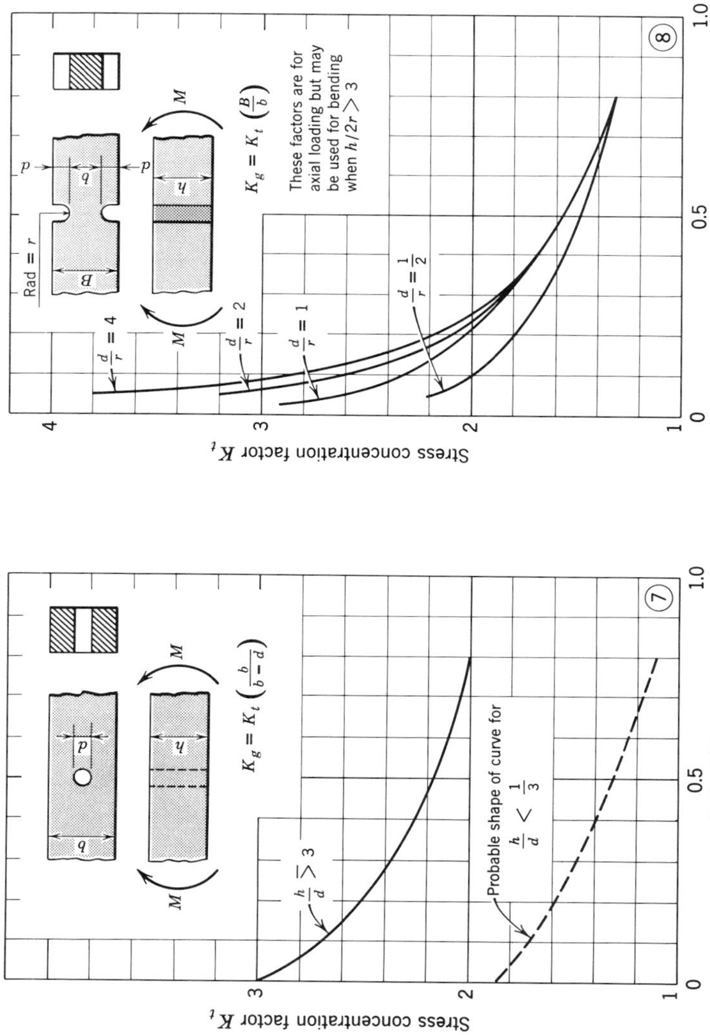


Fig. 10–10b. Stress concentration fatcors for flexural loading.

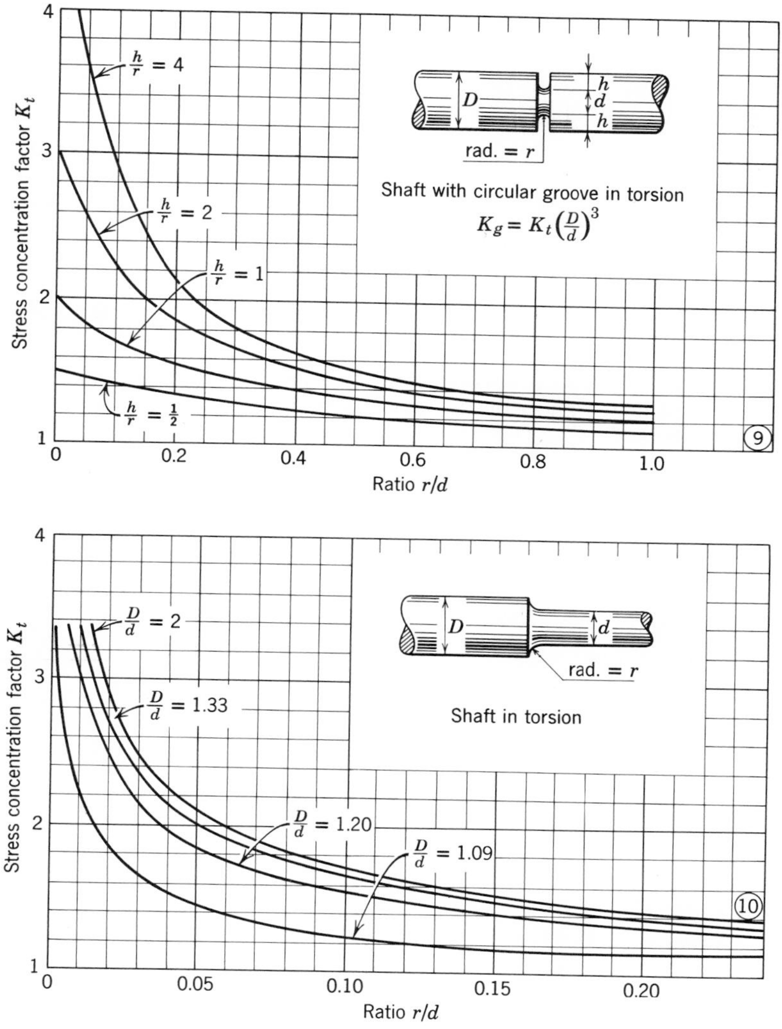


Fig. 10–11. Stress concentration factors for torsional loading.

The remarks in Art. 1–8 concerning the use of the stress concentration factor in design will bear repeating here. Stress concentration is ignored in a ductile material subjected to static loading. However, if the loading is repeated, stress concentration must not be disregarded, regardless of whether the material is brittle or ductile. Example 10–1 will serve to illustrate this point. For the problems in this chapter, unless otherwise indicated, assume that the details of construction are such that stress concentration is negligible at supports and points of application of loads.

**Example 10–1:** The simple beam *AB* of Fig. 10–12*a* is constructed of aluminum alloy 2024-T4 and has a square cross section 2 in. wide by 2 in. deep with a pair of notches (detail in insert) at the center. Determine the maximum permissible magnitude for the load *P*:

(*a*) If applied as a completely reversed repeated load with a factor of safety of 3.

(*b*) If applied as a steady load with a factor of safety of 2 with respect to failure by slip.

*Solution:* (*a*) The critical stress is the maximum tensile stress, which occurs at the surface and is given by the flexure formula modified for stress concentration as necessary. The free-body and bending moment diagrams of Fig. 10–12*b* indicate a maximum moment at *C*; however, the stress concentration at the notches may produce a greater flexural stress at *D* than at *C*. Hence it is necessary to obtain the value for *K* from curve ⑧ of Fig. 10–10 as follows: $h/(2r)$ is 4, $d/r$ is $\frac{1}{2}$, and $r/b$ is 0.143, from which $K_t$ is about 1.9 which is the factor based on the section 1.75 in. wide by 2 in. deep. In this instance, the work is expedited by converting $K_t$ to $K_g$, the factor based on the gross section, since it will then be possible to compare $KM$ at sections *C* and *D*. Thus $K_g$ is $1.9(2)/1.75$ or 2.17. The values for $KM$ are $1(3/2)P$ or $1.5P$ ft-lb at *C* and $2.17P$ ft-lb at *D*. This analysis indicates that *D* is the critical sec-

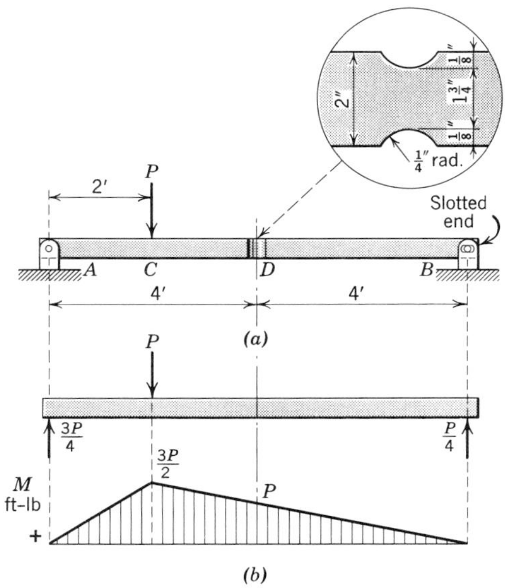


Fig. 10–12

tion, and, when the endurance limit from App. A with a factor of safety, $k$, of 3 is used, the expression for the stress is

$$\frac{\sigma}{k} = \frac{KMc}{I} = \frac{18{,}000}{3} = \frac{2.17P(12)(1)}{2(2^3)/12},$$

from which $$P = \underline{307 \text{ lb.}}$$ Ans.

(*b*) Since the material has an ultimate elongation of 19 per cent (see App. A), it is a ductile material, and under static loading the stress concentration is to be neglected. Comparing $M/I$ at $C$ and $D$, we obtain

$$\frac{1.5P}{2(8)/12} > \frac{P}{1.75(8)/12};$$

therefore, the critical section is now at $C$. Since the mode of failure is slip, the elastic strength will be used; thus,

$$\frac{48{,}000}{2} = \frac{(3P/2)(12)(1)}{2(2^3)/12},$$

from which $$P = \underline{1780 \text{ lb.}}$$ Ans.

## PROBLEMS

**10–1.*** A machine part of 1020 steel is accidentally subjected to a completely reversed repeated stress of 40 ksi at 300 cycles per min. Estimate the probable service life of the member based on Fig. 10–7.

**10–2.*** If a panel of aluminum alloy 2024-T4 in an airplane were subjected to a completely reversed stress (due to vibration) of 28 ksi, what would be the probable service life of the panel, based on Fig. 10–7, if it vibrated at 250 cycles per min?

**10–3.*** If the estimated service life of a part made of T-1 steel in a piece of army ordnance is 700,000 cycles, determine a suitable working stress with a factor of safety of 1.3. Use Fig. 10–7.

**10–4.*** The machine part shown in Fig. P10–4 is 1.0 in. thick and is made of structural steel (see App. A). Determine the maximum safe axial load:

(*a*) If the load is a completely reversed repeated load (use a factor of safety of 3).

(*b*) If the load is static (use a factor of safety of 2 with respect to failure by slip).

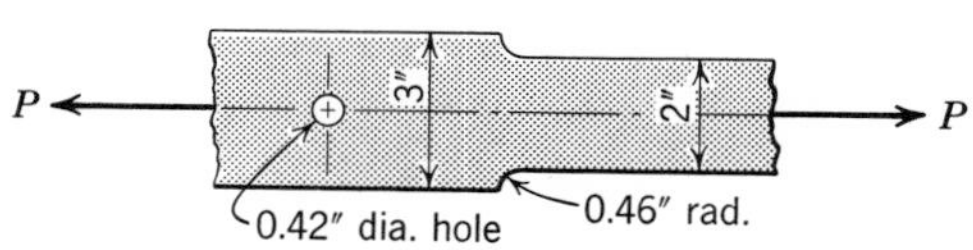


Fig. P10–4

**10–5.*** Solve Prob. 10–4 if the material is aluminum alloy 2024-T4.

**10–6.*** The bar of Fig. P10–6 is $\frac{1}{4}$ in. thick and is made of annealed stainless steel (18–8). Determine the maximum permissible axial load that may be applied to the bar:

(*a*) If the load is completely reversed and repeated and if a factor of safety of 2.8 is specified.

(*b*) If the load is static and if a factor of safety of 2 with respect to failure by slip is specified.

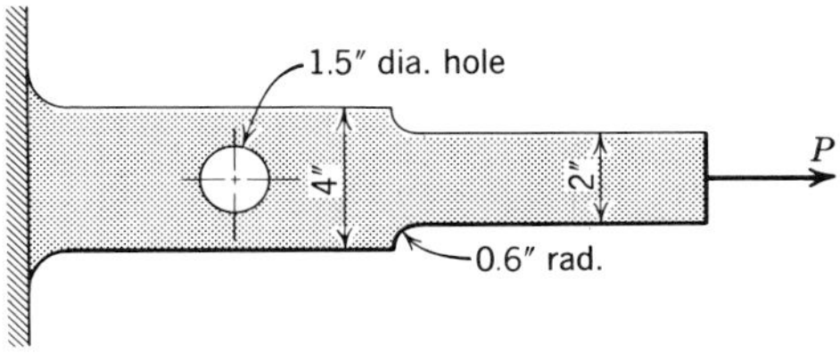


Fig. P10–6

**10–7.** Solve Prob. 10–6 if the material is heat-treated SAE 4340 steel.

**10–8.*** The structural steel plate of Fig. P10–8 is $\frac{3}{8}$ in. thick and is subjected to a tensile load of 20 kips uniformly distributed across the plate. Using a factor of safety of 2.5 with respect to failure by fracture or slip, determine the minimum allowable width of the plate:

(*a*) If the load is completely reversed and repeated.

(*b*) If the load is static.

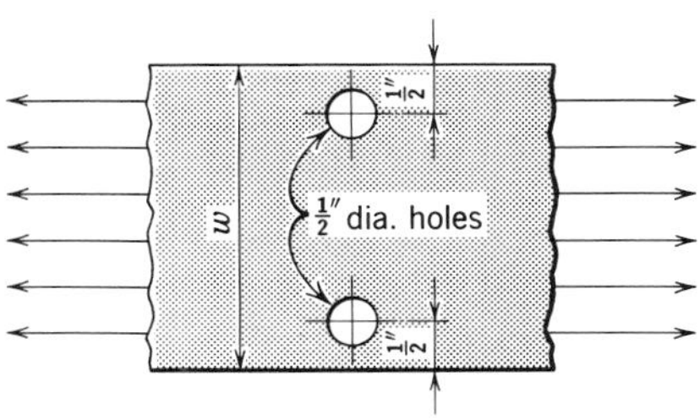


Fig. P10–8

**10–9.** A 3-in. diameter shaft which is subjected only to bending has a U-shaped groove 0.3 in. deep (radius of groove is 0.2 in.) turned in it at the section of maximum moment. The shaft is made of structural steel. Determine the maximum permissible bending moment that may be applied if:

(*a*) The load is static and the factor of safety is 2 with respect to failure by slip.

(*b*) The load is completely reversed and repeated and the factor of safety is 3.

**10–10.*** The cantilever beam shown in Fig. P10–10 is made of structural steel. Determine the maximum safe value of the load $P$:

(*a*) If applied as a steady load. Use a factor of safety of 2.2 with respect to failure by slip.

(*b*) If applied as a completely reversed repeated load. Use a factor of safety of 2.8.

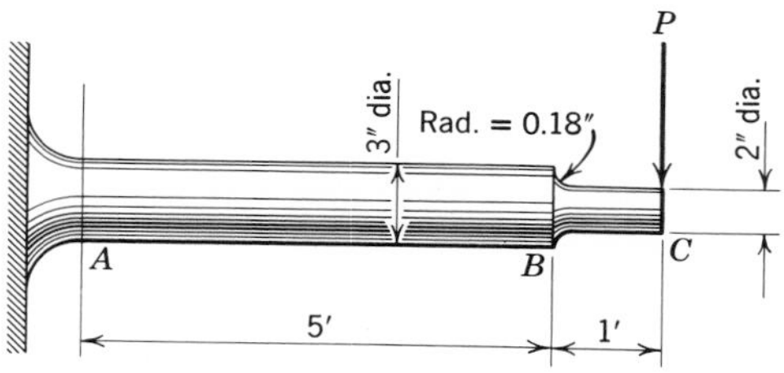


Fig. P10–10

**10–11.** Figure P10–11 shows a cantilever spring which is 2 in. wide. If $P$ is a completely reversed repeated load of 38 lb and the material is heat-treated SAE 4340 steel, determine the maximum safe values for the distances $a$ and $b$ ($b$ in combination with the maximum value for $a$). Use a factor of safety of 2.5.

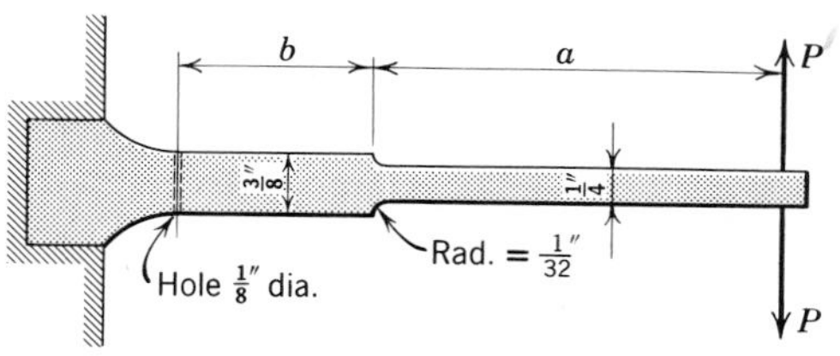


Fig. P10–11

**10–12.*** A simple beam is 6.0 in. wide by 2.0 in. deep and has a span of 10 ft. It is loaded with a completely reversed repeated concentrated load, $P$, at the center of the span and has a vertical hole 0.6 in. in diameter through the beam 1.5 ft from one end. Determine the maximum safe value for $P$ using a factor of safety of 3. The endurance limit for the material is 40,000 psi, and the ultimate strength is 85,000 psi.

**10–13.** The notched cantilever beam of Fig. P10–13 is 2 in. thick and made of aluminum alloy 2024-T4. Determine the maximum completely reversed repeated load, $P$, which may be applied if a factor of safety of 3 is specified.

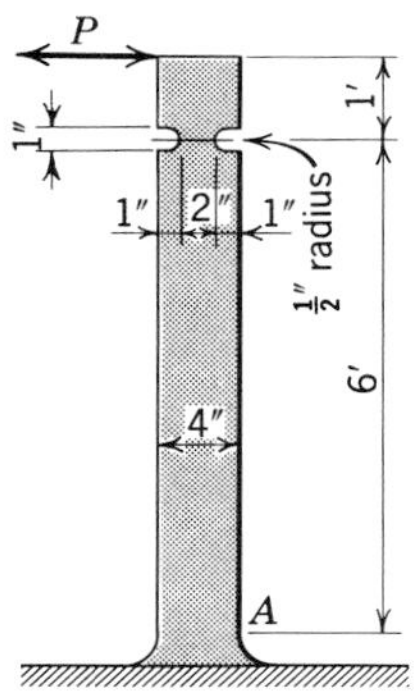


Fig. P10–13

**10–14.*** A 4-in. diameter shaft is to have one end turned down to a 3.33-in. diameter. The shaft is to be made of 1.2 per cent carbon steel normalized (use 25 ksi for the endurance limit in torsion). If the shaft is to be subjected to a completely reversed repeated torque of 3000 ft-lb and a factor of safety of 3 is specified, what should be the minimum radius of the fillet at the junction of the two sections?

**10–15.** A 4-in. diameter shaft has one end turned down to a 2-in. diameter. The fillet at the junction of the two sections has a ⅛-in. radius. The shaft is made of 3.5 per cent nickel steel, hardened and tempered (use 31.5 ksi for the endurance limit in torsion). A factor of safety of 3 is specified.

(*a*) What maximum completely reversed repeated torque may be applied?

(*b*) If a U-shaped groove 0.2 in. wide by 0.4 in. deep (radius of bottom of groove = 0.1 in.) is turned in the 4-in. section, what would be the maximum completely reversed repeated torque?

**10–4. Resistance to Failure under Fluctuating Load.** The maximum stress necessary to cause fracture under a fluctuating load is usually larger

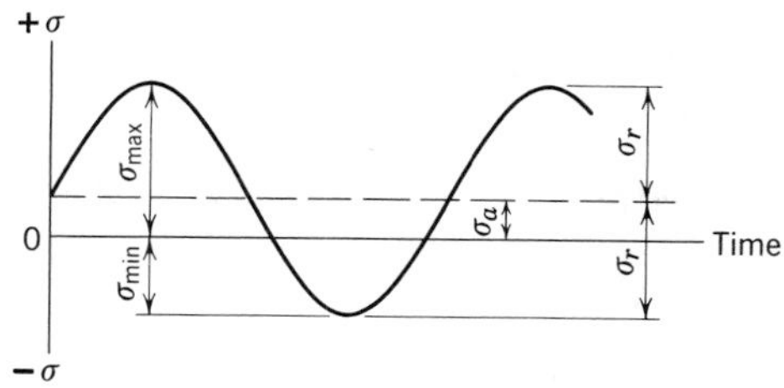


Fig. 10–13

than the endurance limit. Considerable data have been accumulated from tests performed in machines similar to that shown schematically in Fig. 10–6, in which the stress in the specimen may vary from some minimum to a different maximum value. However, the task of experimentally determining the ranges of stress which will cause fatigue failure for all possible (or reasonable) combinations of maximum and minimum stresses and for the numerous materials in use today would be unending. Hence, some method of predicting the strength based on available data is essential. The approach to the problem is made through the interpretation of data from a limited number of tests in various ranges and is based on the concept that a fluctuating stress can be considered the sum of a steady stress and a completely reversed (or alternating) stress as indicated for normal stresses in Fig. **10–13**, where $\sigma_a$, the steady stress, is one-half the sum of $\sigma_{max}$ and $\sigma_{min}$ and where $\sigma_r$, the alternating stress, is one-half the difference between these two stresses (compressive stresses are negative). The steady and alternating stresses may not be the same kind of stress; for example, one could have a steady torsional shearing stress combined with an alternating flexural stress. However, for the restricted scope of this book, the discussion will be limited to the case where the steady and alternating stresses at a point are both normal (or both shearing) stresses on the same plane.

The data from tests can be plotted on a diagram, and a curve can be drawn to represent the strength of the material for various combinations of steady and alternating stresses. Two curves which have received widespread attention are shown in Fig. 10–14, in which $\sigma_E$ is the endurance limit, $\sigma_Y$ is the static elastic[6] tensile strength, and $\sigma_U$ is the static ultimate tensile strength. Any point with coordinates ($\sigma_a$, $\sigma_r$), such as $A$ on the diagram, would represent a particular combina-

[6] Yield point, yield strength, or proportional limit.

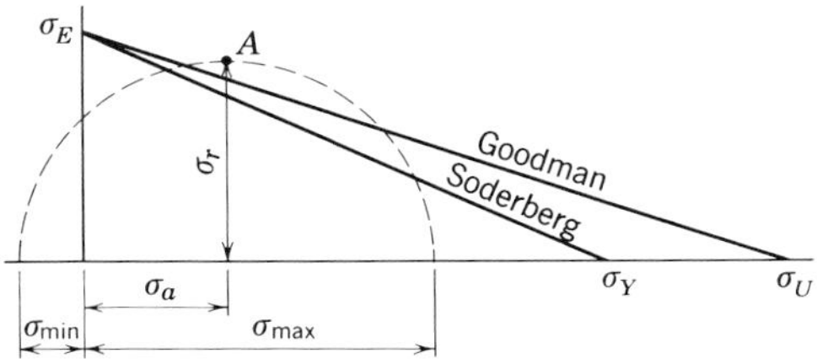


Fig. 10–14

tion of steady and repeated stresses, and, if the point is above the straight line (Soderberg[7] or Goodman,[8] depending on which line is accepted as the failure criterion), failure is presumed to result. Most of the available test data plot above the two lines shown, indicating that both curves are conservative.

When a design problem involves machine elements which will not function properly if the static elastic strength of the material is exceeded (a slip failure), the Soderberg diagram should be regarded as the criterion of failure. The reason for using the more conservative diagram is that $\sigma_{max}$ might exceed the static elastic strength on the first application of the maximum load, whereupon the usefulness of the mechanism would be impaired or destroyed. If fracture is the only mode of failure to be considered, slip is of no concern, and the less conservative Goodman diagram should be used. Needless to say, the approximate criteria of strength are to be used only if fluctuating strength data from mechanical tests are unavailable. For designs that warrant the expense, fatigue tests should be conducted on structural components loaded to simulate actual service conditions (for example, the landing gear cylinder of Fig. 10–2). When the approximate failure criteria are applied, if the endurance limit is unknown *and the material is ferrous,* one-half of the ultimate tensile strength will give a reasonable value for the endurance limit. The use of the Soderberg and Goodman diagrams is illustrated by Ex. 10–2 and 10–3. The choice of diagrams is usually indicated by a statement concerning the mode of failure. For example, a factor of safety with respect to failure by slip or fracture means that both slip (due to the application of the maxi-

[7] See "Working Stresses," C. R. Soderberg, *Trans. A.S.M.E.*, Vol. 55, APM-55-16, p. 131, 1933.

[8] Goodman's diagram does not resemble the one of Fig. 10–14. However, since the two diagrams yield the same results, it has become customary to identify the upper line of Fig. 10–14 as the Goodman criterion.

mum load) and fracture (due to the alternating load) must be avoided; hence, the Soderberg criterion should be used.

**Example 10–2:** A machine element of heat-treated SAE 4340 steel is subjected to stresses that fluctuate from 50 ksi $T$ to 10 ksi $C$. Determine the factor of safety with respect to failure by slip or fracture.

*Solution:* Failure by slip or fracture indicates that the Soderberg diagram should be used. The diagram is shown in Fig. 10–15, for which the properties were obtained from App. A. The first step in the solution is to compute the ratio of $\sigma_r$ to $\sigma_a$ as follows:

$$\sigma_{a1} = \frac{50 + (-10)}{2} = 20 \text{ ksi } T$$

and

$$\sigma_{r1} = \frac{50 - (-10)}{2} = 30 \text{ ksi } T \text{ and } C,$$

from which

$$\sigma_r = \frac{3\sigma_a}{2}.$$

The equation of the line in Fig. 10–15 is

$$\sigma_r = 76 - \left(\frac{76}{132}\right)\sigma_a,$$

and, since

$$\sigma_r = \frac{3\sigma_a}{2},$$

$$\frac{3\sigma_a}{2} = 76 - \left(\frac{76}{132}\right)\sigma_a,$$

from which

$$\sigma_a = 36.6 \text{ ksi}.$$

This value represents the limit of the steady stress. Because of the fixed ratio between $\sigma_a$ and $\sigma_r$, 36.6 ksi can be used as a strength index; hence, the factor of safety is

$$k = \frac{36.6}{20} = \underline{1.83.} \qquad \text{Ans.}$$

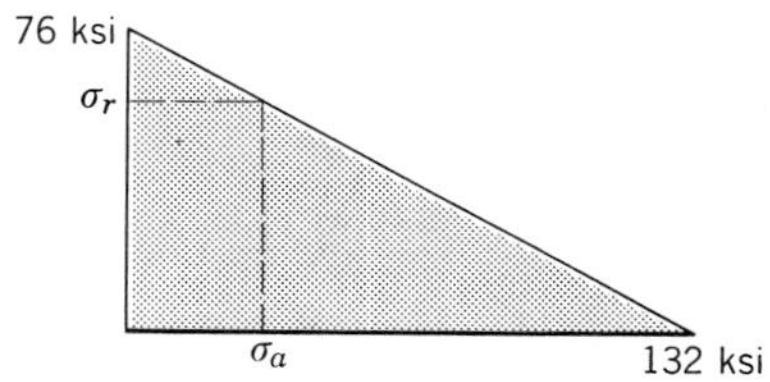


**Fig. 10–15**

Observe that this value is obtained regardless of whether one uses the steady, the alternating, the maximum, or the minimum stress as the strength index; for example,

$$\sigma_{\max} = \sigma_a + \sigma_r = \sigma_a + \frac{3\sigma_a}{2} = \frac{5\sigma_a}{2},$$

and the factor of safety is

$$k = \frac{36.6}{20} = \frac{(3/2)(36.6)}{(3/2)(20)} = \frac{(5/2)(36.6)}{(5/2)(20)} = \cdots$$

**Example 10–3:** The load on the beam of Ex. 10–1 fluctuates from $3P$ downward to $2P$ upward. Determine the maximum permissible magnitude for $P$ if a factor of safety of 3 with respect to failure by fracture is specified.

*Solution:* Since the only criterion of failure is fracture,[9] the Goodman diagram will be used with $\sigma_E$ of 18 ksi and $\sigma_U$ of 68 ksi. Instead of these strength values being used, they may be divided by the factor of safety, thus converting the failure diagram to a working stress diagram. Whether the factor of safety is introduced in the diagram or later is optional. To get the ratio of stresses,

$$P_a = \frac{3P + (-2P)}{2} = \frac{P}{2}$$

and

$$P_r = \frac{3P - (-2P)}{2} = \frac{5P}{2},$$

from which

$$P_r = 5P_a.$$

Therefore, since the stresses will be below the proportional limit,

$$\sigma_r = 5\sigma_a,$$

and, when the Goodman diagram is used with working stresses instead of strengths,

$$\sigma_{r_w} = \frac{18}{3} - \frac{18}{68}\sigma_{a_w} = 5\sigma_{a_w},$$

from which

$$\sigma_{a_w} = 1.14 \text{ ksi}.$$

The loading is repeated; therefore the stress concentration will not be ignored, and the analysis of part *a* of Ex. 10–1 is valid. Thus

$$\frac{\sigma_a}{k} = \frac{KMc}{I} = 1140 = \frac{2.17P_a(12)(1)}{2(2^3)/12},$$

from which

$$P_a = 58.4$$

and

$$P = 2P_a = \underline{117 \text{ lb.}}$$ Ans.

[9] So far as the static strength is concerned, fracture here denotes that the maximum static load is reached when the maximum flexural stress is equal to the ultimate tensile strength. Actually, under a load of this magnitude, the beam would not collapse since the material is ductile (see Art. 5–12).

Observe that the correct solution is obtained whether $\sigma_a$, $\sigma_r$, $\sigma_{max}$, or $\sigma_{min}$ is used, as long as the corresponding value for the load is used. For example, if $\sigma_r$ is used, the above equation becomes

$$1140(5) = 2.17(5P_a)(9),$$

giving the same result as above.

When this result is compared with Ex. 10–1, the fluctuating load is seen to vary from −234 lb to +350 lb, while a completely reversed load varies from −307 to +307 lb.

## PROBLEMS

**10–16.*** Determine the factor of safety with respect to failure by slip or fracture for a machine part for which the stress varies from 10,000 psi *T* to 6000 psi *C*. The material is aluminum alloy 2024-T4.

**10–17.*** A machine member is to be subjected to a repeated load which varies from 3000 lb in tension to 7000 lb in tension. The material is 3.5 per cent nickel steel hardened and tempered with the following properties: endurance limit 63,000 psi, proportional limit 86,400 psi, and ultimate strength 118,000 psi. Using Soderberg's method, determine the combination of maximum and minimum stresses to which the member could probably be subjected millions of times without fracture.

**10–18.*** An axial load varies from 4000 lb compression to 8000 lb tension. Using Soderberg's method, determine the combination of maximum and minimum stresses to which the member could probably be subjected millions of times without fracture if the proposed material is 0.93 per cent carbon spring steel hardened and tempered. The following properties apply: endurance limit 56,000 psi, proportional limit 84,800 psi, and ultimate strength 97,100 psi.

**10–19.** A machine element is subjected to a bending load which varies from +5000 lb to −2000 lb. The material is structural steel (see App. A). Determine the magnitude of a suitable working stress using a factor of safety of 2.5 with respect to failure by slip or fracture.

**10–20.*** Determine the maximum and minimum values for the fluctuating axial load, *P*, of Fig. P10–20 if the magnitude of *P* in tension is twice the magnitude in compression. The material is structural steel (see App. A), and a factor of safety of 3 with respect to failure by slip or fracture is to be maintained.

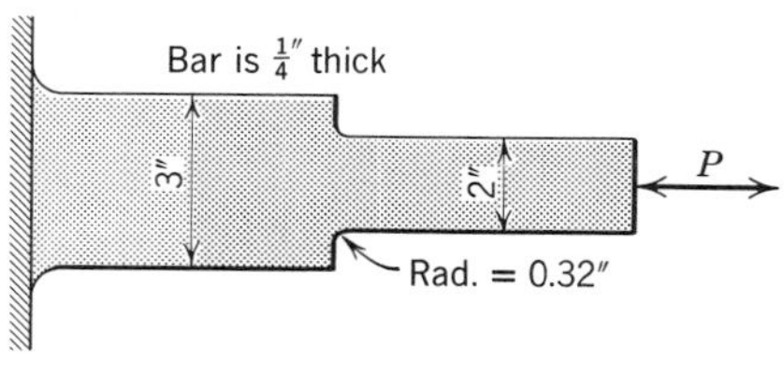


Fig. P10–20

**10–21.** In Fig. P10–21 is shown a connecting rod made of 0.57 per cent carbon steel which is normalized. The rod is of rectangular cross section and is 2 in. thick. Determine the maximum completely reversed repeated load, $P$, that may be applied with a factor of safety of 3 with respect to failure by slip or fracture. The following properties apply: ultimate strength 72,000 psi, proportional limit 35,000 psi, and endurance limit 33,000 psi. Note that the tension on the net section at the pin fluctuates from zero to a maximum.

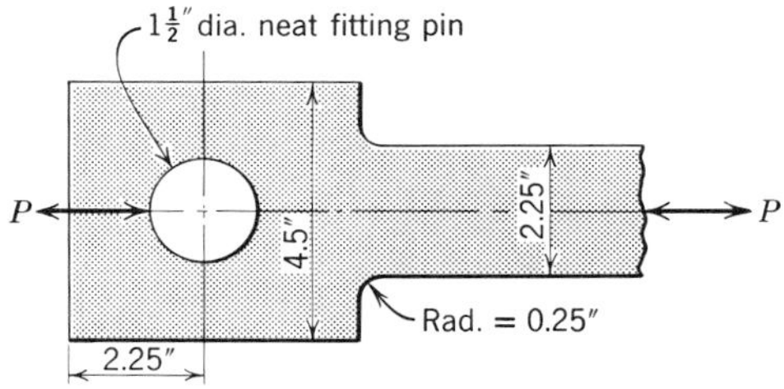


Fig. P10–21

**10–22.*** The bar of Fig. P10–6 is made of ⅜-in. thick structural steel and is subjected to an axial load which fluctuates from $4P$ in tension to $P$ in compression. Determine the maximum permissible value for $P$ to be applied with a factor of safety of 2.5 with respect to failure by slip or fracture.

**10–23.** A simple beam of aluminum alloy 6061-T6 is 4 in. wide by 2 in. deep, has a span of 12 ft, and has a ½-in. vertical hole drilled through it 1.5 ft from one support. The beam is loaded at the center of the span with a concentrated load which varies from $P$ upward to $4P$ downward. Determine the magnitude of $P$ if a factor of safety of 2 with respect to failure by slip or fracture is specified.

**10–24.*** A simple beam of structural steel is 4 in. wide by 2 in. deep, has a span of 10 ft, and has a ¼-in. vertical hole drilled through it 2 ft from one support. The beam is loaded at the center of the span with a fluctuating concentrated load. Determine the maximum and minimum values of the load, $P$, if the magnitude of $P$ downward is twice the magnitude upward. Use a factor of safety of 2.5 with respect to failure by slip or fracture.

**10–25.** A simple beam of aluminum alloy 2014-T4 is 6 in. wide by 2 in. deep and has a span of 14 ft. At the center of the span there is a pair of symmetrical notches for which the stress concentration factor is 1.5 based on the gross section. A concentrated fluctuating load is applied 3 ft from one support. Determine the maximum and minimum values of the load if the magnitude in one direction is three-halves that in the other direction. Use a factor of safety of 2 with respect to failure by slip or fracture.

**10–26.*** The cantilever beam of Fig. P10–26 is made of heat-treated spring steel with the properties of Prob. 10–18. The beam is 2 in. wide with depths and stress concentration factors as shown. The load at $C$ varies from $3P$ downward to $P$ upward. Determine the maximum permissible magnitude for $P$ using a factor of safety of 3 with respect to failure by slip or fracture.

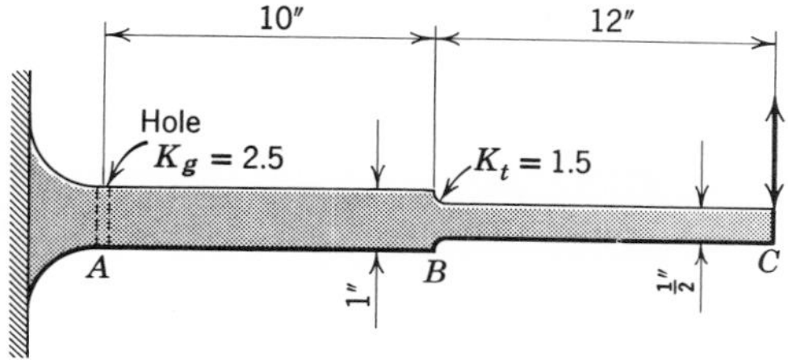


Fig. P10–26

**10–27.** If the load of Prob. 10–13 fluctuates from $5P$ to the right to $P$ to the left and a factor of safety of 2.5 with respect to failure by slip or fracture is specified, what is the maximum permissible value of $P$?

## 10–5. Review Problems

**10–28.*** The unbalanced disk of Fig. P10–28 weighs 300 lb, and the eccentricity of the mass center induces an effective force of 75 lb directed outward from the shaft which is attached to and rotates with the disk. Neglect the weight of the shaft and any torsional moment; assume the end bearings to be simple supports, and consider bending only in a vertical plane. Determine the minimum permissible diameter for the shaft if it is to be made of heat-treated SAE 4340 steel and a factor of safety of 3 with respect to failure by slip or fracture is specified.

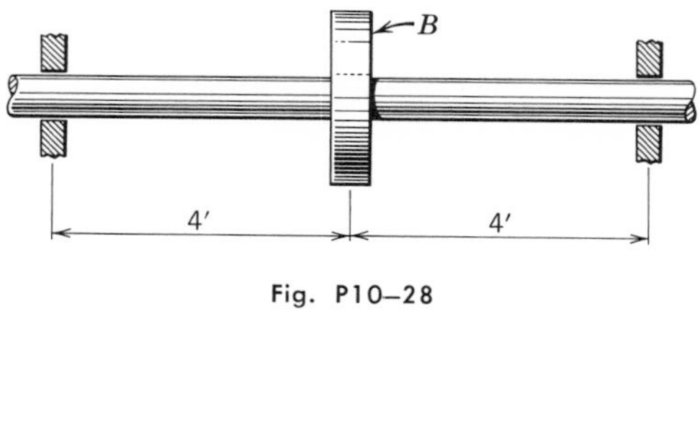


Fig. P10–28

**10–29.** The steel cantilever spring of Fig. P10–29 is deflected repeatedly from zero to 1/4 in. by means of the rotating cam. The spring has a rectangular cross section 0.2 in. thick and 1.2 in. wide. The material is heat-treated SAE 4340 steel. Determine the factor of safety with respect to failure by slip or fracture.

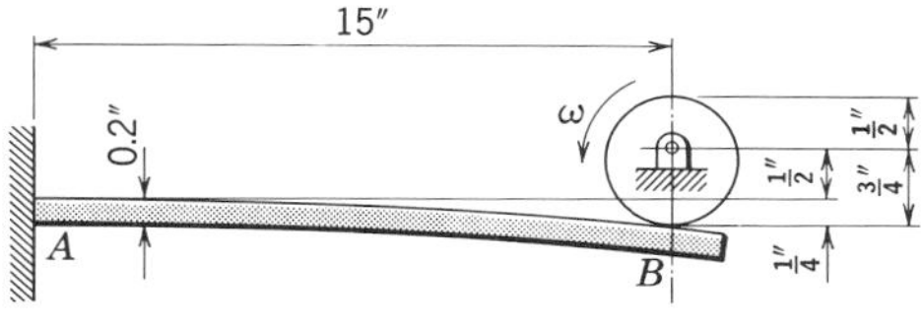


Fig. P10–29

**10–30.*** A spring for a latch mechanism may be approximated by a cantilever beam with a rectangular cross section 0.30 in. wide by 0.03 in. thick with a span of 2.0 in. The free end of the spring is deflected from zero to 0.200 in. as the latch is operated. If the spring is made of cold-rolled 18–8 stainless steel, what is the factor of safety with respect to failure by fracture?

**10–31.** Solve Prob. 10–28 when the disk rotates on a shaft which does not rotate.

**10–32.*** The 18–8 cold-rolled stainless steel beam shown in Fig. P10–32 is 3 in. wide. It is subjected to a completely reversed and repeated uniformly distributed load of 160 lb per in. Determine the factor of safety.

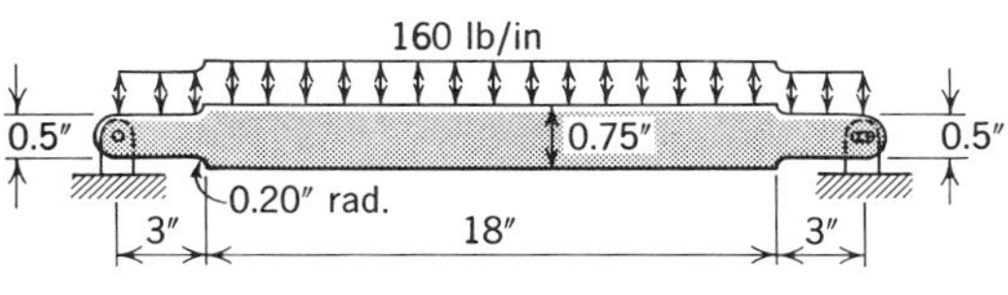


Fig. P10–32

**10–33.** The machine part shown in Fig. P10–33 is 0.5 in. thick and is made of structural steel (see App. A). Determine the maximum safe axial load:

(*a*) If the load is completely reversed and repeated. Use a factor of safety of 2.5.

(*b*) If the load is static. Use a factor of safety of 2 with respect to failure by slip.

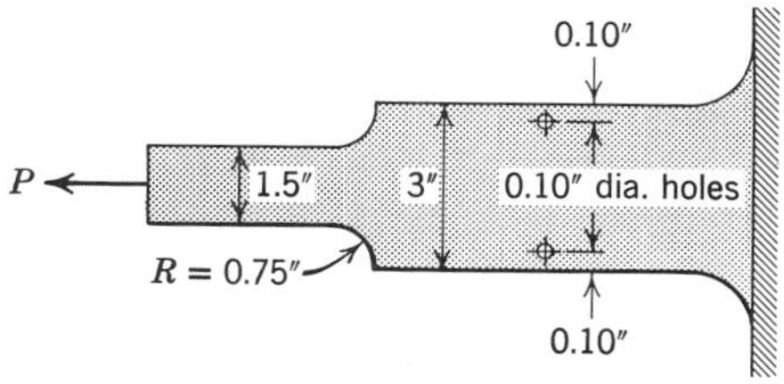


Fig. P10–33

**10–34.*** The 0.2 per cent $C$ hardened steel strut in Fig. P10–34 is subjected to a completely reversed and repeated load of 1800 lb. Determine the minimum thickness of the rectangular cross section that may be used if the factor of safety is to be 2 with respect to failure by slip or fracture. Note that the tension on the net cross section at the pin fluctuates from zero to a maximum.

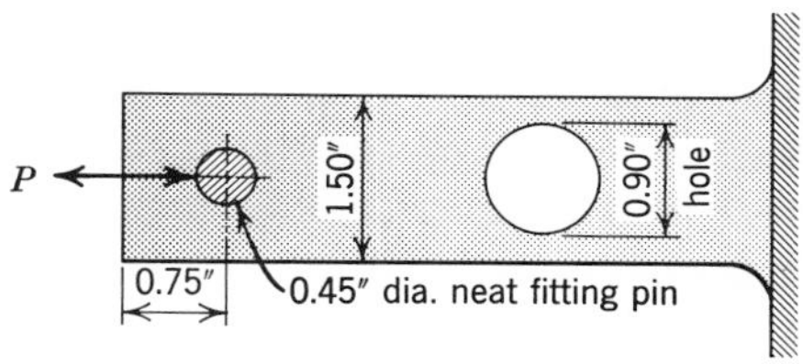


Fig. P10–34

**10–35.** Solve Prob. 10–29 if a $\frac{1}{16}$-in. diameter hole is drilled vertically through the spring at a point 5 in. from $A$.

CHAPTER 11

# Connections

**11–1. General Discussion of Riveted Connections.** One of the most vexing problems confronting the structural designer is the proportioning and detailing of connections to transmit forces from one structural element to another. Among the common types of fastenings are rivets, bolts, welds, pins, and threaded couplings. This article is concerned with riveted (or bolted) connections such as the joints of Fig. 11–1.

A riveted joint can fail as a result of the cross shearing stresses in the rivets, the bearing stresses between the rivets and the plates, the tensile stresses in the plates, or, perhaps, a combination of these stresses.

Before discussing the analysis of these connections, it is best to define, with the aid of Fig. 11–2, some of the terms used. Figure 11–2*a* represents a *double-riveted butt joint* with one *cover (or splice) plate.* The term *double riveted* (or *single riveted,* etc.) refers to the number of rows of rivets which transfer the total load from one element to another (in this case from main plate to cover plate, or from cover plate to main plate). A butt joint may also be fabricated with two cover plates, one on each side of the main plates. In some butt joints, both cover plates engage all rows of rivets; in others, one plate may engage only the inner rows and the other plate all rows. (See Fig. 11–2*b*.) Figure 11–2*c* represents a *triple-riveted lap joint.* The various terms used are indicated in the diagrams. Note carefully the designation of inner and outer rows in the two different types of joints. The term *pitch* alone always denotes spacing between rivets in a row.

When analyzing long joints such as those in boilers, water tanks, and air compressor tanks, a representative length of the joint, called the *repeating group,* is investigated rather than the entire joint. The repeating group of rivets is the shortest pattern which repeats itself along the joint, and the length of the group is the least common multiple of the rivet pitches in the joint. It is recommended that the rivet

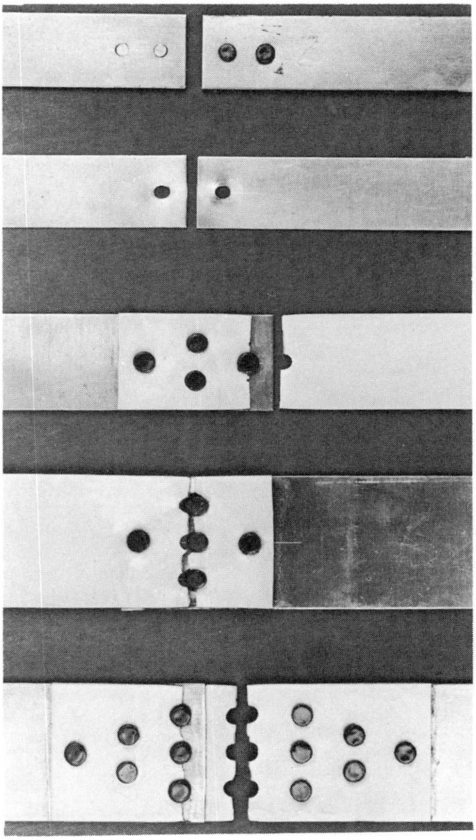

(a) Double-riveted lap joint. Shear failure.

(b) Single-riveted lap joint. Bearing failure.

(c) Triple-riveted lap joint. Tension failure in outer row.

(d) Triple-riveted lap joint. Tension failure in inner row.

(e) Triple-riveted butt joint. Tension failure in inner rows of cover plates (top plate on left side of joint and bottom plate on right side).

Fig. 11–1

pattern be sketched as an aid to obtaining the correct number of rivets (or shear areas) in the repeating group. Such a sketch is given in Fig. 11–3 which is an example of a triple-riveted joint with pitches of 2, 3, and 4 in. It will be observed that there are thirteen rivets in the length of 12 in. This sketch is extended to show that the patterns for two consecutive 12-in. lengths are identical.

The analysis of a riveted or bolted connection based on elastic action[1] is unsatisfactory because the solution is so laborious and the re-

[1] For a discussion of elastic analysis, see "Work of Rivets in Riveted Joints," A. Hrennikoff, *Trans. Am. Soc. of Civil Engrs.*, Vol. 99, p. 437, 1934. The accompanying discussions include a rather extensive bibliography on the subject.

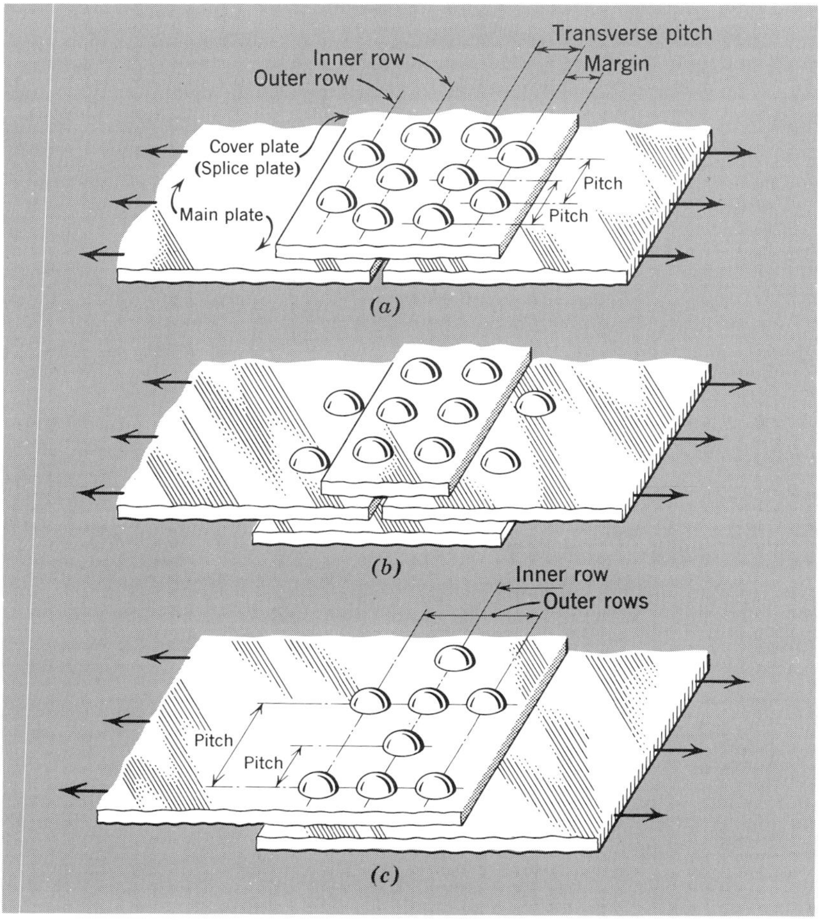


Fig. 11–2

sults of the computations (which are based on certain assumptions) are usually considerably different from experimental results, which of necessity involve indeterminate variables due to fabrication methods. For a brief exploration into the field of elastic analysis, consider section *a–a* of the quadruple-riveted lap joint of Fig. 11–4 in which the rivet distortions are exaggerated. If it is assumed that the distortions in rivets 1 and 2 are equal (equivalent to assuming that the forces on the two rivets are equal) and if there is no slip in the joint (the rivets fill the holes completely and elastic action prevails), the axial deforma-

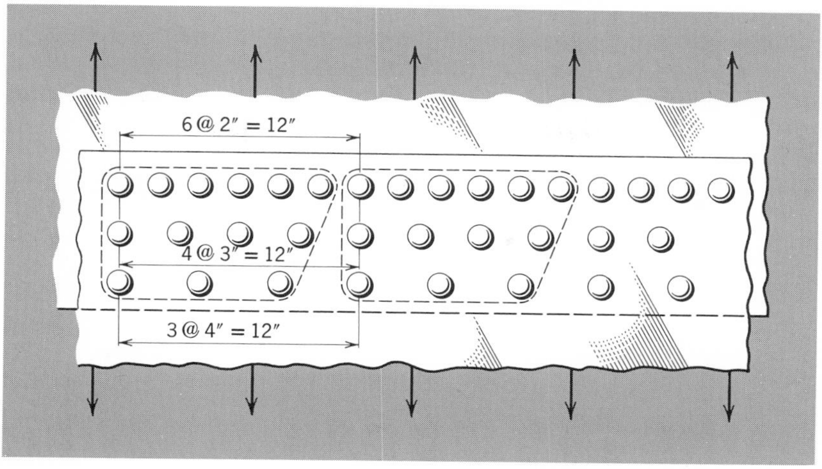


Fig. 11–3

tion between rivets 1 and 2 in plates $A$ and $B$ is the same. Therefore the average axial strain in $A$ is equal to that in $B$; and, if Hooke's law applies and the plates are of the same size and material, the tensile force in $A$ equals that in $B$, equals $P/2$. When this result is applied to the free-body diagrams of Fig. 11–5, the condition of equilibrium in-

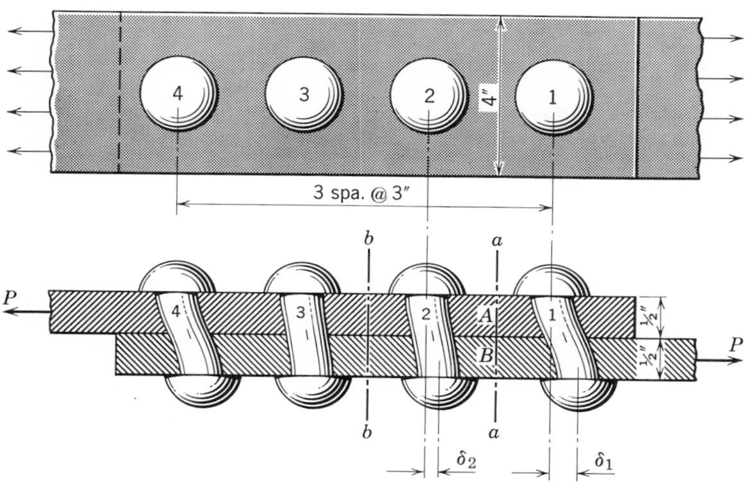


Fig. 11–4

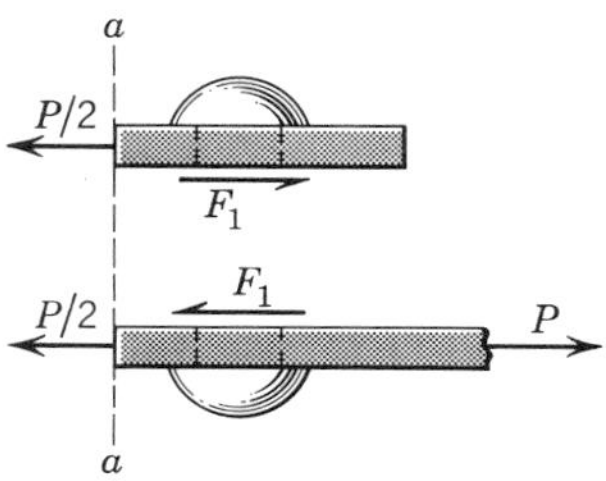


Fig. 11–5

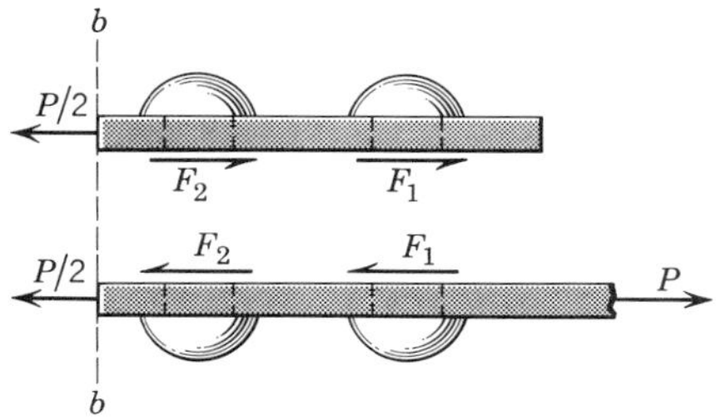


Fig. 11–6

dicates that $F_1$ is equal to $P/2$. Now, when the same reasoning is applied to section $b$–$b$ and the free-body diagrams of Fig. 11–6 are drawn, it becomes apparent that, if $F_1$ is equal to $P/2$, $F_2$ must be zero. This statement is not consistent with the original assumption, indicating that, if elastic action prevails throughout the joint and the rivets completely fill the holes, a uniform distribution of load to rivets is impossible (except for the special case of the double-riveted joint with balanced plate areas).

One possible elastic analysis of the joint of Fig. 11–4 based on the assumptions of negligible friction, all rivets bearing on the plates immediately upon application of load, and the force on any rivet being a constant[2] times the rivet distortion, $\delta$, would indicate that, for a steel joint having the dimensions shown, at working load, rivets 1 and 4 would each carry about 35 per cent of the total force, $P$, and rivets 2 and 3 would each carry about 15 per cent. These results, although closer to the truth than those obtained under the first assumption, are still in error because experiments[3,4,5] show that slip may begin upon application of even a small load, indicating that, although friction may be negligible, the rivets do not completely fill the holes (which is to be

[2] This constant is experimentally determined and varies considerably with materials, joints, loads, and investigators. For this analysis the constant $12.5(10^6)$ lb per in. was taken from reference 3.

[3] "Riveted and Pin-connected Joints of Steel and Aluminum Alloys," L. S. Moisseiff, E. C. Hartmann, and R. L. Moore, *Trans. Am. Soc. of Civil Engrs.,* Vol. 109, p. 1359, 1944.

[4] "Tension Tests of Large Riveted Joints," R. E. Davis, G. B. Woodruff, and H. E. Davis, *Trans. Am. Soc. of Civil Engrs.,* Vol. 105, p. 1193, 1940.

[5] "Tests of Joints in Wide Plates," W. M. Wilson, J. Mather, and C. O. Harris, *Bulletin No.* 239, Engr. Exp. Sta., Univ. of Ill., 1931.

expected if the cooling of the rivet after driving is considered); therefore the assumption of no slip is invalidated. Furthermore, bending of the plates was neglected; and, as can be seen from Fig. 11–4, the applied forces, $P$, cannot become collinear until the plates are bent.

The foregoing discussion will serve to indicate the need for some simplifying assumption or concept regarding the action of a riveted or bolted connection leading to a reliable easily applied method of analysis. Two widely used methods are outlined in the two succeeding articles. For both methods, the following approximations are made:

1. *The nominal size of the rivet or bolt* is used for *shear* and *bearing* stresses in the design of steel bridges and buildings. In the design of pressure vessels and aircraft structures, the actual diameter of the hole in the plate is used (the rivet when driven is assumed to fill the hole).
2. The net section for computing the tensile stress in the plates is calculated using the nominal size of the rivet or bolt plus ⅛ in. for steel bridge and building design. The actual (specified) size of the hole is used to calculate the net section in the design of aircraft structures and pressure vessels.
3. *Friction* and *bending* are neglected for riveted joints but friction may be considered in bolted joints using high-strength steel bolts.

A term frequently encountered in discussions of riveted connections is the *efficiency* of a joint, here defined as *the maximum permissible load on a joint divided by the maximum permissible axial load on the solid plate.* Efficiency may also be defined as *the strength of the joint divided by the strength of the solid plate.* One may speak of the efficiency in shear, bearing, or tension as the permissible load (or strength) in shear, bearing, or tension divided by the safe load capacity (or strength) of the solid plate; however, the *efficiency of the joint* is as defined above.

**11–2. Analysis of Riveted or Bolted Joints—Uniform Shear Method.** One of the common assumptions made regarding load distribution in riveted and bolted connections is that the cross shearing stress in the rivets or bolts is uniformly distributed over all the shear areas (the cross-sectional areas of the rivets or bolts). After the discussion in Art. 11–1, this may appear to be a rather indefensible assumption; however, tests have been conducted which seem to indicate that such

TABLE 11–1. Allowable Stresses for Riveted Joints

| Specification | Tensile Stress, Net Section of Plate, psi | Shearing Stress in Rivet, psi | Bearing Stress on Plate, psi |
|---|---|---|---|
| Am. Soc. of Mech. Engrs. Boiler Construction Code for Unfired Pressure Vessels, 1965 SA-7 plate, SA-31 rivets | 12,650 | 9,000 | 20,250 |
| Am. Inst. of Steel Const. for Struct. Steel Bldgs., 1963 A7 steel, A141 rivets, A325 bolts[a] | 20,000 | 15,000 | 44,500 |
| Am. Assoc. of State Highway Officials for Steel Bridges, 1961 A7 steel, A141 rivets, A325 bolts[a] | 18,000 | 13,500 | 27,000 |
| Am. Ry. Engrs. Assoc., 1965 A36 steel, A141 rivets | 20,000 | 13,500 | 27,000[b]<br>36,000 |
| Am. Soc. of Civil Engrs. Spec. for Alum. Alloy 2014-T6 (14S-T6), 1952 | 22,000 | 10,000[c]<br>8,000 | 36,000[d] |

[a] The shearing stress applies and the bearing stress is not restricted for friction-type joints with A325 bolts.

[b] The smaller value for rivets in single shear, the other for rivets in double shear.

[c] The larger value for alloy 2117-T3 driven cold, the smaller value for alloy 6061-T43 driven hot. If the rivet is used in relatively thin plates, the allowable shear stress shall be reduced (see spec.).

[d] If the margin is less than two rivet diameters, the stress shall be reduced (see spec.).

a condition is possibly attained with only a small amount of slip in the joint.[6] The tensile stress in the plate is assumed to be uniformly distributed over the net area of the plate at each row of rivets. The

[6] "Riveted and Pin-connected Joints of Steel and Aluminum Alloys," L. S. Moisseiff, E. C. Hartmann, and R. L. Moore, *Trans. Am. Soc. of Civil Engrs.*, Vol. 109, p. 1359, 1944. Included in these tests were symmetrical triple-riveted butt

bearing stress between the rivets and the plates is assumed to be uniformly distributed over the projection of the contact area, the plate thickness multiplied by the rivet diameter.

In order to design or check the safety of riveted joints, working stresses must be specified. Table 11–1 lists allowable stresses from some widely used specifications.

To illustrate the analysis of riveted joints by assuming a uniform shearing stress, the joint of Fig. 11–1*d* and a boiler joint will be discussed in Ex. 11–1 and 11–2.

**Example 11–1:** The triple-riveted lap joint of Fig. 11–1*d* is composed of 1⅜-in. wide by 0.025-in. thick aluminum alloy 2024-T4 plates connected with 5/32-in. aluminum alloy 2117-T4 rivets. The holes and rivets are essentially the same size for this joint. The joint ruptured under an axial load of 1640 lb. Determine for this loading:

(*a*) The average shearing stress in the rivets.
(*b*) The average bearing stress between rivet and plate.
(*c*) The maximum average tensile stress, and state where it occurs.

*Solution:* (*a* and *b*) From the free-body diagram of Fig. 11–7*a* and the equilibrium equation $\Sigma F_x = 0$, the force on each rivet is

$$F_s = 1640/5 = 328 \text{ lb},$$

and, from Fig. 11–7*c*, the average shearing stress is

$$\tau = \frac{328}{(\pi/4)(5/32)^2} = \underline{17{,}100 \text{ psi.}} \qquad \text{Ans.}$$

---

joints having ¾-in. main plates and two ⅜-in. cover plates with twelve ⅞-in. rivets in each half. At design loads (average shearing stress = 13,500 psi), the rivets in the inner row were found to carry 33 to 42 per cent of the load, those in the intermediate row 16 to 26 per cent, and those in the outer row 36 to 43 per cent. However, the ultimate loads for the joints designed to fail in shear were within 2 per cent of the predicted loads based on the uniform shearing stress assumption and on shearing strengths obtained from auxiliary control specimens having one and two rivets in each half. These results indicate that, although at working loads, there is definitely a non-uniform distribution of load to the rivets, the inequality disappears by the time the ultimate load is reached. Joints designed to fail in tension averaged within 4 per cent of the loads predicted on the basis of the tensile properties of the plate material (as obtained from control specimens) on the net section. The writers of this paper concluded that, "The amount of plastic yielding necessary to level off high stress concentrations and to effect a uniform distribution of load among the rivets must be relatively small." Of course, this means that a reasonably ductile material is necessary. (See also references 4 and 5 on p. 508.)

From Fig. 11–7*b* and *c*, the average bearing stress is

$$\sigma_b = \frac{328}{0.025(5/32)} = \underline{84{,}000 \text{ psi.}} \qquad \text{Ans.}$$

The specified strengths of these materials in shear and bearing are 27,000 psi and 90,000 psi respectively;[7] therefore, this joint should not fail in shear or bearing under the given load.

[7] ANC-5, Strength of Aircraft Elements, December, 1942. Note that specified strengths are guaranteed minimum values and are usually less than typical values.

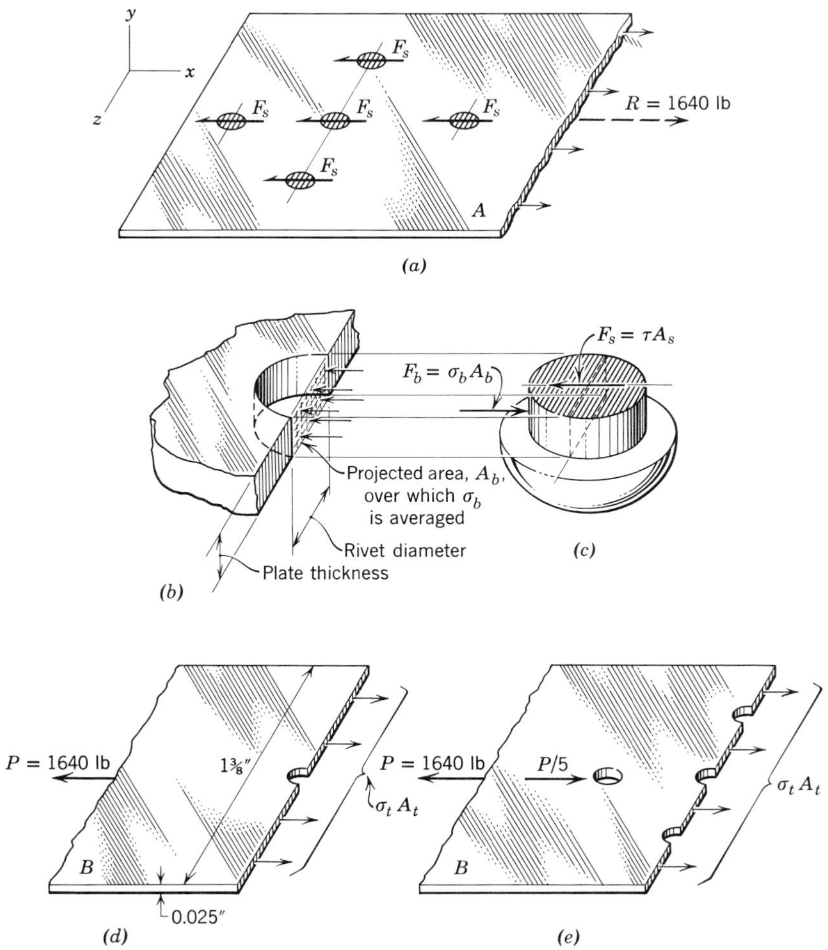


Fig. 11–7

(*c*) The average tensile stress (axial stress) may be maximum at either the outer row where there is only one hole or the inner row where there are three holes and a reduction in the load. The section at the outer row will be checked first with the aid of the free-body diagram of Fig. 11–7*d*. Summing forces in the horizontal direction gives

$$P = \sigma_t A_t = \sigma_t \left(\frac{11}{8} - \frac{5}{32}\right)(0.025)$$

$$1640 = \sigma_t \left(\frac{39}{32}\right)(0.025),$$

from which $$\sigma_t = 53{,}900 \text{ psi } T \text{ in the outer row.}$$

Next the section through the inner row will be checked with the aid of Fig. 11–7*e*. Note that, in order to separate the joint at the inner row, not only must the plate be cut at this section, but the rivet ahead of the section must also be removed (or cut), and, according to the assumption of uniform shearing stress, the load on this rivet is one-fifth of the total load. Again summing forces in the horizontal direction gives

$$P = \frac{P}{5} + \sigma_t A_t \quad \text{or} \quad \frac{4P}{5} = \sigma_t \left[\frac{11}{8} - 3\left(\frac{5}{32}\right)\right](0.025)$$

$$\frac{4(1640)}{5} = \sigma_t \left(\frac{29}{32}\right)(0.025),$$

from which $$\sigma_t = \underline{57{,}900 \text{ psi } T \text{ at the inner row.}} \qquad \text{Ans.}$$

There is no need to check the next row because the situation at the outer row of the other plate is identical with that already checked. The results of these computations indicate that, in tension, the joint is weaker at the inner row than at the outer row; this fact is verified by Fig. 11–1*d* which shows the plate ruptured at the inner row. The specified tensile strength of this material is 62,000 psi (see reference 7, p. 512); therefore, the average stress obtained was about 7 per cent below the specified strength.[8]

**Example 11–2:** A 50-in. diameter boiler is made of ¾-in. plate. The longitudinal seam is a double-riveted butt joint with two 9/16-in. cover plates and with ⅞-in. diameter rivets in 15/16-in. holes. The outer rows of rivets have a

[8] This tends to verify the conclusions in references 4 and 5, p. 508, that protecting the minimum net section by increasing the pitch in the outer rows is not as effective as the analysis would indicate. This joint would very likely have carried a larger load if it had been fabricated with two ⅛-in. rivets in each of the three rows.

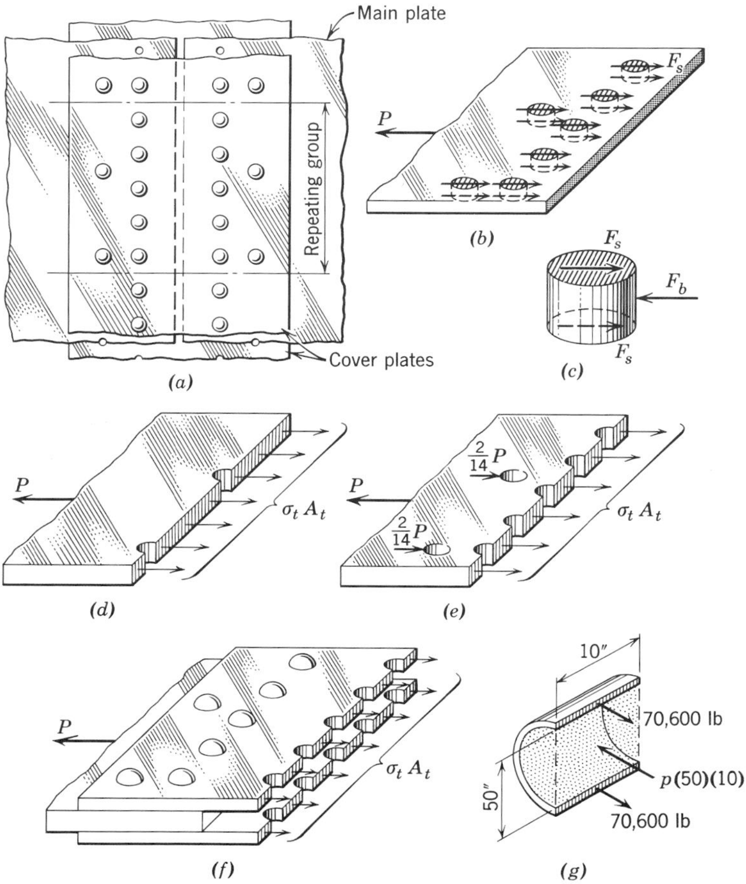


Fig. 11–8

pitch of 5 in. and the inner rows a pitch of 2 in. Using the ASME Boiler Code, determine:

(*a*) The maximum allowable internal pressure.

(*b*) The efficiency of the joint.

*Solution:* (*a*) From the sketch of the rivet pattern shown in Fig. 11–8*a*, it is evident that the repeating group has two rivets in each outer row and five rivets in each inner row. The free-body diagram of Fig. 11–8*b* shows a section of the main plate with a repeating group of rivets and indicates the fourteen shearing forces, $F_s$, two forces for each rivet. The free-body diagram in Fig. 11–8*c* indicates that $F_b = 2F_s$ since each rivet is in double shear. The maximum allowable values of $F_b$ and $F_s$ are

$$(F_b)_{\max} = \sigma_b A_b = 20{,}250\left(\frac{3}{4}\right)\left(\frac{15}{16}\right) = 14{,}240 \text{ lb},$$

and

$$(F_s)_{\max} = \tau A_s = 9{,}000(\pi)\left(\frac{15}{32}\right)^2 = 6{,}210 \text{ lb}.$$

It is evident that the bearing capacity of the rivets and cover plate is greater than that of the main plate because the thickness of the two cover plates exceeds that of the main plate. Since twice the allowable shearing load is less than the maximum allowable bearing load, the shearing strength of the rivets is the critical factor as far as shear and bearing are concerned. The maximum load which can be carried by the rivets is, from equilibrium (see Fig. 11–8*b*),

$$P_s = 7(2F_s) = 7(2)(6{,}210) = 86{,}900 \text{ lb}.$$

Three different sections must be checked for tensile capacity. They are: (1) the main plate at the outer row of rivets, (2) the main plate at the inner row of rivets with some of the load already transferred to the cover plates, and (3) the cover plates at the inner row of rivets. The summation of forces in the direction of the load for the free-body diagram of Fig. 11–8*d* gives

$$P_t(\text{outer row}) = \left[10 - 2\left(\frac{15}{16}\right)\right]\left(\frac{3}{4}\right)12{,}650 = 77{,}100 \text{ lb}.$$

The same equilibrium equation for the free-body diagram of Fig. 11–8*e* for tension in the inner row with $(2/7)P$ already transferred to the cover plates gives

$$P_t(\text{inner row}) = \left[10 - 5\left(\frac{15}{16}\right)\right]\left(\frac{3}{4}\right)12{,}650 + \left(\frac{2}{7}\right)P_t,$$

from which $$P_t = 70{,}600 \text{ lb}$$

as the maximum allowable load for tension at the inner row of the main plate. The summation of forces in the direction of the load for the free-body diagram of Fig. 11–8*f*, which shows the tension in the cover plates at the inner row, gives

$$P_t(\text{cover}) = (2)\left[10 - 5\left(\frac{15}{16}\right)\right]\left(\frac{9}{16}\right)12{,}650 = 75{,}600 \text{ lb}.$$

The 70,600-lb load which can be carried by the main plate at the inner row of rivets is the smallest of the four loads computed and is therefore the maximum allowable load on the 10-in. portion of the joint investigated.

The free-body diagram of Fig. 11–8*g* shows a 10-in. length of one-half of the boiler where $p$ is the internal pressure in the boiler. The applicable equation of equilibrium gives

$$p500 = 2(70{,}600)$$

or $$p = \underline{282 \text{ psi}}. \qquad \text{Ans.}$$

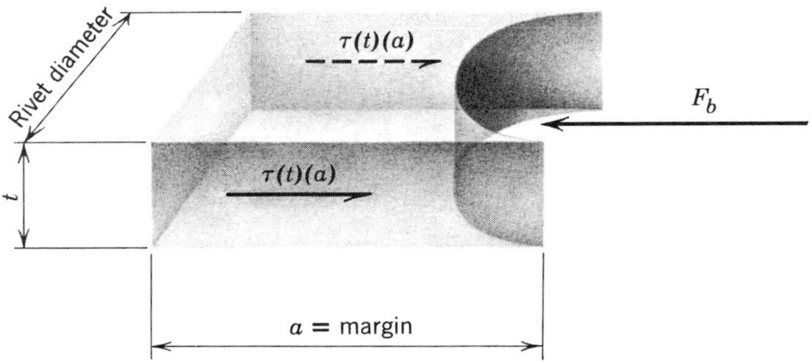


Fig. 11–9

(*b*) The efficiency of the joint is the allowable load on the joint divided by the allowable load on the main plate or

$$\eta = \frac{70{,}600}{(3/4)(10)12{,}650} = 0.744 = \underline{74.4 \text{ per cent.}} \qquad \text{Ans.}$$

A type of failure not yet considered here is a *marginal failure,* frequently called *tearout.* If a joint is designed to satisfy the usual specifications, the margins will be wide enough ($1\frac{1}{2}$ to 2 rivet diameters is usually specified) to prevent tearout failures. Should a narrow margin be needed, the shear tearout investigation should be made as indicated in Fig. 11–9. For the problems in this book, the margins will be assumed to be wide enough to preclude marginal failures, and the transverse pitch great enough to prevent failure along a diagonal (or zig zag) line between rivets in different rows.

## PROBLEMS

In the following problems, when the size of the hole is given, use the hole diameter for shear, bearing, and tension investigations. When the size of the rivet or bolt is given and no hole diameter is specified, use the nominal diameter of the rivet (or bolt) for shear and bearing investigations, and the nominal diameter plus $\frac{1}{8}$ in. for the size of the hole in tension investigations.

**11–1.*** A bolted joint from a truss is shown in Fig. P11–1. Member $A$ consists of two $2\frac{1}{2}$ by $\frac{1}{4}$-in. bars; and $B$ and $C$ each consist of two $2\frac{1}{2}$ by $2\frac{1}{2}$ by $\frac{1}{4}$-in. angles. The diameter of the bolts is $\frac{3}{4}$ in. The total tensile force $P_1 = 12{,}600$ lb. Determine:

(a) The highest average tensile stress in $A$, and state where it occurs.
(b) The maximum average bearing stress on the connection of $B$.
(c) The average shearing stress in the bolts at $C$.

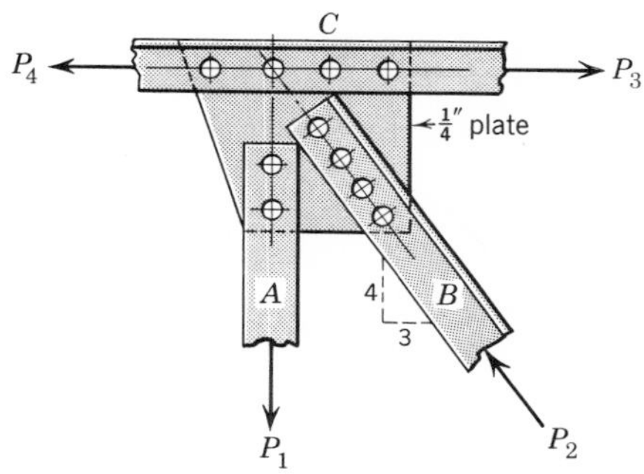


Fig. P11–1

**11–2.*** A 10-in., 25.0-lb steel channel is riveted to a gusset plate as shown in Fig. P11–2. The rivets are $\frac{3}{4}$ in. in diameter. Determine:

(a) The average shearing stress in the rivets.
(b) The average bearing stress between rivets and channel web.
(c) The highest average tensile stress in the channel, and state where it occurs.

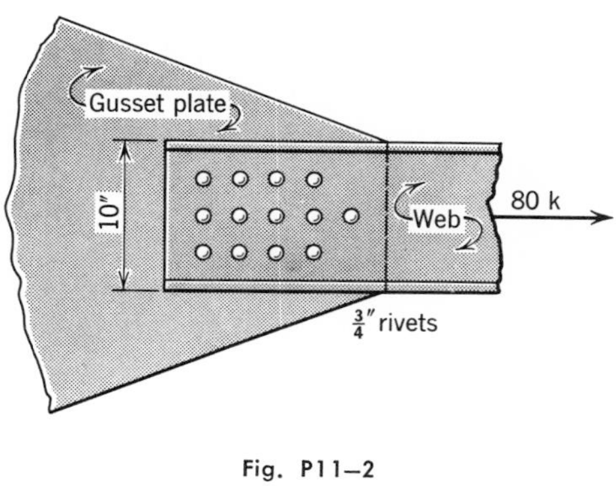


Fig. P11–2

**11–3.** Figure P11–3 represents a compression member in a pin-connected truss. The member is composed of two steel 12 ⊔ 30 latticed together; and, to provide enough bearing area on the pin, two $\frac{3}{8}$-in. pin plates (*a* and *b*) are riveted to each channel web with $\frac{3}{4}$-in. diameter rivets as shown. The channel web is $\frac{1}{2}$-in. thick. The total axial load on the member (two channels) is 230 kips. Assume the bearing stress on the pin to be uniformly distributed, and compute the average shearing stress in the rivets:

(*a*) On the plane of contact between the pin plates and channel web.
(*b*) On the plane of contact between the two pin plates.

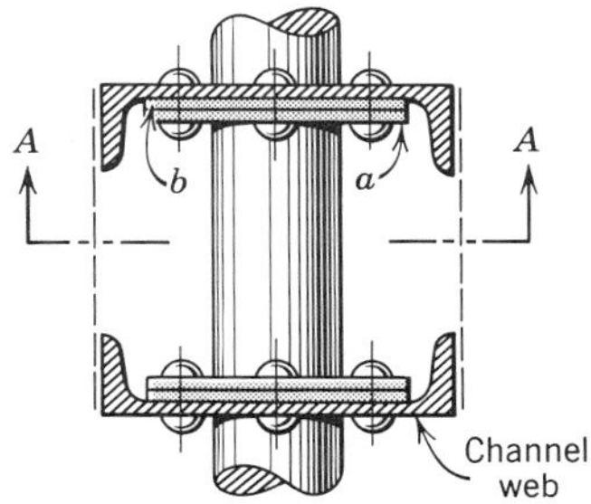


Transverse section

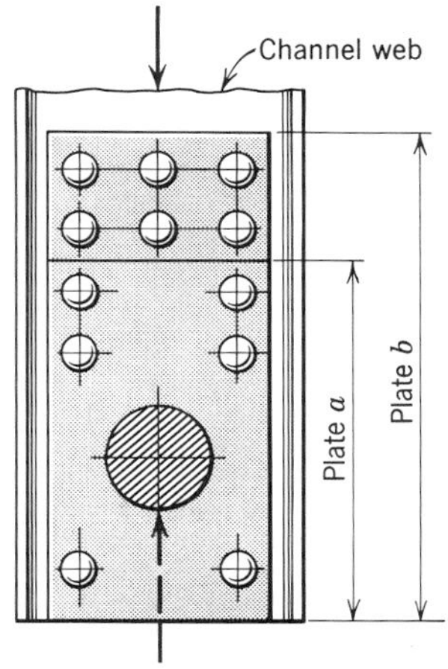


Section *A–A*

**Fig. P11–3**

**11–4.*** A long steel conduit 6 ft in diameter is subjected to an internal pressure of 90 psi. It is made of $\frac{3}{8}$-in. steel plates joined with a longitudinal, double-riveted lap joint. The rivets are $\frac{3}{4}$ in. in diameter in $\frac{13}{16}$-in. holes, and the pitch in each row is 3 in. Determine:

(*a*) The average shearing stress in the rivets.
(*b*) The average bearing stress on the plates.
(*c*) The highest average tensile stress in the plates.

**11–5.*** A triple-riveted lap joint (Fig. P11–5) unites two $\frac{3}{8}$-in. plates. The rivets are $\frac{7}{8}$ in. in diameter and are spaced 6 in. in the outer rows and 3 in. in the inner row. When the tensile stress is 10,000 psi in the gross section of the plate, determine:

(*a*) The average intensity of shearing stress in the rivets.
(*b*) The average bearing stress on the plates.
(*c*) The highest average tensile stress, stating where it occurs.

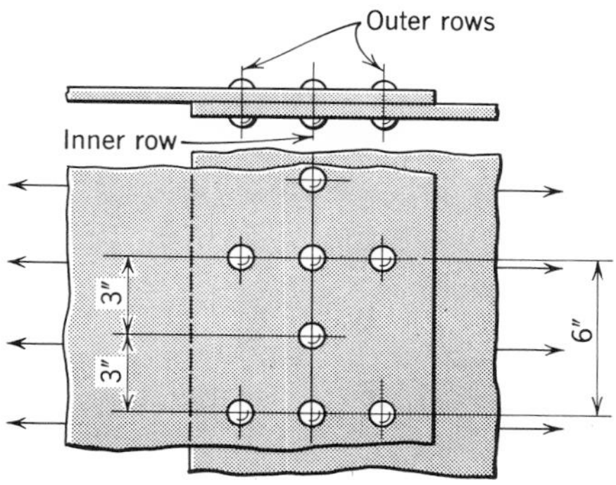


Fig. P11–5

**11–6.** Figure P11–6 shows a riveted lap splice in the stressed skin of an airplane wing. The material is aluminum alloy. The rivets are $\frac{5}{32}$ in. in diameter with 0.159-in. holes. Determine:

(*a*) The average shearing stress in the rivets.

(*b*) The maximum average bearing stress between rivets and sheet.

(*c*) The maximum average tensile stress, stating where it occurs.

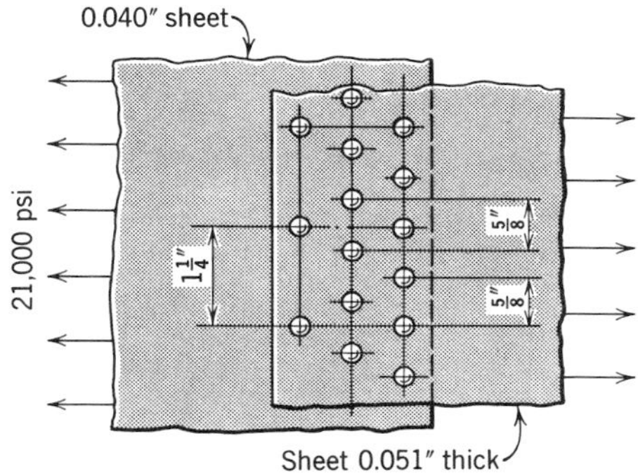


Fig. P11–6

**11–7.*** A cylindrical pressure vessel with an inside diameter of 36 in. is made of $\frac{1}{2}$-in. plate and is subjected to an internal pressure of 250 psi. The longitudinal seam is a double-riveted butt joint with two $\frac{5}{16}$-in. cover plates. The $\frac{3}{4}$-in. diameter rivets in $\frac{13}{16}$-in. holes are spaced at $2\frac{1}{2}$ in. in the inner rows and 5 in. in the outer rows. Determine:

(*a*) The average shearing stress in the rivets.

(*b*) The highest average bearing stress between rivets and any plate.

(*c*) The highest average tensile stress in the plates, and state where it occurs.

**11–8.** Two $\frac{7}{16}$-in. steel plates are connected by a triple-riveted butt joint with two $\frac{1}{4}$-in. cover plates. The $\frac{5}{8}$-in. diameter rivets are spaced at 4 in. in the outer rows, 2 in. in the intermediate rows, and 3 in. in the inner rows. The joint is loaded in tension, and each shear area carries a total force of 3000 lb. Determine the highest average tensile stress in the joint, and state where it occurs.

**11–9.*** Two 8 by ¾-in. plates are connected by a triple-riveted butt joint with unequal length cover plates as shown in Fig. P11–9. The rivets are ⅞ in. in diameter. When $P = 55{,}000$ lb, determine:

(*a*) The average shearing stress in the rivets.

(*b*) The average bearing stress on the rivets in row 3.

(*c*) The average tensile stress in the main plate at each row of rivets.

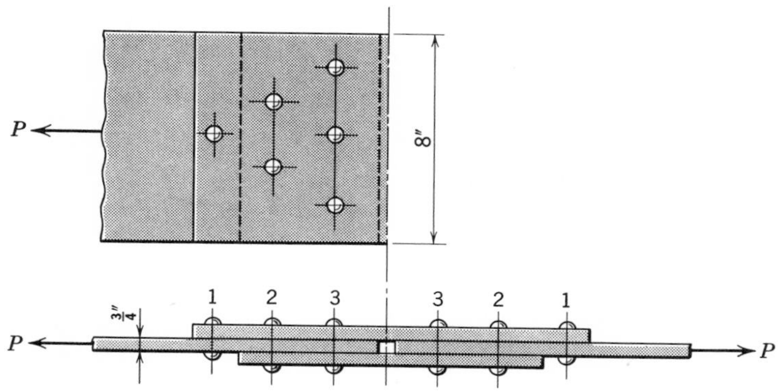


Fig. P11–9

**11–10.** A boiler with a diameter of 40 in. is made of ½-in. plate. The longitudinal seam is a double-riveted lap joint having ⅞-in. diameter rivets in 15/16-in. holes, at a pitch of 3 in. in each row. Using the ASME Boiler Code, determine:

(*a*) The maximum allowable internal pressure.

(*b*) The efficiency of the riveted joint.

**11–11.*** A triple-riveted lap joint is used to connect two ⅜-in. plates with ¾-in. rivets in 13/16-in. holes. The pitch in the outer rows is 6 in. and the pitch in the center row is 2 in. Determine the efficiency of the joint, using the ASME Boiler Code.

**11–12.** A triple-riveted lap joint (see Fig. P11–5) is to be made of steel plates and ¾-in. rivets at pitches of 4 in. in the outer rows and 2 in. in the inner row. Determine the minimum permissible thickness of the plate for the maximum allowable load on the joint, assuming the AISC Building Code can be applied to this joint.

**11–13.*** A cylindrical pressure tank with a diameter of 40 in. is made of $5/16$-in. steel plate. The longitudinal seam is a triple-riveted butt joint with one $3/8$-in. cover plate and $3/4$-in. diameter rivets in $13/16$-in. holes, at pitches of $2\frac{1}{2}$ in. in the inner and intermediate rows and 5 in. in the outer rows. Using the AISC specification of 1963, determine:

(*a*) The maximum allowable pressure in the tank.

(*b*) The efficiency of the joint.

**11–14.** A 4-ft diameter boiler with an internal steam pressure of 200 psi is made of $1/2$-in. plate. The double-riveted longitudinal butt joint is made with two splice plates. Pitch in the inner row of rivets is 3 in. and in the outer row is 5 in. The rivet diameter is $7/8$ in. ($15/16$-in. holes), and the splice plates are each $5/16$ in. thick. Determine the efficiency of the joint based on the following allowable stresses:

| | |
|---|---|
| Shear on rivets | 9,000 psi |
| Bearing | 20,000 psi |
| Tension in plate | 12,000 psi. |

**11–15.*** The triple-riveted butt joint of Fig. 11–1*e* is composed of 0.040-in. main plates spliced with two 0.025-in. cover plates and $5/32$-in. diameter rivets (assume $5/32$-in. holes). The plate material is aluminum alloy 2024-T4, and the rivets are 2117-T6. If the guaranteed ultimate strengths of these materials are 27,000 psi shear, 90,000 psi bearing, and 62,000 psi tension, determine the probable ultimate (failure) load for the specimen, and the efficiency of the joint. The plates are 1.375 in. wide.

**11–16.** Two 6 by $5/8$-in. steel plates are spliced with $3/8$-in. plates and $3/4$-in. diameter rivets as shown in Fig. P11–16. Using the AASHO specification of 1961, determine the maximum allowable load, *P*.

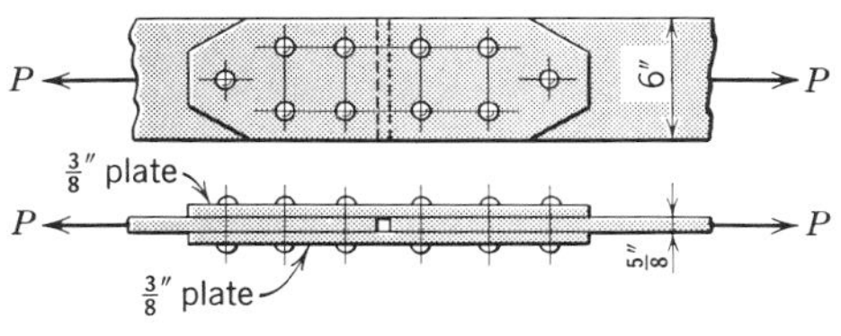


Fig. P11–16

**11–17.** If the cover plates for the joint of Fig. P11–9 are each $1/2$ in. thick, the rivets are $7/8$ in. in diameter, and the material is structural steel, determine the maximum permissible load on the joint if the AASHO specification of 1961 is followed.

**11–18.*** Determine the efficiency of the double-riveted butt joint in Fig. P11–18. The rivets are $7/8$ in. in diameter with $15/16$-in. holes and allowable stresses are: 8800 psi shear, 11,000 psi tension, and 19,000 psi bearing.

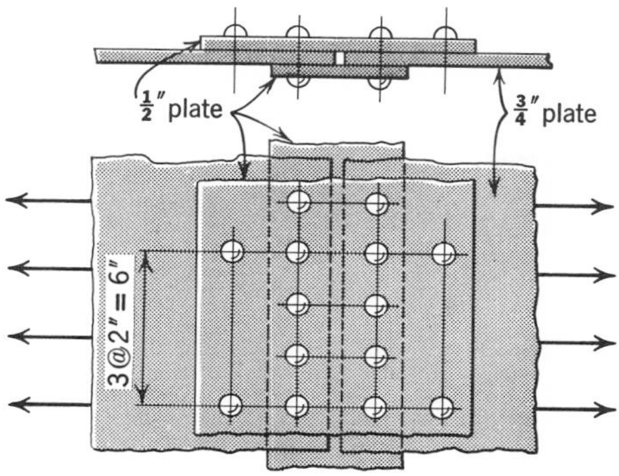


Fig. P11–18

**11–19.** Two 9 by ¾-in. plates are to be connected by a riveted butt joint with two cover plates. The plates are to carry an axial tensile load of 65,000 lb. The rivets are 1 in. in diameter. The allowable stresses are: $\tau = 9000$ psi, $\sigma_t = 13{,}000$ psi, and $\sigma_b = 20{,}000$ psi. Determine:

(*a*) The minimum allowable number of rivets.

(*b*) The minimum allowable number of rows of rivets.

(*c*) The best distribution of rivets in the rows.

(*d*) The minimum allowable thickness for each cover plate, using the distribution of part *c*.

**11–3. Analysis of Riveted Joints—Ultimate Strength Method.** A widely used method for the design and investigation of riveted connections is based on the concept that, if the load necessary to produce excessive inelastic action in a joint (ultimate load) can be predicted, a safe working load can then be determined by introducing an appropriate factor of safety. The determination of stresses at design loads (working loads) is of no consequence; hence, the question of distribution of load to the various elements of a connection at design loads merits no consideration.

The ultimate strength of the joint is determined by considering the various methods and combinations of failure which could occur. Any rivet can fail in either shear or bearing, and these possibilities are usually investigated first. When a joint contains some rivets that are in single shear and some in double shear, it is possible that the shearing strength of the material could be the controlling factor for the rivets in single shear and that the bearing strength could limit the load carried by the rivets in double shear.

The joint can fail in tension at the outer row of rivets in a lap joint, at the outer row of rivets in the main plate of a butt joint, or at the inner row of rivets in the cover plate of a butt joint. Failure can also result from a tensile failure of the main plate, at a row other than the outer row, combined with rivet failure (shear or bearing) which would have to occur if the joint were to separate. All possible combinations

**TABLE 11–2. Ultimate Strengths for Some Selected Materials**

| Specification | Tensile Strength of Plate, psi | Bearing Strength of Plate, psi | Shearing Strength of Rivet, psi |
|---|---|---|---|
| Am. Soc. of Mech. Engrs. Boiler Construction Code for Power Boilers, 1965; for temps to 650 F | | | |
| SA 30 flange steel | 55,000 | 95,000 | |
| SA 31, Grade A rivets | | | 44,000 |
| MIL-HDBK-5 Met. Matls. and Elements for Flight Vehicle Struct., Aug., 1962 | | | |
| 2024-T3 sheet | 68,000[a] | 129,000[b] | |
| 2117-T3 rivets | | | 30,000[a,c] |
| 2024-T31 rivets | | | 41,000[a,c] |
| Structural steel | 55,000 | 90,000[b] | 35,000 |
| Stainless steel, AISI 301 | | | |
| Annealed | 75,000 | 150,000[b] | 40,000 |
| Sheet or strip, ¼ hard | 125,000 | 250,000[b] | |
| Monel rivets | | | 49,000 |

[a] Probability (not guaranteed) values.

[b] For ratio of margin to rivet diameter = 2, and ratio of rivet diameter to sheet thickness $<5.5$.

[c] For ratio of rivet diameter to sheet thickness $\leq 3$ for single shear and 1.5 for double shear. See specification for reduction factors.

of tension failure with shear or bearing failure must be visualized and checked to determine the weakest combination. The smallest of the computed loads is the strength of the joint.

Table 11–2 lists a few representative values of ultimate strengths which may be used in the analysis of riveted connections. Although the preceding discussion concerned ultimate loads on joints, the same analysis can be made using working stresses, instead of ultimate strengths, to obtain the maximum working load on the joint. Such a procedure amounts to dividing by the factor of safety before calculating the joint strength instead of afterward.

The ultimate strength method of analyzing riveted joints is illustrated in Ex. 11–3 and 11–4.

**Example 11–3:** The triple-riveted butt joint of Fig. 11–1*e* is composed of 1⅜ by 0.040-in. main plates spliced with two 0.025-in. cover plates and ⁵⁄₃₂-in. diameter rivets. The diameter of the hole is essentially the same as that of the rivet. The plate material is aluminum alloy 2024-T4, and the rivets are 2117-T6. If the guaranteed[9] ultimate strengths of these materials are 27,000 psi shear, 90,000 psi bearing, and 62,000 psi tension, determine the ultimate load for the specimen.

*Solution:* The modes of failure to be investigated are: shearing of all rivets, bearing on all rivets, and tension at any net section in combination with shear or bearing failure of any rivets assisting the net section.

Ultimate shearing and bearing loads will be computed first with the aid of Fig. 11–10*a* and *b* (note that in Fig. 11–10*a* there are six additional forces, $F_s$, not shown, on the opposite side of the plate) as follows:

$$2F_s = 2(27{,}000)\left(\frac{\pi}{4}\right)\left(\frac{5}{32}\right)^2 = 1035 \text{ lb shear per rivet,}$$

$$F_b = 90{,}000\left(\frac{5}{32}\right)(0.040) = 562 \text{ lb bearing per rivet.}$$

The cover plates will carry more load in bearing than the main plate since the bearing area is greater. The results above indicate that the shearing load is not critical; therefore, as far as shear and bearing are concerned, the ultimate load on the joint is

$$P_b = 6F_b = 6(562) = 3370 \text{ lb.}$$

Next the ultimate tensile load will be determined with the aid of the free-body diagrams of Fig. 11–10*c, d, e,* and *f*. The load for each rivet assisting the net section in the main plate, Fig. 11–10*d* and *e*, is 562 lb since the previous shear and bearing analysis indicates that the rivet will fail in bearing before

[9] ANC-5, Strength of Aircraft Elements, December 1942. Note that guaranteed values are minimum acceptable values and are usually less than typical values.

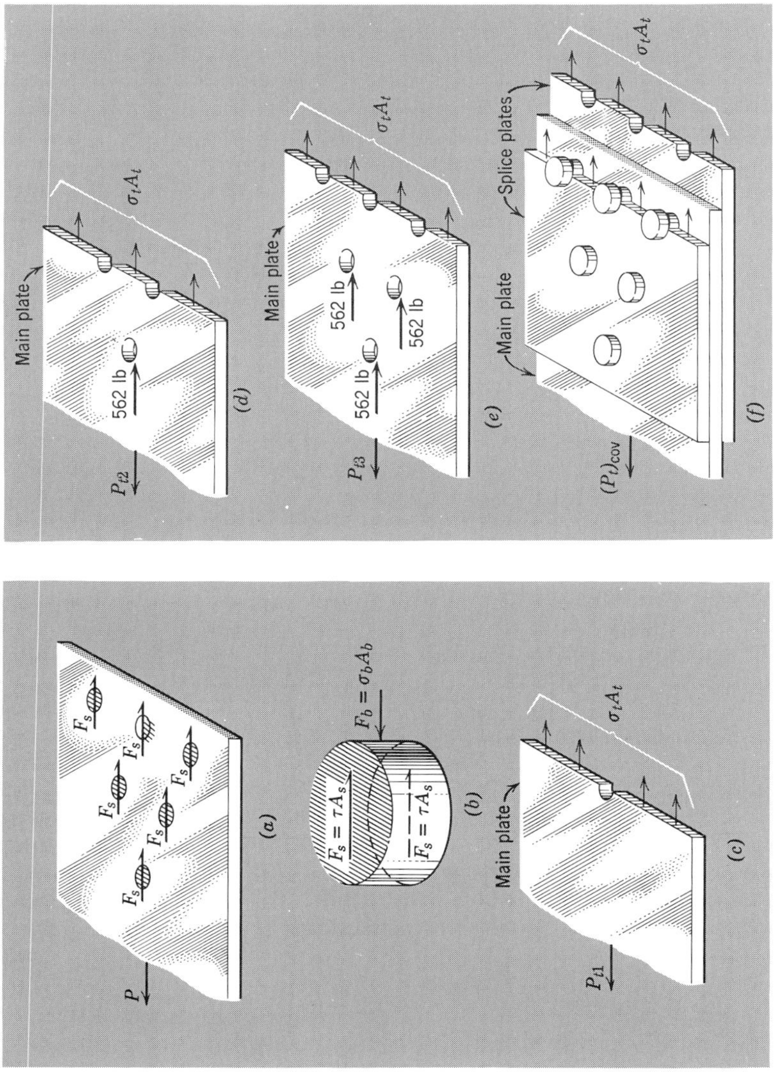

P
$F_s$
$F_s$
$F_s$
$F_s$
$F_s$
$F_s$
(a)
$F_s = \tau A_s$
$F_s = \tau A_s$
$F_b = \sigma_b A_b$
(b)
Main plate
$P_{t1}$
$\sigma_t A_t$
(c)
Main plate
$P_{t2}$
562 lb
$\sigma_t A_t$
(d)
Main plate
$P_{t3}$
562 lb
562 lb
562 lb
$\sigma_t A_t$
(e)
Main plate
Splice plates
$(P_t)_{cov}$
$\sigma_t A_t$
(f)


Fig. 11–10

the ultimate shear load is attained. Summing forces in the horizontal direction on each of the free-body diagrams in order yields the following:

$$(c)\quad P_{t1} = \sigma_t A_t = 62{,}000\left(\frac{11}{8} - \frac{5}{32}\right)(0.040) = 3020 \text{ lb}$$

$$(d)\quad P_{t2} = 562 + 62{,}000\left[\frac{11}{8} - 2\left(\frac{5}{32}\right)\right](0.040) = 3200 \text{ lb}$$

$$(e)\quad P_{t3} = 562(3) + 62{,}000\left[\frac{11}{8} - 3\left(\frac{5}{32}\right)\right](0.040) = 3930 \text{ lb}$$

$$(f)\quad (P_t)_{\text{cov}} = 62{,}000\left(\frac{11}{8} - \frac{15}{32}\right)(0.025)(2) = \underline{2810 \text{ lb.}} \qquad \text{Ans.}$$

As the load on the cover plates is the least of the five loads computed, the joint might be expected to fracture at a load of 2810 lb. The average of two such joints, one of which is shown in Fig. 11–1*e*, loaded to fracture was 2790 lb.

**Example 11–4:** Figure 11–11*a* represents a riveted lap splice in the stressed skin of an airplane wing. The sheet material is 2024-T3, and the 5⁄32-in. diameter rivets in 0.159-in. holes are 2024-T31. Use the MIL-HDBK-5 specifications with the shearing strength reduced to 0.964 times the value given (because the rivet diameter to plate thickness ratio is greater than 3), and determine:

(*a*) The maximum permissible load per linear inch of joint if a factor of safety of 1.5 is specified.

(*b*) The efficiency of the joint.

*Solution:* (*a*) From Fig. 11–11*a* the length of a repeating group of rivets is seen to be 5⁄4 in. with one rivet in the left row and two rivets in each of the other rows. Each rivet is in single shear, and the free-body diagram in Fig. 11–11*b* indicates that $F_b = F_s$. The bearing load is limited by the lower plate since the thickness of this plate is smaller. The maximum allowable loads per rivet in shear and bearing are

$$(F_s)_{\max} = (\pi/4)(0.159^2)(41{,}000)(0.964) = 785 \text{ lb}$$

and

$$(F_b)_{\max} = (0.040)(0.159)(129{,}000) = 820 \text{ lb.}$$

There are five rivets in each repeating group, and the free-body diagram in Fig. 11–11*c* indicates that the maximum allowable load in the joint, as far as shear and bearing are concerned, is

$$P_s = 5(785) = 3920 \text{ lb.}$$

There are three possibilities of joint failure involving tensile failure of the plates to be investigated. The joint might fail by tearing the lower plate at row 1; the joint might fail by tearing the lower plate at row 2 and shearing the rivet in row 1; or the joint might fail by tearing the top plate at row 3.

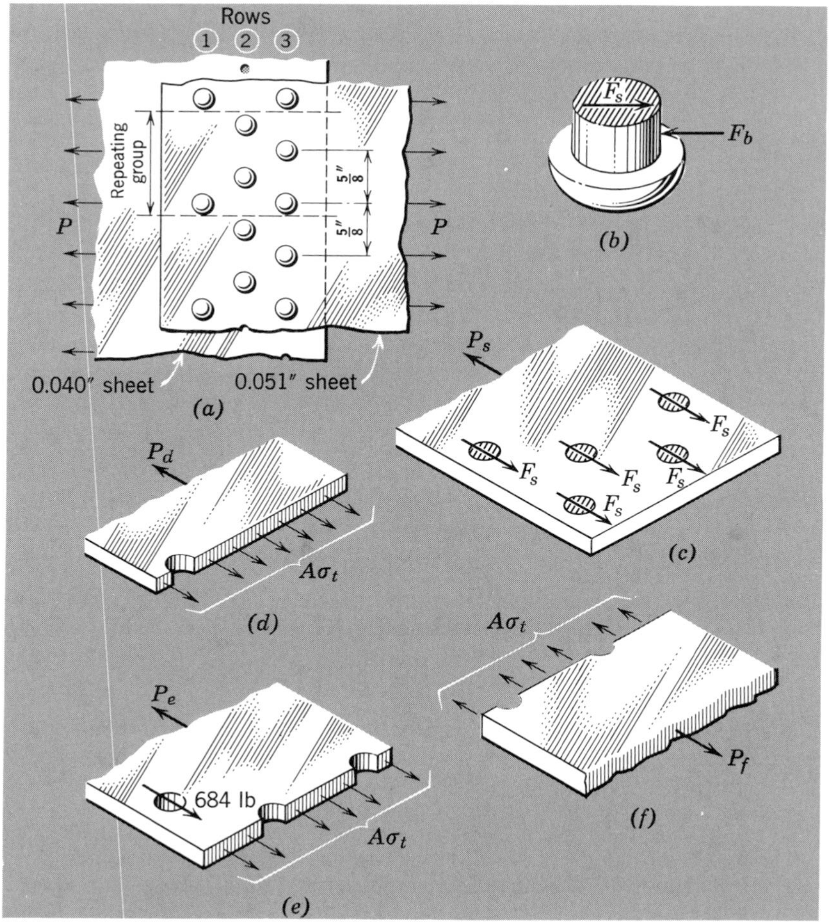


Fig. 11–11

These three possibilities are shown in the free-body diagrams in Fig. 11–11*d*, *e*, and *f*. The equations of equilibrium for these free-body diagrams give

From (*d*) $P_d = 0.040(1.25 - 0.159)68{,}000 = 2970 \text{ lb.}$

From (*e*) $P_e = 785 + 0.040[1.25 - 2(0.159)]68{,}000 = 3320 \text{ lb.}$

From (*f*) $P_f = 0.051[1.25 - 2(0.159)]68{,}000 = 3230 \text{ lb.}$

It is not necessary to check other failure combinations as the plates have as much or more areas at all other rows and more rivets would have to shear for other combinations. The controlling load on a repeating group is the 2970-lb load from Fig. 11–11*d*. With a factor of safety of 1.5 the load per inch becomes

$$P = \left(\frac{4}{5}\right)\frac{(2970)}{1.5} = \underline{1584 \text{ lb per in.}} \qquad \text{Ans.}$$

(*b*) The efficiency of the joint is

$$\eta = \frac{2970}{(5/4)(0.040)(68{,}000)} = 0.874 = \underline{87.4 \text{ per cent.}} \qquad \text{Ans.}$$

## PROBLEMS

**In the following problems, when the size of the hole is given, use the hole diameter for shear, bearing, and tension investigations. When the size of the rivet or bolt is given and no hole diameter is specified, use the nominal diameter of the rivet (or bolt) for shear and bearing investigations, and the nominal diameter plus 1/8 in. for the size of the hole in tension investigations.**

**11–20.*** A boiler with a diameter of 40 in. is made of 1/2-in. plate. The longitudinal seam is a double-riveted lap joint having 1-in. diameter rivets in 1 1/16-in. holes, at a pitch of 3 in. in each row. Using ultimate strengths from the ASME Boiler Code and a factor of safety of 5, determine:

(*a*) The maximum allowable internal pressure.

(*b*) The efficiency of the riveted joint.

**11–21.** The triple-riveted lap joint in Fig. P11–21 connects two 3/8-in. structural steel plates with 3/4-in. diameter rivets. Determine the efficiency of the joint and the allowable load if a factor of safety of 3 is specified. Use the ultimate strengths given in the MIL-HDBK-5 specifications.

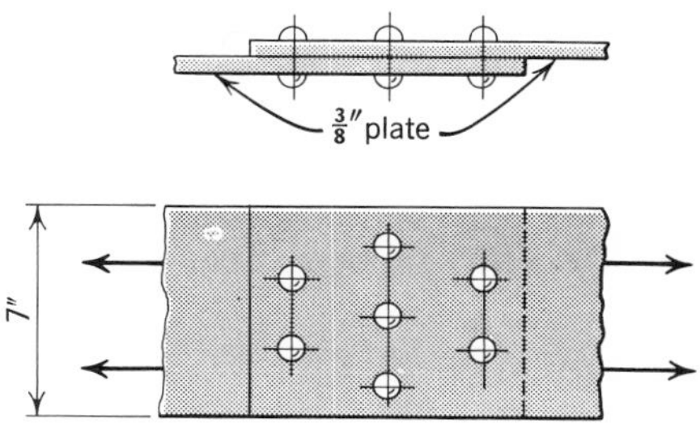


Fig. P11–21

**11–22.*** A triple-riveted lap joint is used to join two 1/4-in. plates with 3/4-in. diameter rivets in 13/16-in. holes. The pitch in the outer rows is 5 in. and that in the center row is 2 in. Using the ultimate strengths from the ASME Boiler Code, determine the efficiency of the joint.

**11–23.*** If the butt joint of Ex. 11–3 were fabricated with 0.040-in. cover plates, one of which engaged only the inner and intermediate rows while the other engaged all three rows of rivets, what would be the ultimate load for the specimen? What would be the efficiency of the joint? Use the strengths given in Ex. 11–3.

**11–24.** A triple-riveted lap joint is to be made of structural steel plates and $\frac{3}{4}$-in. diameter rivets at pitches of 4 in. in the outer rows and 2 in. in the inner row. Using the MIL-HDBK-5 specifications and a factor of safety of 3, determine the minimum permissible thickness of the plate for the maximum allowable load on the rivets.

**11–25.** A 4-ft diameter boiler with an internal steam pressure of 200 psi is made of $\frac{1}{2}$-in. plate. The double-riveted longitudinal butt joint is made with two $\frac{5}{16}$-in. splice plates and $\frac{3}{4}$-in. diameter rivets in $\frac{13}{16}$-in. holes, at pitches of 3 in. in the inner and 5 in. in the outer rows. Determine the factor of safety of the joint using the ASME Boiler Code.

**11–26.*** A double-riveted butt joint with two equal cover plates has a 3.0-in. pitch in the inner rows and a 4.5-in. pitch in the outer rows. The ultimate strengths are: 40,000 psi shear, 64,000 psi tension, and 96,000 psi bearing.

(*a*) Determine the rivet diameter and main plate thickness which will result in equal strength in shear, bearing, and tension in the outer row of rivets in the main plate. Assume rivets and holes are the same size.

(*b*) Using the results of part (*a*), determine the ratio of the joint strength computed on the basis of a tension failure at the inner row of rivets coupled with rivet failure, to the joint strength in bearing. Assume the cover plates are thick enough to prevent tearing of the cover plate.

**11–27.** A 50-in. diameter boiler is made of $\frac{3}{4}$-in. plate. The longitudinal seam is a double-riveted butt joint with two $\frac{9}{16}$-in. cover plates and with $\frac{7}{8}$-in. diameter rivets in $\frac{15}{16}$-in. holes. The outer rows of rivets have a pitch of 5 in. and the inner rows a pitch of 2 in. Using the ASME Boiler Code and a factor of safety of 5, determine the maximum allowable internal pressure.

**11–28.*** Two 5 by $\frac{1}{2}$-in. stainless steel plates are spliced with $\frac{5}{16}$-in. plates and $\frac{3}{4}$-in. rivets as shown in Fig. P11–28. Using the MIL-HDBK-5 specifications, determine the maximum allowable load, *P*, if a factor of safety of 2 is specified. All material is AISI 301 annealed.

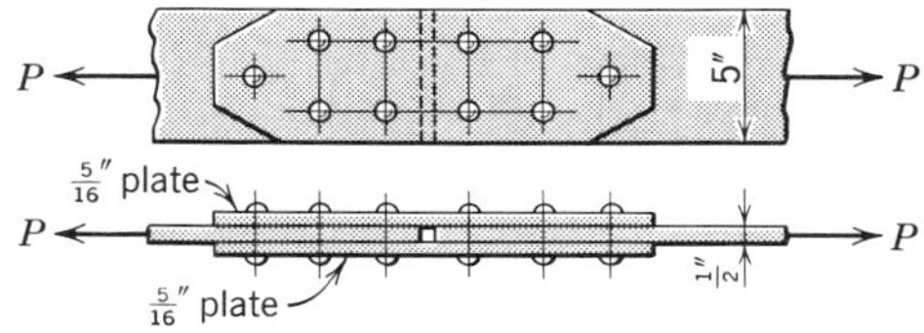


Fig. P11–28

**11–29.*** If the cover plates in Fig. P11–29 are each $\frac{7}{16}$ in. thick, the rivets are $\frac{7}{8}$ in., and the material is structural steel, determine the maximum permissible load on the joint. Use the MIL-HDBK-5 specification and a factor of safety of 2.5.

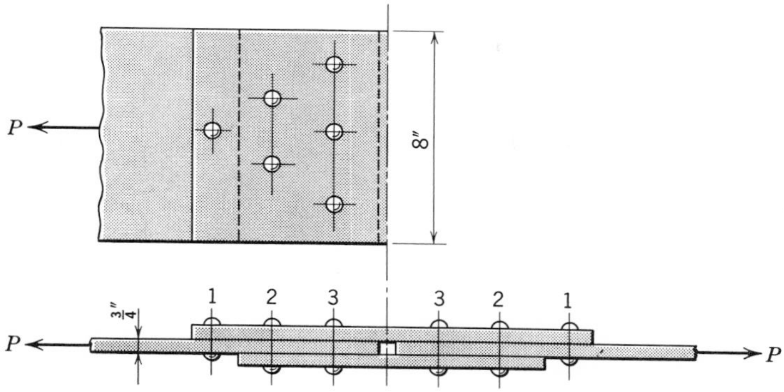


Fig. P11–29

**11–30.** Two 10 by $\frac{3}{4}$-in. steel plates are to be connected by a riveted butt joint with two cover plates and 1-in. rivets. The joint is to carry an axial load of 65,000 lb. Using the ASME Boiler Code and a factor of safety of 5, determine:

(*a*) The minimum allowable number of rivets.
(*b*) The minimum allowable number of rows of rivets.
(*c*) The best distribution of rivets in the rows.
(*d*) The minimum allowable thickness for each cover plate.

**11–4. Welded Connections.** Connections between metal plates, angles, pipes, and other structural elements are frequently made by welding. Fusion welds are made by melting portions of the materials to be joined with an electric arc, a gas flame, or with thermit. In fusion welds, additional welding material is usually added to the melted metal to fill the space between the two parts to be joined or to form a fillet. A gas shield is provided when welding certain metals to prevent rapid oxidation of the molten metal. A good welded joint will usually develop the full strength of the material being joined unless the high temperature necessary for the process changes the properties of the materials.

Metals may also be joined by resistance welding in which a small area or spot is heated under high localized pressure. The material is not melted with this type of welding. Other joining methods for metals include brazing and soldering in which the joining metal is melted, but the parts to be connected are not melted. Such connections are usually much weaker than the materials being connected. Fusion welding is the most effective method when high strength is an important factor, and it will be discussed more in detail.

Figure 11–12 illustrates the two general types of fusion welded joints. The joints illustrated in Fig. 11–12*a*, *b*, and *c* are called *butt* joints and are made with groove welds. The weld may be square, for

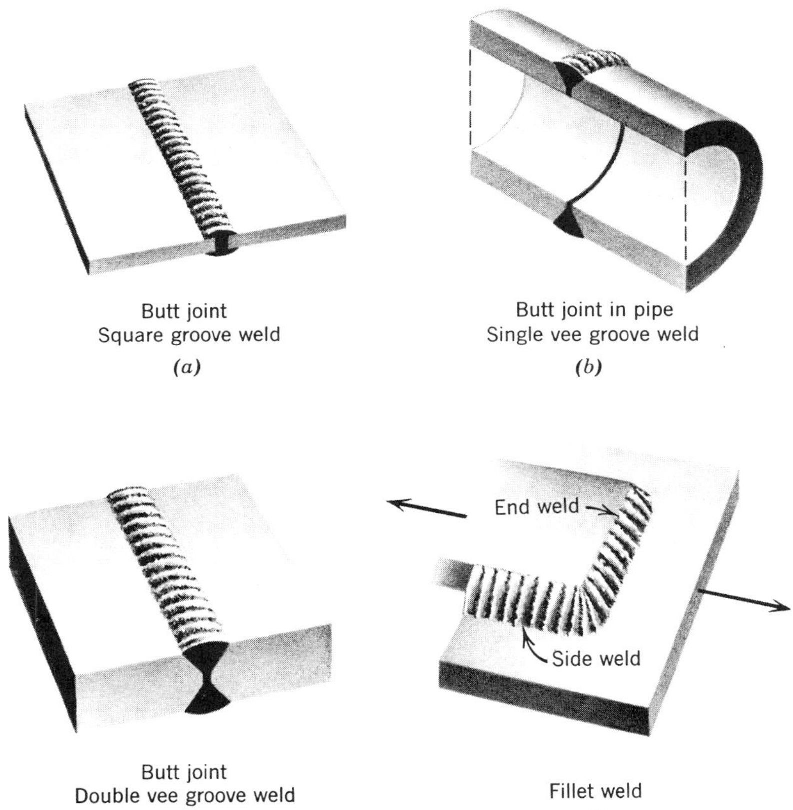


Butt joint
Square groove weld
(*a*)

Butt joint in pipe
Single vee groove weld
(*b*)

Butt joint
Double vee groove weld
(*c*)

Fillet weld
(*d*)

Fig. 11–12

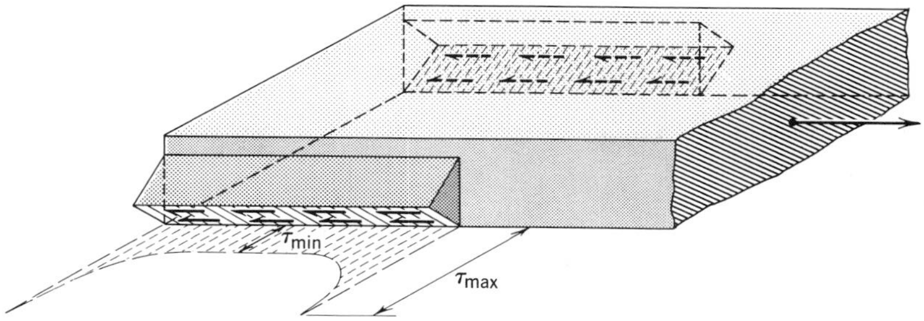


Fig. 11–13

fairly thin materials, or there may be one or two vees for thicker metals. When approved welding techniques and materials are used and complete penetration of the material is obtained, the strength of the joint is assumed to be the same as that of the material being joined.

A fillet weld is illustrated by Fig. 11–12*d* and is used to join a plate, angle, or other structural shape to another structural member. The end weld may be omitted in some applications. Ideally, the fillet should be slightly concave, rather than convex or straight as shown, to reduce possible stress concentrations at reentrant corners.

From an analysis similar to that for the riveted connection (see Art. 11–1), it can be shown that the shearing stress in the side weld cannot be uniformly distributed along the weld as long as the action is elastic, and tests[10] indicate the distribution is somewhat as indicated in Fig. 11–13. However, if the material is ductile (the usual condition for welded connections), a slight amount of inelastic action at the ends of the welds will result in essentially a uniform shearing stress distribution; therefore, in the design of welded connections, a uniform distribution is assumed. Fillet welds are designed on the assumption that failure will occur by shearing the minimum section of the weld, which is called the throat of the weld, section *AB* in Fig. 11–14*a*, whether the weld is a side (longitudinal) weld or an end (transverse) weld. The strength of side fillet welds can be determined from Fig. 11–14*a* and *b*. Although it is desirable to make fillet welds slightly concave, calculations are usually based on the assumption that the weld surface is straight and makes an angle of 45° with the two legs. The maximum

[10] "Distribution of Shear in Welded Connections," H. W. Troelsch, *Trans. Am. Soc. of Civil Engrs.*, Vol. 99, p. 409, 1934.

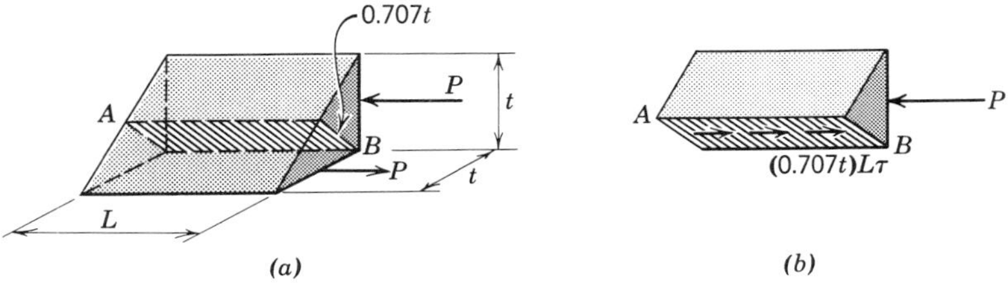


Fig. 11–14

allowable load $P$ which can be carried by a side fillet weld with a length $L$ and legs $t$ (it is common practice to designate the size of fillet welds by the length of leg, $t$) can be obtained from the free-body diagram in Fig. 11–14$b$ and is

$$P = 0.707tL\tau$$

where $\tau$ is the allowable shearing stress which is permitted in the weld material. The allowable shearing stress recommended by the AISC specifications for steel is 13.6 ksi using A233 class E60 series electrodes or 15.8 ksi using A233 class E70 series electrodes.

Tests indicate that the strength of a transverse fillet weld is about 30 per cent greater than that of a side fillet weld with the same dimensions. It is common practice, however, to assume that the strength of a transverse weld is the same as that of a side weld with the same dimensions. This practice will be followed in this book.

The following example illustrates the analysis of a centrically loaded fillet weld.

**Example 11–5:** An 8 by 6 by ½-in. steel angle is to be welded to a flat plate with the long side of the angle against the plate as shown in Fig. 11–15$a$ and $b$.

($a$) Determine the minimum lengths $L_1$ and $L_2$ which will cause the angle to carry the maximum allowable axial load. The allowable tensile stress for the material in the angle is 18,000 psi, and the allowable shearing stress in the weld material is 13,600 psi. Each leg of the weld is 0.50 in.

($b$) Solve part $a$ if a transverse weld is placed across the end of the angle.

*Solution:* ($a$) The maximum longitudinal force in the angle will be developed when the tensile stress is uniformly distributed and will act through the centroid of the cross section of the angle. The area of the cross section and the location of its centroid can be obtained from the table of properties of sections of structural shapes in App. C. The area is 6.75 sq in., and the location of the centroid is shown on Fig. 11–15$c$ which is a free-body diagram of a part of the

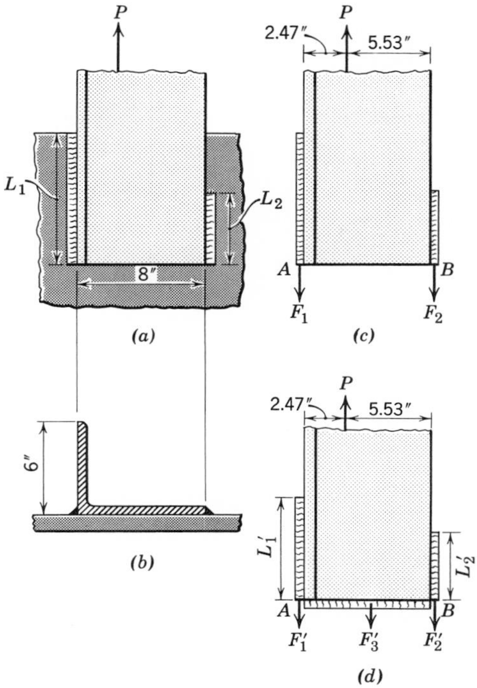


Fig. 11–15

angle. The forces $F_1$ and $F_2$ are assumed to act at the edges of the angle. The force $P$ is

$$P = \sigma A = 18{,}000(6.75) = 121{,}500 \text{ lb.}$$

The equation $\Sigma M_A = 0$, applied to the free-body diagram, gives

$$\circlearrowleft + \qquad \Sigma M_A = 8F_2 - 2.47P = 0,$$

from which $\qquad F_2 = 37{,}500$ lb as shown.

Similarly $\qquad F_1 = 84{,}000$ lb as shown.

The shear area of the throat that must carry the load $F_2$ is $L_2(0.707)(0.50)$, and the shearing stress multiplied by this area must equal the applied load. Thus

$$F_2 = 37{,}500 = \tau A = 13{,}600L_2(0.707)(0.50),$$

from which $\qquad L_2 = \underline{7.80 \text{ in.}}$ $\qquad$ Ans.

In the same manner

$$L_1 = \underline{17.45 \text{ in.}} \qquad \text{Ans.}$$

(*b*) When a weld is placed across the end of the angle, an additional force must be shown on the free-body diagram as in Fig. 11–15*d*. The force $F_3'$ is

$$F_3' = \tau A = 13{,}600(8)(0.707)(0.50) = 38{,}500 \text{ lb.}$$

Note that the transverse weld is assumed to carry the same load per inch of weld as the side weld. The equation $\Sigma M_A = 0$ applied to Fig. 11–15*d* gives

$$(\curvearrowleft +) \qquad \Sigma M_A = 8F_2' + 4(38{,}500) - 2.47P = 0,$$

from which $F_2' = 18{,}250 \text{ lb} = 13{,}600L_2'(0.707)(0.50)$

as in part *a* of the solution. This equation gives

$$L_2' = \underline{3.80 \text{ in.}} \qquad \text{Ans.}$$

Similarly $L_1' = \underline{13.45 \text{ in.}}$ Ans.

## PROBLEMS

**11–31.*** Design a fillet-welded joint, for attaching a 4 by 4 by $\frac{3}{8}$-in. steel angle to a gusset plate, that will develop the full working tensile strength of the angle. Use a tensile working stress of 15,000 psi in the angle and a maximum allowable force per linear inch of weld of 3000 lb. Weld along the end of the angle in order to shorten the gusset plate. See App. C for additional data.

**11–32.** Solve Prob. 11–31, omitting the weld along the end of the angle.

**11–33.*** Design a fillet-welded joint, for attaching a 6 by 4 by $\frac{3}{8}$-in. steel angle to a gusset plate with the 6-in. side against the plate, that will develop the full working tensile strength of the angle. Use a tensile working stress of 24 ksi in the angle and an allowable shearing stress in the $\frac{5}{16}$-in. weld of 13.6 ksi. Weld along the end of the angle in order to shorten the gusset plate. See App. C for additional data.

**11–34.** A cylindrical pressure tank is to be constructed by welding a $\frac{3}{8}$-in. thick plate on the ends of a 6-ft piece of steel pipe. The pipe has an outside diameter of 14 in. and a thickness of 0.20 in. Determine the size of a fillet weld to develop the full tensile working strength of the pipe. The working stress in the pipe is 20,000 psi, and the allowable force per inch of weld is 10,600$t$ where $t$ is the size of the weld.

**11–35.*** An improvised tension member is to be made by splicing two lengths of 6 by 4 by $\frac{9}{16}$-in. angle. Since the angles are too short to permit groove welding, a joint composed of plate $A$ and bar $B$ fillet welded to the angles as shown in Fig. P11–35 is proposed. The working stress for the angles is 18 ksi, and the maximum allowable load on each inch of fillet weld is 4 kips. Determine the length of each weld $a$, $b$, and $c$ necessary to develop the full axial tensile capacity of the angles with no bending about any axis in the joint.

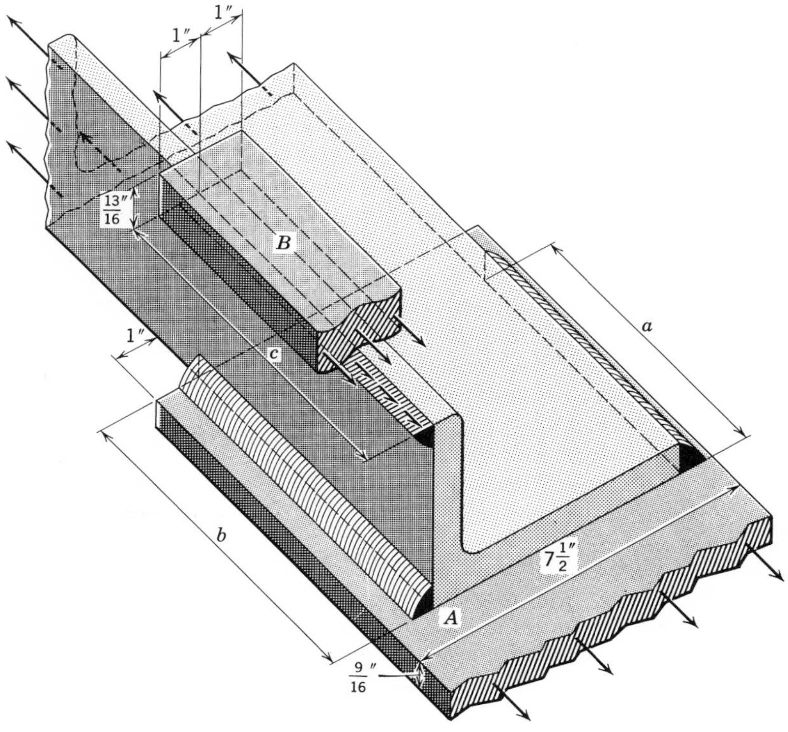


Fig. P11–35

**11–36.*** A cylindrical pressure tank is fabricated by butt welding ½-in. plate with a spiral seam as shown in Fig. P11–36. The pressure in the tank is 200 psi, and an axial load of $4800\pi$ lb is applied to the end of the tank through a rigid bearing plate. Determine the normal and shearing stresses in the weld on the plane of the weld.

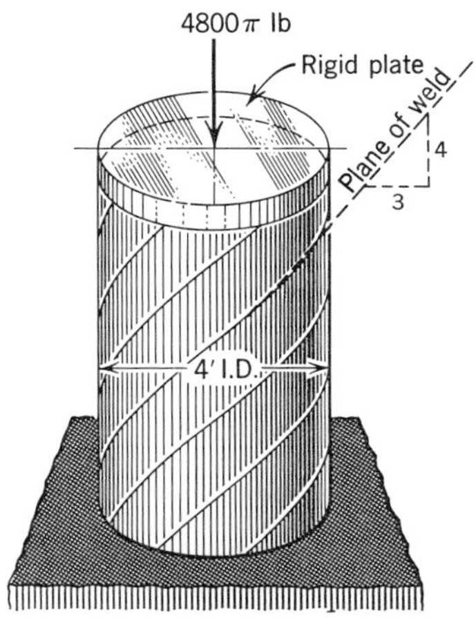


Fig. P11–36

**11–37.** A steel pipe 36 in. in diameter has a welded spiral seam as shown in Fig. P11–37. The pressure inside the pipe is 90 psi. Determine the normal and shearing forces which must be carried by each linear inch of the weld. Assume that there is no longitudinal stress in the pipe.

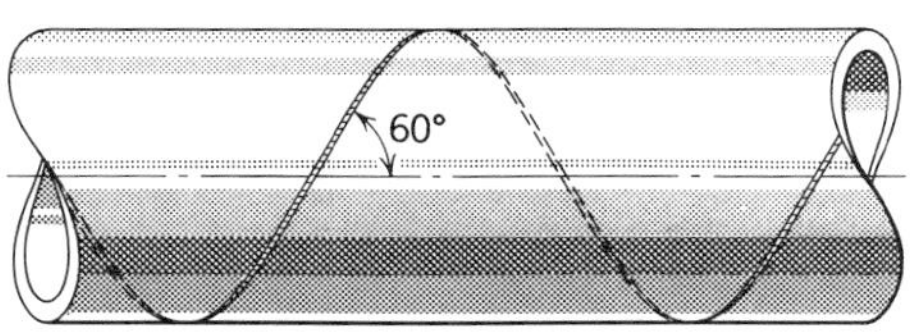


Fig. P11–37

**11–38.*** The flanges of the steel cantilever beam $AB$ of Fig. P11–38 are to be connected to the column with 1 by 14-in. butt welds, and the web connection is to be $\frac{3}{8}$ by 2-in. intermittent fillet welds on both sides of the web. Recall that the 80-kip load at $B$ may be replaced with a vertical force at $A$ and a couple. Assume that the couple is resisted only by the flange welds and that the vertical force is resisted only by the web welds, and determine:

(*a*) The average tensile stress in the flange weld (on the vertical plane).

(*b*) The number of $\frac{3}{8}$ by 2-in. welds to be used on each side of the web, using class E70 series electrodes and AISC specifications.

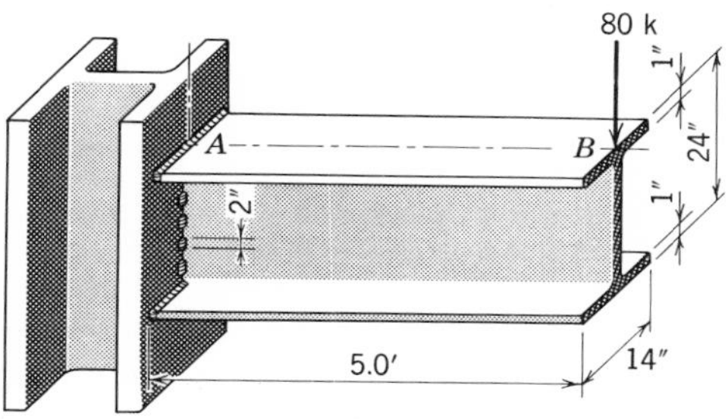


Fig. P11–38

**11–39.** Prove that when a structural angle is to be welded to a plate in such a manner that the maximum tensile strength of the angle is developed, the use of a transverse end weld has the effect of shortening each of the side welds by equal amounts.

**11–5. Connections under Torsional Loading.** Power-transmitting shafts are frequently connected by flange couplings (see, for example, Fig. 4–2), and the design of the coupling is of course important. The flange coupling is not the only device called upon to transmit a torque. Many structural connections are subjected to torsional loading, often combined with centric loading. The latter case is taken up in the next article. This article is limited to pure torsion.

When a torque is applied to the plate $C$ of Fig. 11–16$a$ which is riveted to another structural element, the plate will rotate (a slight amount for elastic action) about some point such as $O$ in the figure. If the assumption is made that any straight line in the plate (such as

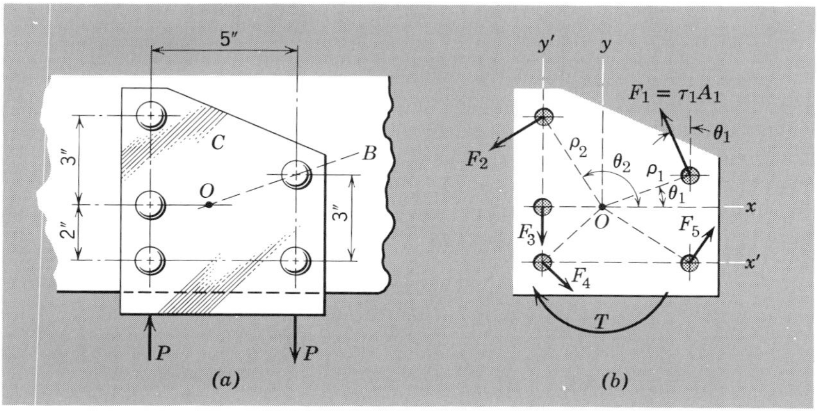


Fig. 11–16

$OB$) before the torque is applied remains straight after application of the torque, the deformations of the rivets, and hence the average cross shearing strain in each rivet, will be proportional to the distance from $O$. If, in addition, stress and strain are proportional, the rivet cross shearing stress will be proportional to the distance from $O$. This means that, for rivets (or bolts) of the same diameter, the force on each rivet is proportional to the distance from $O$ and is in the direction of the strain which is perpendicular to the radius from $O$.

The equations of equilibrium can be used to locate the point of rotation, $O$. The resultant of the resisting forces, $F_1$, $F_2$, . . . , in Fig. 11–16$b$ is a couple; therefore $\sum_{i=1}^{n} F_i$ is equal to zero in any direction, where the $F_i$'s are the $n$ resisting forces. The summation of forces in the $y$ direction gives

$$\Sigma F_y = F_1 \cos \theta_1 + F_2 \cos \theta_2 + \cdots = \sum_{i=1}^{n} F_i \cos \theta_i = 0.$$

Also $\quad F_i = \tau_i A_i = (\tau_i/\rho_i)(\rho_i A_i) = k\rho_i A_i$

because the cross shearing stress is proportional to the radius, $\rho$. The summation thus becomes

$$\Sigma F_y = k \sum_{i=1}^{n} \rho_i \cos \theta_i A_i = k \sum_{i=1}^{n} x_i A_i = k\bar{x}A_t = 0$$

where $\bar{x}$ is the $x$ coordinate of the centroid of the rivet areas and $A_t$ is the sum of the rivet areas.

This expression proves that $\bar{x}$ is zero because neither $k$ nor $A_t$ can be zero. The summation of forces in the $x$ direction will indicate that $\bar{y}$ must also be zero which means that the center of rotation of the plate and the centroid of the rivet areas coincide. The following is an example of a riveted joint loaded in pure torsion.

**Example 11–6:** The plates and rivets of Fig. 11–16*a* are of structural steel, the rivets are all ½ in. in diameter, and the applied torque is 2000 ft-lb. Determine the magnitude of the maximum cross shearing stress in the rivets.

*Solution:* The free-body diagram of Fig. 11–16*b* will be used for this problem. The centroid of the rivet areas, which is the same as the center of rotation, is located by applying the principle of moments of areas with respect to the $x'$ and $y'$ axes. The coordinates of the centroid are

$$\bar{x} = 2 \text{ in.} \quad \text{and} \quad \bar{y} = 2 \text{ in.}$$

The summation of moments of forces with respect to the $O$ axis gives

$$\circlearrowleft + \quad \Sigma M_O = \sqrt{10}\,F_1 + \sqrt{13}\,F_2 + 2F_3 + \sqrt{8}\,F_4 + \sqrt{13}\,F_5 - 2000(12) = 0.$$

Since the rivet areas are all equal, the forces on the rivets are proportional to the radial distances from $O$; thus

$$\frac{F_1}{\sqrt{10}} = \frac{F_2}{\sqrt{13}} = \cdots.$$

The maximum forces are $F_2$ and $F_5$; and, if the forces are written in terms of $F_2$, they become

$$F_1 = \frac{\sqrt{10}\,F_2}{\sqrt{13}}, \qquad F_3 = \frac{2F_2}{\sqrt{13}}, \qquad F_4 = \frac{\sqrt{8}\,F_2}{\sqrt{13}}, \qquad \text{and} \quad F_5 = F_2.$$

When these values are substituted in the moment equation, it becomes

$$\sqrt{10}\left(\frac{\sqrt{10}\,F_2}{\sqrt{13}}\right) + \sqrt{13}\,F_2 + 2\left(\frac{2F_2}{\sqrt{13}}\right) + \sqrt{8}\left(\frac{\sqrt{8}\,F_2}{\sqrt{13}}\right) + \sqrt{13}\,(F_2) = 24{,}000,$$

from which $F_2 = 1803$ lb. The maximum cross shearing stress is

$$\tau_{\max} = \frac{F_2}{A} = \frac{1803}{(\pi/4)(1/2)^2} = \underline{9180 \text{ psi.}} \qquad \text{Ans.}$$

## PROBLEMS

**11–40.*** A hollow circular shaft is subjected to a torque of 10,000 ft-lb. The inside diameter is 2 in., and the outside diameter is 4 in. Determine:

(*a*) The maximum shearing stress in the shaft.

(*b*) The number of $\frac{3}{4}$-in. bolts needed for a coupling on a bolt circle of 6-in. diameter for a cross shearing stress in the bolts of 10,000 psi.

**11–41.** Four equal-sized bolts, placed as shown in Fig. P11–41, are used to transmit a 10,000-ft-lb torque from one machine part to another. Determine:

(*a*) The magnitude of the force on bolt *A*.

(*b*) The minimum required diameter of the bolts if the allowable cross shearing stress is 15,000 psi.

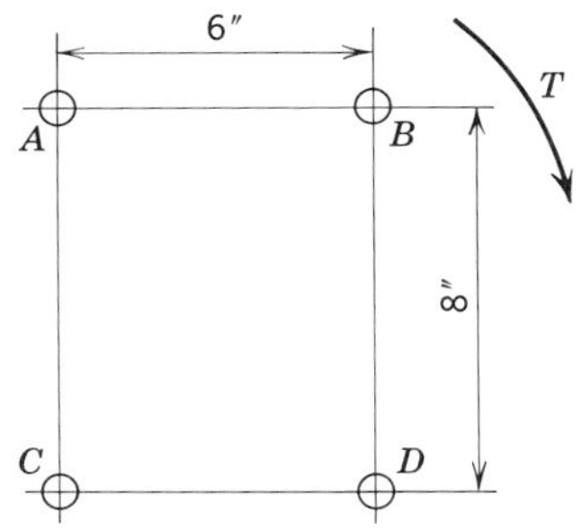


Fig. P11–41

**11–42.*** A shaft coupling has six $\frac{1}{2}$-in. diameter bolts on a 4-in. diameter bolt circle. Determine the maximum horsepower the coupling can transmit at 300 rpm if the cross shearing stress in the bolts is limited to 8 ksi.

**11–43.*** Six bolts placed as shown in Fig. P11–43 are used to transmit a torque of 30,000 ft-lb from one plate to another. Determine:

(*a*) The magnitude of the force on bolt *A*.

(*b*) The magnitude of the force on bolt *B*.

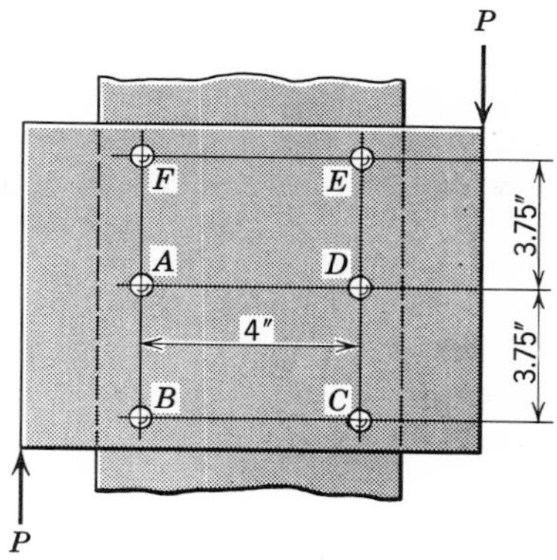


Fig. P11–43

**11–44.** Two plates are riveted together as shown in Fig. P11–44. The couple produced by the forces $P$ ($P = 3500$ lb) on one plate is transmitted to the other plate by the nine ½-in. diameter rivets. Determine the cross shearing stress in each rivet.

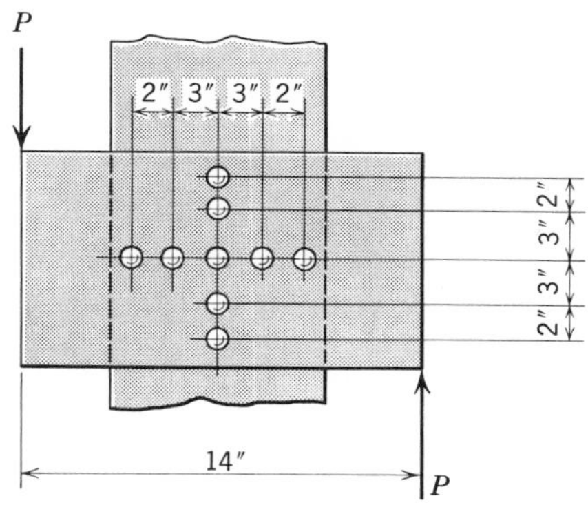


Fig. P11–44

**11–45.*** A motor supplies the torque $Q$ to the driving pulley $B$ which is mounted on the bronze shaft of Fig. P11–45. Pulleys $A$ and $C$ are for power take-off with 200 hp being taken off at $C$ and the remainder at $A$. The allowable shearing stress in the hollow steel segment $AB$ is 12 ksi. The angle of twist at $A$ with respect to $C$ is not to exceed 0.01 rad counterclockwise looking from $A$ toward $C$. The bronze ($G = 6000$ ksi) and steel segments are rigidly connected by a rigid flange coupling with a bolt circle of 8-in. diameter. When the shaft is rotating at 200 rpm, determine:

(*a*) The maximum permissible power that can be supplied to pulley $B$.

(*b*) The number of $\frac{7}{8}$-in. diameter bolts required for the flange coupling with an allowable cross shearing stress of 9 ksi in the bolts.

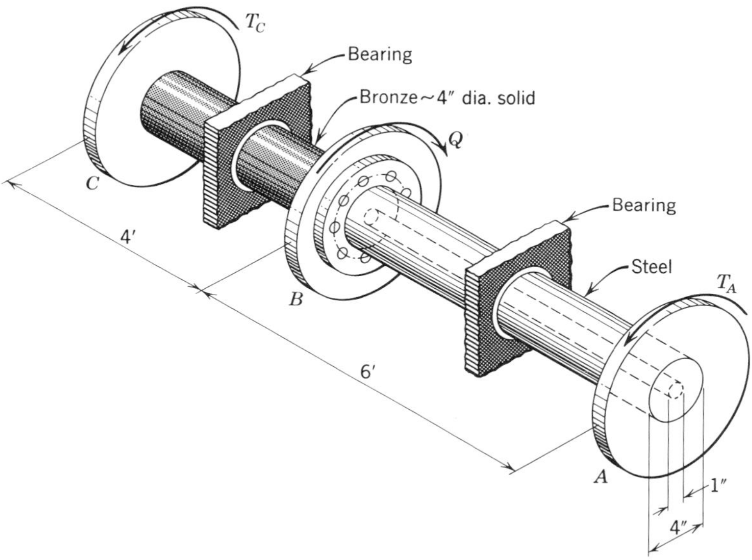


Fig. P11–45

**11–46.** A hollow steel shaft has diameters of 6 in. and 4 in. and an allowable shearing stress of 9000 psi. Determine:

(*a*) The horsepower transmitted at 25 rad per sec.

(*b*) The magnitude of the angle of twist in a length of 10 ft.

(*c*) The required number of $\frac{3}{4}$-in. diameter bolts for a flange coupling in which two-thirds of the bolts are on a 12-in. diameter bolt circle and one-third are on a 9-in. diameter circle. The allowable cross shearing stress in the bolts is 12,000 psi.

**11–47.*** The riveted structural connection in Fig. P11–47 is subjected to a torque as shown. All rivets are the same size.

(*a*) Which rivet transmits the largest force?

(*b*) Determine the magnitude of the force of part *a*.

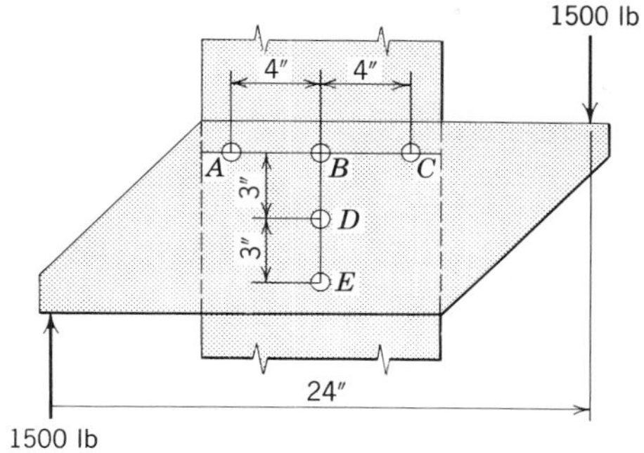


Fig. P11–47

**11–48.** A riveted structural connection is subjected to the torque shown in Fig. P11–48. All rivets are the same size.

(*a*) Which rivet transmits the largest force?

(*b*) Determine the magnitude of the force of part *a*.

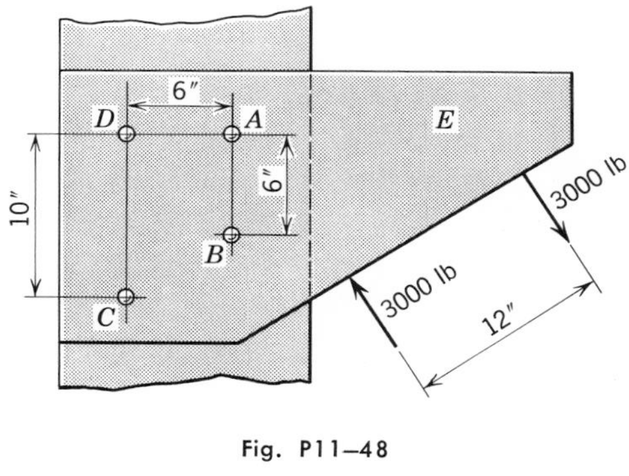


Fig. P11–48

**11–6. Eccentric Load on a Riveted Joint.** A common and important combination of torsional and centric loads occurs when a riveted or bolted joint is subjected to an eccentric load. The presentation in Art. 11–1 through 11–3 was limited to centric loads on riveted or bolted joints. The present discussion will be limited to eccentric loads in the plane of the shear areas. In Art. 11–5 bolted or riveted connections subjected to torsional loads were analyzed. The proof that a riveted joint

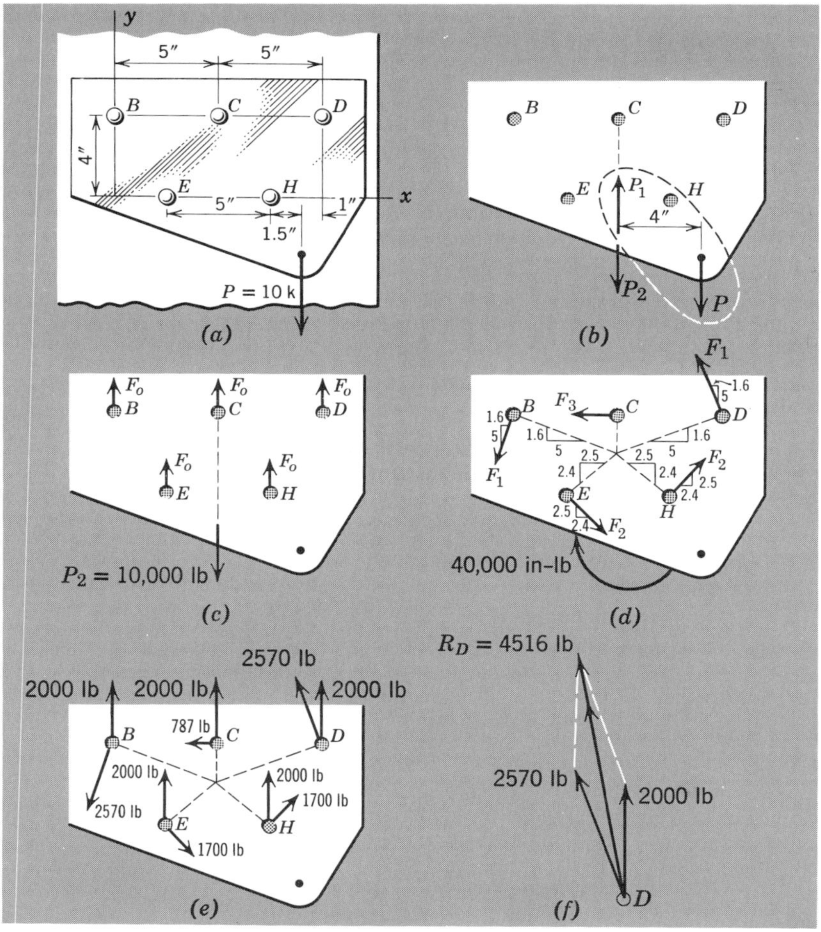


Fig. 11–17

rotates or tends to rotate about the centroid of the rivet areas when the joint is subjected to a couple was given in Art. 11–5.

Any load on a riveted or bolted joint that does not act through the centroid of the rivet areas is defined as an eccentric load. Fortunately any eccentric load can be replaced by a centric load and a couple in the plane of the shear areas. Furthermore, the force acting on any given shear area can be readily obtained by superposition of the component forces due to the torsional and centric loads. The following example illustrates the procedure.

**Example 11–7:** Determine the cross shearing stress in the rivet of the joint of Fig. 11–17*a* carrying the greatest load. The diameter of each rivet is $\frac{3}{4}$ in.

*Solution:* By symmetry the $x$ coordinate of the centroid of the rivet areas is $\bar{x} = 5.00$ in. By the principal of moments the $y$ coordinate of the centroid of the rivet areas is

$$\bar{y} = \frac{\Sigma M_x}{\Sigma A} = \frac{(3A)4}{5A} = 2.40 \text{ in.},$$

where $A$ is the cross-sectional area of one rivet.

Two equal opposite centric loads $P$ (labeled $P_1$ and $P_2$) are introduced as shown in Fig. 11–17*b*. By this means the eccentric load $P$ is replaced by a centric load $P_2$ of 10,000 lb and a clockwise couple $P_1P$ of 40,000 in-lb as shown in Fig. 11–17*b*. An upward resisting force of $F_o$, equal to 2000 lb, is required on each shear area, as shown in Fig. 11–17*c*, to hold the centric load, $P_2$, in equilibrium. The five resisting forces shown in Fig. 11–17*d* are required to hold the 40,000-in-lb couple $PP_1$ in equilibrium. As shown in Art. 11–5,

$$\frac{F_1}{\sqrt{27.56}} = \frac{F_2}{\sqrt{12.01}} = \frac{F_3}{1.6},$$

and

$$2F_1\sqrt{27.56} + 2F_2\sqrt{12.01} + 1.6F_3 = 40{,}000.$$

The simultaneous solution of these equations gives

$$F_1 = 2570 \text{ lb}, \quad F_2 = 1700 \text{ lb}, \quad \text{and} \quad F_3 = 787 \text{ lb all as shown.}$$

The component forces acting on each rivet as shown in Fig. 11–17*e* reveals that rivet $D$ is carrying the greatest load. The two resisting forces on rivet $D$ are added vectorially to give the resultant resisting force, $R_D$ of 4516 lb, as shown in Fig. 11–17*f*. The cross shearing stress on rivet $D$ is

$$\tau = \frac{4516}{\pi(3/8)^2} = \underline{10{,}240 \text{ psi.}} \qquad \text{Ans.}$$

## PROBLEMS

**11–49.*** A gusset plate carrying a load of 2400 lb is held by three ½-in. diameter bolts as shown in Fig. P11–49.

(*a*) Which bolt carries the greatest load?

(*b*) Determine the cross shearing stress in the bolt carrying the greatest load.

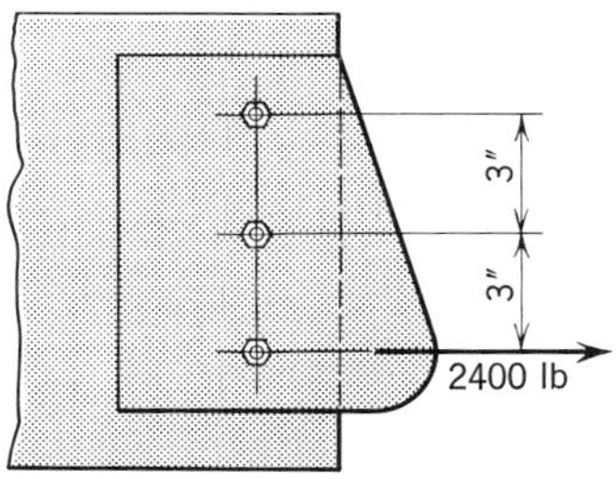


Fig. P11–49

**11–50.** Given the bolted connection in Fig. P11–50, determine the cross shearing stress in bolt *A*. All bolts are 1 in. in diameter.

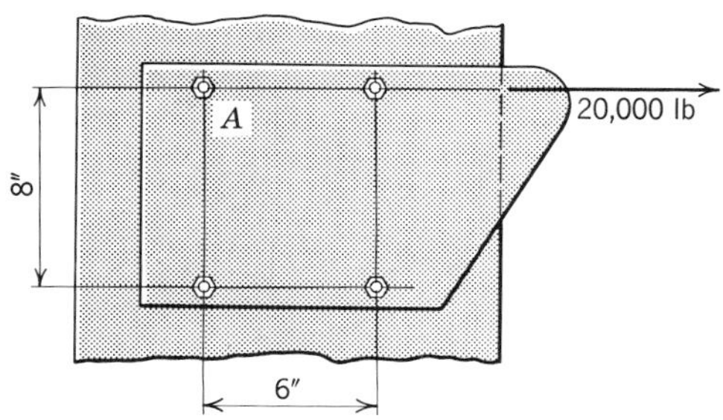


Fig. P11–50

**11–51.*** The diameter of each rivet in the joint of Fig. P11–51 is $\frac{7}{8}$ in. Determine the cross shearing stresses in rivets $A$ and $B$.

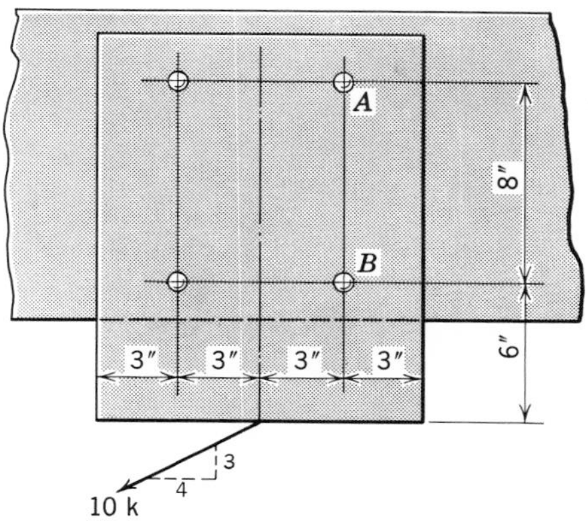


Fig. P11–51

**11–52.** The diameter of each rivet in the joint in Fig. P11–52 is $\frac{3}{4}$ in. Determine the cross shearing stresses in the top and bottom rivets.

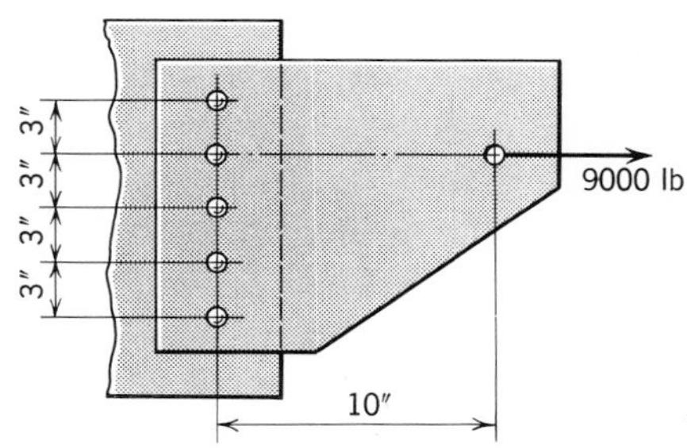


Fig. P11–52

**11–53.*** Solve Prob. 11–52 if the 9000-lb load acts vertically downward instead of to the right as shown.

**11–54.** The bolts in the joint in Fig. P11–54 each have a diameter of $\frac{5}{8}$ in. Determine the maximum cross shearing stress in any of the bolts.

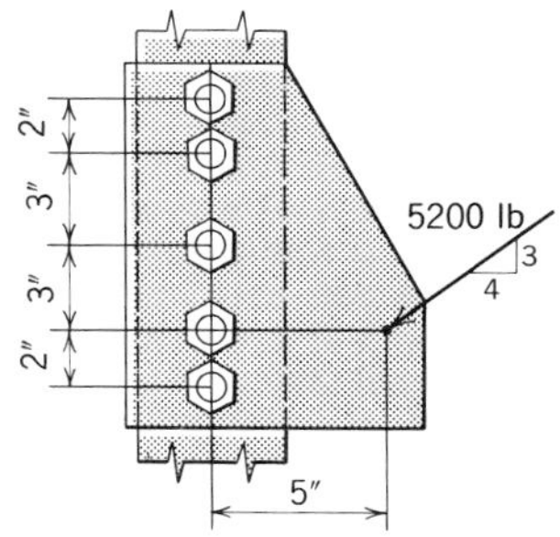


**Fig. P11–54**

**11–55.*** In the joint of Fig. P11–55 the diameter of each bolt is $\frac{1}{2}$ in. Determine the cross shearing stresses in bolts $A$ and $B$.

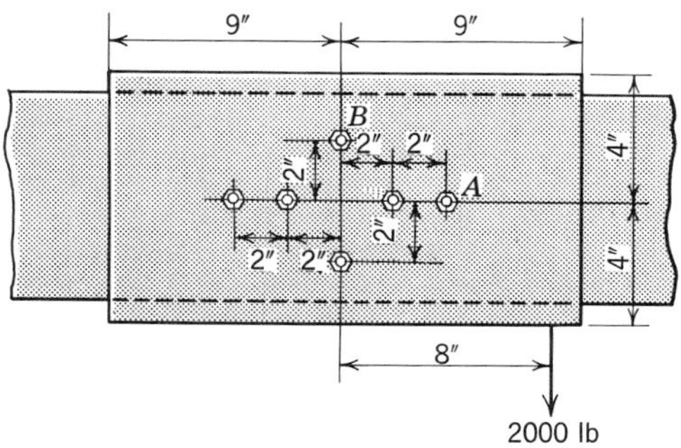


**Fig. P11–55**

**11–56.** Each of the rivets in the joint shown in Fig. P11–56 has a diameter of 1 in.

(*a*) Which of the three rivets would have the maximum cross shearing stress? Why?

(*b*) Determine the cross shearing stress in this rivet.

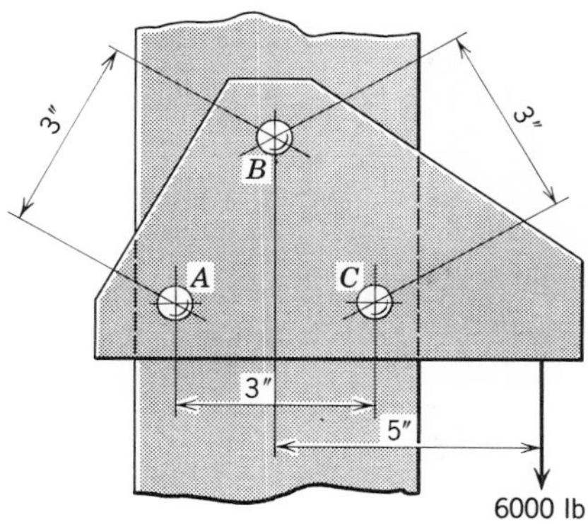


Fig. P11–56

**11–57.** A load $P$ is applied to the plate riveted to a support as shown in Fig. P11–57. Determine the maximum value of $P$ if the maximum allowable cross shearing stress in the rivets is 12,000 psi. All rivets are 1 in. in diameter.

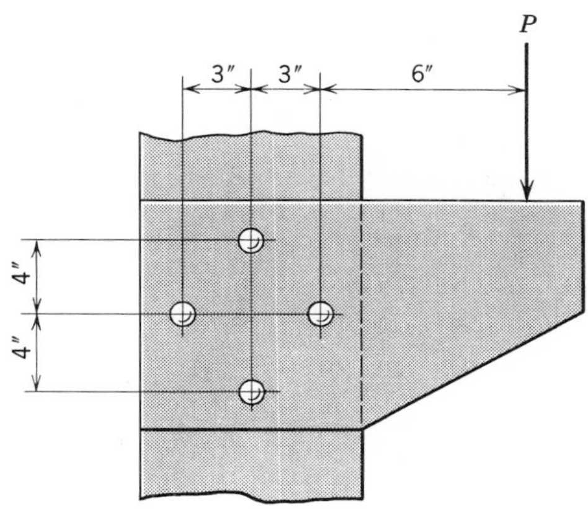


Fig. P11–57

**11–58.*** The bolts in Fig. P11–58 are $\frac{3}{8}$ in. in diameter. If the allowable cross shearing stress in the bolts is 12,000 psi determine the maximum permissible value for the load $P$.

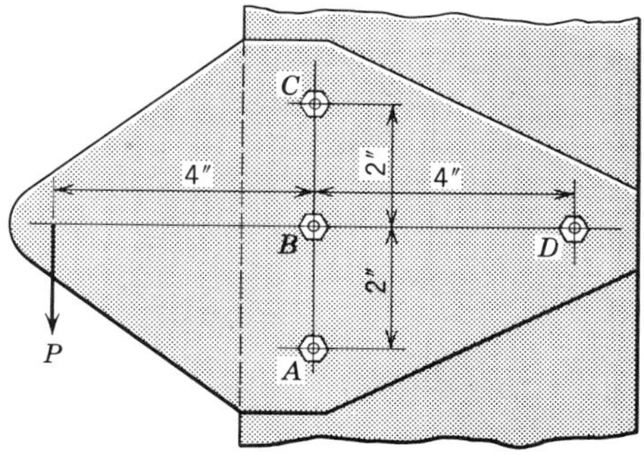


Fig. P11–58

**11–59.*** The bolted joint in Fig. P11–59 is to be designed to support the 2400-lb eccentric load as shown. The diameter of the bolts is $\frac{1}{2}$ in. and the allowable cross shearing stress in the bolts is 8000 psi. Determine the minimum permissible distance $a$.

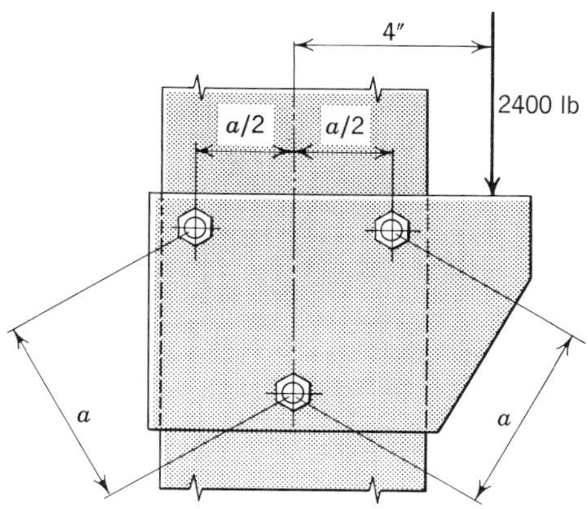


Fig. P11–59

APPENDIX **A**

# *1. Average Properties of Selected Materials*

# *2. Stress-Strain Diagrams for Typical Metals*

## TABLE A–1. Average Properties of Selected Engineering Materials

Exact values may vary widely with changes in composition, heat treatment, and mechanical working. More precise information can be obtained from manufacturers.

| Materials | Specific Weight | Elastic Strength* | | | Ultimate Strength | | | Endurance Limit*** | Modulus of Elasticity | Modulus of Rigidity | Per Cent Elongation in 2″ | Coefficient of Thermal Expansion |
|---|---|---|---|---|---|---|---|---|---|---|---|---|
| | | Tension | Comp. | Shear | Tension | Comp. | Shear | | | | | |
| | lb/cu in. | ksi | ksi | ksi | ksi | ksi | ksi | ksi | 1000 ksi | 1000 ksi | | $10^{-6}$ in. per in. per °F |
| Ferrous metals | | | | | | | | | | | | |
| Wrought iron | 0.278 | 30 | ** | | 48 | ** | 25 | 23 | 28 | | 30† | 6.7 |
| Structural steel | 0.284 | 36 | ** | | 66 | ** | | 28 | 29 | 11.0 | 28† | 6.6 |
| Steel, 0.2 % C hardened | 0.284 | 62 | ** | | 90 | ** | | | 30 | 11.6 | 22 | 6.6 |
| Steel, 0.4 % C hot-rolled | 0.284 | 53 | ** | | 84 | ** | | 38 | 30 | 11.6 | 29 | |
| Steel, 0.8 % C hot-rolled | 0.284 | 76 | ** | | 122 | ** | | | 30 | 11.6 | 8 | |
| Cast iron—gray | 0.260 | | | | 25 | 100 | | 12 | 15 | | 0.5 | 6.7 |
| Cast iron—malleable | 0.266 | 32 | ** | | 50 | ** | | | 25 | | 20 | 6.6 |
| Cast iron—nodular | 0.266 | 70 | | | 100 | | | | 25 | | 4 | 6.6 |
| Stainless steel (18–8) annealed | 0.286 | 36 | ** | | 85 | ** | | 40 | 28 | 12.5 | 55 | 9.6 |
| Stainless steel (18–8) cold-rolled | 0.286 | 165 | ** | | 190 | ** | | 90 | 28 | 12.5 | 8 | 9.6 |
| Steel, SAE 4340, heat-treated | 0.283 | 132 | 145 | | 150 | ** | 95 | 76 | 29 | 11.0 | 19 | |

| | | | | | | | | | | | | |
|---|---|---|---|---|---|---|---|---|---|---|---|---|
| Non-ferrous metal alloys | | | | | | | | | | | | |
| Aluminum, cast, 195-T6 | 0.100 | 24 | 25 | | 36 | | 30 | 7 | 10.3 | 3.8 | 5 | |
| Aluminum, wrought, 2014-T4 | 0.101 | 41 | 41 | 24 | 62 | ** | 38 | 18 | 10.6 | 4.0 | 20 | 12.5 |
| Aluminum, wrought, 2024-T4 | 0.100 | 48 | 48 | 28 | 68 | ** | 41 | 18 | 10.6 | 4.0 | 19 | 12.5 |
| Aluminum, wrought, 6061-T6 | 0.098 | 40 | 40 | 26 | 45 | ** | 30 | 13.5 | 10.0 | 3.8 | 17 | 12.5 |
| Magnesium, extrusion, AZ80X | 0.066 | 35 | 26 | | 49 | ** | 21 | 19 | 6.5 | 2.4 | 12 | 14.4 |
| Magnesium, sand cast, AZ63-HT | 0.066 | 14 | 14 | | 40 | ** | 19 | 14 | 6.5 | 2.4 | 12 | 14.4 |
| Monel, wrought, hot-rolled | 0.319 | 50 | ** | | 90 | ** | | 40 | 26 | 9.5 | 35 | 7.8 |
| Red brass, cold-rolled | 0.316 | 60 | | | 75 | | | | 15 | 5.6 | 4 | 9.8 |
| Red brass, annealed | 0.316 | 15 | ** | | 40 | ** | | | 15 | 5.6 | 50 | 9.8 |
| Bronze, cold-rolled | 0.320 | 75 | | | 100 | | | | 15 | 6.5 | 3 | 9.4 |
| Bronze, annealed | 0.320 | 20 | ** | | 50 | ** | | | 15 | 6.5 | 50 | 9.4 |
| Titanium alloy, annealed | 0.167 | 135 | ** | | 155 | ** | | | 14 | 5.3 | 13 | |
| Invar, annealed | 0.292 | 42 | ** | | 70 | ** | | | 21 | 8.1 | 41 | 0.6 |
| Non-metallic materials | | | | | | | | | | | | |
| Douglas fir, green†† | 0.022 | 4.8 | 3.4 | | | 3.9 | 0.9 | | 1.6 | | | |
| Douglas fir, air dry†† | 0.020 | 8.1 | 6.4 | | | 7.4 | 1.1 | | 1.9 | | | |
| Red oak, green†† | 0.037 | 4.4 | 2.6 | | | 3.5 | 1.2 | | 1.4 | | | 1.9 |
| Red oak, air dry†† | 0.025 | 8.4 | 4.6 | | | 6.9 | 1.8 | | 1.8 | | | |
| Concrete, medium strength | 0.087 | | 1.2 | | | 3.0 | | | 3.0 | | | 6.0 |
| Concrete, fairly high str. | 0.087 | | 2.0 | | | 5.0 | | | 4.5 | | | 6.0 |

* Elastic strength may be represented by proportional limit, yield point, or yield strength at a specified offset (usually 0.2 per cent for ductile metals).

** For ductile metals (those which have an appreciable ultimate elongation) it is customary to assume the properties in compression have the same values as those in tension.

*** Rotating beam.

† Elongation in 8 in.

†† All timber properties are parallel to the grain.

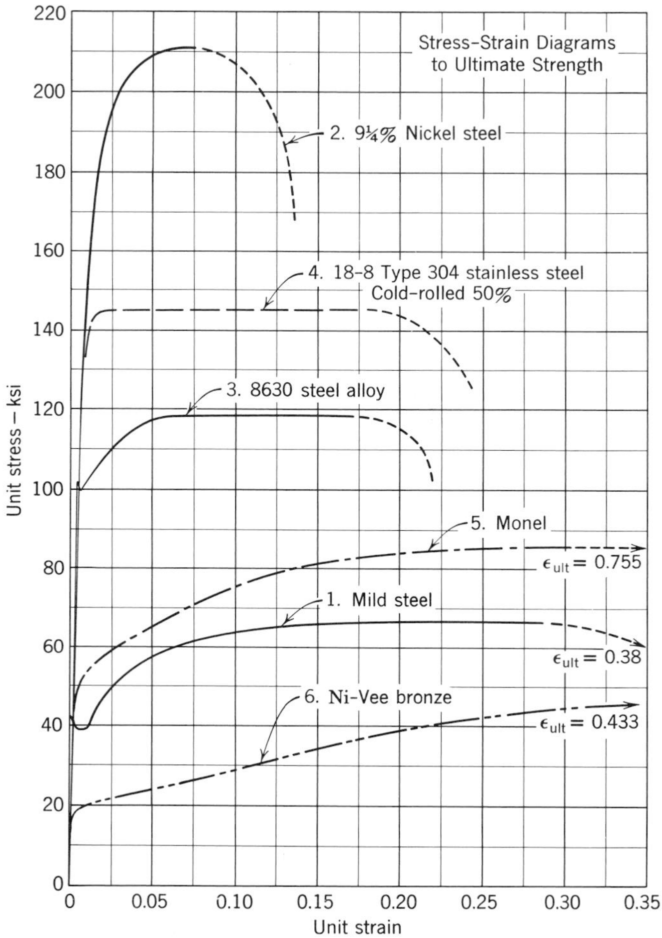


**Stress-strain diagrams. Courtesy: International Nickel Company.**

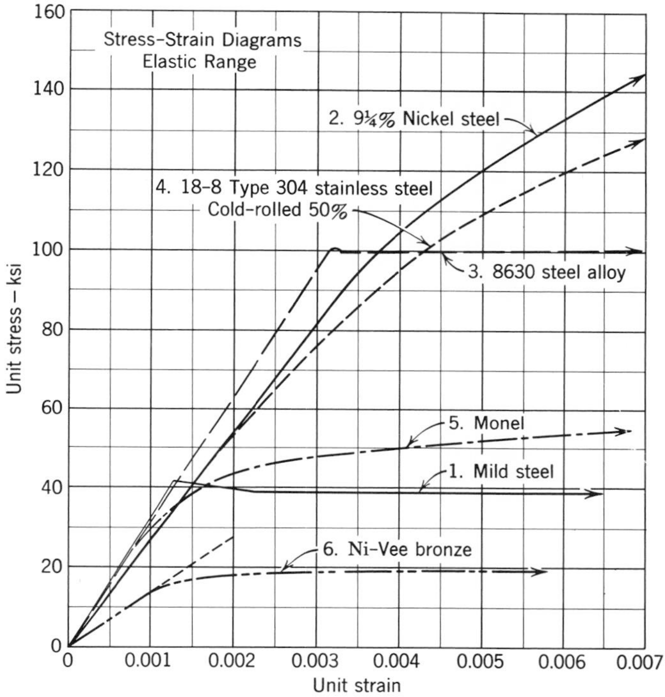


**Stress-strain diagrams. Courtesy: International Nickel Company.**

APPENDIX B

# *Properties of Selected Areas*

### Rectangle

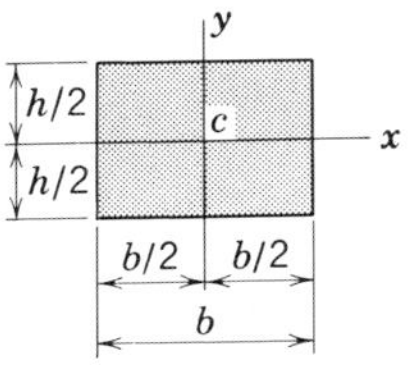


$A = bh$

$I_x = bh^3/12$

$I_y = b^3h/12$

### Segment of 2nd-Degree Parabola

$y = kx^2 = (h/b^2)x^2$

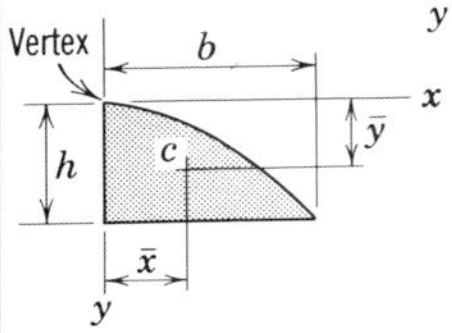


$A = 2bh/3$

$\bar{x} = 3b/8$

$\bar{y} = 3h/5$

### Circle

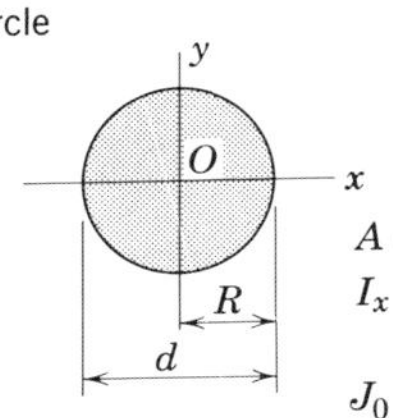


$A = \pi R^2 = \pi d^2/4$

$I_x = I_y = \pi R^4/4$
$= \pi d^4/64$

$J_0 = \pi R^4/2 = \pi d^4/32$

### Spandrel of 2nd-Degree Parabola

$y = kx^2 = (h/b^2)x^2$

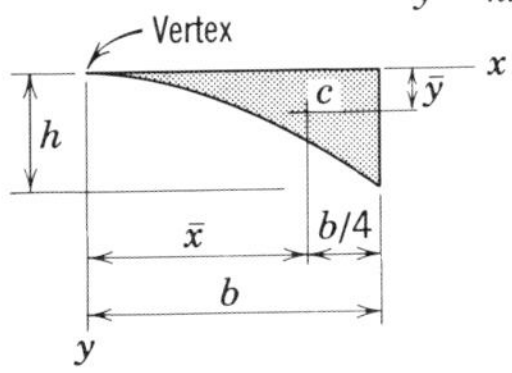


$A = bh/3$

$\bar{x} = 3b/4$

$\bar{y} = 3h/10$

### Semicircle

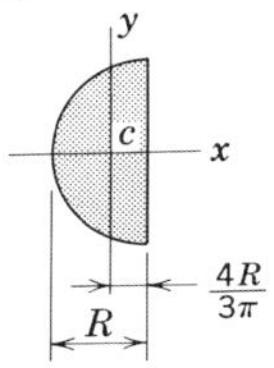


$A = \pi R^2/2$

$I_x = \pi R^4/8$

$I_y = \pi R^4\left(\frac{1}{8} - \frac{8}{9\pi^2}\right)$

### Spandrel of 3rd-Degree Parabola

$y = kx^3 = (h/b^3)x^3$

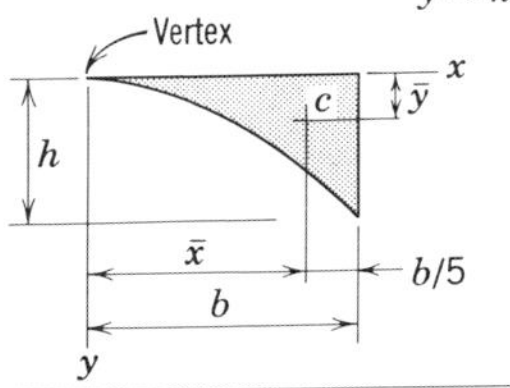


$A = bh/4$

$\bar{x} = 4b/5$

$\bar{y} = 2h/7$

### Right Triangle

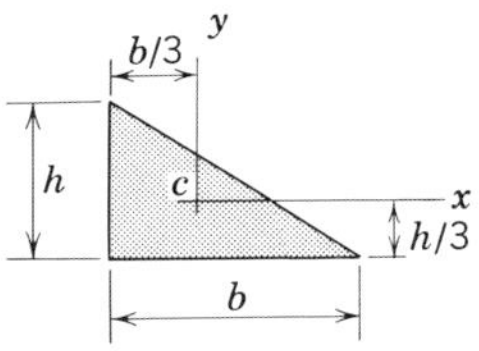


$A = bh/2$

$I_x = bh^3/36$

$I_y = hb^3/36$

### Spandrel of $n$th-Degree Parabola

$y = kx^n = (h/b^n)x^n$

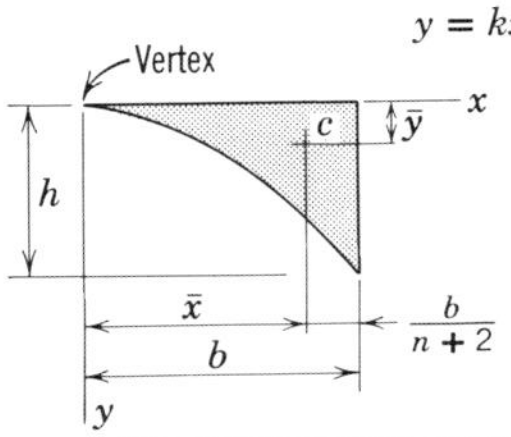


$A = \frac{bh}{n+1}$

$\bar{x} = \frac{n+1}{n+2}b$

$\bar{y} = \frac{n+1}{2(2n+1)}h$

APPENDIX C

# *Properties of Rolled Steel Shapes*

## WF SHAPES

PROPERTIES FOR DESIGNING

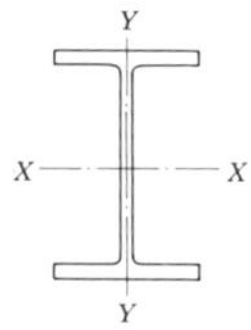


| Nominal Size | Weight per Foot | Area | Depth | Flange Width | Flange Thickness | Web Thickness | Axis X-X I | Axis X-X Z | Axis X-X r | Axis Y-Y I | Axis Y-Y Z | Axis Y-Y r |
|---|---|---|---|---|---|---|---|---|---|---|---|---|
| in. | lb | sq in. | in. | in. | in. | in. | in.$^4$ | in.$^3$ | in. | in.$^4$ | in.$^3$ | in. |
| 36 × 16½ | 230 | 67.73 | 35.88 | 16.475 | 1.260 | .765 | 14988.4 | 835.5 | 14.88 | 870.9 | 105.7 | 3.59 |
| 36 × 12 | 160 | 47.09 | 36.00 | 12.000 | 1.020 | .653 | 9738.8 | 541.0 | 14.38 | 275.4 | 45.9 | 2.42 |
| 33 × 15¾ | 200 | 58.79 | 33.00 | 15.750 | 1.150 | .715 | 11048.2 | 669.6 | 13.71 | 691.7 | 87.8 | 3.43 |
| 33 × 11½ | 130 | 38.26 | 33.10 | 11.510 | .855 | .580 | 6699.0 | 404.8 | 13.23 | 201.4 | 35.0 | 2.29 |
| 30 × 15 | 172 | 50.65 | 29.88 | 14.985 | 1.065 | .655 | 7891.5 | 528.2 | 12.48 | 550.1 | 73.4 | 3.30 |
| 30 × 10½ | 108 | 31.77 | 29.82 | 10.484 | .760 | .548 | 4461.0 | 299.2 | 11.85 | 135.1 | 25.8 | 2.06 |
| 27 × 14 | 145 | 42.68 | 26.88 | 13.965 | .975 | .600 | 5414.3 | 402.9 | 11.26 | 406.9 | 58.3 | 3.09 |
| 27 × 10 | 94 | 27.65 | 26.91 | 9.990 | .747 | .490 | 3266.7 | 242.8 | 10.87 | 115.1 | 23.0 | 2.04 |
| 24 × 14 | 130 | 38.21 | 24.25 | 14.000 | .900 | .565 | 4009.5 | 330.7 | 10.24 | 375.2 | 53.6 | 3.13 |
| 24 × 12 | 100 | 29.43 | 24.00 | 12.000 | .775 | .468 | 2987.3 | 248.9 | 10.08 | 203.5 | 33.9 | 2.63 |
| 24 × 9 | 84 | 24.71 | 24.09 | 9.015 | .772 | .470 | 2364.3 | 196.3 | 9.78 | 88.3 | 19.6 | 1.89 |
| 21 × 13 | 112 | 32.93 | 21.00 | 13.000 | .865 | .527 | 2620.6 | 249.6 | 8.92 | 289.7 | 44.6 | 2.96 |
| 21 × 9 | 82 | 24.10 | 20.86 | 8.962 | .795 | .499 | 1752.4 | 168.0 | 8.53 | 89.6 | 20.0 | 1.93 |
| 21 × 8¼ | 62 | 18.23 | 20.99 | 8.240 | .615 | .400 | 1326.8 | 126.4 | 8.53 | 53.1 | 12.9 | 1.71 |
| 18 × 11¾ | 96 | 28.22 | 18.16 | 11.750 | .831 | .512 | 1674.7 | 184.4 | 7.70 | 206.8 | 35.2 | 2.71 |
| 18 × 8¾ | 77 | 22.63 | 18.16 | 8.787 | .831 | .475 | 1286.8 | 141.7 | 7.54 | 88.6 | 20.2 | 1.98 |
| 18 × 7½ | 60 | 17.64 | 18.25 | 7.558 | .695 | .416 | 984.0 | 107.8 | 7.47 | 47.1 | 12.5 | 1.63 |
| | 50 | 14.71 | 18.00 | 7.500 | .570 | .358 | 800.6 | 89.0 | 7.38 | 37.2 | 9.9 | 1.59 |
| 16 × 11½ | 88 | 25.87 | 16.16 | 11.502 | .795 | .504 | 1222.6 | 151.3 | 6.87 | 185.2 | 32.2 | 2.67 |
| 16 × 8½ | 64 | 18.80 | 16.00 | 8.500 | .715 | .443 | 833.8 | 104.2 | 6.66 | 68.4 | 16.1 | 1.91 |
| 16 × 7 | 50 | 14.70 | 16.25 | 7.073 | .628 | .380 | 655.4 | 80.7 | 6.68 | 34.8 | 9.8 | 1.54 |
| | 36 | 10.59 | 15.85 | 6.992 | .428 | .299 | 446.3 | 56.3 | 6.49 | 22.1 | 6.3 | 1.45 |

Courtesy American Institute of Steel Construction

# WF SHAPES

PROPERTIES FOR DESIGNING

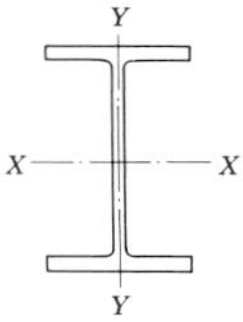


| Nominal Size | Weight per Foot | Area | Depth | Flange |  | Web Thickness | Axis X–X |  |  | Axis Y–Y |  |  |
|---|---|---|---|---|---|---|---|---|---|---|---|---|
|  |  |  |  | Width | Thickness |  | $I$ | $Z$ | $r$ | $I$ | $Z$ | $r$ |
| in. | lb | sq in. | in. | in. | in. | in. | in.$^4$ | in.$^3$ | in. | in.$^4$ | in.$^3$ | in. |
| 14 × 16 | 202 | 59.39 | 15.63 | 15.750 | 1.503 | .930 | 2538.8 | 324.9 | 6.54 | 979.7 | 124.4 | 4.06 |
| 14 × 14½ | 119 | 34.99 | 14.50 | 14.650 | .938 | .570 | 1373.1 | 189.4 | 6.26 | 491.8 | 67.1 | 3.75 |
|  | 87 | 25.56 | 14.00 | 14.500 | .688 | .420 | 966.9 | 138.1 | 6.15 | 349.7 | 48.2 | 3.70 |
| 14 × 12 | 78 | 22.94 | 14.06 | 12.000 | .718 | .428 | 851.2 | 121.1 | 6.09 | 206.9 | 34.5 | 3.00 |
| 14 × 10 | 68 | 20.00 | 14.06 | 10.040 | .718 | .418 | 724.1 | 103.0 | 6.02 | 121.2 | 24.1 | 2.46 |
| 14 × 8 | 53 | 15.59 | 13.94 | 8.062 | .658 | .370 | 542.1 | 77.8 | 5.90 | 57.5 | 14.3 | 1.92 |
|  | 43 | 12.65 | 13.68 | 8.000 | .528 | .308 | 429.0 | 62.7 | 5.82 | 45.1 | 11.3 | 1.89 |
| 14 × 6¾ | 30 | 8.81 | 13.86 | 6.733 | .383 | .270 | 289.6 | 41.8 | 5.73 | 17.5 | 5.2 | 1.41 |
| 12 × 12 | 106 | 31.19 | 12.88 | 12.230 | .986 | .620 | 930.7 | 144.5 | 5.46 | 300.9 | 49.2 | 3.11 |
|  | 72 | 21.16 | 12.25 | 12.040 | .671 | .430 | 597.4 | 97.5 | 5.31 | 195.3 | 32.4 | 3.04 |
| 12 × 10 | 53 | 15.59 | 12.06 | 10.000 | .576 | .345 | 426.2 | 70.7 | 5.23 | 96.1 | 19.2 | 2.48 |
| 12 × 8 | 40 | 11.77 | 11.94 | 8.000 | .516 | .294 | 310.1 | 51.9 | 5.13 | 44.1 | 11.0 | 1.94 |
| 12 × 6½ | 27 | 7.97 | 11.95 | 6.500 | .400 | .240 | 204.1 | 34.1 | 5.06 | 16.6 | 5.1 | 1.44 |
| 10 × 10 | 100 | 29.43 | 11.12 | 10.345 | 1.118 | .685 | 625.0 | 112.4 | 4.61 | 206.6 | 39.9 | 2.65 |
|  | 77 | 22.67 | 10.62 | 10.195 | .868 | .535 | 457.2 | 86.1 | 4.49 | 153.4 | 30.1 | 2.60 |
|  | 60 | 17.66 | 10.25 | 10.075 | .683 | .415 | 343.7 | 67.1 | 4.41 | 116.5 | 23.1 | 2.57 |
|  | 49 | 14.40 | 10.00 | 10.000 | .558 | .340 | 272.9 | 54.6 | 4.35 | 93.0 | 18.6 | 2.54 |
| 10 × 8 | 45 | 13.24 | 10.12 | 8.022 | .618 | .350 | 248.6 | 49.1 | 4.33 | 53.2 | 13.3 | 2.00 |
|  | 33 | 9.71 | 9.75 | 7.964 | .433 | .292 | 170.9 | 35.0 | 4.20 | 36.5 | 9.2 | 1.94 |
| 10 × 5¾ | 25 | 7.35 | 10.08 | 5.762 | .430 | .252 | 133.2 | 26.4 | 4.26 | 12.7 | 4.4 | 1.31 |
|  | 21 | 6.19 | 9.90 | 5.750 | .340 | .240 | 106.3 | 21.5 | 4.14 | 9.7 | 3.4 | 1.25 |
| 8 × 8 | 67 | 19.70 | 9.00 | 8.287 | .933 | .575 | 271.8 | 60.4 | 3.71 | 88.6 | 21.4 | 2.12 |
|  | 40 | 11.76 | 8.25 | 8.077 | .558 | .365 | 146.3 | 35.5 | 3.53 | 49.0 | 12.1 | 2.04 |
|  | 31 | 9.12 | 8.00 | 8.000 | .433 | .288 | 109.7 | 27.4 | 3.47 | 37.0 | 9.2 | 2.01 |
| 8 × 6½ | 28 | 8.23 | 8.06 | 6.540 | .463 | .285 | 97.8 | 24.3 | 3.45 | 21.6 | 6.6 | 1.62 |
|  | 24 | 7.06 | 7.93 | 6.500 | .398 | .245 | 82.5 | 20.8 | 3.42 | 18.2 | 5.6 | 1.61 |
| 8 × 5¼ | 17 | 5.00 | 8.00 | 5.250 | .308 | .230 | 56.4 | 14.1 | 3.36 | 6.7 | 2.6 | 1.16 |

Courtesy American Institute of Steel Construction

# AMERICAN STANDARD BEAMS

PROPERTIES FOR DESIGNING

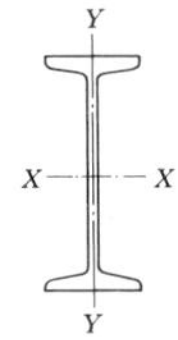


| Nominal Size | Weight per Foot | Area | Depth | Flange | | Web Thickness | Axis X–X | | | Axis Y–Y | | |
|---|---|---|---|---|---|---|---|---|---|---|---|---|
| | | | | Width | Thickness | | *I* | *Z* | *r* | *I* | *Z* | *r* |
| in. | lb | sq in. | in. | in. | in. | in. | $in.^4$ | $in.^3$ | in. | $in.^4$ | $in.^3$ | in. |
| 24 × 7⅞ | 120.0 | 35.13 | 24.00 | 8.048 | 1.102 | .798 | 3010.8 | 250.9 | 9.26 | 84.9 | 21.1 | 1.56 |
| 24 × 7 | 100.0 | 29.25 | 24.00 | 7.247 | .871 | .747 | 2371.8 | 197.6 | 9.05 | 48.4 | 13.4 | 1.29 |
| | 90.0 | 26.30 | 24.00 | 7.124 | .871 | .624 | 2230.1 | 185.8 | 9.21 | 45.5 | 12.8 | 1.32 |
| | 79.9 | 23.33 | 24.00 | 7.000 | .871 | .500 | 2087.2 | 173.9 | 9.46 | 42.9 | 12.2 | 1.36 |
| 20 × 7 | 95.0 | 27.74 | 20.00 | 7.200 | .916 | .800 | 1599.7 | 160.0 | 7.59 | 50.5 | 14.0 | 1.35 |
| | 85.0 | 24.80 | 20.00 | 7.053 | .916 | .653 | 1501.7 | 150.2 | 7.78 | 47.0 | 13.3 | 1.38 |
| 20 × 6¼ | 75.0 | 21.90 | 20.00 | 6.391 | .789 | .641 | 1263.5 | 126.3 | 7.60 | 30.1 | 9.4 | 1.17 |
| | 65.4 | 19.08 | 20.00 | 6.250 | .789 | .500 | 1169.5 | 116.9 | 7.83 | 27.9 | 8.9 | 1.21 |
| 18 × 6 | 70.0 | 20.46 | 18.00 | 6.251 | .691 | .711 | 917.5 | 101.9 | 6.70 | 24.5 | 7.8 | 1.09 |
| | 54.7 | 15.94 | 18.00 | 6.000 | .691 | .460 | 795.5 | 88.4 | 7.07 | 21.2 | 7.1 | 1.15 |
| 15 × 5½ | 50.0 | 14.59 | 15.00 | 5.640 | .622 | .550 | 481.1 | 64.2 | 5.74 | 16.0 | 5.7 | 1.05 |
| | 42.9 | 12.49 | 15.00 | 5.500 | .622 | .410 | 441.8 | 58.9 | 5.95 | 14.6 | 5.3 | 1.08 |
| 12 × 5¼ | 50.0 | 14.57 | 12.00 | 5.477 | .659 | .687 | 301.6 | 50.3 | 4.55 | 16.0 | 5.8 | 1.05 |
| | 40.8 | 11.84 | 12.00 | 5.250 | .659 | .460 | 268.9 | 44.8 | 4.77 | 13.8 | 5.3 | 1.08 |
| 12 × 5 | 31.8 | 9.26 | 12.00 | 5.000 | .544 | .350 | 215.8 | 36.0 | 4.83 | 9.5 | 3.8 | 1.01 |
| 10 × 4⅝ | 35.0 | 10.22 | 10.00 | 4.944 | .491 | .594 | 145.8 | 29.2 | 3.78 | 8.5 | 3.4 | .91 |
| | 25.4 | 7.38 | 10.00 | 4.660 | .491 | .310 | 122.1 | 24.4 | 4.07 | 6.9 | 3.0 | .97 |
| 8 × 4 | 23.0 | 6.71 | 8.00 | 4.171 | .425 | .441 | 64.2 | 16.0 | 3.09 | 4.4 | 2.1 | .81 |
| | 18.4 | 5.34 | 8.00 | 4.000 | .425 | .270 | 56.9 | 14.2 | 3.26 | 3.8 | 1.9 | .84 |
| 7 × 3⅝ | 20.0 | 5.83 | 7.00 | 3.860 | .392 | .450 | 41.9 | 12.0 | 2.68 | 3.1 | 1.6 | .74 |
| | 15.3 | 4.43 | 7.00 | 3.660 | .392 | .250 | 36.2 | 10.4 | 2.86 | 2.7 | 1.5 | .78 |
| 6 × 3⅜ | 17.25 | 5.02 | 6.00 | 3.565 | .359 | .465 | 26.0 | 8.7 | 2.28 | 2.3 | 1.3 | .68 |
| | 12.5 | 3.61 | 6.00 | 3.330 | .359 | .230 | 21.8 | 7.3 | 2.46 | 1.8 | 1.1 | .72 |
| 5 × 3 | 14.75 | 4.29 | 5.00 | 3.284 | .326 | .494 | 15.0 | 6.0 | 1.87 | 1.7 | 1.0 | .63 |
| | 10.0 | 2.87 | 5.00 | 3.000 | .326 | .210 | 12.1 | 4.8 | 2.05 | 1.2 | .82 | .65 |
| 4 × 2⅝ | 9.5 | 2.76 | 4.00 | 2.796 | .293 | .326 | 6.7 | 3.3 | 1.56 | .91 | .65 | .58 |
| 3 × 2⅜ | 7.5 | 2.17 | 3.00 | 2.509 | .260 | .349 | 2.9 | 1.9 | 1.15 | .59 | .47 | .52 |

Courtesy American Institute of Steel Construction

# AMERICAN STANDARD CHANNELS

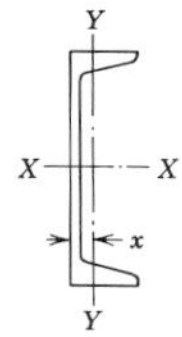


## PROPERTIES FOR DESIGNING

| Nominal Size | Weight per Foot | Area | Depth | Flange |  | Web Thickness | Axis X–X |  |  | Axis Y–Y |  |  |  |
|---|---|---|---|---|---|---|---|---|---|---|---|---|---|
|  |  |  |  | Width | Average Thickness |  | *I* | *Z* | *r* | *I* | *Z* | *r* | *x* |
| in. | lb | sq in. | in. | in. | in. | in. | in.⁴ | in.³ | in. | in.⁴ | in.³ | in. | in. |
| *18 × 4 | 58.0 | 16.98 | 18.00 | 4.200 | .625 | .700 | 670.7 | 74.5 | 6.29 | 18.5 | 5.6 | 1.04 | .88 |
|  | 51.9 | 15.18 | 18.00 | 4.100 | .625 | .600 | 622.1 | 69.1 | 6.40 | 17.1 | 5.3 | 1.06 | .87 |
|  | 45.8 | 13.38 | 18.00 | 4.000 | .625 | .500 | 573.5 | 63.7 | 6.55 | 15.8 | 5.1 | 1.09 | .89 |
|  | 42.7 | 12.48 | 18.00 | 3.950 | .625 | .450 | 549.2 | 61.0 | 6.64 | 15.0 | 4.9 | 1.10 | .90 |
| 15 × 3⅜ | 50.0 | 14.64 | 15.00 | 3.716 | .650 | .716 | 401.4 | 53.6 | 5.24 | 11.2 | 3.8 | .87 | .80 |
|  | 40.0 | 11.70 | 15.00 | 3.520 | .650 | .520 | 346.3 | 46.2 | 5.44 | 9.3 | 3.4 | .89 | .78 |
|  | 33.9 | 9.90 | 15.00 | 3.400 | .650 | .400 | 312.6 | 41.7 | 5.62 | 8.2 | 3.2 | .91 | .79 |
| 12 × 3 | 30.0 | 8.79 | 12.00 | 3.170 | .501 | .510 | 161.2 | 26.9 | 4.28 | 5.2 | 2.1 | .77 | .68 |
|  | 20.7 | 6.03 | 12.00 | 2.940 | .501 | .280 | 128.1 | 21.4 | 4.61 | 3.9 | 1.7 | .81 | .70 |
| 10 × 2⅝ | 30.0 | 8.80 | 10.00 | 3.033 | .436 | .673 | 103.0 | 20.6 | 3.42 | 4.0 | 1.7 | .67 | .65 |
|  | 25.0 | 7.33 | 10.00 | 2.886 | .436 | .526 | 90.7 | 18.1 | 3.52 | 3.4 | 1.5 | .68 | .62 |
|  | 15.3 | 4.47 | 10.00 | 2.600 | .436 | .240 | 66.9 | 13.4 | 3.87 | 2.3 | 1.2 | .72 | .64 |
| 9 × 2½ | 20.0 | 5.86 | 9.00 | 2.648 | .413 | .448 | 60.6 | 13.5 | 3.22 | 2.4 | 1.2 | .65 | .59 |
|  | 15.0 | 4.39 | 9.00 | 2.485 | .413 | .285 | 50.7 | 11.3 | 3.40 | 1.9 | 1.0 | .67 | .59 |
|  | 13.4 | 3.89 | 9.00 | 2.430 | .413 | .230 | 47.3 | 10.5 | 3.49 | 1.8 | .97 | .67 | .61 |
| 8 × 2¼ | 18.75 | 5.49 | 8.00 | 2.527 | .390 | .487 | 43.7 | 10.9 | 2.82 | 2.0 | 1.0 | .60 | .57 |
|  | 13.75 | 4.02 | 8.00 | 2.343 | .390 | .303 | 35.8 | 9.0 | 2.99 | 1.5 | .86 | .62 | .56 |
|  | 11.5 | 3.36 | 8.00 | 2.260 | .390 | .220 | 32.3 | 8.1 | 3.10 | 1.3 | .79 | .63 | .58 |
| 7 × 2⅛ | 14.75 | 4.32 | 7.00 | 2.299 | .366 | .419 | 27.1 | 7.7 | 2.51 | 1.4 | .79 | .57 | .53 |
|  | 12.25 | 3.58 | 7.00 | 2.194 | .366 | .314 | 24.1 | 6.9 | 2.59 | 1.2 | .71 | .58 | .53 |
|  | 9.8 | 2.85 | 7.00 | 2.090 | .366 | .210 | 21.1 | 6.0 | 2.72 | .98 | .63 | .59 | .55 |
| 6 × 2 | 13.0 | 3.81 | 6.00 | 2.157 | .343 | .437 | 17.3 | 5.8 | 2.13 | 1.1 | .65 | .53 | .52 |
|  | 10.5 | 3.07 | 6.00 | 2.034 | .343 | .314 | 15.1 | 5.0 | 2.22 | .87 | .57 | .53 | .50 |
|  | 8.2 | 2.39 | 6.00 | 1.920 | .343 | .200 | 13.0 | 4.3 | 2.34 | .70 | .50 | .54 | .52 |
| 5 × 1¾ | 9.0 | 2.63 | 5.00 | 1.885 | .320 | .325 | 8.8 | 3.5 | 1.83 | .64 | .45 | .49 | .48 |
|  | 6.7 | 1.95 | 5.00 | 1.750 | .320 | .190 | 7.4 | 3.0 | 1.95 | .48 | .38 | .50 | .49 |
| 4 × 1⅝ | 7.25 | 2.12 | 4.00 | 1.720 | .296 | .320 | 4.5 | 2.3 | 1.47 | .44 | .35 | .46 | .46 |
|  | 5.4 | 1.56 | 4.00 | 1.580 | .296 | .180 | 3.8 | 1.9 | 1.56 | .32 | .29 | .45 | .46 |
| 3 × 1½ | 6.0 | 1.75 | 3.00 | 1.596 | .273 | .356 | 2.1 | 1.4 | 1.08 | .31 | .27 | .42 | .46 |
|  | 4.1 | 1.19 | 3.00 | 1.410 | .273 | .170 | 1.6 | 1.1 | 1.17 | .20 | .21 | .41 | .44 |

* Car and Shipbuilding Channel; not an American Standard.

Courtesy American Institute of Steel Construction

## ANGLES
## EQUAL LEGS

PROPERTIES FOR DESIGNING

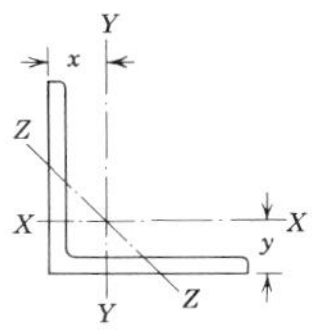


| Size | Thickness | Weight per Foot | Area | Axis X–X and Axis Y–Y | | | | Axis Z–Z |
|---|---|---|---|---|---|---|---|---|
| | | | | $I$ | $Z$ | $r$ | $x$ or $y$ | $r$ |
| in. | in. | lb | sq in. | in.⁴ | in.³ | in. | in. | in. |
| 8 × 8 | 1 | 51.0 | 15.00 | 89.0 | 15.8 | 2.44 | 2.37 | 1.56 |
| | 3/4 | 38.9 | 11.44 | 69.7 | 12.2 | 2.47 | 2.28 | 1.57 |
| | 5/8 | 32.7 | 9.61 | 59.4 | 10.3 | 2.49 | 2.23 | 1.58 |
| | 9/16 | 29.6 | 8.68 | 54.1 | 9.3 | 2.50 | 2.21 | 1.58 |
| | 1/2 | 26.4 | 7.75 | 48.6 | 8.4 | 2.50 | 2.19 | 1.59 |
| 6 × 6 | 1 | 37.4 | 11.00 | 35.5 | 8.6 | 1.80 | 1.86 | 1.17 |
| | 3/4 | 28.7 | 8.44 | 28.2 | 6.7 | 1.83 | 1.78 | 1.17 |
| | 5/8 | 24.2 | 7.11 | 24.2 | 5.7 | 1.84 | 1.73 | 1.18 |
| | 1/2 | 19.6 | 5.75 | 19.9 | 4.6 | 1.86 | 1.68 | 1.18 |
| | 3/8 | 14.9 | 4.36 | 15.4 | 3.5 | 1.88 | 1.64 | 1.19 |
| 5 × 5 | 7/8 | 27.2 | 7.98 | 17.8 | 5.2 | 1.49 | 1.57 | .97 |
| | 3/4 | 23.6 | 6.94 | 15.7 | 4.5 | 1.51 | 1.52 | .97 |
| | 5/8 | 20.0 | 5.86 | 13.6 | 3.9 | 1.52 | 1.48 | .98 |
| | 7/16 | 14.3 | 4.18 | 10.0 | 2.8 | 1.55 | 1.41 | .98 |
| | 5/16 | 10.3 | 3.03 | 7.4 | 2.0 | 1.57 | 1.37 | .99 |
| 4 × 4 | 3/4 | 18.5 | 5.44 | 7.7 | 2.8 | 1.19 | 1.27 | .78 |
| | 1/2 | 12.8 | 3.75 | 5.6 | 2.0 | 1.22 | 1.18 | .78 |
| | 3/8 | 9.8 | 2.86 | 4.4 | 1.5 | 1.23 | 1.14 | .79 |
| | 1/4 | 6.6 | 1.94 | 3.0 | 1.1 | 1.25 | 1.09 | .80 |
| 3 1/2 × 3 1/2 | 3/8 | 8.5 | 2.48 | 2.9 | 1.2 | 1.07 | 1.01 | .69 |
| | 1/4 | 5.8 | 1.69 | 2.0 | .79 | 1.09 | .97 | .69 |
| 3 × 3 | 1/2 | 9.4 | 2.75 | 2.2 | 1.1 | .90 | .93 | .58 |
| | 3/8 | 7.2 | 2.11 | 1.8 | .83 | .91 | .89 | .58 |
| | 1/4 | 4.9 | 1.44 | 1.2 | .58 | .93 | .84 | .59 |
| 2 1/2 × 2 1/2 | 1/2 | 7.7 | 2.25 | 1.2 | .72 | .74 | .81 | .49 |
| | 1/4 | 4.1 | 1.19 | .70 | .39 | .77 | .72 | .49 |
| | 3/16 | 3.07 | .90 | .55 | .30 | .78 | .69 | .49 |
| 2 × 2 | 1/4 | 3.19 | .94 | .35 | .25 | .61 | .59 | .39 |
| | 1/8 | 1.65 | .48 | .19 | .13 | .63 | .55 | .40 |
| 1 1/2 × 1 1/2 | 1/4 | 2.34 | .69 | .14 | .13 | .45 | .47 | .29 |
| | 1/8 | 1.23 | .36 | .08 | .07 | .47 | .42 | .30 |
| 1 × 1 | 1/4 | 1.49 | .44 | .04 | .06 | .29 | .34 | .20 |
| | 1/8 | .80 | .23 | .02 | .03 | .30 | .30 | .20 |

Courtesy American Institute of Steel Construction

L

## ANGLES
## UNEQUAL LEGS

PROPERTIES FOR DESIGNING

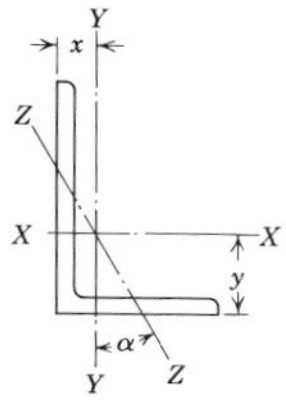


| Size | Thickness | Weight per Foot | Area | Axis X–X | | | | Axis Y–Y | | | | Axis Z–Z | |
|---|---|---|---|---|---|---|---|---|---|---|---|---|---|
| | | | | $I$ | $Z$ | $r$ | $y$ | $I$ | $Z$ | $r$ | $x$ | $r$ | tan $\alpha$ |
| in. | in. | lb | sq in. | in.$^4$ | in.$^3$ | in. | in. | in.$^4$ | in.$^3$ | in. | in. | in. | |
| 8 × 6 | 1 | 44.2 | 13.00 | 80.8 | 15.1 | 2.49 | 2.65 | 38.8 | 8.9 | 1.73 | 1.65 | 1.28 | .543 |
| | 3/4 | 33.8 | 9.94 | 63.4 | 11.7 | 2.53 | 2.56 | 30.7 | 6.9 | 1.76 | 1.56 | 1.29 | .551 |
| | 1/2 | 23.0 | 6.75 | 44.3 | 8.0 | 2.56 | 2.47 | 21.7 | 4.8 | 1.79 | 1.47 | 1.30 | .558 |
| 8 × 4 | 1 | 37.4 | 11.00 | 69.6 | 14.1 | 2.52 | 3.05 | 11.6 | 3.9 | 1.03 | 1.05 | .85 | .247 |
| | 3/4 | 28.7 | 8.44 | 54.9 | 10.9 | 2.55 | 2.95 | 9.4 | 3.1 | 1.05 | .95 | .85 | .258 |
| | 1/2 | 19.6 | 5.75 | 38.5 | 7.5 | 2.59 | 2.86 | 6.7 | 2.2 | 1.08 | .86 | .86 | .267 |
| 7 × 4 | 7/8 | 30.2 | 8.86 | 42.9 | 9.7 | 2.20 | 2.55 | 10.2 | 3.5 | 1.07 | 1.05 | .86 | .318 |
| | 9/16 | 20.0 | 5.87 | 29.6 | 6.5 | 2.24 | 2.44 | 7.2 | 2.4 | 1.11 | .94 | .87 | .332 |
| | 3/8 | 13.6 | 3.98 | 20.6 | 4.4 | 2.27 | 2.37 | 5.1 | 1.6 | 1.13 | .87 | .88 | .339 |
| 6 × 4 | 7/8 | 27.2 | 7.98 | 27.7 | 7.2 | 1.86 | 2.12 | 9.8 | 3.4 | 1.11 | 1.12 | .86 | .421 |
| | 9/16 | 18.1 | 5.31 | 19.3 | 4.8 | 1.90 | 2.01 | 6.9 | 2.3 | 1.14 | 1.01 | .87 | .438 |
| | 3/8 | 12.3 | 3.61 | 13.5 | 3.3 | 1.93 | 1.94 | 4.9 | 1.6 | 1.17 | .94 | .88 | .446 |
| | 5/16 | 10.3 | 3.03 | 11.4 | 2.8 | 1.94 | 1.92 | 4.2 | 1.4 | 1.17 | .92 | .88 | .449 |
| 6 × 3½ | 3/8 | 11.7 | 3.42 | 12.9 | 3.2 | 1.94 | 2.04 | 3.3 | 1.2 | .99 | .79 | .77 | .350 |
| 5 × 3½ | 5/8 | 16.8 | 4.92 | 12.0 | 3.7 | 1.56 | 1.70 | 4.8 | 1.9 | .99 | .95 | .75 | .472 |
| | 1/2 | 13.6 | 4.00 | 10.0 | 3.0 | 1.58 | 1.66 | 4.1 | 1.6 | 1.01 | .91 | .75 | .479 |
| | 3/8 | 10.4 | 3.05 | 7.8 | 2.3 | 1.60 | 1.61 | 3.2 | 1.2 | 1.02 | .86 | .76 | .486 |
| | 1/4 | 7.0 | 2.06 | 5.4 | 1.6 | 1.61 | 1.56 | 2.2 | .83 | 1.04 | .81 | .76 | .492 |
| 5 × 3 | 1/2 | 12.8 | 3.75 | 9.5 | 2.9 | 1.59 | 1.75 | 2.6 | 1.1 | .83 | .75 | .65 | .357 |
| | 5/16 | 8.2 | 2.40 | 6.3 | 1.9 | 1.61 | 1.68 | 1.8 | .75 | .85 | .68 | .66 | .368 |
| | 1/4 | 6.6 | 1.94 | 5.1 | 1.5 | 1.62 | 1.66 | 1.4 | .61 | .86 | .66 | .66 | .371 |
| 4 × 3½ | 1/2 | 11.9 | 3.50 | 5.3 | 1.9 | 1.23 | 1.25 | 3.8 | 1.5 | 1.04 | 1.00 | .72 | .750 |
| | 5/16 | 7.7 | 2.25 | 3.6 | 1.3 | 1.26 | 1.18 | 2.6 | 1.0 | 1.07 | .93 | .73 | .757 |
| | 1/4 | 6.2 | 1.81 | 2.9 | 1.0 | 1.27 | 1.16 | 2.1 | .81 | 1.07 | .91 | .73 | .759 |
| 4 × 3 | 3/8 | 8.5 | 2.48 | 4.0 | 1.5 | 1.26 | 1.28 | 1.9 | .87 | .88 | .78 | .64 | .551 |
| | 1/4 | 5.8 | 1.69 | 2.8 | 1.0 | 1.28 | 1.24 | 1.4 | .60 | .90 | .74 | .65 | .558 |
| 3½ × 3 | 3/8 | 7.9 | 2.30 | 2.7 | 1.1 | 1.09 | 1.08 | 1.9 | .85 | .90 | .83 | .62 | .721 |
| | 1/4 | 5.4 | 1.56 | 1.9 | .78 | 1.11 | 1.04 | 1.3 | .59 | .91 | .79 | .63 | .727 |
| 3½ × 2½ | 3/8 | 7.2 | 2.11 | 2.6 | 1.1 | 1.10 | 1.16 | 1.1 | .59 | .72 | .66 | .54 | .496 |
| | 1/4 | 4.9 | 1.44 | 1.8 | .75 | 1.12 | 1.11 | .78 | .41 | .74 | .61 | .54 | .506 |
| 3 × 2½ | 5/16 | 5.6 | 1.62 | 1.4 | .69 | .94 | .93 | .90 | .49 | .74 | .68 | .53 | .680 |
| 3 × 2 | 5/16 | 5.0 | 1.47 | 1.3 | .66 | .95 | 1.02 | .47 | .32 | .57 | .52 | .43 | .435 |
| 2½ × 2 | 5/16 | 4.5 | 1.31 | .79 | .47 | .78 | .81 | .45 | .31 | .58 | .56 | .42 | .620 |
| 2½ × 1½ | 1/4 | 3.19 | .94 | .59 | .36 | .79 | .88 | .16 | .14 | .41 | .38 | .32 | .357 |
| 2 × 1½ | 3/16 | 2.12 | .62 | .25 | .18 | .63 | .64 | .12 | .11 | .44 | .39 | .32 | .551 |

Courtesy American Institute of Steel Construction

APPENDIX D

# *Table of Beam Deflections and Slopes*

| Case | Load and Support (Length $L$) | Slope at End (+  ) | Maximum Deflection (+ upward) |
|---|---|---|---|
| 1 | 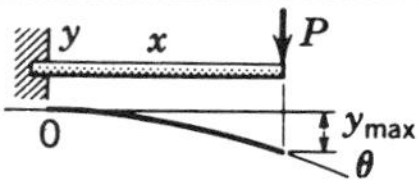 | $\theta = -\dfrac{PL^2}{2EI}$ at $x = L$ | $y_{\max} = -\dfrac{PL^3}{3EI}$ at $x = L$ |
| 2 | 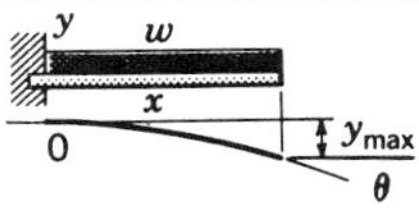 | $\theta = -\dfrac{wL^3}{6EI}$ at $x = L$ | $y_{\max} = -\dfrac{wL^4}{8EI}$ at $x = L$ |
| 3 | 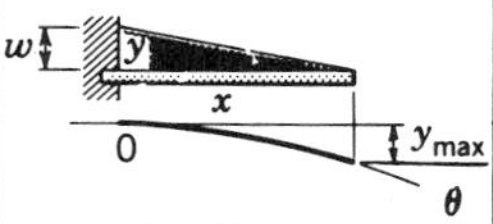 | $\theta = -\dfrac{wL^3}{24EI}$ at $x = L$ | $y_{\max} = -\dfrac{wL^4}{30EI}$ at $x = L$ |
| 4 | 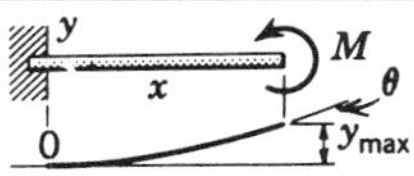 | $\theta = +\dfrac{ML}{EI}$ at $x = L$ | $y_{\max} = +\dfrac{ML^2}{2EI}$ at $x = L$ |
| 5 | 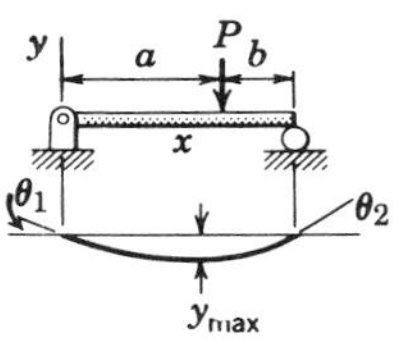 | $\theta_1 = -\dfrac{Pb(L^2 - b^2)}{6LEI}$ at $x = 0$<br>$\theta_2 = +\dfrac{Pa(L^2 - a^2)}{6LEI}$ at $x = L$ | $a > L/2 > b$<br>$y_{\max} = -\dfrac{Pb(L^2 - b^2)^{3/2}}{9\sqrt{3}\,LEI}$<br>at $x = \sqrt{(L^2 - b^2)/3}$<br>$y_{\text{center}} = -\dfrac{Pb(3L^2 - 4b^2)}{48EI}$ not max |
| 6 | 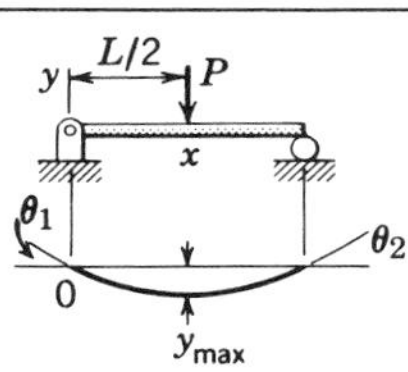 | $\theta_1 = -\dfrac{PL^2}{16EI}$ at $x = 0$<br>$\theta_2 = +\dfrac{PL^2}{16EI}$ at $x = L$ | $y_{\max} = -\dfrac{PL^3}{48EI}$ at $x = L/2$ |
| 7 | 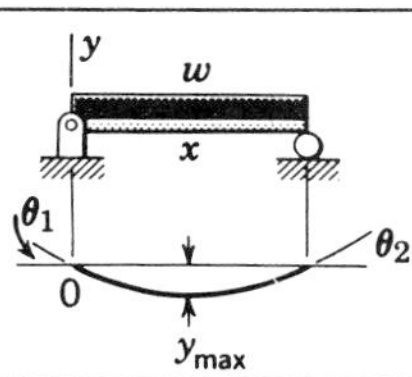 | $\theta_1 = -\dfrac{wL^3}{24EI}$ at $x = 0$<br>$\theta_2 = +\dfrac{wL^3}{24EI}$ at $x = L$ | $y_{\max} = -\dfrac{5wL^4}{384EI}$ at $x = L/2$ |
| 8 | 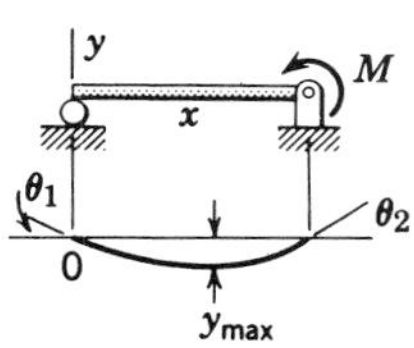 | $\theta_1 = -\dfrac{ML}{6EI}$ at $x = 0$<br>$\theta_2 = +\dfrac{ML}{3EI}$ at $x = L$ | $y_{\max} = -\dfrac{ML^2}{9\sqrt{3}\,EI}$<br>at $x = L/\sqrt{3}$<br>$y_{\text{center}} = -\dfrac{ML^2}{16EI}$ not max |

# Answers

## Chapter 1

**1–1.** (*a*) 26.0 ksi $T$.
(*b*) 13.0 ksi $T$.
**1–2.** (*a*) 3.92 ksi $C$.
(*b*) 2.50 ksi $C$.
(*c*) 3.81 ksi $C$.
**1–3.** 11.61 kips as shown.
**1–5.** (*a*) 0.876 sq in.
(*b*) 2.40 sq in.
**1–6.** $A_A = 0.975$ sq in.
$A_B = 1.125$ sq in.
**1–8.** 12,990 psi $T$.
**1–10.** 1.490 sq in.
**1–12.** 52,800 lb.
**1–14.** (*a*) 24,500 psi $T$.
(*b*) 10,200 psi.
**1–16.** (*a*) 0.300 sq in.
(*b*) 0.1546 sq in.
**1–19.** 13,240 psi $C$.
**1–21.** (*a*) 0.938 sq in.
(*b*) 0.620 in.
**1–22.** (*a*) $\sigma = 3470$ psi $T$.
$\tau = 6010$ psi.
(*b*) $\sigma = 13{,}890$ psi $T$.
$\tau = 6940$ psi.
**1–23.** (*a*) 109,600 lb.
(*b*) 1099 psi.
(*c*) 1710 psi.
**1–24.** 2250 lb.
**1–25.** 3.49 kips.
**1–27.** 12,010 psi $T$.
**1–30.** 6250 psi $C$.
**1–32.** 7050 psi.
**1–33.** 960 psi $T$.
**1–36.** 17.44 ksi $T$.
**1–38.** 3675 psi $T$.
**1–40.** $\theta = 16.85°$
$\sigma = 3.30$ ksi $T$.
**1–41.** Partial Results:
$\sigma_{p1} = 13.66$ ksi $T$.
$\sigma_{p2} = 9.66$ ksi $C$.
$\tau_{max} = 11.66$ ksi.
$\theta = -15.48°$.
**1–42.** Partial Results:
$\sigma_{p1,p2} = 22.0$ ksi $T$, 12.0 ksi $C$.
$\tau_{max} = 17.0$ ksi.
$\theta = -14.03°$.
**1–43.** Partial Results:
$\sigma_{p1,p2} = 2.19$ ksi $T$, 3.19 ksi $C$.
$\tau_{max} = 2.69$ ksi.
$\theta = -10.9°$.
**1–45.** (*a*) 800 psi.
(*b*) 15° and −75°.
(*c*) 693 psi.
**1–46.** Partial Results:
$\sigma_{p1,p2} = 2.24$ ksi $T$, 6.24 ksi $C$.
$\tau_{max} = 4.24$ ksi.
$\theta = -22.5°$.
**1–50.** Partial Results:
$\sigma_{p1} = 7000$ psi $T$.
$\sigma_{p2} = 13{,}000$ psi $C$.
$\tau_{max} = 10{,}000$ psi.
$\theta = 18.44°$.
**1–52.** Partial Results:
$\sigma_{p1} = 7000$ psi $C$.
$\sigma_{p2} = 17{,}000$ psi $C$.
$\tau_{max} = 8500$ psi.
$\theta = -18.44°$.
**1–54.** Partial Results:
$\tau_{xy} = \pm 16{,}970$ psi.
$\sigma_{p1} = 22{,}000$ psi $T$.
$\tau_{max} = 17{,}000$ psi.
$\theta = \pm 43.3°$.
**1–56.** Partial Results:
$\sigma_x = 6270$ psi $T$.
$\sigma_y = 2810$ psi $T$.
$\sigma_{p2} = 1080$ psi $T$.
$\tau_{max} = 4000$ psi.

**1–57.** See 1–41.
**1–58.** See 1–42.
**1–59.** See 1–43.
**1–61.** See 1–52.
**1–63.** See 1–27.
**1–65.** See 1–33.
**1–70.** (*a*) $\sigma = 9.20$ ksi $T$.
$\tau = 7.39$ ksi.
(*b*) $\sigma_{p1} = 11.42$ ksi $T$.
$\sigma_{p2} = 15.42$ ksi $C$.
$\tau = 13.42$ ksi.
**1–72.** Partial Results:
$\tau_{xy} = 3.00$ ksi.
$\tau_{\max} = 5.00$ ksi.
$\sigma_{p1,p2} = 2.00$ ksi $T$, 8.00 ksi $C$.
**1–74.** Partial Results:
$\tau_{xy} = 12.0$ ksi.
$\sigma_x = 20.0$ ksi $C$.
$\sigma_{p2} = 28.0$ ksi $C$.
$\theta = -33.7°$.
**1–76.** Partial Results:
$\sigma_x = 20.0$ ksi $C$.
$\sigma_y = 4.00$ ksi $T$.
$\tau_{\max} = 13.00$ ksi.
$\theta = 11.3°$.
**1–78.** Partial Results:
$\sigma_x = 15{,}000$ psi $T$.
$\sigma_{p1} = 15{,}200$ psi $T$.
$\tau_{\max} = 8200$ psi.
$\theta = 6.35°$.
**1–79.** 0.300.
**1–80.** 0.00803.
**1–81.** $\epsilon_D = \epsilon_B = 0.979\epsilon_C + 0.020$.
**1–83.** 0.00118.
**1–85.** 0.00090.
**1–86.** $\epsilon_{p1} = 411\mu$.
$\epsilon_{p2} = -301\mu$.
$\gamma_{\max} = \pm 712\mu$.
**1–88.** $\epsilon_{p1} = 201\mu$.
$\epsilon_{p2} = 78.8\mu$.
$\gamma_{\max} = \pm 122.4\mu$.
**1–90.** $\epsilon_{p1} = 627\mu$.
$\epsilon_{p2} = -307\mu$.
$\gamma_{\max} = \pm 934\mu$.

## Chapter 2

**2–1.** 0.0140 in. longer.
**2–2.** 0.1857 in. longer.
**2–3.** (*a*) 16,300 ksi.
(*b*) 0.333.
(*c*) 24.4 ksi $T$.
**2–4.** $\mu = 0.350$; $E = 25(10^6)$ psi.
**2–6.** (*a*) 0.333.
(*b*) $15(10^6)$ psi.
(*c*) $5.62(10^6)$ psi.
**2–8.** $\dfrac{WL}{2AE}$.
**2–10.** $\dfrac{3\gamma}{8E} L^2$.
**2–12.** $\dfrac{5\gamma}{6E} L^2$.
**2–13.** 39,300 lb.
**2–14.** 2320 lb.
**2–15.** (*a*) 12,500 psi.
(*b*) 8.62 $(10^{-4})$.
(*c*) 2.64.
**2–16.** 85,000 lb $T$.
**2–18.** 32.0 kips.
**2–20.** 2.09.
**2–22.** For $B$: 2.94.
For $A$: 3.33.
**2–24.** 0.03144 in. left.
**2–25.** 0.059 in. upward.
**2–27.** (*a*) 93,000 lb $C$.
(*b*) 3472 psi.
**2–29.** (*a*) 10,500 psi $T$.
(*b*) 5250 psi.
(*c*) Decrease $3.30(10^{-3})$ in.
**2–31.** (*a*) 214°F.
(*b*) 44.5 ksi $T$.
(*c*) 1.348.
**2–32.** (*a*) $\sigma_{st} = 30.0$ ksi $C$.
$\sigma_{oak} = 1.500$ ksi $C$.
(*b*) 234 kips.
(*c*) 1.308 in. to the right.
**2–33.** (*a*) $\sigma_A = 7000$ psi $T$.
$\sigma_B = 2100$ psi $T$.
(*b*) $1.400(10^{-3})$ in. downward.
(*c*) 1.362 in. to the right of $A$.
**2–34.** $A_A = 2.00$ sq in.
$A_C = 1.244$ sq in.
**2–35.** 20,000 lb.
**2–37.** $\sigma_A = 40.0$ ksi $T$.
$\sigma_B = 42.1$ ksi $C$.
$\delta_C = 0.1000$ in. downward.
**2–39.** 33,000 lb.

2–41. 50.0 kips.
2–43. 4880 lb.
2–45. 21,500 lb.
2–46. 2570 lb.
2–48. 30,800 psi $T$.
2–50. 468 lb.
2–53. 780 lb upward.
2–54. 13,530 psi $T$.
2–56. 7430 lb.
2–58. $\sigma_B = 36.0$ ksi $T$.
$\sigma_A = 37.1$ ksi $T$.
$\delta_C = 0.1696$ in. downward.
2–60. $\sigma_A = 37.1$ ksi $T$.
$\sigma_B = 36.0$ ksi $T$.
$\delta_C = 0.220$ in. downward.
2–62. 16,670 lb.
2–64. 120,000 lb.
2–66. (a) 15,590 psi $T$.
(b) 4.49
2–67. 0.600 in. or 5/8 in.
2–69. 15,750 lb $T$.
2–70. 15,840 lb $T$.
2–73. $f_r = 1700$ lb per in. $T$.
$f_t = 600$ lb per in.
2–75. $\sigma = 6400$ psi $T$.
$\tau = 2400$ psi.

## Chapter 3

3–1. $\epsilon_x = -\dfrac{My}{EI}$; $\epsilon_y = \dfrac{\mu My}{EI}$;

$\gamma_{xy} = 0$; $\sigma_x = -\dfrac{My}{I}$; $\sigma_y = 0$;

$\tau_{xy} = 0$ in the plate.

On $y = \pm c$, $\epsilon_x = \mp \dfrac{Mc}{EI}$;

$\epsilon_y = \pm \dfrac{\mu Mc}{EI}$; $\gamma_{xy} = 0$;

$\sigma_x = \mp \dfrac{Mc}{I}$; $\sigma_y = \tau_{xy} = 0$.

On $x = 0, L$, $\epsilon_x = -\dfrac{My}{EI}$;

$\epsilon_y = \dfrac{\mu My}{EI}$; $\gamma_{xy} = 0$;

$\sigma_x = -\dfrac{My}{I}$; $\sigma_y = \tau_{xy} = 0$.

3–2. $\epsilon_x = \dfrac{\gamma}{E}(L - x)$;

$\epsilon_y = -\dfrac{\mu\gamma}{E}(L - x)$; $\gamma_{xy} = 0$;

$\sigma_x = \gamma(L - x)$; $\sigma_y = \tau_{xy} = 0$.
The case of a bar stretched by its own weight.

3–10. Partial Results:
$\sigma_x = 0$; $\sigma_y = 6a_2xy$; $\tau_{xy} = -3a_2x^2$.
On $x = 0$, forces are zero.
On $x = b$, $R_x = 0$; $R_y = 3a_2b^2ct$ downward. On $y = 0$, $R_y = 0$; $R_x = a_2b^3t$ to the right. On $y = c$, $R_x = a_2b^3t$ to the left; $R_y = 3a_2b^2ct$ upward.

3–11. Partial Results:
$\sigma_x = -6a_1x^2$; $\sigma_y = 6a_1(2x^2 - y^2)$; $\tau_{xy} = 12a_1xy$. On $x = 0$, $R_x = R_y = 0$. On $x = b$, $R_x = 6a_1b^2ct$ to the left; $R_y = 6a_1bc^2t$ upward. On $y = 0$, $R_x = 0$; $R_y = 4a_1b^3t$ downward. On $y = c$, $R_x = 6a_1b^2ct$ to the right; $R_y = 2a_1bt(2b^2 - 3c^2)$ upward.

3–13. $\sigma_x = -6a_2x^2y$; $\sigma_y = 2a_2y(6x^2 - y^2)$; $\tau_{xy} = -2a_2x(2x^2 - 3y^2)$.
On $x = 0$, $R_x = R_y = 0$. On $x = b$, $R_x = 12a_2c^4t$ to the left; $R_y = 28a_2c^4t$ downward. On $y = 0$, $R_y = 0$; $R_x = 16a_2c^4t$ to the right. On $y = c$, $R_x = 4a_2c^4t$ to the left; $R_y = 28a_2c^4t$ upward.

3–16. (a) $\sigma_x = -\dfrac{2Px^2y}{\pi(x^2 + y^2)^2}$;

$\sigma_y = -\dfrac{2Py^3}{\pi(x^2 + y^2)^2}$;

$\tau_{xy} = -\dfrac{2Pxy^2}{\pi(x^2 + y^2)^2}$.

On $y = 0$, $\sigma_x = \sigma_y = \tau_{xy} = 0$ except at $x^2 + y^2 = 0$.

## Chapter 4

4–1. 0.00090.
4–2. (a) 672 lb.
(b) 806 in-lb.

**4–3.** 1508 in-lb.

**4–4.** 2390 in-kips.

**4–6.** (*a*) 7500 psi.
(*b*) 128.6 per cent increase in area; 46.3 per cent increase in torque.

**4–8.** (*a*) 0.054 rad.
(*b*) 4050 psi.
(*c*) $5.63(10^{-4})$.

**4–9.** (*a*) 2.68 in.
(*b*) 0.036 rad.

**4–11.** 125,700 in-lb.

**4–14.** (*a*) Arr. *a*:
$T_{BC} = 5000$ in-lb,
$T_{AB} = 12{,}000$ in-lb;
Arr. *b*:
$T_{AB} = 7000$ in-lb,
$T_{AC} = 5000$ in-lb.
(*b*) Arr. *b* produces the smaller maximum stress.

**4–17.** (*a*) 94,250 in-lb.
(*b*) 0.016 rad .

**4–18.** 87,800 in-lb.

**4–20.** 20,500 in-lb.

**4–22.** (*a*) 1.910°.
(*b*) 10,000 psi. .

**4–24.** (*a*) 5730 psi.
(*b*) 0.00955 rad .

**4–26.** (*a*) 14,300 psi.
(*b*) 0.0179 rad .

**4–28.** (*a*) $\dfrac{9.22}{\pi d^4}$ rad .
(*b*) 2.73 in.

**4–30.** $T_1 = 226.5$ in-kips,
$T_2 = 650.0$ in-kips.

**4–31.** $\dfrac{32TL(D_2^2 + D_1D_2 + D_1^2)}{3\pi G D_2^3 D_1^3}$.

**4–33.**
$$\frac{8TL}{\pi G D_3^3(D_2 - D_1)}\left[\ln\left(\frac{D_2 - D_3}{D_2 + D_3}\cdot\frac{D_1 + D_3}{D_1 - D_3}\right) - 2\tan^{-1}\left(\frac{D_2}{D_3}\right) + 2\tan^{-1}\left(\frac{D_1}{D_3}\right)\right].$$

**4–35.** Partial Results:
(*a*) $\tau = 1042$ psi, $\sigma = 5920$ psi *C*.
(*b*) 6000 psi *T* and *C*.

**4–36.** 14,730 in-lb.

**4–38.** (*a*) 3760 psi.
(*b*) 0.1354 in.
(*c*) 2.52 in.

**4–40.** (*a*) 32.0 in-kips.
(*b*) 20.4 ksi *T*.

**4–41.** Partial Results:
(*a*) $\sigma_{p1} = 18.00$ ksi *T*,
$\sigma_{p2} = 8.00$ ksi *C*,
$\tau_{max} = 13.00$ ksi, $\theta = 33.7°$.
(*b*) $\sigma = 17.00$ ksi *T*, $\tau = 5.00$ ksi.

**4–42.** Partial Results:
$\sigma_{p1} = 2.85$ ksi *T*,
$\sigma_{p2} = 0.351$ ksi *C*,
$\tau_{max} = 1.601$ ksi, $\theta = 19.3°$.

**4–44.** Partial Results:
(*a*) Any surface point of the left 4 ft of the shaft.
(*b*) $\sigma_{p1} = 4.00$ ksi *T*,
$\sigma_{p2} = 1.000$ ksi *C*,
$\tau_{max} = 2.50$ ksi,
$\theta = 26.6°$.

**4–46.** 125.0 kips.

**4–48.** 109.6 kips.

**4–49.** (*a*) 179.5 hp.
(*b*) 0.0240 rad.

**4–50.** (*a*) 170.3 hp.
(*b*) 136.3 hp.

**4–51.** (*a*) 2.86 in.
(*b*) 342 rpm.

**4–53.** 4.37 in.

**4–55.** 3.69 in.

**4–56.** (*a*) 1212 in-lb.
(*b*) 4.27.

**4–57.** (*a*) $4.89(10^{-3})$.
(*b*) 1.318.

**4–58.** 10.8 in-kips.

**4–60.** (*a*) 458 in-kips.
(*b*) 0.09 rad per ft.

**4–62.** $\tau = 34{,}400$ psi,
$\theta = 0.0689$ rad .

**4–63.** 16,340 psi.

**4–64.** 0.00356 rad .

**4–65.** $\tau_{st} = 18{,}890$ psi,
$\tau_B = 15{,}720$ psi,
$\theta = 0.0315$ rad .

**4–67.** 530 in-kips.

**4–70.** 1.780 ksi.

**4–72.** (*a*) 5330 psi.
(*b*) $2.05(10^{-3})$ rad .

**4–73.** 2180 in-lb.

**4–75.** 325,000 in-lb.
**4–77.** 35.9 in-kips.
**4–79.** $\tau_{st} = 10.00$ ksi, $\tau_B = 7.50$ ksi.
**4–81.** (c) $\frac{15\sqrt{3}\,T}{2a^3}$. It occurs at the centers of the sides.
**4–82.** $\tau_A = \frac{C}{2a}(4a^2 - b^2)$; $\tau_B = 0$; $\tau_D = 2C(2a - b) = \tau_{max}$.
**4–83.** (a) $T_1 = 48.9$ in-kips; $T_2 = 38.4$ in-kips; $T_3 = 28.8$ in-kips.
(b) $\theta_1 = 0.0393$ rad; $\theta_2 = 0.0500$ rad; $\theta_3 = 0.0667$ rad.
**4–84.** (a) $T_{\bigcirc} = 137.5$ in-kips; $T_{\square} = 108.0$ in-kips; $T_{\Delta} = 83.1$ in-kips.
(b) $\theta_{\bigcirc} = 0.0262$ rad; $\theta_{\square} = 0.0333$ rad; $\theta_{\Delta} = 0.0433$ rad.
**4–86.** (a) 1000 in-kips.
(b) $1.520(10)^{-3}$ rad per ft.
**4–88.** (a) 358 in-kips.
(b) 10.00 ksi, 6.25 ksi.
(c) 0.00310 rad.
**4–90.** $T = 403$ in-kips; $\tau_{0.050} = 6.90$ ksi; $\tau_{0.040} = 8.61$ ksi.

## Chapter 5

**5–1.** 18,710 psi $C$.
**5–2.** 13,330 psi $C$.
**5–3.** 43,200 lb $T$; 0.833 in. below top of beam.
**5–6.** $54.4\sigma_T$ in-lb.
**5–8.** 1200 psi $T$ and $C$.
**5–10.** 89.8 per cent.
**5–12.** See 5–6.
**5–13.** See 5–8.
**5–14.** See 5–10.
**5–15.** 120 psi $C$.
**5–16.** 550 psi $C$.
**5–18.** (a) 503 psi $T$.
(b) 2010 psi $C$.
**5–20.** 4040 psi $T$ and $C$.
**5–22.** 8000 psi $T$ and $C$.
**5–24.** 4620 lb.
**5–27.** (a) $V = -2000 + 1000x$; $M = -2000(x + 2) + 500x^2$.
(b) Partial Results: $M_{4'} = -6000$ ft-lb; $M_{15'} = 30{,}000$ ft-lb; inflection point at 7.46′ from left end.
**5–28.** (a) $V = 5000 - 500x$; $M = 2000(12 - x) - 250(14 - x)^2$.
(b) Partial Results: $M_{0'} = -25{,}000$ ft-lb; $M_{12'} = -1000$ ft-lb; inflection point at $x = 10$ ft.
**5–29.** (a) $V = -5(4 + x)$; $M = -2.5(4 + x)^2$.
(b) Partial Results: $M_{0'} = -40$ ft-kip; $M_{+2'} = -30$ ft-kip; $M_{10'} = -190$ ft-kip.
(c) 114 in.$^3$
**5–32.** (a) $V = 5$ for $0 \leq x \leq 12$; $M_1 = 5x$ for $0 \leq x < 5$; $M_2 = 5x - 60$ for $5 < x \leq 12$.
(b) Partial Results: $R_L = 5$ kips upward; $R_R = 5$ kips downward; $M_{5'-} = 25$ ft-kips; $M_{5'+} = -35$ ft-kips.
**5–34.** (a) $V = 5200 - 800x$; $M = 26{,}000 + 5200x - 400x^2$.
(b) Partial Results: $R_L = 9600$ lb upward; $R_R = 16{,}800$ lb upward; $M_{8'}(\max) = 38{,}400$ ft-lb; $M_{15'} = 14{,}000$ ft-lb; $M_{20'} = -30{,}000$ ft-lb; inflection point at 3.15 ft left of right support.
**5–35.** (a) $V = 100(x^2/3) - 1600$; $M = 1600x - 100x^3/9$.
(b) Partial Results: $R_L = 3200$ lb upward; $R_R = 1600$ lb upward;

$M_{0'} = M_{12'} = 0$;
$M_{6.94'}$(max) = 7390 ft-lb.

**5–37.** (*a*) $V = 3x^2/8 + 2x - 9.2$;
$M = -x^3/8 - x^2 + 9.2x$.
(*b*) Partial Results:
$R_L$ = 20.8 kips upward;
$R_R$ = 9.2 kips upward;
$M_{2.97'}$(max) = 18.12 ft-kips; $M_{4'}$ = 12.8 ft-kips;
$M_{7'}$ = −1.6 ft-kips;
$M_{10'}$ = −25 ft-kips;
inflection point at $x$ = 6.67 ft.

**5–39.** (*a*) Partial Results:
$M_{4'}$ = −3000 ft-lb;
$M_{8'}$ = −1400 ft-lb;
$M_{10'}$(max) = −600 ft-lb;
$M_{12'}$ = −1400 ft-lb;
$M_{14'}$ = 1000 ft-lb;
$M_{17'}$ = 3400 ft-lb;
inflection point at 1.17 ft right of right support.
(*b*) Origin at $D$ with $x$ positive left; $V = 400x - 800$;
$M = -200x^2 + 800x - 1400$.

**5–41.** Partial Results:
$V_A = +0.45wL$;
$V_B = -0.55wL$;
$M_{0.45L} = +0.0512wL^2$;
inflection points at $0.130L$ and $0.770\,L$ from $A$.

**5–43.** (*a*) 66 ft.
(*b*) Partial Results:
$M_{16'}$ = −69,700 ft-lb;
$M_{24.74'}$(max) = −104,900 ft-lb; $M_{44'}$ = 75,100 ft-lb;
$M_{48.62'}$(max) = 118,300 ft-lb; $M_{52'}$ = 95,200 ft-lb;
inflection point at 39.4 ft. from left end.
(*c*) With origin at $B$, $x$ positive to right, $M = -67{,}900 - 8500x + 485x^2$.

**5–45.** Partial Results:
$M_{0'}$ = −1500 ft-lb;
$M_{5'}$ = −4000 ft-lb;
$M_{9'-}$ = 800 ft-lb;
$M_{9'+}$ = 3800 ft-lb;
$M_{13'}$(max) = 5400 ft-lb;
$M_{15'}$ = 5000 ft-lb; inflection point at 3.10 ft right of left support; with origin at $B$, $x$ positive to right, $M = -4000 + 1600x - 100x^2$.

**5–47.** Partial Results:
$M_{0'}$ = 1600 ft-lb;
$M_{4'}$ = 0; $M_{9'-}$ = 7500 ft-lb;
$M_{9'+}$ = 3000 ft-lb;
$M_{14'}$ = 6750 ft-lb; with origin at $C$, $x$ positive to right,
$M = 3000 + 1500x - 150x^2$.

**5–50.** Partial Results:
$M_{2'} = M_{4'-}$ = −1000 ft-lb;
$M_{3'}$(max) = −750 ft-lb;
$M_{4'+}$ = 6000 ft-lb;
$M_{7'}$ = 4500 ft-lb;
$M_{9'}$ = −3500 ft-lb;
$M_{11'}$ = −7500 ft-lb;
$M_{13'-}$ = −3500 ft-lb;
$M_{13'+}$ = −8000 ft-lb; inflection point at 1.125 ft right of 3500 lb load; with origin at $E$, $x$ positive to right, $M = -3500 - 4000x + 1000x^2$.

**5–55.** 8390 psi $T$ at the bottom of section $B$.

**5–57.** 5530 psi $T$ at the bottom at section $C$.

**5–62.** (*a*) $M_2$ = 1800 ft-lb counterclockwise; $F_1$ = 3360 lb as shown; $F_2$ = 4800 lb as shown.
(*b*) 1440 lb to the left.
(*c*) 40.0 psi.

**5–63.** (*a*) 2880 lb to the left.
(*b*) 120.0 psi.

**5–64.** (*a*) 960 lb to the left.
(*b*) 40.0 psi.

**5–67.** (*a*) 84.4 psi.
(*b*) 150.0 psi.

**5–68.** (*a*) 91.4 psi at the neutral axis.
(*b*) 76.5 psi.

**5–69.** (*a*) 151.6 psi.
(*b*) 202 psi at neutral axis at each support.

**5–71.** (*a*) 750 psi $T$ and $C$.
(*b*) 91.9 psi.

**5–73.** (*a*) 63.7 psi.
(*b*) 4190 psi $C$.

**5–75.** 2220 lb.
**5–77.** 1600 lb per ft.
**5–79.** 800 lb.
**5–81.** 1889 lb per ft.
**5–83.** (*a*) 73.5 psi.
(*b*) 529 psi $T$.
**5–85.** (*a*) 1858 psi $C$.
(*b*) 136.2 psi.
(*c*) 710 psi $T$.
**5–87.** 249 lb per ft.
**5–89.** Partial Results:
$\sigma_p$ = 18,390 psi $T$, 1970 psi $C$;
$\tau_{max}$ = 10,180 psi.
**5–90.** Partial Results:
$\sigma_p$ = 15.00 ksi $T$, 4.10 ksi $C$;
$\tau_{max}$ = 9.55 ksi.
**5–91.** Partial Results:
$\sigma_p$ = 558 psi $C$, 31 psi $T$;
$\tau_{max}$ = 297 psi.
**5–93.** $\sigma_{p\max}$ = 13.95 ksi $T$ and $C$;
$\tau_{max}$ = 7.16 ksi.
**5–95.** 15.25 kips per ft.
**5–97.** Partial Results:
$a = -3P/(4ct)$;
$b = P/(4c^3t)$;
All boundary conditions satisfied but $\tau_{xy}$ at $x = 0$ must be distributed parabolically; this solution coincides with results from elementary theory.
**5–99.** (*a*) $\sigma_x = \dfrac{4w}{(4 - \pi)} \left[\dfrac{\pi}{4} - \dfrac{xy}{x^2 + y^2} - \tan^{-1}\dfrac{y}{x}\right]$;
at $A$, $\sigma_x = 3.66w$;
at $B$, $\sigma_x = -0.1007w$;
at $C$, $\sigma_x = -2.33w$.
(*b*) at $A$, $\sigma_x = 3w$; at $B$, $\sigma_x = 0$; at $C$, $\sigma_x = -3w$.
**5–101.** 4220 in-kips.
**5–102.** 10,980 in-kips.
**5–104.** (*a*) 2.56.
(*b*) 2.35.
**5–106.** (*a*) 3.33 in. above bottom of beam.
(*b*) 3840 in-kips.
(*c*) 1.795.
**5–108.** (*a*) 825 in-kips.
(*b*) 1757 in-kips.
**5–109.** 1.505 in. to left of outside of web.
**5–111.** 7.94 in.
**5–113.** $\sigma_w$ = 1000 psi $T$ and $C$;
$\sigma_{al}$ = 10,000 psi $T$ and $C$.
**5–114.** $\sigma_w$ = 802 psi $T$ and $C$;
$\sigma_{st}$ = 20,050 psi $T$ and $C$;
**5–116.** 9350 psi $T$.
**5–118.** $\sigma_w$ = 1555 psi $C$;
$\sigma_{st}$ = 14,640 psi $T$.
**5–120.** (*a*) 8960 psi $T$;
(*b*) 5280 lb.

## Chapter 6

**6–1.** (*a*) $\rho$ = 6,670 in.,
$\sigma$ = 9000 psi $T$ and $C$.
(*b*) $\rho$ = 11,360 in.,
$\sigma$ = 8,590 psi $T$.
**6–2.** (*a*) 5000 in.
(*b*) arc of a circle.
**6–3.** 0.900 in.
**6–4.** $\rho$ = 3330 in.,
$y$ = 0.540 in. upward.
**6–6.** (*a*) $EIy = Mx^2/2 - MLx/2$ with the origin at the left support.
(*b*) The parabola of Prob. 6-6*a* is an approximation due to the approximate expression used for the radius of curvature.
**6–7.** $EIy = \dfrac{wLx^3}{12} - \dfrac{wx^4}{24} - \dfrac{wL^3x}{12} + \dfrac{wL^4}{24}$.
**6–8.** $EIy = \dfrac{wLx^3}{30} - \dfrac{wx^4}{24} + \dfrac{wL^3x}{120}$ with the origin at the left support.
**6–9.** $EIy = \dfrac{7wLx^3}{96} - \dfrac{wx^4}{24} - \dfrac{wL^3x}{32}$ with the origin at the left support.
**6–12.** $EIy = wLx^3/12 - wL^2x^2/16 - wx^4/24 + wL^3x/48$

**6–14.** $\dfrac{11wL^4}{120EI}$ downward.

**6–16.** (*a*) 0.338 in. upward.
(*b*) 0.405 in. downward.

**6–18.** $\dfrac{wL^4}{6EI}$ upward.

**6–19.** $EIy = -\dfrac{wL^2x^2}{12} + \dfrac{7wLx^3}{72} - \dfrac{wx^4}{24} + \dfrac{wL^3x}{36}$
with the origin at the center support.

**6–21.** $\dfrac{0.0251wL^4}{EI}$ upward.

**6–23.** $\dfrac{PL^3}{9EI}$ downward.

**6–26.** $\dfrac{23PL^3}{24EI}$ downward.

**6–28.** $\dfrac{2wL^4}{3EI}$ downward.

**6–31.** (*a*) $EIy = \dfrac{wax^3}{6} - \dfrac{w(x-a)^4}{24} - \dfrac{11wa^3x}{6}$
with the origin at the left support.
(*b*) 0.342 in. downward.

**6–32.** $-wL^2x^2 + \dfrac{wLx^3}{3} - \dfrac{w}{12}\langle x-L\rangle^4 - \dfrac{wL^2}{2}\langle x - L\rangle^2$.

**6–33.** See 6–28.

**6–36.** (*a*) $EIy = \dfrac{wax^3}{6} - \dfrac{w}{24}\langle x-a\rangle^4 + \dfrac{w}{24}\langle x-3a\rangle^4 - \dfrac{11wa^3x}{6}$
with the origin at the left support.
(*b*) 0.342 in. downward.

**6–38.** $\dfrac{61wL^4}{96EI}$ downward.

**6–40.** 0.251 in. upward.

**6–42.** $EIy = -\dfrac{wax^3}{6} - \dfrac{wa^2x^2}{2} + \dfrac{wa}{2}\langle x-0\rangle^3 + \dfrac{wa^2}{2}\langle x-a\rangle^2 - \dfrac{wa}{3}\langle x-3a\rangle^3 - \dfrac{w}{24}\langle x-a\rangle^4 + \dfrac{w}{6}\langle x-2a\rangle^4 - \dfrac{w}{40a}\langle x-a\rangle^5 + \dfrac{w}{40a}\langle x-2a\rangle^5 - \dfrac{691wa^3}{-240}x$
with the origin at the left support, $y = \dfrac{559wa^4}{120EI}$ downward.

**6–49.** $\dfrac{17PL^3}{240EI}$ downward.

**6–50.** $\dfrac{25PL^3}{96EI}$ upward.

**6–52.** $\dfrac{PL^3}{2EI}$ downward.

**6–54.** $\dfrac{35wL^4}{6144EI}$ downward.

**6–56.** (*a*) 3660 lb.
(*b*) 26,200 psi *T* and *C*.

**6–58.** 15,180 lb.

**6–63.** $\dfrac{wL^4}{6EI}$ upward.

**6–65.** $0.441L$ from the left end.

**6–67.** (*a*) 0.784 in. downward.
(*b*) 938 psi *T* and *C*.
(*c*) 39.1 psi.

**6–69.** (*a*) 0.321 in. downward.
(*b*) 0.0470 in. upward.

**6–71.** $0.472L$ from the left support.

**6–72.** 2.50 in. downward.

**6–73.** 0.595 in. downward.

**6–74.** 1.171 in. downward.

**6–76.** $\dfrac{2PL^3}{3EI}$ downward.

**6–78.** $\dfrac{29wL^4}{48EI}$ downward.

**6–81.** $\dfrac{7wL^4}{10EI}$ downward.

**6–82.** See 6–72.

**6–83.** 3030 lb upward.

**6–84.** $\dfrac{29wL^4}{24EI}$ downward.

**6–86.** $\dfrac{17PL^3}{240EI}$ downward.
**6–88.** See 6–74.
**6–89.** $\dfrac{73wL^4}{120EI}$ downward.
**6–90.** See 6–38.
**6–91.** See 6–78.
**6–92.** See 6–81.
**6–94.** 0.0229 rad
**6–96.** 1045 lb per ft.
**6–100.** $\dfrac{3wL^4}{16EI}$ downward.
**6–102.** $\dfrac{wL^4}{24EI}$ upward.
**6–104.** $\dfrac{2wL^4}{15EI}$ downward.
**6–108.** $\dfrac{2.828ML^2}{81EI}$ upward.
**6–110.** $\dfrac{5PL^3}{48EI}$ downward at $L/2$ from the left end.
**6–111.** $\dfrac{49wL^4}{240EI}$ downward.

## Chapter 7

**7–1.** $wL/2$ lb upward.
**7–2.** $wL^2/4$ ft-lb.
**7–3.** (*a*) $EIy = \dfrac{wLx^3}{12} - \dfrac{Qx^3}{18} - \dfrac{wx^4}{24} - \dfrac{wL^3}{24}x + \dfrac{QL^2}{18}x$ with the origin at the left support.
(*b*) $3wL/8$ lb downward.
**7–4.** 400 lb downward.
**7–6.** $wL^2/12$ ft-lb clockwise.
**7–8.** $V_{\text{Left}} = wL/3$ lb upward; $M_{\text{Left}} = wL^2/36$ ft-lb counterclockwise; $R_{\text{Right}} = 13wL/9$ lb upward.
**7–11.** (*a*) $14wL/9$ lb;
(*b*) $\dfrac{46wL^4}{27EI}$ ft downward;
(*c*) $\dfrac{191wL^4}{108EI}$ ft downward.
**7–13.** $P/4$ lb upward.
**7–15.** $64wL/125$ lb upward.
**7–17.** $wL/3$ lb downward.
**7–19.** $V = wL/4$ lb upward; $M = 0$.
**7–21.** $7wL/32$ lb upward.
**7–23.** $wL/4$ lb downward.
**7–24.** (*a*) $wL^2/6$ ft-lb clockwise.
(*b*) $\dfrac{wL^4}{24EI}$ ft downward.
**7–25.** $P/2$ lb upward.
**7–28.** (*a*) 2500 lb downward.
(*b*) 7.5 ft from the right support.
**7–30.** $V_{\text{Left}} = 960$ lb upward; $M_{\text{Left}} = 4320$ ft-lb counterclockwise; $V_{\text{Right}} = 2280$ lb upward; $M_{\text{Right}} = 6480$ ft-lb clockwise.
**7–33.** (*a*) $1.163wL$ lb upward;
(*b*) $2.85L$ ft from the right end.
**7–35.** $R_C = 3440$ lb upward;
$R_L = R_R = 2280$ lb upward.
**7–36.** (*a*) $R_C = 6790$ lb upward;
$R_L = R_R = 3810$ lb upward;
(*b*) 5260 psi $T$ and $C$.
**7–37.** 3700 lb.
**7–38.** $wL/2$ lb upward.
**7–40.** 391 lb.
**7–42.** Partial Results:
$R_C = 3700$ lb upward,
$R_L = R_R = 1350$ lb upward,
$V = 0$ at 3.38 ft from each end.
$V_{\max} = \pm 1850$ lb at center.
**7–44.** 1481 lb upward.
**7–46.** (*a*) $6wL/17$ lb upward.
(*b*) 0.1220 in. downward.
**7–48.** 13,040 psi $T$.
**7–50.** 133.3 lb $C$.
**7–51.** See 7–2.
**7–52.** See 7–6.
**7–53.** See 7–1.
**7–54.** See 7–3*b*.
**7–55.** See 7–8.
**7–56.** $3wL^2/4$ ft-lb counterclockwise.
**7–58.** See 7–19.
**7–65.** 324 lb upward.
**7–66.** 16*WF*50.
**7–67.** 2.80 kips per ft.
**7–69.** 3.50 kips per ft.
**7–71.** 5.62 kips per ft.
**7–72.** 1.520 kips per ft.
**7–74.** $wL/24$ lb upward.

**7–76.** $13wL^2/54$ ft-lb clockwise.
**7–78.** (*a*) $-wL/30$ lb;
(*b*) $0.0253wL^4/(EI)$ ft upward.
**7–80.** $5wL/12$ lb upward.
**7–82.** 0.1453 in. downward.

## Chapter 8

**8–1.** (*a*) 15,630 psi $C$.
(*b*) 15,720 psi $C$.
**8–2.** 27.7 ksi $C$ at bottom 20 in. to the right of $B$.
**8–5.** $\sigma_A = 1100$ psi $T$;
$\sigma_B = 500$ psi $T$;
$\sigma_C = 1500$ psi $C$;
$\sigma_D = 900$ psi $C$; Line of zero normal stress through point 1.5 in. from $B$ toward $C$ and through point 3.3 in. from $A$ toward $D$.
**8–7.** 1961 psi $C$.
**8–8.** 1.888 in.
**8–10.** 9600 lb.
**8–12.** $P = 23.7$ kips;
$\tau = 1052$ psi.
**8–14.** Partial Results:
*at G* — $\sigma_{p1} = 1260$ psi $C$;
$\sigma_{p2} = 0$; $\tau_{\max} = 630$ psi.
*at H* — $\sigma_{p1} = 565$ psi $T$;
$\sigma_{p2} = 2.50$ psi $C$; $\tau_{\max} = 284$ psi.
**8–16.** Partial Results:
(*a*) $P = 5600$ lb; $T = 0$.
(*b*) $\sigma_{p1} = 5000$ psi $C$;
$\sigma_{p2} = 200$ psi $T$;
$\tau_{\max} = 2600$ psi.
**8–18.** $P = 22{,}400$ lb;
$a = 12.33$ in.
**8–20.** Partial Results:
$\tau_{\max} = 261$ psi; $\sigma_{p1} = 520$ psi $C$;
$\sigma_{p2} = 2$ psi $T$.
**8–22.** Partial Results:
$\tau_{\max} = 12.42$ ksi; $\sigma_{p1} = 23.9$ ksi $T$;
$\sigma_{p2} = 0.955$ ksi $C$.
**8–23.** Partial Results:
$\tau_{\max} = 636$ psi; $\sigma_{p1} = 1146$ psi $T$;
$\sigma_{p2} = 127$ psi $C$.
**8–24.** (*a*) Top surface at wall.
(*b*) Partial Results:
$\tau_{\max} = 5660$ psi;
$\sigma_{p1} = 9660$ psi $T$;
$\sigma_{p2} = 1656$ psi $C$.
**8–26.** 3620 lb.
**8–30.** 1.655 kips.
**8–31.** (*a*) 1.28 in. downward.
(*b*) 154 lb downward.
**8–32.** 9520 lb.
**8–33.** (*a*) $4.24(10^4)$ lb.
(*b*) 0.034 in. downward.
**8–34.** 71.3 in.
**8–36.** (*a*) 40.4 lb.
(*b*) 60.6 lb.
**8–38.** 1920 lb.
**8–40.** (*a*) 0.106 in. downward.
(*b*) 1589 lb.
(*c*) 15,890 psi $T$ and $C$.
**8–42.** (*a*) 9210 psi.
(*b*) 8240 psi.
**8–44.** 12,350 psi.
**8–46.** (*a*) 82.4 percent.
(*b*) 3.00.
**8–48.** 21.6 in.
**8–51.** $h = 6.80$ in.; load factor $= 12.0$;
$U_A = 96.0$ in-lb;
$U_B = 48.0$ in-lb.
**8–52.** (*a*) 1.20 in.
(*b*) $U_A = 3.20$ in-lb;
$U_B = 6.40$ in-lb;
$U_C = 6.40$ in-lb.
**8–54.** $(\sigma_{BC})_{\text{max fiber}} = 12{,}070$ psi $T$ and $C$; $(\sigma_p)_{AB} = 2880$ psi $T$ and $C$.
**8–57.** (*a*) For maximum normal stress 1.800; For maximum shearing stress 1.800. For maximum normal strain 2.20. For maximum distortion-energy 2.06.
(*b*) Same order as in (*a*) 1.800, 1.125, 1.525, 1.286.
**8–58.** (*a*) $\sigma_x = 16.97$ ksi $T$ and $C$;
$\tau_{xy} = 13.58$ ksi.
(*b*) $\sigma_{p1} = 24.5$ ksi $T$;
$\sigma_{p2} = 7.53$ ksi $C$;
$\sigma_{p3} = 0$;
$\tau_{\max} = 16.02$ ksi.
(*c*) 1.87.
**8–60.** 339 kips.
**8–62.** 3.88 in.
**8–64.** 4.68 kips.
**8–66.** 4.27 kips.

## Chapter 9

**9–2.** 96.0.
**9–3.** 124.7.
**9–4.** 72.5.
**9–6.** (*a*) 78.0.
(*b*) 92.3.
(*c*) 133.3.
**9–8.** 103.9.
**9–10.** $ec/r^2 = 0.36$; $e = 0.030$ in.
**9–12.** 124.8 in.
**9–14.** $L/r = 169.7$; $E = 29.6(10^6)$ psi.
**9–16.** 24,600 lb.
**9–18.** 4120 lb.
**9–21.** (*a*) 41.3 kips.
(*b*) 206 kips.
**9–22.** (*a*) 64.2 kips.
(*b*) 16.60 kips.
**9–23.** (*a*) 219 kips.
(*b*) 14.77 kips.
**9–24.** (*a*) 90.2 kips.
(*b*) 36.5 kips.
**9–26.** 101.6 kips.
**9–28.** 63.0 kips.
**9–30.** 73.2 kips.
**9–32.** 57.1 kips.
**9–33.** 254 kips.
**9–36.** (*a*) 128.0 kips.
(*b*) 53.7 kips.
**9–37.** (*a*) 7.68 in.
(*b*) 194.5 kips.
**9–50.** 41.0 kips.
**9–51.** 38.3 kips.
**9–53.** 47,900 lb.
**9–54.** 59.3 kips.
**9–56.** 205 kips.
**9–57.** (*a*) 21.2 kips.
(*b*) 44,800 in-lb.
**9–59.** 56.2 kips.
**9–61.** 4.22 in.
**9–63.** 42.3 kips.
**9–66.** $\dfrac{n}{\sqrt{P/(EI)}} \tan L\sqrt{P/(EI)} - nL.$
**9–68.** $y = \dfrac{nEI \sin(\sqrt{P/(EI)})\, x}{PL \sin(\sqrt{P/(EI)})L} + \dfrac{nx^3}{6L^2} - n\left(\dfrac{1}{6} + \dfrac{EI}{PL^2}\right)x.$
**9–70.** $1.553\pi^2 EI/L^2$ lb.

## Chapter 10

**10–1.** 1000 minutes.
**10–2.** 10,000 minutes.
**10–3.** 55.7 ksi.
**10–4.** (*a*) 9.3 kips.
(*b*) 36.0 kips.
**10–5.** (*a*) 6.0 kips.
(*b*) 48.0 kips.
**10–6.** (*a*) 4.03 kips.
(*b*) 9.0 kips.
**10–8.** (*a*) 16.67 in.
(*b*) 4.70 in.
**10–10.** (*a*) 602 lb.
(*b*) 358 lb.
**10–12.** 1778 lb.
**10–14.** 0.23 in.
**10–16.** 2.05.
**10–17.** $\sigma_{max} = 78.1$ ksi $T$;
$\sigma_{min} = 33.5$ ksi $T$.
**10–18.** $\sigma_{max} = 61.2$ ksi $T$;
$\sigma_{min} = 30.6$ ksi $C$.
**10–20.** $P_{max} = 2.74$ kips $T$;
$P_{min} = 1.37$ kips $C$.
**10–22.** 1.29 kips.
**10–24.** $P_{max} = 976$ lb downward;
$P_{min} = 438$ lb upward.
**10–26.** 32.5 lb.
**10–28.** 1.49 in.
**10–30.** 1.94.
**10–32.** 1.77.
**10–34.** 0.372 in.

## Chapter 11

**11–1.** (*a*) 15,510 psi $T$ at lower bolt.
(*b*) 21,000 psi $C$.
(*c*) 2670 psi.
**11–2.** (*a*) 13,900 psi.
(*b*) 15,600 psi $C$.
(*c*) 12,400 psi $T$.
at second row from right.
**11–4.** (*a*) 9380 psi.
(*b*) 15,950 psi $C$.
(*c*) 11,840 psi $T$.
**11–5.** (*a*) 9360 psi.
(*b*) 17,140 psi $C$.
(*c*) 12,000 psi $T$ at outer row.

**11–7.** (*a*) 7230 psi.
(*b*) 18,420 psi *C*.
(*c*) 10,740 psi *T* in outer row of main plate.

**11–9.** (*a*) 8320 psi.
(*b*) 15,240 psi *C*.
(*c*) 10,475 psi *T* in row 1; 11,110 psi *T* in row 2; 8,000 psi *T* in row 3.

**11–11.** 75.7 per cent.

**11–13.** (*a*) 253 psi.
(*b*) 81.0 per cent.

**11–15.** 2810 lb; 82.4 per cent.

**11–18.** 62.0 per cent.

**11–20.** (*a*) 177.8 psi.
(*b*) 64.7 per cent.

**11–22.** 83.7 per cent.

**11–23.** 3020 lb; 88.6 per cent.

**11–26.** (*a*) $d = 0.948$ in.; $t = 0.620$ in.
(*b*) 1.27.

**11–28.** 76,200 lb.

**11–29.** 90,300 lb.

**11–31.** Lengths of side welds are 2.07 in. and 8.22 in.

**11–33.** Lengths of side welds are 6.35 in. and 16.51 in.

**11–35.** $a = 8.00$ in.; $b = 9.86$ in.; $c = 6.03$ in.

**11–36.** $\sigma = 7800$ psi $T$; $\tau = 2400$ psi.

**11–38.** (*a*) 14.9 ksi.
(*b*) 5.

**11–40.** (*a*) 10,100 psi.
(*b*) 10.

**11–42.** 89.9 hp.

**11–43.** (*a*) 8980 lb.
(*b*) 19,100 lb.

**11–45.** (*a*) 597 hp.
(*b*) 6.

**11–47.** (*a*) *A* and *C*.
(*b*) 2600 lb.

**11–49.** (*a*) The bottom bolt.
(*b*) 10,200 psi.

**11–51.** $\tau_A = 6780$ psi; $\tau_B = 10{,}800$ psi.

**11–53.** $\tau_T = \tau_B = 14.2$ ksi.

**11–55.** $\tau_A = 8480$ psi. $\tau_B = 3790$ psi.

**11–58.** 1870 lb.

**11–59.** 6.71 in.

# Index